Differential Equations

Differential Equations

Second Edition

JOHN A. TIERNEY

wcb

Wm. C. Brown Publishers
Dubuque, Iowa

To
John, Frankie, Terri, and Steven

Contents

3

Applications of First-Order Differential Equations 60

4

Linear Differential Equations 116

5

Applications of Second-Order Equations   162

6

Systems of Differential Equations   213

7

The Laplace Transform   258

10

Partial Differential Equations; Boundary-Value Problems; Fourier Series   379

Answers to Selected Odd-Numbered Problems   434

Index   454

Preface

This second edition employs the same approach as the original text and is based on the same underlying philosophy. It is intended primarily for a sophomore or a junior level one-semester introductory course in differential equations. Addressed to students who have completed a basic calculus sequence, it is not too difficult for a good student who has completed two semesters of calculus. The level of mathematical sophistication is appropriate for average students whose interests lie in mathematics, science, or engineering. It is also suitable for those who expect to encounter differential equations in such diverse fields as economics, biology, medicine, or demography.

Although theory is not slighted, the main emphasis is on applications. These are taken not only from the physical sciences but also from the life sciences, the social sciences, and geometry. A perusal of the Table of Contents gives a fairly specific indication of the variety of applications covered. Chapters 3 and 5 are devoted entirely to applications; other chapters include applications whenever appropriate. Few applications receive long, detailed treatment since time limitations in most courses render such an approach impractical. Attention is focused on those applications that are readily understood by students who do not have specialized knowledge of the fields from which applications are selected. More applications are included than can be covered in the usual course. This permits flexibility and allows the instructor to consider the interests, preparation, and mathematical maturity of the students.

Various other options are available. Chapter 7 on the Laplace transform and Chapter 10 on partial differential equations may be omitted, treated lightly, or covered completely. Sections 4.8, 4.9, 4.11, 6.2, 9.2, and 9.7 through 9.10 are not essential to the main development and may be omitted or given minor emphasis in a short course.

The main objective of the textbook is to present simply and clearly the important concepts of the theory of differential equations and to show how differential equations are used in mathematical models describing real-life situations. This is accomplished by using simple yet modern terminology and notation, including an unusually large number of examples, and presenting extensive, carefully graded problem lists. Numerous expositions illustrate the manner in which reasonable physical assumptions lead to differential equations governing concrete situations. The text emphasizes that the same mathematical model often governs many seemingly unrelated physical situations.

The importance of numerical methods of solving differential equations is stressed. Although the computer's importance is emphasized, the material is developed without assuming that the student has access to a high-speed computer. The basic objective of Chapter 9 is to give the student an idea of what is important in numerical solutions and to present a few simple numerical methods.

The existence and the uniqueness of solutions are considered, with illustrations. This is accomplished without including the numerous proofs that would be essential in a more theoretical development. Students are encouraged to seek properties of solutions of differential equations without obtaining explicit solutions. Geometric and physical interpretations are exploited in this direction.

Numerous references are given at the end of each chapter. These furnish sources for additional proofs, more detailed theoretical approaches, and expanded treatments of applications.

This revision differs from the first edition in several respects. In Section 2.4 an example is presented in which the function Q in the differential equation $y' + P(x) y = Q(x)$ is discontinuous at a point $x = a$ inside the interval under consideration. Section 2.7 includes three graphs of direction fields obtained using computer graphics. These graphs will give students an awareness of the manner in which a computer, without actually solving a differential equation, can furnish information about properties of the solutions of the equation. Section 2.7 also includes several examples illustrating the Peano existence theorem and the Picard uniqueness theorem.

Chapter 4 on Linear Differential Equations has been expanded considerably. The Wronskian in Section 4.3 is given more prominence, the topic Reduction of Order is treated separately in Section 4.5, and in Section 4.6 Homogeneous Second-Order Linear Differential Equations and the Euler Identity receive more detailed treatments. The chapter contains two new sections, Section 4.10 on Higher-Order Linear Equations and Section 4.11 on the Euler Equation.

Section 6.2 on Plane Autonomous Systems, Critical Points, and Stability is new. In courses where time permits, this section will give students a better understanding of the qualitative aspects of the theory of differential equations.

In Section 8.3 the Method of Frobenius has been expanded slightly. Chapter 9 on Numerical Methods now contains separate sections on Runge-Kutta methods and the Classical Runge-Kutta Method. Section 10.7 includes two illustrations based on computer graphic treatments of the Heat Equation.

In addition to these specific changes and additions, a number of general modifications have been incorporated. To render the overall presentation more instructive, interesting, and challenging, numerous illustrations, drawings, and problems have been added. The exposition has been modified in various places in an effort to achieve greater clarity.

It is our hope that this new edition will give students an understanding and appreciation of the important role that differential equations play in modern life.

I am indebted to Professor Mahlon F. Stilwell and Marion G. Tierney for their careful reading of the manuscript and their many highly valuable suggestions.

I also express my appreciation to Raeia Maes for coordinating the production of this text.

Finally, I am deeply grateful to Allyn and Bacon editors Gary Folven and Carol Nolan-Fish. Their constant encouragement and assistance have been invaluable.

J.A.T.

Differential Equations

1

Preliminary
Concepts

1.1 Introduction

In the latter half of the seventeenth century, Newton and Leibniz systematized the calculus into a unified body of mathematical knowledge. Their exploitation of the fact that differentiation and integration are inverse processes led to the development of the theory of differential equations [abbreviated DE henceforth, for "differential equation(s)"]. Mathematicians and scientists were quick to realize that many basic physical laws were best expressed by equations involving not only the underlying variables but also their derivatives or instantaneous rates of change. Interest in the theory and application of DE has persisted to the present day. Efforts to resolve various theoretical questions concerning DE have resulted in the enrichment of mathematical analysis, the study of infinite processes. Investigators continue to discover new applications of DE, not only in the physical sciences but also in such diverse fields as biology, physiology, medicine, statistics, sociology, psychology, and economics. Both theoretical and applied DE are active fields of current research.

Definition A differential equation is an equation involving unknown functions and their derivatives.

If the functions are real functions of one real variable, the derivatives occurring are ordinary derivatives, and the equation is called an *ordinary* DE.

If the functions are real functions of more than one real variable, the derivatives occurring are partial derivatives, and the equation is called a *partial* DE. When we refer to an equation as a DE, we shall mean an ordinary DE. Until we state otherwise, we shall restrict ourselves to DE involving a single unknown function.

Definition By a solution of the DE

$$F\left(x, y, \frac{dy}{dx}, \frac{d^2y}{dx^2}, \ldots, \frac{d^ny}{dx^n}\right) = 0 \qquad (1.1)$$

we mean a real function f, denoted by $y = f(x)$, defined on a set S of real numbers, where S is the union of nonoverlapping intervals such that

$$F(x, f(x), f'(x), f''(x), \ldots, f^{(n)}(x)) \equiv 0$$

for all x in S.

If x_0 is a left (right) endpoint of an interval in S, derivatives on the right (left) of f at x_0 are intended. NOTE: The variables x and y need not appear in (1.1); however, at least one derivative must appear if (1.1) is to be termed a DE.

EXAMPLE 1. The function defined by $y = f(x) = e^{2x} \equiv \exp(2x)$ is a solution of the DE $y' - 2y = 0$ on $(-\infty, +\infty)$, since $2e^{2x} - 2(e^{2x}) \equiv 0$ for all x. The function defined by $y = g(x) = e^{2x}$ having domain $[0, 1]$ is also a solution of $y' - 2y = 0$.

In many applications we seek a solution of a DE where the domain of the solution is a specified interval. For example, if t denotes time in a DE, a solution is often sought on the interval $t \geq 0$. If no domain is specified, we seek a solution (or solutions) having the largest possible domain, consisting of one or more intervals.

EXAMPLE 2. The function defined by $y = \sqrt{1 - x^2}$ is a solution of the DE $yy' + x = 0$ on $(-1, 1)$, since

$$\sqrt{1 - x^2}\left(\frac{-2x}{2\sqrt{1 - x^2}}\right) + x \equiv 0$$

for all x in $(-1, 1)$. Note that y is undefined on $(-\infty, -1) \cup (1, +\infty)$, and that y' is undefined at $x = +1$ and at $x = -1$.

EXAMPLE 3. Find solutions of the DE $y' = x^{-2}$.

Solution: Since $d/dx(-x^{-1} + C) = x^{-2}$ for all $x \neq 0$, then $y = -x^{-1} + C$ defines a family of solutions on $S = (-\infty, 0) \cup (0, +\infty)$, where C is an arbitrary constant. Thus, the DE has an infinite number of solutions, each having domain S.

EXAMPLE 4. Find solutions of $y' = \dfrac{x}{\sqrt{x^2 - 1}} - \dfrac{x}{\sqrt{9 - x^2}}$.

Solution: Since

$$\frac{d}{dx}\left(\sqrt{x^2 - 1} + \sqrt{9 - x^2} + C\right) = \frac{x}{\sqrt{x^2 - 1}} - \frac{x}{\sqrt{9 - x^2}}$$

then

$$y = \sqrt{x^2 - 1} + \sqrt{9 - x^2} + C$$

defines a family of solutions, each with domain $S = (-3, -1) \cup (1, 3)$.

EXAMPLE 5. Find solutions of $y' = \dfrac{3\sqrt{x}}{2}$.

Solution: Since

$$\frac{d}{dx}(x^{3/2} + C) = \frac{3\sqrt{x}}{2}$$

then $y = f(x) = x^{3/2} + C$ defines a family of solutions, each having domain $S = [0, +\infty)$. By $f'(0) = 0$ we mean the derivative on the right of f at $x = 0$.

EXAMPLE 6. The DE $(dy/dx)^2 + x^2y^2 + 1 = 0$ has no real-valued solution, since the left member is positive for all differentiable real functions of a real variable. This example shows us that there is no guarantee that a given DE has even one solution.

EXAMPLE 7. The DE $(dy/dx)^2 + y^2 = 0$ has the *unique* solution f with specified domain $(-\infty, +\infty)$ where $y = f(x) \equiv 0$; that is, the DE has *one and only one* solution. If a solution g existed on $(-\infty, +\infty)$ where $g(x_0) \neq 0$ at some $x = x_0$, the left member of the DE would be positive for $x = x_0$. Note that the solution of this DE involves no arbitrary constants.

EXAMPLE 8. Find solutions of $y' = x^{-1}$ on $(-\infty, 0)$.

Solution: Since $d/dx(\ln|x| + C) = x^{-1}$ on $(-\infty, 0)$, solutions are given by $y = \ln|x| + C$. Since $|x| = -x$ on $(-\infty, 0)$, the solutions are also given by $y = \ln(-x) + C$. The DE has an infinite number of solutions, each having the specified domain $(-\infty, 0)$. By $\ln u$ we mean the natural, or Napierian, logarithm of u.

Definition The order of a differential equation is the order of the highest-order derivative appearing in the equation.

The DE of Examples 1–8 are first-order equations.

EXAMPLE 9. The DE $y'' + xy' = 0$ is a second-order equation.

EXAMPLE 10. The DE $y''' - x^2 = 0$ is a third-order equation.

Definition   An ordinary DE is said to be *linear* if and only if it can be written in the form

$$a_0(x)y^{(n)} + a_1(x)y^{(n-1)} + \cdots + a_{n-1}(x)y' + a_n(x)y = f(x)$$

where f and the coefficients $a_0, a_1, \ldots, a_n$ are continuous functions of x. All other DE are called *nonlinear*. (NOTE: $y^{(n)}$ denotes $d^n y/dx^n$.)

Thus a linear DE is linear in y and its derivatives.

EXAMPLE 11. The DE $y'' + xy' + x^2 y - e^x = 0$ is linear, since it is linear in y, y', and y''.

EXAMPLE 12. The DE $y'' + \cos y = 0$ is nonlinear, since it is not linear in y.

EXAMPLE 13. The DE $y'' + yy' + x = 0$ is nonlinear, since y, the coefficient of y', denotes an unknown function of x instead of a specific function of x.

In Chapter 5 we will encounter the linear DE

$$L\frac{d^2 i}{dt^2} + R\frac{di}{dt} + \frac{1}{C}i = \frac{d}{dt}E(t)$$

and the nonlinear DE

$$\frac{d^2\theta}{dt^2} + \frac{g}{l}\sin\theta = 0$$

In the first DE, which governs the flow of current in an electric circuit, L, R, and C are constants, i denotes the current in amperes, and E is the voltage at time t.

In the second DE, which governs the motion of a simple pendulum, g and l are constants and θ denotes the angle between the pendulum rod and the vertical.

The first-order DE $y' = \cos x$ has a solution given by $y = \sin x$, valid for all x. It also has an infinite number of solutions given by $y = \sin x + C$, where C is an arbitrary constant.

The second-order DE $y'' = 6x$ has solutions given by $y = x^3 + C_1 x + C_2$, valid for all x, where C_1 and C_2 are arbitrary constants. These solutions are found by integrating $y'' = 6x$ to obtain $y' = 3x^2 + C_1$, and then integrating a second time.

In general, the solution of an nth-order DE contains n arbitrary constants. The n constants are said to be essential if it is not possible to write the solution in a form involving fewer than n constants. For example, the function given by Ae^{B+x} appears to contain two essential arbitrary constants, but in fact contains only one. This is readily seen by writing

$$Ae^{B + x} = Ae^B e^x = Ce^x$$

where Ae^B is replaced by the single arbitrary constant C. (For a more complete discussion of essential arbitrary constants see Reference 1.1.) We shall assume that any constants appearing in a solution are essential unless we state otherwise.

Definition   By a *general solution* of an nth-order DE, we mean a solution containing n essential arbitrary constants.

If *every* solution of a DE can be obtained by assigning particular values to the n arbitrary constants in a general solution, that general solution is called a *complete solution*. A solution of the DE that cannot be obtained from a general solution by assigning particular values to the arbitrary constants is called a *singular solution*.

The term *general solution* is unsatisfactory since a general solution may or may not be a complete solution. Some authors use the term "general solution" only when treating linear DE. The reason for this will appear in Chapter 4. More theoretical developments prefer to focus attention on differential systems that include DE and to prove existence and uniqueness theorems for such systems. (This approach will be discussed in Sections 1.2 and 2.7.)

EXAMPLE 14.   The solution given by $y = \sin x + C$ is a general solution of the DE $y' = \cos x$. This solution is also a complete solution, since two functions having a derivative given by $\cos x$ can differ by at most a constant.

EXAMPLE 15.   The solution given by $y = Cx - C^2$, where C is an arbitrary constant, is a general solution of the first-order DE $(dy/dx)^2 - x\, dy/dx + y = 0$. This is not a complete solution, since the DE also has the singular solution given by $y = x^2/4$. The singular solution cannot be obtained from $y = Cx - C^2$ by assigning a particular value to C.

Definition   Any solution of a DE that can be obtained from a general solution by assigning values to the essential arbitrary constants is called a *particular solution*.

For example, by setting $C = 0$ in Example 14, we obtain the particular solution given by $y = \sin x$ of the DE $y' = \cos x$.

EXAMPLE 16.   Solutions of the DE $y' = 2x^{-3}$ are given by $y = C - x^{-2}$, and the domain of each solution is the set $S = (-\infty, 0) \cup (0, +\infty)$.

The following example illustrates a partial DE. (We shall consider partial DE in Chapter 10.)

EXAMPLE 17. Show that the function f defined by $z = f(x, t) = (x - 4t)^2$ is a solution of the *partial* DE

$$\frac{\partial^2 z}{\partial t^2} = 16 \frac{\partial^2 z}{\partial x^2}$$

Solution: From

$$\frac{\partial z}{\partial x} = 2(x - 4t), \quad \frac{\partial^2 z}{\partial x^2} = 2$$

$$\frac{\partial z}{\partial t} = -8(x - 4t), \quad \frac{\partial^2 z}{\partial t^2} = 32$$

we obtain $32 = 16(2)$, true for all x and all t. The domain of f is the set of all ordered pairs (x, t) of real numbers.

A special type of DE is classified according to degree.

Definition If the DE

$$F(x, y, y', y'', \ldots, y^{(n)}) = 0 \qquad (1.2)$$

can be expressed as a polynomial in $y, y', y'', \ldots, y^{(n)}$, the exponent of the highest-order derivative is called the *degree* of the DE.

EXAMPLES. The differential equations

$$y''' - x^2 = 0 \qquad y'' - y'^2 + x = 0 \qquad y'' - 3y' + 2y - e^x = 0$$

$$x^2 y'' + xy' + x^2 = 0 \quad \text{and} \quad \frac{\partial^2 z}{\partial x^2} + \frac{\partial^2 z}{\partial y^2} = 0$$

are first-degree equations. The DE

$$(y')^2 - xy' + y = 0 \quad \text{and} \quad (y')^2 - xy^3 = 0$$

are second-degree equations. The DE $y'' = \pm\sqrt{1 + y'}$ is also a second-degree equation since it can be written in the form $(y'')^2 - y' - 1 = 0$. The DE $y'' - \ln y = 0$ and $y' - \cos y = 0$ have no degree since neither can be written in the form (1.2).

We make the usual assumption that the DE

$$F(x, y, y', y'', \ldots, y^{(n)}) = 0 \qquad (1.2)$$

can be solved for the highest-order derivative appearing; that is, that it can be written in the normal form

$$y^{(n)} = G(x, y, y', y'', \ldots, y^{(n-1)}) \qquad (1.3)$$

which is a DE of degree one.

The reason for this assumption is that DE (1.2) may have the same solutions as two or more DE of the form (1.3). The assumption enables us to treat one DE at a time.

For example, the DE $(y')^2 - 4x^2 = 0$ has the same solutions as the two DE

$$y' = 2x \quad \text{and} \quad y' = -2x$$

having solutions given by $y = f(x) = x^2 + C_1$ and $y = g(x) = -x^2 + C_2$. It is easy to verify that f and g define solutions of $(y')^2 - 4x^2 = 0$ on $(-\infty, +\infty)$. Under our assumption, we would treat $y' = 2x$ and $y' = -2x$ separately.

Problem List 1.1

1. Verify that the following expressions define functions that are solutions of the given DE.
   (a) $y = \sin x$, $y'' + y = 0$
   (b) $y = e^{3x}$, $y'' - 9y = 0$
   (c) $y = \dfrac{x^3}{2} + 5x$, $xy' - x^3 - y = 0$
   (d) $y = \tan x$, $y' - y^2 - 1 = 0$
   (e) $y = \cos x$, $y'^2 + y^2 = 1$
   (f) $y = a \sin x + b \cos x$, $y'' + y = 0$
   (g) $y = \tan x$, $y' - y^2 - 1 = 0$
   (h) $y = x^3 + \exp(x^2)$, $y' - 2xy = 3x^2 - 2x^4$

2. Find a general solution of each of the following DE.
   (a) $y' = x^2 - 2x$ (b) $y' = x^3 + x - 2$
   (c) $y' = 3x^4 - 4x^2$ (d) $y' = 2x^3 - x^2 + x - 4$

3. Find a general solution of each of the following DE.
   (a) $y' - e^x = 0$ (b) $y'' - e^x = 0$
   (c) $y'' - \cos x = 0$ (d) $y''' - x + 1 = 0$
   (e) $y' - x \ln x = 0$

4. For what values of x does $y = \sqrt{x^2 - 1}$ define a solution of $x - y(dy/dx) = 0$?

5. Show that the function defined by $y = A \sin 2x + B \cos 2x$ is a solution of $y'' + 4y = 0$, where A and B are arbitrary constants.

6. Show that the function defined by $y = x \int_0^x e^{-t^2}\, dt$ is a solution of $xy' - y - x^2 e^{-x^2} = 0$.

7. Give the order of each of the following DE and state whether it is linear or nonlinear.
   (a) $\dfrac{d^2 y}{dx^2} + xy = 0$ (b) $\dfrac{dy}{dt} + t^2 y = 0$ (c) $\dfrac{d^3 x}{dy^3} + \cos y = 0$

(d) $\dfrac{d^3y}{dx^3} + \cos y = 0$ (e) $\dfrac{d^2r}{d\theta^2} + r\dfrac{dr}{d\theta} = 0$ (f) $x^2y'' + xy' + x^2 - 1 = 0$

(g) $\dfrac{dy}{dx} = x + y^2$ (h) $\dfrac{dy}{dx} = x^2 + y$

8. State the degree of each of the following DE.
   (a) $y'' - 5y' + 4y = \sin x$
   (b) $y'' = (y')^{2/3}$
   (c) $(y'')^2 + 3(y')^3 = 0$
   (d) $y' + \ln y - 1 = 0$
   (e) $(y')^{1/2} = (1 + y'')^{1/3}$
   (f) $\dfrac{\partial z}{\partial t} = 4\dfrac{\partial^2 z}{\partial x^2}$

9. Find by inspection a particular solution of $y'' + 2y' - 8y = 16$.

10. For what values of m does $y = \exp(mx)$ define a solution of $y'' + y' - 12y = 0$?

11. Find a DE having the function defined by $y = e^t + e^{-t}$ as a solution.

12. Given that the equations $y = f(x)$ and $y = g(x)$ define solutions of $y'' + y = 0$, show that $y = Af(x) + Bg(x)$, where A and B are arbitrary constants, also defines solutions of the same DE.

13. Show that the three arbitrary constants A, B, and C in the expression $y = A\sin^2 x + B\cos^2 x + C\cos 2x$ are not essential; that is, show that y can be expressed by using fewer than three arbitrary constants.

14. The DE $(dy/dx)^2 - 4x\,dy/dx + 4y = 0$ has a general solution defined by $y = cx - c^2/4$. Use the function given by $y = x^2$ to show that this general solution is not a complete solution.

15. Find the unique solution of $y'^2 + 4y^2 = 0$.

16. Find a solution of $y' = \ln x$ on $(0, +\infty)$.

17. Show that $y = 2xe^{x-1}$ defines a solution of $y'' - 2y' + y = 0$, satisfying $y(0) = 0$ and $y(1) = 2$.

18. Show that $|dy/dx| + |y| + 3 = 0$ has no solution.

19. Find the unique solution of $y'^2 + 3|y| = 0$ on $(-\infty, +\infty)$.

20. Find solutions of $dy/dx = 1/x$ on $(0, +\infty)$.

21. Show that the function defined by $z = x^3 - 3xy^2$ is a solution of the partial DE $z_{xx} + z_{yy} = 0$.

22. Determine a so that $y = \sin ax$ defines a solution of $y'' + 4y = 0$.

23. Show that $y = x^2 + \cos 3x$ defines a solution of $y'' + 9y = 9x^2 + 2$.

24. Find a solution by inspection:

(a) $y' - y = 0$ (b) $y'' + y = 0$ (c) $\left|\dfrac{dy}{dx}\right| + |x| = 0$

25. (a) A DE of the form

$$y = x\left(\dfrac{dy}{dx}\right) + f\left(\dfrac{dy}{dx}\right)$$

is known as *Clairaut's equation*, named after the French mathematician Alexis Claude Clairaut (1713–1765). Show that $y = cx + f(c)$ defines a solution of the equation.

(b) Find two solutions of $y = xy' - (y')^2$ for which $y = -9$ when $x = 0$.

26. Find a DE for which the function given by $y = |x|$ is a solution.
    (a) on $-2 < x < -1$; (b) on $1 < x < 2$
27. Find the particular solution in each of the following cases.
    (a) $y' - \sec^2 x = 0$; $x = 0$, $y = 1$
    (b) $y' + e^x = 0$; $x = 1$, $y = 1$
    (c) $y'' - 1 = 0$; $x = 0$, $y = 1$, $y' = 2$
    (d) $y'' = 0$; $x = 1$, $y = 2$, $y' = -1$
28. Show that any differentiable function y defined implicitly by $x^2 - 2xy - y^2 = c$ is a solution of the DE $dy/dx = (x - y)/(x + y)$.
29. Show that any differentiable function defined by the parametric equations $x = \cos^3 t$, $y = \sin^3 t$ is a solution of the DE $x(dy/dx)^3 + y = 0$.
30. Find a solution of the form $x = g(y)$ of the DE $dy/dx = ky$. Solve $x = g(y)$ for y in terms of x to obtain a solution of the form $y = f(x)$.
31. Show that

$$y = f(x) = \begin{cases} 0 & \text{for } x \le 0 \\ x^2 & \text{for } x > 0 \end{cases}$$

defines a solution of $y'^2 - 4y = 0$ on $(-\infty, +\infty)$.

32. Let

$$f(x) = \begin{cases} 0 & \text{for } -1 < x < 0 \\ 1 & \text{for } 0 \le x < 1 \end{cases}$$

Show that no solution of $y' = f(x)$ having domain $(-1, 1)$ exists.

1.2 Existence and Uniqueness of Solutions

The basic problem of the quantitative theory of differential equations is to find all solutions of a given DE. This is a formidable task, since most DE do not have solutions that can be expressed as combinations of the trigonometric, inverse trigonometric, exponential, logarithmic, and algebraic functions encountered in the calculus.

An important problem in the qualitative theory of DE consists of proving that under certain conditions, solutions of a given DE exist. Theorems of this nature, termed *existence theorems*, were introduced by the French mathematician Augustin-Louis Cauchy (1789–1857) in the 1820s. His investigations marked the beginning of the modern theory of DE. Existence theorems were later studied by such prominent mathematicians as Lipschitz (1832–1903), Picard (1856–1941), Poincaré (1854–1912), Peano (1858–1932), and Liapounoff (1857–1918). Another important qualitative problem consists of proving that a particular solution satisfying certain given conditions is unique; that is, that the solution is the only solution. Theorems of this kind are termed *uniqueness theorems*. When a DE is used as a mathematical model to describe a physical situation, it is gratifying to know in advance that the DE has a unique solution satisfying the given conditions of the physical problem.

If the DE did not have a unique solution, its appropriateness as a model would be in serious question.

Another aspect of the qualitative theory of DE involves the study of properties of solutions without actually solving the given DE. For example, we can investigate the boundedness of solutions, the periodicity of solutions, the existence of asymptotes of the solution curves, and we can try to construct approximate graphs of solutions. We can also use numerical methods to approximate values of solutions of DE to any required degree of accuracy. Computing machines are very useful in this type of attack.

We now establish an existence and uniqueness theorem for DE of the form $y' = f(x)$.

Theorem 1-I Let f be continuous on an interval I containing a and let k be a given constant. Then the DE $y' = f(x)$ has a unique solution ϕ on I such that $\phi(a) = k$.

Proof: Since f is continuous on I, the indefinite integral F of f, given by $F(x) = \int_a^x f(t)\,dt$, exists for all $x \in I$. By the first form of the fundamental theorem of calculus, $F'(x) = f(x)$ on I. Let ϕ be the function defined on I by

$$y = \phi(x) = k + \int_a^x f(t)\,dt$$

Since $\phi'(x) = f(x)$ on I and $\phi(a) = k$, therefore ϕ is a solution guaranteed by the theorem. This completes the existence portion of the theorem.

To prove that ϕ is the only solution of $y' = f(x)$, let ψ denote an *arbitrary* solution on I of $y' = f(x)$ for which $\psi(a) = k$. Since ϕ and ψ are primitives, or antiderivatives, of f on I,

$$\psi(x) - \phi(x) = c = \text{constant on } I$$

and hence

$$\psi(a) - \phi(a) = k - k = 0 = c$$

Thus, $\psi(x) \equiv \phi(x)$ on I; that is, no solution different from ϕ exists. In other words, the solution ϕ is unique.

EXAMPLE 1. Solve the DE

$$y' = f(x) = 3x^2 + 2x - 1 \tag{1.4}$$

Solution: Since the antiderivative of a sum is the sum of the antiderivatives, and since any two antiderivatives of f differ by at most a constant, a complete solution is given by

$$y = x^3 + x^2 - x + C$$

Specific or particular solutions can be obtained by assigning proper values to C and every solution has domain $(-\infty, +\infty)$.

Example 2. Solve the DE $y' = e^{-x^2}$, given that $y = 1$ when $x = 0$.

Solution: By Theorem 1-I the unique solution is given by

$$y = \phi(x) = 1 + \int_0^x e^{-t^2}\, dt$$

Since the function f given by $f(x) = e^{-x^2}$ is continuous on $(-\infty, +\infty)$, the solution ϕ has domain $(-\infty, +\infty)$. The DE is regarded as "solved," even though we cannot readily evaluate $\phi(x_0)$ for a given x_0. We could estimate $\phi(x_0)$ by means of an approximating technique. The approximate integration can be performed on a computer and the results presented in a table. A DE is customarily considered solved when its solution has been reduced to one or more integrations (sometimes referred to as *quadratures*), and the integrals involved are known to exist. This explains why a solution of a DE is sometimes referred to as an *integral* of the DE.

If we employ the extensively tabulated error function defined by

$$\text{erf } x = \frac{2}{\sqrt{\pi}} \int_0^x e^{-t^2}\, dt$$

we can write our solution as

$$y = 1 + \frac{\sqrt{\pi}}{2} \text{ erf } x$$

The important point is not that erf is a known function, but rather that $\int_0^x e^{-t^2}\, dt$ defines a function.

Example 3. Show that any differentiable function defined implicitly on an interval I by the equation

$$\ln (x^2 + y^2) = \text{Tan}^{-1} \frac{y}{x} + C \tag{1.5}$$

is a solution on I of the DE

$$\frac{dy}{dx} = \frac{2x + y}{x - 2y}$$

Solution: Differentiating implicitly, we obtain

$$\frac{2x + 2yy'}{x^2 + y^2} = \frac{1}{1 + y^2 x^{-2}} \cdot \frac{xy' - y}{x^2} = \frac{xy' - y}{x^2 + y^2}$$

Multiplying both sides by $x^2 + y^2$, where $x^2 + y^2 \neq 0$, and solving for y' yields

$$y' = \frac{dy}{dx} = \frac{2x + y}{x - 2y}$$

The given DE is regarded as "solved," with solution (or solutions) defined implicitly by Equation (1.5). See Reference 1.8 for a discussion of the function defined by $\phi(x) = \tan^{-1} x$.

The question of whether or not an implicit equation defines one or more differentiable functions belongs to the theory of implicit functions. (See Reference 1.7.)

We also regard a DE as solved if we can express a solution (or solutions) parametrically by equations of the form $y = \phi(t)$, $x = \psi(t)$.

EXAMPLE 4. Show that the parametric equations

$$y = 5 - e^{-t}; \quad x = 2e^t; \quad -\infty < t < +\infty$$

define a solution of the DE

$$\frac{dy}{dx} = \frac{5 - y}{x} \qquad \text{for } 0 < x < +\infty$$

Solution:

$$\frac{dy}{dx} = \frac{dy/dt}{dx/dt} = \frac{e^{-t}}{2e^t} = \frac{5 - (5 - e^{-t})}{2e^t} = \frac{5 - y}{x}$$

It is interesting to note that DE often give rise to, or define, new functions. For example, the DE

$$x^2 y'' + xy' + (x^2 - p^2)y = 0 \qquad \textbf{(1.6)}$$

where p is a nonnegative constant, is known as *Bessel's equation*, after the German astronomer Friedrich Wilhelm Bessel (1784–1846), who encountered it in a problem on planetary motion. The same equation arises in problems involving heat flow in cylinders, propagation of electric currents in cylindrical conductors, vibration of membranes, vibration of chains, and in numerous other important investigations. This is one of the many instances in which the same mathematical model is appropriate for describing several different physical situations. The solutions of (1.6) have been investigated in detail, have been extensively tabulated, and are termed *Bessel functions*. (We consider them in Chapters 8 and 10.)

Another example is furnished by the DE $d^2 y/d\theta^2 + y = 0$, from which it is possible to develop analytic trigonometry. The solution for which $y = 0$ when $\theta = 0$ is given by $y = \sin \theta$, the solution for which $y = 1$ when $\theta = 0$ is given by $y = \cos \theta$, and a complete solution is given by $y = C_1 \sin \theta + C_2 \cos \theta$.

A complete solution is also given by $y = A \sin (\theta + \phi)$, where $A = \sqrt{C_1^2 + C_2^2}$ and $\phi = \mathrm{Tan}^{-1}\left(\dfrac{C_2}{C_1}\right)$.

EXAMPLE 5. Solutions of $y' = 2x^{-3}$ are given by $y = C - x^{-2}$, each solution having domain $S = (-\infty, 0) \cup (0, +\infty)$. By Theorem 1-I, $y = C - x^{-2}$ defines a complete solution on $(-\infty, 0)$ and also on $(0, +\infty)$. On $S = (-\infty, 0) \cup (0, +\infty)$, however, $y = C - x^{-2}$ does not define a complete solution. The solution g given by

$$y = g(x) = \begin{cases} 1 - x^{-2} & \text{on } (-\infty, 0) \\ 2 - x^{-2} & \text{on } (0, +\infty) \end{cases}$$

is a solution on S and yet it cannot be obtained by assigning a single value to C. This difficulty does not arise in an application in which the domain of a solution consists of a single interval.

Problem List 1.2

1. Find a complete solution of
   (a) $y' = 6x^2 - 2x$ on $(-\infty, +\infty)$
   (b) $y' = 4 \tan x \sec^2 x$ on $\left(-\dfrac{\pi}{2}, \dfrac{\pi}{2}\right)$
   (c) $y' = \ln x$ on $(0, +\infty)$

2. Find the solution of $y' = 2 \exp(-x^2)$ for which $y = 3$ when $x = 1$. State the domain of the solution.

3. Why doesn't Theorem 1-I apply to the DE $y' = e^{1/x}$ on $(-1, +1)$?

4. Show that

$$y = \begin{cases} C_1 x^2 & \text{for } x \leqslant 0 \\ C_2 x^2 & \text{for } x > 0 \end{cases}$$

   defines a solution of $xy' - 2y = 0$ for all C_1, C_2.

5. Show that any differentiable function defined implicitly on an interval I by

$$x^3 y - 3x^2 + y^3 + C = 0$$

   is a solution on I of the DE

$$(3x^2 y - 6x) + (3y^2 + x^3)y' = 0$$

6. Show that the parametric equations

$$x = C(\theta - \sin \theta), \quad y = C(1 - \cos \theta); \quad 0 < \theta < 2\pi$$

   define functions that are solutions of the DE

$$1 + \left(\frac{dy}{dx}\right)^2 + 2y\frac{d^2y}{dx^2} = 0$$

7. Solve $y' = |x|$ on $(-\infty, +\infty)$ given that (a) $y = 6$ when $x = 2$ and (b) $y = 6$ when $x = -2$.

8. Find a solution ϕ of $y' = 2x^{-1}$ on $(-\infty, 0) \cup (0, +\infty)$ given that $\phi(2) = 4$ and $\phi(-2) = 6$.

1.3 Initial-Value and Two-Point Boundary-Value Problems

When we solve DE, we often find their solutions in forms from which specific values of these solutions can be determined.

EXAMPLE 1. Solve the DE $y' = x^2$, given that $y = 2$ when $x = 1$.

Solution I: A general solution, which is also a complete solution, is given by $y = (x^3/3) + C$. Substituting $x = 1$ and $y = 2$, we obtain $2 = \frac{1}{3} + C$, and hence C must equal $\frac{5}{3}$ to satisfy the given condition. Hence the required solution, unique by Theorem 1-I, is given by $y = (x^3/3) + \frac{5}{3}$ and has domain $(-\infty, +\infty)$.

Solution II: $y = F(x) = C + \int_1^x t^2 \, dt$, since the right member is the general anti-derivative of x^2. Setting $x = 1$, we obtain $2 = F(1) = C + 0$. Hence,

$$y - 2 = \int_2^y du = \int_1^x t^2 \, dt \tag{1.7}$$

$$y - 2 = \left. \frac{t^3}{3} \right]_1^x = \frac{x^3}{3} - \frac{1}{3}$$

or

$$y = \frac{x^3}{3} + \frac{5}{3}$$

Solution II differs from Solution I mainly in notation. When the notation of (1.7) is used, the lower limits on the two integral signs are corresponding values of y and x, the upper variable limits also correspond, and the variables u and t are dummy variables.

Variables x and y in Example 1 can be made to satisfy one condition, since our general solution contains one arbitrary constant. A problem of this type is called an *initial-value* problem, because in many applications one variable is the time, and the specified condition gives the value of the other variable at the initial time $t = 0$.

EXAMPLE 2. Solve the second-order initial-value problem $y'' = 6x$; $y = 3$, $y' = 2$ when $x = 0$.

Solution: In $y' = 3x^2 + c$, we set $x = 0$ and $y' = 2$ to obtain $y' = 3x^2 + 2$.
In $y = x^3 + 2x + k$, we set $x = 0$ and $y = 3$ to obtain $y = x^3 + 2x + 3$.

EXAMPLE 3. Solve the DE $y'' = 12x$, given that $y = 0$ when $x = 0$, and $y = 6$ when $x = 2$.

Solution: Forming antiderivatives, we obtain $y' = 6x^2 + c$, and

$$y = 2x^3 + cx + k \tag{1.8}$$

Applying the given conditions to (1.8) yields $0 = k$ and $6 = 16 + 2c + k$.
The required solution $y = 2x^3 - 5x$ is obtained by substituting $k = 0$ and $c = -5$ in (1.8).

In Example 3, variables x and y were made to satisfy two conditions and (1.8) contained two arbitrary constants. This type of problem is called a *two-point boundary-value problem*, since the given conditions usually involve the endpoints of the interval that is of interest and importance in the problem.

The solution of many important problems in applied mathematics involves DE. This is because physical laws are usually stated as DE.

EXAMPLE 4. The motion of a particle of mass m moving on the x axis is governed by Newton's second law, which states that the force f acting on the particle equals the mass m times the acceleration a of the particle. Since $a = d^2x/dt^2$, Newton's second law is expressed as a second-order DE. When we say that the motion is governed by the DE, we mean that the displacement function, denoted by $x = h(t)$, is a solution of the DE. If the force f depends on the time t, or the velocity dx/dt, or the displacement x, or some combination of these three quantities, the DE will have the form

$$F\left(t, x, \frac{dx}{dt}, \frac{d^2x}{dt^2}\right) = 0 \tag{1.9}$$

which is of the type (1.1), with t replacing x and x replacing y.

A general solution of (1.9) would contain two essential arbitrary constants. These constants could be determined from initial conditions of the form

$$t = 0, \quad x = x_0 \qquad t = 0, \quad \frac{dx}{dt} = v_0$$

The resulting particular solution would be given by the displacement function denoted by $x = h(t)$. The domain of h would be the time interval of interest in the problem. In summary, if we know the initial displacement and velocity of the particle, and also the DE governing the motion, a solution of the problem consists of the displacement function, which gives the position of the particle at variable time t. The DE and the initial conditions furnish a mathematical model for the physical situation.

EXAMPLE 5. A particle moves on the x axis with acceleration $a = 6t - 4$ ft/sec^2. Find the position and velocity of the particle at $t = 3$ if the particle is at the origin and has velocity 10 ft/sec when $t = 0$.

Solution I: We first solve the initial-value problem:

$$a = \frac{dv}{dt} = \frac{d^2x}{dt^2} = 6t - 4; \quad v = 10, \quad x = 0, \quad t = 0$$

$$v = 3t^2 - 4t + c$$

$$10 = 0 - 0 + c$$

and

$$\frac{dx}{dt} = v = 3t^2 - 4t + 10 \qquad (1.10)$$

$$x = t^3 - 2t^2 + 10t + k$$

$$0 = 0 - 0 + 0 + k$$

$$x = t^3 - 2t^2 + 10t \qquad (1.11)$$

Substituting $t = 3$ in (1.10) and (1.11), we obtain $v_3 = 25$ ft/sec and $x_3 = 39$ ft.

Solution II:

$$\int_{10}^{v} dz = \int_{0}^{t} (6w - 4)\, dw$$

$$v - 10 = 3t^2 - 4t$$

$$v_3 = 27 - 12 + 10 = 25 \text{ ft/sec}$$

and

$$\frac{dx}{dt} = 3t^2 - 4t + 10$$

$$\int_{0}^{x_3} du = \int_{0}^{3} (3t^2 - 4t + 10)\, dt$$

$$x_3 = t^3 - 2t^2 + 10t]_{0}^{3} = 39 \text{ ft}$$

EXAMPLE 6. A ball is released from a balloon that is 192 ft above the ground and rising at 64 ft/sec. Find the distance over which the ball continues to rise and the time elapsed before the ball strikes the ground. The flight of the ball is depicted in Figure 1.1.

Solution: Let the displacement s of the ball be measured from the point at which the ball is released and let s be positive upward. Also let the time t be measured from the instant the ball is released. Assume that the ball has constant acceleration $a = dv/dt = -32$ ft/sec^2. The acceleration, known as the *acceleration of gravity*, is negative, since

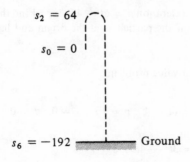

Fig. 1.1

the velocity decreases with time. The initial velocity of the ball is the same as the velocity of the balloon, and we must solve the initial-value problem:

$$a = \frac{dv}{dt} = \frac{d^2s}{dt^2} = -32; \quad \begin{Bmatrix} t = 0 \\ v = 64 \end{Bmatrix}, \quad \begin{Bmatrix} t = 0 \\ s = 0 \end{Bmatrix}$$

From $v = -32t + c$, we obtain $64 = 0 + c$, and hence

$$v = \frac{ds}{dt} = -32t + 64$$

From $s = -16t^2 + 64t + k$, we obtain $0 = 0 + 0 + k$, and hence

$$s = -16t^2 + 64t$$

The ball continues to rise until $v = -32t + 64 = -32(t - 2) = 0$, or $t = 2$ sec. The distance traveled in the first two seconds is

$$s_2 = -16t^2 + 64t|_{t=2} = 64 \text{ ft}$$

To find the time before the ball strikes the ground, we set $s = -192$ and solve for t:

$$-192 = -16t^2 + 64t$$

$$t^2 - 4t - 12 = 0$$

$$(t - 6)(t + 2) = 0$$

The answer is 6 sec, since $t \geqslant 0$.

Remark: If we decided to measure s positive downward, v_0 would have been -64 and a would have been $+32$. We could also have chosen $s = 0$ at the ground level. The important point is that we must first choose our coordinate system and then use it consistently throughout the problem.

Problem List 1.3

1. Solve the following initial-value problems.
   (a) $y' = 6x; x = 0, y = 4$
   (b) $y' = -2x; x = 0, y = 7$
   (c) $y' = 4; x = 0, y = -3$
   (d) $y' = x - 3; x = 0, y = 8$
   (e) $y' = x^2 - 2x - 3; x = 3, y = 1$
   (f) $y' = 3x^2 + 4; x = -2, y = 3$
2. Solve the following initial-value problems.
   (a) $y'' = 3x; y = 5, y' = 2$ when $x = 0$
   (b) $y'' = 6x - 8; y = 2, y' = -1$ when $x = 0$
   (c) $y'' = 6x + 10; y = 8, y' = 13$ when $x = 1$
   (d) $y'' = 12x^2; y = -1, y' = -7$ when $x = -1$
   (e) $y''' = 24x + 6; y = 1, y' = 0, y'' = 2$ when $x = 0$
   (f) $y''' = 60x^2; y = 2, y' = 3, y'' = 18$ when $x = 0$

3. Solve the initial-value problem

$$xy' - 2 = 0; \quad x = 1, \quad y = 2 \ln 3$$

Find (a) y when $x = 2$, and (b) x when $y = 0$.

4. Find a solution of $y'' = 6x - 8$ such that $y(2) = 9$ and $y'(1) = 2$.

5. Solve the following two-point boundary-value problems.
   (a) $y'' = 2; x = 0, y = -1; x = 1, y = 5$
   (b) $y'' = 12x; x = 0, y = -3; x = 1, y = 0$
   (c) $y'' = -6x + 2; x = 1, y = -3; x = 3, y = -21$
   (d) $y'' = 12x^2; x = -1, y = -3; x = 2, y = 21$

6. The variables x and y satisfy the DE $y' = -14x + 14$. For $y = -1$ when $x = 0$, find the maximum value of y.

7. A particle moves on the x axis with velocity $v = dx/dt = 2t + 3$ ft/sec. If it starts from the point $(-4, 0)$ at time $t = 0$, what is its position and velocity at time $t = 2$?

8. A particle moves on the y axis with velocity $v = dy/dt = 3t^2 - 4$. If it is at $(0, 3)$ at time $t = 2$, what is its position and velocity at time $t = 3$?

9. Solve illustrative Example 6 with s positive downward and $s = 0$ at ground level.

10. A particle moves in a straight line with acceleration $a = 2t - 3$ ft/sec². For $s = 0$ and $v = -4$ when $t = 0$, find (a) the time at which the particle reverses its direction, and (b) the total distance the particle moves in the first 5 sec.

11. A particle moves on the x axis with acceleration $a = 5 - 2t$. If the particle starts from the origin with $v = 0$, how far does it move in the positive direction? Assume $t \geq 0$.

12. A particle moves in a straight line with constant acceleration a. This type of motion is called *uniformly accelerated rectilinear motion*. For $s = 0$ and $v = v_0$ when $t = 0$, show that:

   (a) $v = v_0 + at$ (b) $s = v_0 t + \dfrac{1}{2} at^2$

   (c) $v^2 = v_0^2 + 2as$ (d) $s = \dfrac{1}{2}(v + v_0)t$

   It is important to remember that these formulas, although useful, are valid only when the acceleration is *constant*.

13. A body is uniformly accelerated from 20 ft/sec to 100 ft/sec in 16 sec. Find the distance traveled in that time.

14. A car moving in a straight line has constant acceleration $a = 12$ ft/sec². Find (a) the time required for the velocity to increase from 20 ft/sec to 40 ft/sec, and (b) the distance the car moves in 3 sec, starting from the time when the velocity is 20 ft/sec.

15. Particle A has acceleration $6t$ ft/sec² and starts from the origin with $v = 0$ at $t = 0$. Particle B has constant acceleration 2 ft/sec² and starts from $(15, 0)$ with $v = 1$ ft/sec at $t = 0$. If A and B move along the x axis, where does A overtake B?

16. A bullet is shot upward at 640 ft/sec. Find the distance the bullet rises and its velocity when it strikes the ground. (Use $g = -32$.)

17. It takes 6 sec for a stone to reach the bottom of a vertical mine shaft when it is dropped from rest. Find the depth of the shaft. (Use $g = 32$.)

18. A ball is released from a balloon that is 320 ft above the ground and rising at 32 ft/sec. Find the length of time the ball continues to rise and the velocity with which it strikes the ground. (Use $g = -32$.)

19. If the brakes on a certain car can produce a constant deceleration of 20 ft/sec^2, what is the maximum allowable velocity if the car is to be stopped in 60 ft or less after the brakes are applied?

20. A box moves on an inclined plane with constant acceleration 12 ft/sec^2. If $v_0 = 6$ ft/sec, how long does it take for the box to slide 120 ft?

21. A particle moves on the x axis with acceleration $a = te^{-t}$, t in sec and x in ft. Find x and $v = dx/dt$ for $t = 1$ if $x = 2$ and $v = 5$ when $t = 0$. Also find

$$\lim_{t \to +\infty} x \quad \text{and} \quad \lim_{t \to +\infty} \frac{dx}{dt}$$

22. Find y in terms of x if:

(a) $\displaystyle\int_2^y du = \int_1^x (3t^2 + 1)\, dt$

(b) $\displaystyle\int_{-1}^y du = \int_3^x (4t + 7)\, dt$

23. Find y when $x = 3$ if:

(a) $\displaystyle\int_1^y du = \int_2^x (t + 1)\, dt$

(b) $\displaystyle\int_2^y du = \int_1^x (9t^2 + t)\, dt$

References

1.1 Agnew, Ralph. 1962. *Differential Equations*, 2nd ed. New York: McGraw-Hill.

1.2 Bell, E. T. 1945. *The Development of Mathematics*. New York: McGraw-Hill.

1.3 Birkhoff, G., and G.-C. Rota. 1962. *Ordinary Differential Equations*. New York: Ginn.

1.4 Ince, E. L. 1959. *Ordinary Differential Equations*. New York: Dover.

1.5 Kaplan, Wilfred. 1962. *Ordinary Differential Equations*. Reading, Mass.: Addison-Wesley.

1.6 Murray, F. J., and K. S. Miller. 1954. *Existence Theorems for Ordinary Differential Equations*. New York: New York University Press.

1.7 Taylor, A. E. 1955. *Advanced Calculus*. New York: Ginn.

1.8 Tierney, J. 1979. *Calculus and Analytic Geometry*, 4th ed. Boston: Allyn and Bacon, Inc.

2

First-Order and
Simple Higher-Order
Differential Equations

2.1 Exact Equations

We consider first-order DE of the form

$$M(x, y) + N(x, y)\frac{dy}{dx} = 0 \qquad (2.1)$$

where M and N denote continuous functions of x and y possessing continuous first partial derivatives in a region R of the xy plane. We also assume that $N(x, y) \neq 0$ in R, thereby ensuring that dy/dx is defined at every point of R. Equation (2.1) is often written in the form

$$M(x, y)\, dx + N(x, y)\, dy = 0 \qquad (2.2)$$

The left member of (2.2) is called a differential form in two variables. An advantage of form (2.2) is that it enables us to regard either y or x as an element of the range of the unknown function or functions. It is sometimes convenient to assume that $M(x, y) \neq 0$ in R and to seek solutions of

$$M(x, y)\frac{dx}{dy} + N(x, y) = 0 \qquad (2.3)$$

given by equations of the form $x = g(y)$. Solutions of (2.1) are then obtained by solving $x = g(y)$ for y in terms of x.

Definition   Equation (2.2) is said to be *exact* in a region R if and only if there exists a function, denoted by $u = u(x, y)$, such that at every point of R,

$$\frac{\partial u}{\partial x} = M \qquad \frac{\partial u}{\partial y} = N$$

Since

$$du = \frac{\partial u}{\partial x}\,dx + \frac{\partial u}{\partial y}\,dy$$

Equation (2.2) is exact if and only if the left member denotes the *total* (*or exact*) *differential* of a function given by $u = u(x, y)$. This is the reason the term *exact* is used.

EXAMPLE 1.   The equation

$$y\,dx + x\,dy = 0$$

is exact, since the left number is seen by inspection to be the differential of the function given by $u = xy$.

The equation

$$(2xy + 3y^3)\,dx + (x^2 + 9xy^2)\,dy = 0$$

is also exact, although the truth of this assertion is not evident by inspection. The following theorem states a necessary condition for an equation to be exact.

Theorem 2-I   If $M(x, y)\,dx + N(x, y)\,dy = 0$ is exact in R, then

$$\frac{\partial M}{\partial y} = \frac{\partial N}{\partial x} \qquad \text{in } R$$

Proof: Since the equation is exact in R, there exists a function given by $u = u(x, y)$ such that $\partial u/\partial x = M$ and $\partial u/\partial y = N$ in R. Therefore,,

$$\frac{\partial M}{\partial y} = \frac{\partial^2 u}{\partial y\,\partial x} \qquad \text{and} \qquad \frac{\partial N}{\partial x} = \frac{\partial^2 u}{\partial x\,\partial y}$$

Since we are assuming that $\partial M/\partial y$ and $\partial N/\partial x$ are continuous in R, a well-known theorem of the calculus guarantees that the mixed partials $\partial^2 u/\partial y\,\partial x$ and $\partial^2 u/\partial x\,\partial y$ are equal in R and hence $\partial M/\partial y = \partial N/\partial x$ in R.

To show that an equation of the form (2.2) is not exact in R, we use the contrapositive of Theorem 2-I. The contrapositive of a theorem is obtained by negating both the hypothesis and the conclusion of the theorem, and then interchanging them. A theorem and its contrapositive are both true or both false. Hence Theorem 2-II is equivalent to Theorem 2-I.

Theorem 2-II   If there is at least one point in R at which $\partial M/\partial y \neq \partial N/\partial x$, then the expression $M(x, y)\,dx + N(x, y)\,dy$ is not exact in R. (Contrapositive of Theorem 2-I.)

EXAMPLE 2.   There is no region in which $x \cos y\,dx + y \cos x\,dy = 0$ is exact since in an arbitrary region R of the xy plane,

$$\frac{\partial}{\partial y}(x \cos y) = -x \sin y \neq \frac{\partial}{\partial x}(y \cos x) = -y \sin x$$

The converse of Theorem 2-I, if true, would furnish a test for exactness. That is, $\partial M/\partial y = \partial N/\partial x$ in R would ensure exactness in R. This converse can be established provided that suitable restrictions are placed on R. For simplicity, we assume R to be the interior of a rectangle with sides parallel to the x and y axes.

Theorem 2-III   If $\partial M/\partial y = \partial N/\partial x$ in the interior of a rectangle with sides parallel to the x and y axes, then the equation $M(x, y)\,dx + N(x, y)\,dy = 0$ is exact in the interior of the rectangle.

Proof: Let (a, b) be any convenient fixed point in the rectangle R and let u be the function defined by

$$u(x, y) = \int_{a}^{x} M(s, b)\,ds + \int_{b}^{y} N(x, t)\,dt \tag{2.4}$$

where (x, y) is an arbitrary point in R. Then

$$\frac{\partial u}{\partial x} = \frac{d}{dx}\int_{a}^{x} M(s, b)\,ds + \frac{\partial}{\partial x}\int_{b}^{y} N(x, t)\,dt$$

Since N and $\partial N/\partial x$ are continuous in a rectangle $b \le t \le y, x_1 \le x \le x_2$,

$$\frac{\partial}{\partial x}\int_{b}^{y} N(x, t)\,dt = \int_{b}^{y} \frac{\partial N(x, t)}{\partial x}\,dt$$

This result, known as Leibniz's rule for differentiating under the integral sign, is proved in Reference 2.2. Hence, since $\partial N/\partial x = \partial M/\partial y$,

$$\frac{\partial u}{\partial x} = \frac{d}{dx} \int_a^x M(s, b)\, ds + \int_b^y \frac{\partial N(x, t)}{\partial x}\, dt$$

$$= M(x, b) + \int_b^y \frac{\partial M(x, t)}{\partial t}\, dt$$

$$= M(x, b) + [M(x, t)]_b^y$$

$$= M(x, b) + M(x, y) - M(x, b) = M(x, y)$$

Also,

$$\frac{\partial u}{\partial y} = \frac{\partial}{\partial y} \int_a^x M(s, b)\, ds + \frac{\partial}{\partial y} \int_b^y N(x, t)\, dt$$

$$= 0 + N(x, y) = N(x, y)$$

This completes the proof of exactness.

The situation is depicted graphically in Figure 2.1. In $\int_a^x M(s, b)\, ds$, the integrand is the function of x assumed by $M(x, y)$ on the line segment between (a, b) and (x, b), y having the constant value b. In $\int_b^y N(x, t)\, dt$, the integrand is the function of y assumed by $N(x, y)$ on the line segment between (x, b) and (x, y), with x regarded as fixed during the integration. Students familiar with line integrals will recognize the right member of (2.4) as the line integral $\int_C M\, dx + N\, dy$ taken over the path C, where C is the "elbow path," consisting of the line segment from (a, b) to (x, b) plus the line segment from (x, b) to (x, y). Line integrals have very important mathematical and physical applications and interpretations.

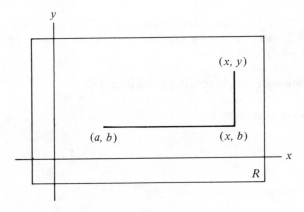

Fig. 2.1

EXAMPLE 3. The equation $(2xy + 3y^3) dx + (x^2 + 9xy^2) dy = 0$ is exact in the entire xy plane, since for all values of x and y,

$$\frac{\partial}{\partial y}(2xy + 3y^3) = 2x + 9y^2 = \frac{\partial}{\partial x}(x^2 + 9xy^2)$$

EXAMPLE 4. The equation $2x \ln y \, dx + x^2 y^{-1} dy = 0$ is exact in any rectangle for which $y > 0$, since

$$\frac{\partial}{\partial y}(2x \ln y) = 2xy^{-1} = \frac{\partial}{\partial x}(x^2 y^{-1})$$

Let $M(x, y) dx + N(x, y) dy = 0$ be exact. Then there exists a function u, which we denote by $u = u(x, y)$, such that $\partial u/\partial x = M$ and $\partial u/\partial y = N$. Solutions of $M(x, y) dx + N(x, y) dy = 0$ are defined implicitly by $u(x, y) = c$, where c is an arbitrary constant. To see this, let $y = f(x)$ denote any differentiable function defined implicitly by $u(x, y) = c$. By the chain rule,

$$\frac{\partial u}{\partial x} + \frac{\partial u}{\partial y}\frac{dy}{dx} = 0$$

$$\frac{\partial u}{\partial x} dx + \frac{\partial u}{\partial y} dy = M(x, y) dx + N(x, y) dy = 0$$

and hence $y = f(x)$ defines a solution of (2.1) or (2.2). To find a solution for which $y_0 = f(x_0)$, one determines the constant c from $u(x_0, y_0) = c$. (For a discussion of the existence and uniqueness of such a solution, see Reference 2.1.)

After applying Theorem 2-III to determine that $M \, dx + N \, dy = 0$ is exact, we still have the problem of obtaining the equation $u(x, y) = c$. The simplest method is to recognize by inspection that $M \, dx + N \, dy$ is an exact differential.

EXAMPLE 5. The equation $2xy \, dx + x^2 \, dy = 0$ is exact since $(\partial/\partial y)(2xy) = 2x = (\partial/\partial x)(x^2)$. Since $d(x^2 y) = 2xy \, dx + x^2 \, dy$, the solutions are given by $x^2 y = c$, or $y = cx^{-2}$, valid on $S = (-\infty, 0) \cup (0, +\infty)$.

The following example requires more ingenuity.

EXAMPLE 6. Solve the initial-value problem

$$\frac{3x^2 y + 1}{y} dx - \frac{x}{y^2} dy = 0; \quad y(2) = 1$$

Solution: The equation is exact since

$$\frac{\partial}{\partial y}\left(\frac{3x^2 y + 1}{y}\right) = -y^{-2} = \frac{\partial}{\partial x}\left(\frac{-x}{y^2}\right)$$

Writing the equation in the form

$$\frac{y\,dx - x\,dy}{y^2} + 3x^2\,dx = 0$$

and remembering that $d(x/y) = (y\,dx - x\,dy)/y^2$, we obtain $x/y + x^3 = c$. Setting $x = 2$ and $y = 1$, we find that $c = 10$. Solving for y, we get the required solution given by $y = x/(10 - x^3)$, valid on $(0, \sqrt[3]{10})$.

The equation $u(x, y) = c$, defining implicitly the solutions of $M\,dx + N\,dy = 0$, can also be obtained directly from (2.4).

EXAMPLE 7. Solve $(2xy + 3y^3)\,dx + (x^2 + 9xy^2)\,dy = 0$, shown in Example 3 to be exact.

Solution: A natural choice for a and b in (2.4) is $a = b = 0$. In this case, however, $N(0, 0) = 0$, so let us choose $a = 1$ and $b = 0$. From (2.4), we obtain

$$u(x, y) = \int_1^x 0\,ds + \int_0^y (x^2 + 9xt^2)\,dt$$

$$= k + [x^2 t + 3xt^3]_0^y = k + x^2 y + 3xy^3$$

Incorporating the arbitrary constant k into an arbitrary constant c, we obtain $x^2 y + 3xy^3 = c$. It is not easy to solve this equation for y in terms of x.

The following example presents a more informal and, in some instances, simpler method of finding $u(x, y)$.

EXAMPLE 8. Solve the equation $(e^x + 4y)\,dx + (4x - \sin y)\,dy = 0$.

Solution: The equation is exact since

$$\frac{\partial}{\partial y}(e^x + 4y) = 4 = \frac{\partial}{\partial x}(4x - \sin y)$$

Remembering that $\partial u/\partial x = e^x + 4y$, we hold y fixed and integrate with respect to x, to obtain $u(x, y) = e^x + 4xy + g(y)$.
We add $g(y)$ since

$$\frac{\partial}{\partial x}[e^x + 4xy + g(y)] = e^x + 4y$$

for an *arbitrary* differentiable function g of y. From

$$\frac{\partial u}{\partial y} = 4x + \frac{dg(y)}{dy} = 4x - \sin y$$

we obtain

$$\frac{dg(y)}{dy} = -\sin y$$

$$g(y) = \cos y + k$$

and

$$u(x, y) = e^x + 4xy + \cos y = c$$

where the arbitrary constant c absorbs the arbitrary constant k.

It is sometimes easier in constructing the function u to hold x fixed and integrate N with respect to y, than to hold y fixed and integrate M with respect to x.

EXAMPLE 9. Solve

$$(x^2 e^{x+y} + 2xe^{x+y} + 2x)\, dx + (x^2 e^{x+y} + 4y)\, dy = 0$$

Solution: The equation is exact since

$$\frac{\partial}{\partial y}(x^2 e^{x+y} + 2xe^{x+y} + 2x) = x^2 e^{x+y} + 2xe^{x+y}$$

$$= \frac{\partial}{\partial x}(x^2 e^{x+y} + 4y)$$

Since the integration of $x^2 e^{x+y} + 2xe^{x+y} + 2x$ with respect to x, with y held fixed, is complicated, we use

$$\frac{\partial u}{\partial y} = x^2 e^{x+y} + 4y$$

We hold x fixed and integrate with respect to y to obtain

$$u(x, y) = x^2 e^{x+y} + 2y^2 + f(x)$$

From

$$\frac{\partial u}{\partial x} = x^2 e^{x+y} + 2xe^{x+y} + \frac{df(x)}{dx}$$

$$= x^2 e^{x+y} + 2xe^{x+y} + 2x$$

we find that

$$\frac{df(x)}{dx} = 2x, \quad f(x) = x^2 + k$$

and

$$u(x, y) = x^2 e^{x+y} + 2y^2 + x^2 = c$$

When the methods of Examples 8 and 9 are combined, $f(x)$ and $g(y)$ can often be found by inspection.

EXAMPLE 10. Solve $(y \cos x - x^2)\, dx + (y + \sin x)\, dy = 0$.

Solution: The equation is exact since

$$\frac{\partial}{\partial y}(y \cos x - x^2) = \cos x = \frac{\partial}{\partial x}(y + \sin x)$$

From $\partial u/\partial x = y \cos x - x^2$, we obtain

$$u(x, y) = y \sin x - \frac{x^3}{3} + g(y)$$

and from $\partial u/\partial y = y + \sin x$, we obtain

$$u(x, y) = \frac{y^2}{2} + y \sin x + f(x)$$

Comparing the two expressions for $u(x, y)$, we see by inspection that $g(y) = y^2/2$ and $f(x) = -x^3/3$. Thus

$$u(x, y) = y \sin x - \frac{x^3}{3} + \frac{y^2}{2} = c$$

Problem List 2.1

1. Show that the following DE are not exact.
   (a) $x^2y\, dx + x^3y^2\, dy = 0$
   (b) $x^2\, dx + y^2x\, dy = 0$
   (c) $-y\, dx + x\, dy = 0$
   (d) $(x^2 + 2xy)\, dx + (y^2 - 2xy)\, dy = 0$
   (e) $(4x \sin y)\, dx + (x^2 \cos y)\, dy = 0$
   (f) $(2x^2y + y^2e^x)\, dx + (x^2 + 2ye^x)\, dy = 0$

2. Show that the following DE are exact.
   (a) $y^2\, dx + 2xy\, dy = 0$
   (b) $\ln y\, dx + xy^{-1}\, dy = 0$
   (c) $(3x^2 \sin y)\, dx + (x^3 \cos y)\, dy = 0$
   (d) $(ye^x - \sin y)\, dx + (e^x - x \cos y)\, dy = 0$

3. Show that the following DE are exact. Solve by inspection.
   (a) $y^2\, dx + 2xy\, dy = 0$
   (b) $e^y\, dx + xe^y\, dy = 0$
   (c) $e^x \ln y\, dx + e^x y^{-1}\, dy = 0$
   (d) $2xy\, dx + (x^2 + 3y^2)\, dy = 0$
   (e) $\sin y\, dx + x \cos y\, dy = 0$

4. Show that the following DE are exact. Solve each DE.
   (a) $2xy\, dx + (x^2 + \cos y)\, dy = 0$
   (b) $(y + 2x)\, dx + (x - 2y)\, dy = 0$
   (c) $3x^2\, dx - 2x^2y\, dy = 2xy^2\, dx$
   (d) $2xy\, dx + (x^2 + \cos y)\, dy = 0$
   (e) $(2xy + x^4)\, dx + (x^2 + 2y)\, dy = 0$
   (f) $2xy^3\, dx + (3x^2y^2 + \cos y)\, dy = 0$
   (g) $(3x^2 + 4y \sin x \cos x)\, dx + (2 \sin^2 x + 3y^2)\, dy = 0$
   (h) $(e^y + y \cos x)\, dx + (\sin x - \sin y + xe^y)\, dy = 0$

(i) $e^x(x + 1) \ln y \, dx + \dfrac{xe^x}{y} \, dy = 0$

(j) $y^2(x \cos x + \sin x) \, dx + 2xy \sin x \, dy = 0$

5. Solve the initial-value problems:

(a) $(y^2 + 2xy^3) \, dx + (2xy + 3x^2y^2) \, dy = 0$, $y(1) = 2$

(b) $(2xy + \cos y) \, dx + (x^2 - x \sin y) \, dy = 0$, $y(2) = 0$

(c) $(x^2 - y) \, dx + (y^2 - x) \, dy = 0$, $y(\frac{4}{3}) = \frac{2}{3}$

6. Solve the initial-value problem $1 \, dx - 3y^2 \, dy = 0$, $y(0) = 0$.
   Hint: Since $N(0, 0) = 0$, seek a solution of the form $x = \phi(y)$.

7. Determine A so that $(3x^2y^2 + Axy) \, dx + (2x^3y + 2x^2) \, dy = 0$ is exact. For $y(1) = 2$ and $y(2) > 0$, find $y(2)$.

8. Show that $(3y^2 + 4xy) \, dx + (2xy + x^2) \, dy = 0$ is not exact. Show also that there exists a positive integer n such that $x^n(3y^2 + 4xy) \, dx + x^n(2xy + x^2) \, dy = 0$ is exact.

9. Solve illustrative Example 9 by the method of Example 8.

10. Show that $du = M(x, y) \, dx + N(x, y) \, dy$ if u is defined by

$$u(x, y) = \int_b^y N(a, t) \, dt + \int_a^x M(s, y) \, ds$$

It is sometimes simpler to construct this expression for $u(x, y)$ than to construct the expression defined by (2.4). In Figure 2.1 this construction of $u(x, y)$ corresponds to an integration from (a, b) to (a, y) followed by an integration from (a, y) to (x, y).

2.2 Separation of Variables

If Equation (2.2) can be written in the form

$$M(x) \, dx + N(y) \, dy = 0 \qquad\qquad \textbf{(2.5)}$$

where M is a function of x alone and N is a function of y alone, the variables x and y in (2.5) are said to be *separable*. Equation (2.5) is exact, since

$$\frac{\partial}{\partial y} M(x) \equiv 0 \equiv \frac{\partial}{\partial x} N(y)$$

Let us assume that $\int M(x) \, dx = G(x) + c_1$ and $\int N(y) \, dy = H(y) + c_2$. Then

$$\frac{dG(x)}{dx} = M(x) \quad \text{and} \quad \frac{dH(y)}{dy} = N(y)$$

and since $M(x) \, dx + N(y) \, dy$ is the differential of $G(x) + H(y)$, a general solution of (2.5) is given implicitly by

$$G(x) + H(y) = c \qquad\qquad \textbf{(2.6)}$$

If $y = f(x)$ denotes any differentiable function defined by (2.6), it follows from the chain rule that f is a solution of (2.5).

If (2.2) can be written in the form

$$A(x)B(y)\, dx + C(x)D(y)\, dy = 0 \qquad\qquad \textbf{(2.7)}$$

the variables are then separable, since division by $B(y)C(x)$ yields the equation

$$\frac{A(x)}{C(x)}\, dx + \frac{D(y)}{B(y)}\, dy = 0 \qquad\qquad \textbf{(2.8)}$$

Definition Two DE are said to be *equivalent* if and only if they have the same solutions.

The solutions of (2.7) could be obtained by solving (2.8) if the two DE were equivalent. They may not be, however, since division by $B(y)C(x)$ may introduce discontinuities or extraneous solutions. Such possibilities are investigated by considering the equations $B(y) = 0$ and $C(x) = 0$.

EXAMPLE 1. The solutions of $x\, dx + y^2\, dy = 0$ are given implicitly by

$$\frac{x^2}{2} + \frac{y^3}{3} = c$$

EXAMPLE 2. Solve $1\, dx + (y^4 + x^2 y^4)\, dy = 0$.

Solution: The given DE is equivalent to the DE

$$\frac{dx}{1 + x^2} + y^4\, dy = 0$$

The solutions are given implicitly by

$$\mathrm{Tan}^{-1}\, x + \frac{y^5}{5} = c$$

EXAMPLE 3. Solve

$$\frac{dy}{dx} = ky, \quad k \neq 0 \qquad\qquad \textbf{(2.9)}$$

Solution: Dividing the equivalent DE

$$ky\ dx - 1\ dy = 0 \tag{2.10}$$

by y, we obtain

$$k\ dx - \frac{dy}{y} = 0 \tag{2.11}$$

If y satisfies (2.11), then

$$kx - \ln|y| = c, \quad \ln|y| = kx - c$$

and

$$|y| = e^{kx-c} = e^{-c}e^{kx} = Pe^{kx}$$

where P is an arbitrary positive constant.

It can now be shown (see Problem 7) that $y = \pm Pe^{kx} = Be^{kx}$ where B is an arbitrary nonzero constant. That is, every *differentiable* y for which $|y| = Pe^{kx}\ (P > 0)$ is of the form $y = Be^{kx}$, $B \neq 0$.

Since we divided by y to obtain (2.11) from (2.10), we must investigate to see whether $y \equiv 0$ defines a solution of (2.10). This is seen by inspection to be the case, and hence *all* solutions of (2.10) must be of the form $y = Be^{kx}$, where B is an arbitrary constant. Conversely, if y is of the form $y = Be^{kx}$, y satisfies (2.9) since

$$Bke^{kx} \equiv k(Be^{kx})$$

for all x. Every solution has domain $(-\infty, +\infty)$.

EXAMPLE 4. Solve the initial-value problem

$$\frac{dy}{dx} = ky; \quad k \neq 0, \quad y(0) = 5$$

Solution: Setting $x = 0$ and $y = 5$ in $y = Be^{kx}$, we obtain $B = 5$ and $y = 5e^{kx}$.

EXAMPLE 5. Solve the initial-value problem

$$\frac{dy}{dx} = 1 + y^2, \quad y\left(\frac{\pi}{4}\right) = 1$$

Solution:

$$\int_1^y \frac{du}{1 + u^2} = \int_{\pi/4}^x dt$$

$$\mathrm{Tan}^{-1}\ y - \mathrm{Tan}^{-1}\ 1 = x - \frac{\pi}{4}$$

$$\mathrm{Tan}^{-1}\ y = x$$

$$y = \tan x$$

The solution has domain $(-\pi/2, \pi/2)$.

EXAMPLE 6. Solve the initial-value problem

$$(1 + 2e^x) \, dx - (1 + e^x)(2y + \cos y) \, dy = 0, \quad y(0) = 0$$

Solution:

$$\int_0^x \left(1 + \frac{e^u}{1 + e^u}\right) du = \int_0^y (2t + \cos t) \, dt$$

$$u + \ln(1 + e^u)]_0^x = t^2 + \sin t]_0^y$$

The solution is given implicitly by

$$x + \ln(1 + e^x) - \ln 2 = y^2 + \sin y$$

Problem List 2.2

1. Solve:
   (a) $dx - 2xy \, dy = 0$
   (b) $x \, dx - 3 \, dy = 0$
   (c) $\dfrac{dx}{y(1 + x^2)} - 2 \, dy = 0$
   (d) $\dfrac{dy}{dx} = e^{y-x}$
   (e) $2x(1 + y^2) \, dx + dy = 0$
   (f) $\sqrt{1 - y^2} \, dx - x\sqrt{x^2 - 1} \, dy = 0$
   (g) $\dfrac{dy}{dx} = \dfrac{y}{x^2}$
   (h) $\dfrac{dy}{dx} = \dfrac{2x + xy^2}{3y + x^2y}$

2. Solve the initial-value problem:
   (a) $(2xy + 2x) \, dx - dy = 0, \, y(0) = 0$
   (b) $dx = 4\sqrt[3]{y(x - 1)} \, dy, \, y(1) = 0$
   (c) $\dfrac{dy}{dx} = 3y, \, y(0) = 1$
   (d) $\dfrac{dy}{dx} = \dfrac{3y - xy}{x}, \, y(1) = 3e^{-1}$
   (e) $\dfrac{dy}{dx} = (3 + y) \cot x, \, y\left(\dfrac{\pi}{2}\right) = 4$
   (f) $(y^2 + 4) \, dx - (x^2 + 4) \, dy = 0, \, y(2) = 1$
   (g) $\dfrac{dy}{dx} = x^2y^2, \, y(1) = 9$

3. Find the particular solution of $x \, dy + y \, dx = 0$, satisfying the condition $y = 4$ when $x = 1$ (a) by separating the variables; (b) by treating the equation as exact.

4. Find the solution of the initial-value problem $dx + (x - 2) \, dy = 0, \, y(3) = 2$. Give the domain of the solution.

5. Given $y = f(x) = \sum_{k=0}^{+\infty} \dfrac{x^k}{k!}$, show that $y = e^x$. *Remark:* $0! = 1$. *Hint:* Use the series for $f'(x)$.

6. Find all differentiable functions that satisfy the functional equation $f(x + h) = f(x)f(h)$ for all x, h.

7. Given that y is a differentiable function satisfying $dy/dx = ky$, $k \neq 0$, and that $|y| = Pe^{kx}$, where $P > 0$, show that y is of the form $\pm Pe^{kx}$.

8. Find the curve through $(3, 4)$ such that the segment of every tangent between the coordinate axes is bisected at the point (x, y) of tangency.

2.3 Integrating Factors

One can scarcely expect the equation $M\, dx + N\, dy = 0$ to be exact since the requirement $M_y = N_x$ is very special. If the equation is not exact, it is still plausible that there may exist a function given by $\mu = \mu(x, y)$, such that

$$\mu(x, y)M(x, y)\, dx + \mu(x, y)N(x, y)\, dy = 0$$

is exact. Such a function μ is called an *integrating factor* of the DE. Euler introduced this useful concept in 1734. In Reference 2.5 it is shown that $M\, dx + N\, dy = 0$ has infinitely many integrating factors whenever a general solution exists.

In Section 2.2 we solved the equation $A(x)B(y)\, dx + C(x)D(y)\, dy = 0$ by considering the equation

$$\frac{A(x)}{C(x)}\, dx + \frac{D(y)}{B(y)}\, dy = 0$$

We effectively made the observation that $\mu(x, y) = [C(x)B(y)]^{-1}$ is an integrating factor.

Integrating factors are sometimes obtained by inspection; the method depends on the introduction of certain formulas for exact differentials.

EXAMPLE 1. The equation $-y\, dx + x\, dy = 0$ is not exact since

$$\frac{\partial}{\partial y}(-y) = -1 \neq \frac{\partial}{\partial x}(x) = +1$$

By recalling the formula $d(y/x) = (x\, dy - y\, dx)/x^2$, we note that $\mu(x, y) = x^{-2}$ is an integrating factor. Hence, $y/x = c$, or $y = cx$, defines a general solution. [Note that $y \equiv 0$ defines a solution on $(-\infty, 0) \cup (0, +\infty)$.]

By recalling the formula for $d(-x/y)$, it is easy to show that y^{-2} is also an integrating factor.

EXAMPLE 2. The equation $(2y + 4)\, dx + x\, dy = 0$ is not exact since $(\partial/\partial y)(2y + 4) = 2 \neq (\partial/\partial x)(x) = 1$. If we are clever enough to notice that $\mu(x, y) = x$ is an integrating factor, we obtain $(2xy + 4x)\, dx + x^2\, dy = 0$, which has solutions given implicitly by $x^2y + 2x^2 = c$.

Example 3. Solve $(x^2 + y^2 - y) dx + x dy = 0$.

Solution: The equation is not exact since

$$\frac{\partial}{\partial y}(x^2 + y^2 - y) = 2y - 1 \neq \frac{\partial}{\partial x}(x) = 1$$

Writing the equation in the form $(x^2 + y^2) dx + x dy - y dx = 0$, and noting that

$$d \operatorname{Tan}^{-1}\left(\frac{y}{x}\right) = \frac{x\,dy - y\,dx}{x^2(1 + y^2/x^2)} = \frac{x\,dy - y\,dx}{x^2 + y^2}$$

we see that $\mu(x, y) = [x^2 + y^2]^{-1}$ is an integrating factor. From

$$dx + \frac{x\,dy - y\,dx}{x^2 + y^2} = 0$$

we obtain $x + \operatorname{Tan}^{-1}(y/x) = c$.

The following formulas are often exploited in similar fashion:

$$d(xy) = x\,dy + y\,dx$$

$$d\left(\ln \frac{x}{y}\right) = \frac{y\,dx - x\,dy}{xy}$$

$$d\left[\frac{1}{2}\ln(x^2 + y^2)\right] = \frac{x\,dx + y\,dy}{x^2 + y^2}$$

$$d\sqrt{x^2 + y^2} = \frac{x\,dx + y\,dy}{\sqrt{x^2 + y^2}}$$

$$d\sqrt{x^2 - y^2} = \frac{x\,dx - y\,dy}{\sqrt{x^2 - y^2}}$$

In the examples presented, the discovery of an integrating factor required considerable ingenuity. Let us consider the possibility of a more direct determination of μ. If

$$\mu M\, dx + \mu N\, dy = 0 \tag{2.12}$$

is exact, then

$$\frac{\partial(\mu M)}{\partial y} = \frac{\partial(\mu N)}{\partial x}$$

or

$$\mu \frac{\partial M}{\partial y} + M \frac{\partial \mu}{\partial y} = \mu \frac{\partial N}{\partial x} + N \frac{\partial \mu}{\partial x}$$

and

$$\frac{1}{\mu}\left(N\frac{\partial \mu}{\partial x} - M\frac{\partial \mu}{\partial y}\right) = \frac{\partial M}{\partial y} - \frac{\partial N}{\partial x} \qquad (2.13)$$

To find even a particular solution μ of this partial DE is a formidable problem. There are, however, two fairly simple special cases of interest.

Suppose that (2.12) has an integrating factor μ that is a function of x alone. Then

$$\frac{\partial \mu}{\partial x} = \frac{d\mu}{dx} \quad \text{and} \quad \frac{\partial \mu}{\partial y} = 0$$

and hence (2.13) reduces to

$$\frac{1}{\mu}\frac{d\mu}{dx} = \frac{\partial M/\partial y - \partial N/\partial x}{N} \qquad (2.14)$$

Since the left member of (2.14) depends only on x, the same is true of the right member, which we denote by $f(x)$.

From

$$\frac{1}{\mu}\frac{d\mu}{dx} = f(x) \quad \text{or} \quad \frac{d\mu}{\mu} = f(x)\,dx$$

we obtain

$$\ln|\mu| = \int f(x)\,dx$$

and

$$\mu = ce^{\int f(x)\,dx} = e^{\int f(x)\,dx}$$

for $c = 1$.

Conversely, if $(M_y - N_x)/N$ depends on x alone, then $\mu = e^{\int f(x)\,dx}$ satisfies (2.14) and hence is an integrating factor for $M\,dx + N\,dy = 0$.

EXAMPLE 4. Solve $-y\,dx + x\,dy = 0$.

Solution: From $(M_y - N_x)/N = (-1 - 1)/x = -2/x$, we obtain

$$\mu = e^{\int -2\,dx/x} = e^{-2\ln|x|} = (e^{\ln|x|})^{-2} = x^{-2}$$

This is the integrating factor we found by inspection in Example 1. Of course, $-y\,dx + x\,dy = 0$ is easily solved by separating the variables.

EXAMPLE 5. Solve $(x^2 + 2x + y)\,dx + 1\,dy = 0$.

Solution: From $(M_y - N_x)/N = (1 - 0)/1 = 1$, we obtain

$$\mu = e^{\int 1\,dx} = e^x$$

It is easy to show that $e^x(x^2 + 2x + y)\,dx + e^x\,dy = 0$ has solutions defined by $x^2e^x + ye^x = c$. See Problem 2(c).

If $(M_y - N_x)/(-M)$ is a function of y alone, given by $g(y)$, an analogous development shows that $\mu = e^{\int g(y)\,dy}$ is an integrating factor. See Problem 3.

Problem List 2.3

1. Solve:
   (a) $y/x\,dx + dy = 0$
   (b) $y\,dx - x\,dy = 0$
   (c) $2x(x^2 + y^2)\,dx - x\,dy + y\,dx = 0$
   (d) $x\,dx + (y + e^y\sqrt{x^2 + y^2})\,dy = 0$
   (e) $x(1 - 4\sqrt{x^2 - y^2})\,dx - y\,dy = 0$
   (f) $2x\,dx + (x^2 + 2y + y^2)\,dy = 0$

2. Find $(M_y - N_x)/N$ and solve:
   (a) $(2y + 4)\,dx + x\,dy = 0$
   (b) $(3y + 8x)\,dx + x\,dy = 0$
   (c) $(x^2 + 2x + y)\,dx + dy = 0$
   (d) $(y + 2x\cos^2 x\sin y)\,dx + (\sin x\cos x + x^2\cos^2 x\cos y)\,dy = 0$

3. Show that if $(M_y - N_x)/(-M)$ is a function of y alone, given by $g(y)$, then $\exp(\int g(y)\,dy)$ is an integrating factor for $M\,dx + N\,dy = 0$.

4. Find $(M_y - N_x)/(-M)$ and solve:
   (a) $2y\,dx + x\,dy = 0$
   (b) $(y^3 - 3x^2y)\,dx + (4xy^2 - 2x^3)\,dy = 0$

5. Show that $y\,dx - x\,dy = 0$ has $y^{-2}, x^{-2}, (xy)^{-1}$, and $(x^2 + y^2)^{-1}$ as integrating factors.

6. Determine m and n so that $\mu = x^m y^n$ is an integrating factor for $(4xy^2 + 3y)\,dx + (5x^2y + 4x)\,dy = 0$.

7. Solve the initial-value problem

$$\frac{dy}{dx} + y\tan x = \cos x, \quad y(0) = 5$$

2.4 First-Order Linear Equations

The general form of a first-order linear DE is

$$A(x)\frac{dy}{dx} + B(x)y + C(x) = 0 \tag{2.15}$$

where the coefficients A, B, and C denote continuous functions of x defined on a set S of one or more intervals. It is customary to assume that $A(x)$ is always positive or always negative on any interval I in S, to divide (2.15) by $A(x)$, and to obtain the form

$$\frac{dy}{dx} + P(x)y = Q(x) \tag{2.16}$$

where P and Q are continuous on a set T of one or more intervals.

Linear DE of this type are important in both pure and applied mathematics.

Writing (2.16) in the form $(Py - Q) dx + 1 dy = 0$, and noting that $(M_y - N_x)/N = (P - 0)/1 = P$, we conclude that $e^{\int P\,dx}$ is an integrating factor. See Section 2.3.

We assume for convenience that $e^{\int P\,dx}$ contains no constant of integration, since any primitive of P will serve our purpose. For example, if $P(x) = 2x$, we take e^{x^2} for $e^{\int P\,dx}$. This amounts to setting $c = 0$ in $x^2 + c$. See Problem 4.

Since

$$e^{\int P(x)\,dx}\frac{dy}{dx} + e^{\int P(x)\,dx}P(x)y = \frac{d}{dx}\left(ye^{\int P(x)\,dx}\right)$$

a general solution y of (2.16) satisfies

$$ye^{\int P(x)\,dx} = \int e^{\int P(x)\,dx}Q(x)\,dx + c \tag{2.17}$$

where, again for convenience, we assume that $\int [\exp(\int P(x)\,dx)]Q(x)\,dx$ contains no constant of integration.

We find y by multiplying both sides of (2.17) by $\exp[-\int P(x)\,dx]$ to obtain

$$y = e^{-\int P(x)\,dx}\left[\int e^{\int P(x)\,dx}Q(x)\,dx + c\right] \tag{2.18}$$

Remark: An alternative method of obtaining the integrating factor $\exp[\int P(x)\,dx]$ is to observe that the left member of (2.16) resembles the derivative of a product. This prompts us to seek a function μ of x such that

$$\mu\left(\frac{dy}{dx} + Py\right) = \frac{d}{dx}(\mu y)$$

Then

$$\mu\frac{dy}{dx} + \mu Py = \mu\frac{dy}{dx} + y\frac{d\mu}{dx}$$

$$\frac{d\mu}{dx} = \mu P$$

$$\frac{d\mu}{\mu} = P\,dx$$

and

$$\mu = e^{\int P\,dx}$$

EXAMPLE 1. Solve $dy/dx + 2y/x = 4x$.

Solution: The integrating factor

$$\mu(x) = \exp\left(\int P(x)\,dx\right) = \exp\left(\int 2\frac{dx}{x}\right)$$

$$= \exp\left(2\ln|x|\right) = \exp\left(\ln|x|^2\right)$$

$$= \exp\left(\ln x^2\right) = x^2$$

since $\exp(\ln u) = u$ and $|x|^2 = x^2$.

Hence from (2.17), the DE has general solution given by

$$yx^2 = \int (4x)x^2\,dx + c = x^4 + c$$

or

$$y = x^2 + cx^{-2}$$

valid on $(-\infty, 0) \cup (0, +\infty)$.

EXAMPLE 2. Solve $dy/dx + y/x = 3x$.

Solution: Since $\mu(x) = \exp\left(\int dx/x\right) = \exp(\ln|x|) = |x|$, we have, from (2.17),

$$y|x| = \int 3x\,|x|\,dx + c$$

On any interval I_1 where $x > 0$, $|x| = x$ and the DE has solutions given by

$$yx = \int 3x^2\,dx + c = x^3 + c \quad \text{or} \quad y = x^2 + cx^{-1}$$

On any interval I_2 where $x < 0$, $|x| = -x$ and the DE has solutions given by

$$-yx = \int 3x(-x)\,dx + k = -x^3 + k \quad \text{or} \quad y = x^2 - kx^{-1}$$

If we set $k = -c$, we note that the solutions on I_1 and I_2 agree.

In Reference 2.2 it is shown that (2.18) is a complete solution of (2.16) on the set T of intervals. That is, every solution of (2.16) can be written in the form (2.18). It is also shown that if $a \in I$ where I is an interval in T, (2.18) yields the unique solution of (2.16) for which $y(a) = b$.

EXAMPLE 3. Given $dy/dx - 5y = 3e^{5x}$ and that $y = 8$ when $x = 0$, find y when $x = 1$.

Solution:

$$\exp\left(\int -5\,dx\right) = \exp(-5x)$$

$$ye^{-5x} = \int 3e^{5x}e^{-5x}\,dx + c$$

$$= 3x + c$$

Setting $x = 0$ and $y = 8$, we obtain

$$8 = c$$

$$y = y(x) = e^{5x}(3x + 8)$$

$$y(1) = 11e^5$$

If a DE is not linear in y and dy/dx, it may be linear in x and dx/dy. To solve a DE of this type, we interchange the roles of x and y and seek solutions of the form $x = g(y)$.

EXAMPLE 4. Solve the initial-value problem

$$\sin y \, dx + (x \sec y - \cos y) \, dy = 0; \quad x = \frac{\pi}{2}, \quad y = \frac{\pi}{4}$$

Assume $\pi/4 < x < 3\pi/4$ and $0 < y < \pi/2$.

Solution: The DE is clearly not linear in y but is linear in x and dx/dy since

$$\frac{dx}{dy} + \frac{\sec^2 y}{\tan y} x = \cot y$$

Using the integrating factor

$$\exp\left(\int \frac{\sec^2 y \, dy}{\tan y}\right) = \exp\left(\ln \tan y\right) = \tan y$$

we obtain

$$x \tan y = \int (\cot y)(\tan y) \, dy + c = y + c$$

Since $x = \pi/2$ when $y = \pi/4$, then $\pi/2 = \pi/4 + c$, $c = \pi/4$, and

$$x \tan y = y + \frac{\pi}{4}$$

or

$$x = \left(y + \frac{\pi}{4}\right) \cot y$$

valid for $0 < y < \pi/2$.

We have assumed that P and Q in the DE $y' + Py = Q$ are continuous in any interval under consideration. In some applications, for example in the study of mechanical and electrical systems, Q (or possibly P) may have a jump discontinuity at a point $x = a$ in an interval $I = (x_1, x_2)$. That is,

$$\lim_{x \to a^+} Q(x) = L_1 \quad \text{and} \quad \lim_{x \to a^-} Q(x) = L_2 \neq L_1$$

This situation is handled by solving the DE on (x_1, a) and also on (a, x_2). We then relate the constants of integration so that y will be continuous at $x = a$. A resulting function ϕ given by $y = \phi(x)$ will satisfy the DE on I except at $x = a$. Although ϕ will be continuous at a, the derivative of ϕ at a will not exist and hence ϕ will not be a solution on I of the DE. In such applications it is customary to refer to ϕ as a solution on I, thereby generalizing the concept of "solution."

To find a particular solution on I for which $\phi(x_3) = y_3$ we proceed in the usual fashion. These ideas are illustrated in the following example.

EXAMPLE 5. Solve

$$\frac{dy}{dx} + \frac{2y}{x} = Q(x) \quad \text{on } (0, +\infty)$$

given that

$$Q(x) = \begin{cases} 0 & \text{for } 0 < x \le 1 \\ 3 & \text{for } 1 < x \end{cases}$$

Solution: We find as in Example 1 that x^2 is an integrating factor.

On $(0, 1]$, $\quad y_1 x^2 = \int (x^2)(0)\, dx + c_1 = c_1 \quad$ or $\quad y_1 = c_1 x^{-2}$

On $(1, +\infty)$, $y_2 x^2 = \int (x^2)(3)\, dx + c_2 = x^3 + c_2 \quad$ or $\quad y_2 = x + c_2 x^{-2}$

Setting $y_1(1) = y_2(1)$, we obtain $c_1 = 1 + c_2$. Solutions on $(0, +\infty)$ are given by

$$y = \begin{cases} (1 + c_2)x^{-2} & \text{for } 0 < x \le 1 \\ x + c_2 x^{-2} & \text{for } 1 < x \end{cases}$$

To find the particular solution for which $y(2) = 5$, we set $x = 2$ in $x + c_2 x^{-2}$. This yields $5 = 2 + \dfrac{c_2}{4}$ or $c_2 = 12$. The required solution is given by

$$y = \phi(x) = \begin{cases} 13x^{-2} & \text{for } 0 < x \le 1 \\ x + 12x^{-2} & \text{for } 1 < x \end{cases}$$

Note that $\phi(1) = 13$. Also, from

$$\phi'(x) = \begin{cases} -26x^{-3} & \text{for } 0 < x < 1 \\ 1 - 24x^{-3} & \text{for } 1 < x \end{cases}$$

we find that

$$\lim_{x \to 1^-} (-26x^{-3}) = -26 \quad \text{and} \quad \lim_{x \to 1^+} (1 - 24x^{-3}) = -23$$

The function ϕ is continuous at $x = 1$ but $\phi'(1)$ does not exist. The curve $y = \phi(x)$ has a "corner" at $(1, 13)$.

Problem List 2.4

1. Solve:

(a) $\dfrac{dy}{dx} + \dfrac{2y}{x} = x^2 + 1$ (b) $\dfrac{dy}{dx} + 2xy = 2x$

(c) $\dfrac{dy}{dx} + \dfrac{y}{x} = x^2$ (d) $\dfrac{dy}{dx} + \dfrac{y}{x} = \dfrac{\sin x}{x}$

(e) $\dfrac{dr}{d\theta} + \dfrac{4r}{\theta} = \theta$ (f) $\dfrac{dy}{dx} + e^x y = e^x$

(g) $\dfrac{dy}{dx} - \dfrac{2y}{x+1} = 3(x+1)^2$ (h) $\dfrac{dy}{dx} + y \cot x = \csc^2 x$

2. Solve (a) $dx/dy + 2xy = y$, and (b) $dx + (2x - y^2)\, dy = 0$.

3. Solve the initial-value problem:

(a) $\dfrac{dy}{dx} + y = e^{-x}$, $y(0) = 5$

(b) $\dfrac{dy}{dx} + \dfrac{2y}{x} = \dfrac{4}{x}$, $y(1) = 6$

(c) $\dfrac{dy}{dx} + \dfrac{y}{x} = e^x$, $y(-1) = e^{-1}$

(d) $\dfrac{dy}{dx} + y \cot x = 2 \cos x$, $y\left(\dfrac{\pi}{2}\right) = 3$

(e) $x\, dy = (2x^4 + x^2 + 2y)\, dx$, $y(1) = 4$

4. Show that no greater generality is obtained by adding a constant of integration to the integrating factor $\mu(x) = \exp(\int P(x)\, dx)$ of Equation (2.16).

5. Solve the initial-value problem

$$\frac{dy}{dx} + ay = b, \quad y(0) = k$$

6. Show that if $f(x)$ denotes a general solution of $y' + P(x)y = 0$, and $g(x)$ denotes any particular solution of $y' + P(x)y = Q(x)$, then a general solution of $y' + P(x)y = Q(x)$ is given by $y = f(x) + g(x)$.

7. Show that if $y = f(x)$ denotes a solution of $y' + P(x)y = 0$, then $y = kf(x)$ also denotes a solution for every constant k.

8. Given $k \geq 0$, $a > 0$, and $b > 0$, solve $y' + ay = be^{-kx}$. Evaluate $\lim_{x \to +\infty} y$ for (a) $k = 0$, and (b) $k > 0$.

9. Given that $y = b$ when $x = a$, solve $A(x)y' + B(x)y = 0$ by noting that $d/dx\,(\ln |y|) = -B(x)/A(x)$.

10. Show that the initial-value problem $xy' - 2y = 0$, $y(1) = 1$, has a solution given by $y_1 = x^2$ on $(-2, 2)$. Show also that the function given by

$$y_2 = \begin{cases} x^2 & \text{for } x \geq 0 \\ 0 & \text{for } x < 0 \end{cases}$$

is also a solution on the same interval. This result does not contradict the uniqueness theorem referred to in the text. The DE $y' - 2x^{-1}y = 0$ is equivalent to the DE $xy' - 2y = 0$ on $(0, +\infty)$ and each has the unique solution y_1 on $(0, +\infty)$ satisfying $y(1) = 1$.

11. Solve $y' + y = |x|$ on $(-1, 1)$ given that $y(0) = 6$.

12. In illustrative Example 5
   (a) Show that ϕ satisfies $y' + 2yx^{-1} = Q(x)$ on $(0, 1) \cup (1, +\infty)$.
   (b) Sketch the graph of ϕ for $1/2 < x < 3/2$.

13. Solve $y' + y = Q(x)$ on $(-\infty, +\infty)$ given that

$$Q(x) = \begin{cases} 0 & \text{on } (-\infty, 1] \\ 1 & \text{on } (1, +\infty) \end{cases}$$

Find the particular solution for which $y(0) = 3e$.

2.5 Differential Equations With One Variable Missing; The Method of Substitution

A differential equation in which y does not appear is converted into a DE of lower order when $y' = dy/dx$ is replaced by a new variable v.

EXAMPLE 1. Solve

$$\frac{d^2y}{dx^2} + 2\frac{dy}{dx} = e^{-2x}$$

Solution: Let $v = dy/dx$; then

$$\frac{dv}{dx} + 2v = e^{-2x}$$

$$\mu(x) = \exp\left(\int 2\,dx\right) = \exp(2x)$$

$$ve^{2x} = \int dx = x + c$$

$$v = \frac{dy}{dx} = xe^{-2x} + ce^{-2x}$$

and, integrating by parts or using a table of integrals,

$$y = \int (xe^{-2x} + ce^{-2x})\,dx = \frac{e^{-2x}}{4}(-2x - 1) - \frac{c}{2}e^{-2x} + k$$

If x is missing from a second-order DE, the replacement of dy/dx by v and the replacement of d^2y/dx^2 by $v(dv/dy)$ is often effective. Note that

$$\frac{d^2y}{dx^2} = \frac{dv}{dx} = \frac{dv}{dy}\cdot\frac{dy}{dx} = \frac{dy}{dx}\cdot\frac{dv}{dy} = v\frac{dv}{dy}$$

Remark: Writing d^2y/dx^2 in the form $v(dv/dy)$, where $v = dy/dx$, is a useful technique. We shall encounter an important illustration when we study simple harmonic motion (Section 5.1).

EXAMPLE 2. Solve the initial-value problem

$$y\frac{d^2y}{dx^2} = \left(\frac{dy}{dx}\right)^2; \quad y(0) = 1, \quad y'(0) = 1$$

Solution: Replacing dy/dx by v and d^2y/dx^2 by $v(dv/dy)$, we obtain

$$yv\frac{dv}{dy} = v^2$$

$$\frac{dv}{v} = \frac{dy}{y}$$

$$\ln|v| = \ln|y| + \ln|c| = \ln|cy|$$

and

$$|v| = |cy|$$

Since $y = 1$ and $v = 1$ when $x = 0$, $1 = |c|$ and $c = +1$ or -1, and $v = dy/dx = +y$. (We discard $v = -y$ since $v = 1$ when $y = 1$.) From $dy/y = dx$, we obtain

$$\ln|y| = x + k$$

$$\ln|1| = 0 + k, \quad k = 0$$

and

$$y = +e^x$$

(We discard $y = -e^x$ since $y = 1$ when $x = 0$.) NOTE: The DE $v = dy/dx = y$ is called a *first integral*, or an *intermediate integral*, of the DE $y(d^2y/dx^2) = (dy/dx)^2$.

The method used in Examples 1 and 2, known as a substitution or transformation or change of variable in a DE, is often very effective and is somewhat analogous to a change of variable in an integral. The objective is to transform a DE into a second DE that we know how to solve. Useful substitutions are discovered by trial and error, through experience, and by sheer ingenuity. If, for example, the expression $(x + y)^3$ appeared in a DE, one might try letting $v = x + y$, that is, letting $y = v - x$. Other common substitutions include $y = v + x$, $y = vx$, $y = v^{-1}$, $y = v/x$, and $y = ve^x$. Sometimes both variables are transformed by equations of the form

$$x = \phi(u, v), \quad y = \psi(u, v)$$

illustrated by $x = u + h$, $y = v + k$, and by $x = r \cos\theta$, $y = r \sin\theta$.

It is possible to lose solutions by a change of variable. For ex
replace y by e^v, the DE obtained will not yield solutions of the orig
which $y \leq 0$. (See Reference 2.2 for a more complete discussion.)

EXAMPLE 3. Solve the initial-value problem

$$y' = (x + y)^2, \quad y(0) = 0$$

Solution: Let $y = v - x$. Then

$$y' = \frac{dv}{dx} - 1 = v^2$$

$$\frac{dv}{1 + v^2} = dx$$

and

$$\text{Tan}^{-1} v = x + c$$

Since $v = x + y$, $v(0) = 0$,

$$\text{Tan}^{-1} 0 = 0 + c, \quad c = 0$$
$$y + x = \text{Tan } x$$

and

$$y = -x + \text{Tan } x$$

EXAMPLE 4. The nonlinear DE $dy/dx + P(x)y = Q(x)y^n$; $n \neq 0$, $n \neq 1$, known as *Bernoulli's equation*, was studied in 1695 by the Swiss mathematician James Bernoulli (1654–1705). Show that the substitution $z = y^{1-n}$ reduces Bernoulli's equation to a linear equation.

Solution:

$$\frac{dz}{dx} = (1 - n)y^{-n}\frac{dy}{dx}$$

$$\frac{dy}{dx} = \frac{y^n}{1 - n}\frac{dz}{dx}$$

$$\frac{y^n}{1 - n}\frac{dz}{dx} + P(x)y = Q(x)y^n$$

$$\frac{dz}{dx} + [(1 - n)P(x)]z = (1 - n)Q(x)$$

Problem List 2.5

1. Solve:

(a) $x\dfrac{d^2y}{dx^2} + \dfrac{dy}{dx} = 0$ (b) $\dfrac{d^2y}{dx^2} + \dfrac{dy}{dx} = 4$

(c) $\dfrac{d^2y}{dx^2} + \dfrac{dy}{dx} = 0$ (d) $y\dfrac{d^2y}{dx^2} + \left(\dfrac{dy}{dx}\right)^2 = 0$

2. Solve the initial-value problem:

(a) $\dfrac{d^2y}{dx^2} + \dfrac{dy}{dx} = 2$; $y(0) = 3$, $y'(0) = 1$

(b) $\dfrac{d^2y}{dx^2} = -\dfrac{dy/dx}{2}$; $y(0) = 0$, $y'(0) = 50$

(c) $x\dfrac{d^2y}{dx^2} + \dfrac{dy}{dx} = 4x$; $y(1) = 2$, $y'(1) = 4$

(d) $\dfrac{d^2y}{dx^2} + \dfrac{1}{x-1}\dfrac{dy}{dx} = 9(x-1)$; $y(2) = 9$, $y'(2) = 5$

3. Solve each Bernoulli equation:

(a) $\dfrac{dy}{dx} + \dfrac{2y}{x} = xy^2$ (b) $\dfrac{dy}{dx} + \dfrac{y}{x} = \dfrac{1}{x^3y^3}$

(c) $\dfrac{dy}{dx} - ay = \dfrac{b}{y}$

4. (a) Solve $dy/dx = x + y$ by making the substitution $y = ve^x$.
   (b) In the DE $dy/dx - y^2 - 4 = 0$, let $y = 1/u$. Show that u is a solution of $du/dx + 4u^2 + 1 = 0$.

5. (a) Show that the substitution $v = ax + by + c$ reduces the DE $dy/dx = f(ax + by + c)$ to a separable equation.
   (b) Solve $dy/dx = (x + y - 1)^2$.

6. Solve $dy/dx = (x - y)/(x + y)$ by using the substitutions $x = r\cos\theta$, $y = r\sin\theta$.

7. A differential equation of the form

$$\frac{dy}{dx} + P(x)y + Q(x)y^2 = R(x) \tag{2.19}$$

is called a *Riccati equation*. Certain special cases of (2.19) were considered by the Italian mathematician Count Jacopo Francesco Riccati (1676–1754).

    (a) Show that if Y is a known solution of (2.19), a general solution can be obtained by substituting $y = Y + v^{-1}$, thereby introducing a first-order linear DE.

    (b) Solve $dy/dx - 2y + y^2 = -1$ by observing that a particular solution is given by $Y \equiv 1$.

8. Show that the general solution of $d^2y/dx^2 + y = 0$ can be written in the form $y = C\sin(x + \phi)$ and also in the form $y = A\sin x + B\cos x$.

2.6 Homogeneous First-Order Equations

    The area A of a rectangle of length x and width y is given by $A = A(x, y) = xy$. The area A_1 of a rectangle of length λx and width λy is given by

$$A_1 = A(\lambda x, \lambda y) = (\lambda x)(\lambda y) = \lambda^2(xy) = \lambda^2 A(x, y)$$

We say that the area is homogeneous of degree 2. A similar argument shows that the volume of a box x by y by z is homogeneous of degree 3. Values of λ satisfying $0 < \lambda < 1$ correspond to a contraction; values of λ satisfying $\lambda > 1$ correspond to an expansion. On the other hand, the ratio of the length of a rectangle to its width is unchanged when the length x and width y are each multiplied by λ. That is,

$$\frac{x}{y} = \frac{\lambda x}{\lambda y} = \lambda^0 \left(\frac{x}{y}\right)$$

These observations suggest the following definitions.

Definition   A function denoted by $f(x, y)$ is called a *homogeneous function* of x and y of degree n if and only if

$$f(\lambda x, \lambda y) = \lambda^n f(x, y), \quad \text{where } \lambda > 0$$

Definition   The DE $M(x, y)\, dx + N(x, y)\, dy = 0$ is called a *homogeneous first-order DE* if and only if M and N are homogeneous functions of the same degree.

EXAMPLE 1.   The function f defined by $f(x, y) = y^2 - xy$ is homogeneous of degree 2 since

$$f(\lambda x, \lambda y) = (\lambda y)^2 - (\lambda x)(\lambda y)$$
$$= \lambda^2(y^2 - xy)$$
$$= \lambda^2 f(x, y)$$

EXAMPLE 2.   The function f defined by $f(x, y) = (xe^{y/x} + 2y)/x^2$ is homogeneous of degree -1 since

$$f(\lambda x, \lambda y) = \frac{\lambda x e^{\lambda y/\lambda x} + 2\lambda y}{(\lambda x)^2}$$

$$= \lambda^{-1} \frac{xe^{y/x} + 2y}{x^2}$$

$$= \lambda^{-1} f(x, y)$$

EXAMPLE 3.   The function f defined by

$$f(x, y) = \frac{\sqrt{x^2 + y^2}}{x} + \sin^2 \frac{y}{x}$$

is homogeneous of degree 0 since

$$f(\lambda x, \lambda y) = \frac{\sqrt{\lambda^2 x^2 + \lambda^2 y^2}}{\lambda x} + \sin^2 \frac{\lambda y}{\lambda x}$$

$$= \frac{\sqrt{x^2 + y^2}}{x} + \sin^2 \frac{y}{x} = \lambda^0 f(x, y)$$

If $M(x, y)\, dx + N(x, y)\, dy = 0$ is a homogeneous DE, then

$$M(\lambda x, \lambda y) = \lambda^n M(x, y), \quad N(\lambda x, \lambda y) = \lambda^n N(x, y)$$

and hence

$$\lambda^{-n} M(\lambda x, \lambda y) \, dx + \lambda^{-n} N(\lambda x, \lambda y) \, dy = 0$$

Multiplying by λ^n, letting $\lambda = 1/x$, and assuming that $x > 0$, we obtain

$$M\left(1, \frac{y}{x}\right) dx + N\left(1, \frac{y}{x}\right) dy = 0 \qquad (2.20)$$

If we now let $y = vx$, then dy becomes $v \, dx + x \, dv$, and (2.20) takes the form $M(1, v) \, dx + N(1, v)(v \, dx + x \, dv) = 0$, or

$$[M(1, v) + vN(1, v)] \, dx + xN(1, v) \, dv = 0$$

or

$$\frac{dx}{x} + \frac{N(1, v) \, dv}{M(1, v) + vN(1, v)} = 0 \qquad (2.21)$$

Equation (2.21), in which the variables x and v are separated, is solved to find $v = v(x)$ in terms of x, whereupon $M \, dx + N \, dy = 0$ has solutions defined by $y = xv(x)$. For an implicit solution $g(v, x) = 0$ of (2.21), v is replaced by y/x. NOTE: To show that the homogeneous DE $M(x, y) \, dx + N(x, y) \, dy = 0$ is separable for $x < 0$, let $\lambda = -x^{-1}$ in

$$M(\lambda x, \lambda y) \, dx + N(\lambda x, \lambda y) \, dy = 0$$

The student should not memorize (2.21) but should employ the procedure of the following example.

EXAMPLE 4.   Solve $(y^2 - xy) \, dx + x^2 \, dy = 0$.

Solution: Since M and N are homogeneous functions of degree 2, we let $y = vx$. Replacing dy by $v \, dx + x \, dv$, we obtain

$$(v^2 x^2 - vx^2) \, dx + x^2(v \, dx + x \, dv) = 0$$

or

$$v^2 \, dx + x \, dv = 0$$

Separating the variables and simplifying yields $dx/x + v^{-2} \, dv = 0$, from which we obtain

$$\ln |x| - v^{-1} = c$$

Replacing v by y/x and simplifying yields the required solution defined by

$$y = \frac{x}{\ln |x| - c}$$

Substituting $y = vx$ is frequently effective in solving DE other than homogeneous first-order DE.

Problem List 2.6

1. Show that each of the following expressions defines a homogeneous function and find the degree.
   (a) $x^2y + 3xy^2$ (b) $4x - 3y$
   (c) $x^3 + y^3 - 3x^2y$ (d) $\ln y - \ln x$
   (e) $\dfrac{x^2e^{x/y}}{y^2}$

2. Find the general solution of:
   (a) $(x + y)\, dx - x\, dy = 0$
   (b) $(x^2 + y^2)\, dx - xy\, dy = 0$
   (c) $\dfrac{dy}{dx} = \dfrac{x}{y} + \dfrac{y}{x} + 2$
   (d) $\dfrac{dy}{dx} = \dfrac{y + x}{y - x}$
   (e) $(y^2 + xy)\, dx - x^2\, dy = 0$
   (f) $y\, dx - (x + 2y)\, dy = 0$
   (g) $(y - \sqrt{x^2 - y^2})\, dx - x\, dy = 0$

3. Solve the initial-value problem:
   (a) $(x + y)\, dx - x\, dy = 0$, $y(2) = 0$
   (b) $(2xy + y^2)\, dx = x^2 dy$, $y(2) = 4$
   (c) $(x^2 + y^2)\, dx - 2xy\, dy = 0$, $y(1) = -3$

4. (a) Show that if $f(v) = v$ for $v = k$, then $y = kx$ defines a solution of $dy/dx = f(y/x)$. Explain how this result applies to Problem 2(d).
   (b) Show that if $M(x, y)$ and $N(x, y)$ are homogeneous of the same degree, then $M(x, y)/N(x, y)$ is homogeneous of degree zero.
   (c) Show that if $f(x, y)$ is homogeneous of degree zero, then f is a function of y/x.

5. Show that the substitutions $x = r \cos \theta$, $y = r \sin \theta$ reduce the homogeneous DE $M(x, y)\, dx + N(x, y)\, dy = 0$ to a separable DE.

6. Show that substituting $x = vy$ reduces the homogeneous DE $M(x, y)\, dx + N(x, y)\, dy = 0$ to a separable DE.

7. Show that substituting $v = yx^{-n}$ separates the variables in the DE $dy/dx = x^{n-1}f(y/x^n)$.

8. (a) Show that if $aB \neq bA$, it is possible to choose h and k so that the substitutions $x = u + h$, $y = v + k$ reduce the DE

$$\frac{dy}{dx} = f\left(\frac{ax + by + c}{Ax + By + C}\right)$$

   to a homogeneous DE.
   (b) Show that if $aB = bA$, the substitution $v = ax + by$ reduces the DE

$$\frac{dy}{dx} = f\left(\frac{ax + by + c}{Ax + By + C}\right)$$

   to a separable DE.
   (c) Solve $\dfrac{dy}{dx} = \dfrac{x - y - 4}{x + y - 2}$.
   (d) Solve $\dfrac{dy}{dx} = \dfrac{x + y + 1}{x + y}$.

2.7 Geometric Interpretation of Differential Equations; Direction Fields

We consider a first-order DE written in the normal form

$$\frac{dy}{dx} = F(x, y) \tag{2.22}$$

The DE (2.22) is not necessarily linear. The DE $y' = x^2 + y$ is linear; the DE $y' = x + y^2$ is nonlinear. We also note that DE (2.2)

$$M(x, y) \, dx + N(x, y) \, dy = 0$$

can be written in the form (2.22) provided we restrict ourselves to a region R in which $N(x, y) \neq 0$.

To interpret (2.22) geometrically, we let $P(x, y)$ be an arbitrary point in the xy plane at which F is defined. We say that (2.22) determines a direction field, since at each point (a, b) of the domain D of F, it defines the slope of a line through (a, b). A direction field is illustrated graphically by drawing short line segments at several points, each line segment (or line element) having slope $F(a, b)$ at (a, b). Let us assume that F is continuous on its domain D. Loosely speaking, this amounts to assuming that if P_1 and P_2, both in D, are sufficiently close together, the slopes at P_1 and P_2 differ by a small number.

An example of a direction field is provided by the classic iron filings experiment of elementary physics. Iron filings are sprinkled on a sheet of paper that has been placed over a magnet. The filings assume the directions of the magnetic field force. See Figure 2.2.

A curve, given by $y = f(x)$, that is tangent at each of its points to the line element associated with that point is called an *integral curve* of the DE (2.22). The function f is a solution of (2.22) since $dy/dx = df/dx = F(x, y)$ at every point in the domain of F. Thus, finding the integral curves of a DE corresponds to solving the DE analytically. The appearance of a direction field gives a good indication of the nature of the integral curves of the associated DE. After a large number of line elements have been drawn, the integral curves begin to take shape. An integral curve may be approximated by starting at a

Fig. 2.2

point and sketching a smooth curve that is tangent to the direction field at every point.

EXAMPLE 1. The direction field determined by $dy/dx = -x/y$ is illustrated in Figure 2.3. The line element at point (x, y) has slope $-x/y$. A general solution is given implicitly by $x^2 + y^2 = c^2$, an equation of the family of circles with centers at the origin. Figure 2.3 also illustrates the integral curve through $(2, 1)$, having equation $y = \sqrt{5 - x^2}$, where $-\sqrt{5} < x < \sqrt{5}$. Note that the direction field is not defined on the x axis.

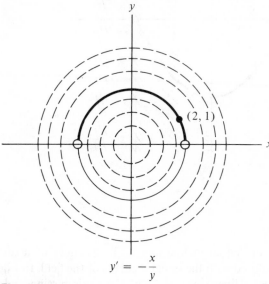

$$y' = -\frac{x}{y}$$

Fig. 2.3

For DE of the form $y' = F(x)$, the direction fields are easily drawn. We note that along a vertical line $x = k = $ constant the slope equals $F(k)$ at every point.

EXAMPLE 2. The direction field associated with the DE $y' = 2x$ is displayed in Figure 2.4. A general solution is given by $y = x^2 + c$; the integral curves are the members of a one-parameter family of parabolas. The particular integral curve through $(1, 0)$ has equation $y = x^2 - 1$.

A similar discussion applies to DE of the form $y' = F(y)$. Slopes are constant along horizontal lines. A simple yet interesting example is afforded by the DE $y' = y$. (See Problem 6(a).)

If $F(x, y)$ in (2.22) is complicated, drawing the associated direction field may be difficult. It does not help much to select a set of arbitrary points and to draw the associated line elements. A more specific and fruitful approach is to set $y' = F(x, y) = k = $ constant, to draw several members of the one-parameter family $F(x, y) = k$ of curves, and to draw line elements of equal slopes along each of these curves. The curves $F(x, y) = k$ are called *isoclines*,

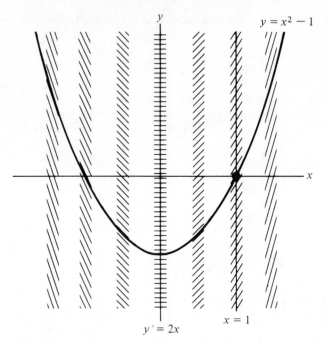

Fig. 2.4

from the Greek for "equal inclinations." The isoclines of a direction field should not be confused with the integral curves of the field. In Figure 2.3, the isoclines are the radial lines $-x/y = k$ with the x axis missing, and the integral curves are the semicircles. In Figure 2.4, the isoclines are the vertical lines $2x = k$ and the integral curves are the parabolas $y = x^2 + c$.

EXAMPLE 3. The direction field associated with the DE $y' = xy$ is displayed in Figure 2.5. The isoclines are the hyperbolas $xy = k$ plus the axes $x = 0$ and $y = 0$. From $dy/y = x\, dx$, $\ln |y| = x^2/2 + \ln |c|$, and $|y/c| = \exp(x^2/2)$, we find that the integral curves consist of the family having equation $y = c \exp(x^2/2)$, plus the x axis.

　　The importance of direction fields cannot be exaggerated. The solutions of many DE that arise in applications cannot be expressed in simple explicit or implicit form. In such cases, the associated direction fields often provide qualitative information concerning solutions and their properties. The direction field is also the basis underlying numerical methods for solving a DE approximately. An introduction to these numerical methods will be presented in Chapter 9.

　　The main disadvantage of the method of direction fields is that it is time-consuming. It is interesting to note that a digital computer can be programmed to print out a direction field. Three computer-generated direction fields are displayed in Figures 2.6, 2.7, and 2.8. The author is indebted to James F. Hall for this application of computer graphics.

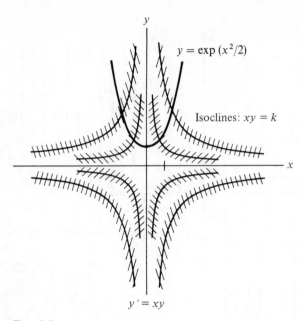

$y = \exp(x^2/2)$

Isoclines: $xy = k$

$y' = xy$

Fig. 2.5

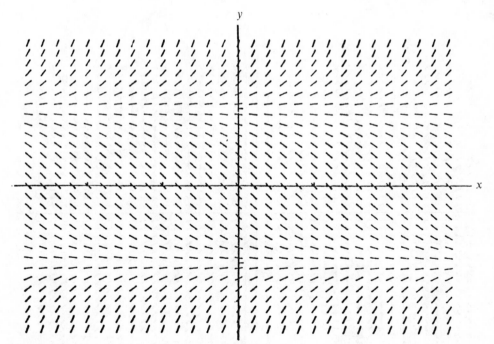

Direction field of $y' = y^2 - 1$

Fig. 2.6

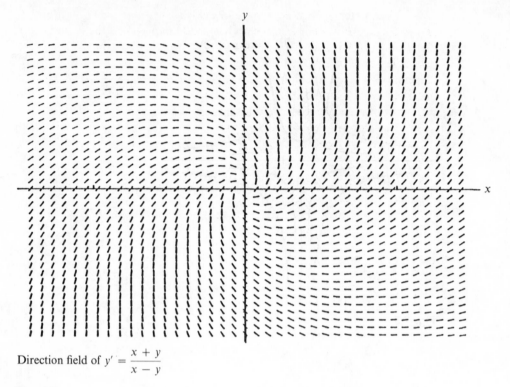

Direction field of $y' = \dfrac{x + y}{x - y}$

Fig. 2.7

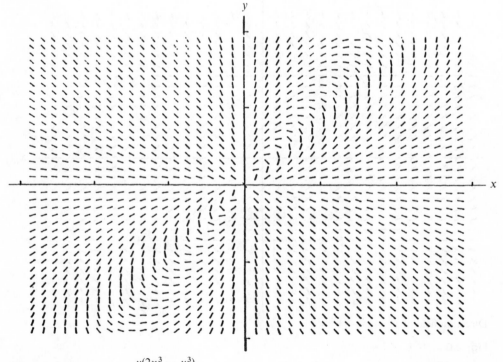

Direction field of $y' = -\dfrac{y(2x^3 - y^3)}{x(2y^3 - x^3)}$
"Folia of Descartes"

Fig. 2.8

The existence of one or more integral curves of $y' = F(x, y)$ through $P(a, b)$, where P is a point in the domain of F, depends on the nature of the function F. If F is continuous in a region R containing P, intuition suggests that at least one integral curve through P exists. For the special case in which F depends on x alone, a unique solution of $y' = F(x, y)$ exists, as we have proved in Theorem 1-I. Interpreted geometrically, this theorem guarantees the existence of a unique integral curve through $P(a, b)$.

The following existence theorem was stated by the Italian mathematician Giuseppe Peano (1858–1932). His proof required later modification.

Theorem 2-IV   *Peano's Existence Theorem:* Let F be continuous on a closed rectangle R: $x_0 \leq x \leq x_1$, $y_0 \leq y \leq y_1$; and let $P(a, b)$ be any interior point of R. Then there exists at least one function f, denoted by $y = f(x)$, defined on an open interval I such that $a \in I$, $y'(x) = F(x, y(x))$ on I, and $y(a) = b$.

The geometric interpretation of the theorem, illustrated in Figure 2.9, is that at least one integral curve through $P(a, b)$ of $y' = F(x, y)$ exists.

EXAMPLE 4. By Peano's Theorem, the DE $y' = F(x, y) = x/y$ has a solution satisfying $y(a) = b$ if $b \neq 0$. The function F is continuous on any closed rectangle R that contains no point at which $y = 0$. From

$$y\, dy = x\, dx \quad \text{and} \quad \frac{y^2}{2} = \frac{x^2}{2} + \frac{c}{2}$$

we obtain $y = \pm\sqrt{x^2 + c}$. A solution f satisfying $f(0) = 1$ is given by $y = f(x) = \sqrt{x^2 + 1}$, with f having domain $(-\infty, +\infty)$. The solution f is unique, although this fact does not follow from Peano's Theorem.

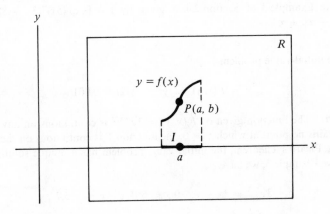

Fig. 2.9

EXAMPLE 5. By Peano's Theorem, the DE $y' = F(x, y) = 3y^{2/3}$ has a solution f satisfying $f(2) = 0$. The function F is continuous on the entire xy plane. Dividing by $y^{2/3}$, we obtain

$$y^{-2/3} \, dy = 3 \, dx, \quad 3y^{1/3} = 3x + c,$$

$$0 = 6 + c, \quad c = -6, \quad \text{and} \quad y = f(x) = (x - 2)^3$$

with f having domain $(-\infty, +\infty)$. In obtaining f we divided by $y^{2/3}$, and hence we must investigate the possibility that $y = g(x) \equiv 0$ may also define a solution. It is readily verified that g is a solution on $(-\infty, +\infty)$. Thus the initial-value problem does not have a unique solution. It can be shown that an infinite number of solutions exist. See Problem 11.

The solution guaranteed by Theorem 2-IV can be shown to be unique provided that appropriate additional conditions are imposed on F. The following uniqueness theorem is due to the French mathematician Emile Picard (1856–1941).

Theorem 2-V *Picard's Uniqueness Theorem:* If F and $\partial F / \partial y$ are continuous on a closed rectangle R: $x_0 \leq x \leq x_1$, $y_0 \leq y \leq y_1$; then the solution f guaranteed by Peano's Theorem is unique.

Picard's Theorem states conditions on F sufficient to ensure uniqueness. The conditions are not necessary; it is possible to prove uniqueness under less restrictive hypotheses. (For proofs of the theorems of Peano and Picard, see References 2.3, 2.4, and 2.5.)

EXAMPLE 6. Apply Picard's Theorem to the initial-value problem

$$y' = F(x, y) = 5y + 3 \exp (5x), \quad y(0) = 8$$

Solution: Noting that F and $\partial F / \partial y = 5$ are continuous on the entire xy plane, we conclude from Picard's Theorem that a unique solution exists. This solution, found in illustrative Example 3 of Section 2.4, is given by $y = [\exp (5x)](3x + 8)$ and has domain $(-\infty, +\infty)$.

EXAMPLE 7. Solve the initial-value problem

$$y' = F(x, y) = 3y^{2/3}, \quad y(2) = 0.001$$

Solution: The function given by $F_y(x, y) = 2y^{-1/3}$ is continuous on any rectangle that contains no point at which $y = 0$; the function F is continuous on the entire xy plane. By Picard's Theorem, the initial-value problem has a unique solution. From illustrative Example 5, we have

$$3y^{1/3} = 3x + c, \quad 0.3 = 6 + c, \quad c = -5.7$$

and

$$y = (x - 1.9)^3, \quad -\infty < x < +\infty$$

Unlike the solution in Example 5, this solution is unique; Picard's Theorem does not apply in Example 5 since the function given by $F_y(x, y) = 2y^{-1/3}$ is not continuous at $(2, 0)$.

In Picard's Theorem the unique solution guaranteed holds on an interval I containing $x = a$. The theorem provides no information relating the size of I and the size of the interval $J = [x_0, x_1]$ determined by the rectangle R.

EXAMPLE 8. Solve the initial-value problem

$$y' = F(x, y) = 1 - y^2, \quad y(0) = 0$$

Solution:

$$\int_0^y \frac{du}{1 - u^2} = \int_0^x dt$$

$$\tanh^{-1} y = x$$

$$y = f(x) = \tanh x = \frac{e^x - e^{-x}}{e^x + e^{-x}}$$

The unique solution f holds on $(-\infty, +\infty)$ and hence holds on $J = [x_0, x_1]$ regardless of the size of the rectangle R in Picard's Theorem. Since $F(x, y) = 1 - y^2$ and $F_y(x, y) = -2y$ are polynomials, the interval $J = [x_0, x_1]$ is unlimited in size.

EXAMPLE 9. Solve the initial-value problem

$$y' = F(x, y) = 1 + y^2, \quad y(0) = 0$$

Solution:

$$\int_0^y \frac{du}{1 + u^2} = \int_0^x dt$$

$$\text{Tan}^{-1} y = x$$

$$y = f(x) = \text{Tan } x$$

Since $F(x, y) = 1 + y^2$ and $F_y(x, y) = 2y$ are polynomials, the interval $J = [x_0, x_1]$ is unlimited in size as in Example 8. The interval I, however, must coincide with, or be contained in, the interval $(-\pi/2, \pi/2)$. The solution f is not defined at $x = \pm\frac{\pi}{2}$ (see Figure 2.10) and hence cannot be valid at either of these points.

EXAMPLE 10. Solve the initial-value problem

$$y' = F(x, y) = y^2, \quad y(0) = k^{-1}$$

where $k > 0$.

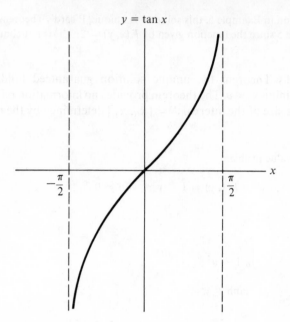

$y = \tan x$

Fig. 2.10

Solution:

$$\int_{k^{-1}}^{y} u^{-2}\, du = \int_{0}^{x} dt$$

$$-u^{-1}]_{k^{-1}}^{y} = t]_{0}^{x}$$

$$-y^{-1} + k = x$$

$$y = f(x) = (k - x)^{-1}$$

As in Example 9, $F(x, y) = y^2$ and $F_y(x, y) = 2y$ are polynomials and $J = [x_0, x_1]$ is unlimited in size. The interval I, however, must coincide with, or be contained in, the interval $(-\infty, k)$ since f is not defined at $x = k$. If the value of k is changed, then the size of the interval $(-\infty, k)$ is changed. See Figure 2.11. In this example the interval I in Picard's Theorem clearly depends on the initial condition.

In Examples 8, 9, and 10, the maximum size of the interval I could not have been anticipated. In each example the functions F and F_y defined simple polynomials and were continuous on the entire xy plane. The explanation lies in the fact that all three DE were nonlinear. When Picard's Theorem is applied to the linear DE $y' = Q(x) - y\, P(x)$, it can be shown that the unique Picard solution holds at least at every point at which P and Q are continuous.

Picard's Theorem is very useful since it covers most cases that arise in applications. If we cannot solve a DE by elementary methods, we can often determine by Picard's Theorem that a unique solution exists. We thereby gain

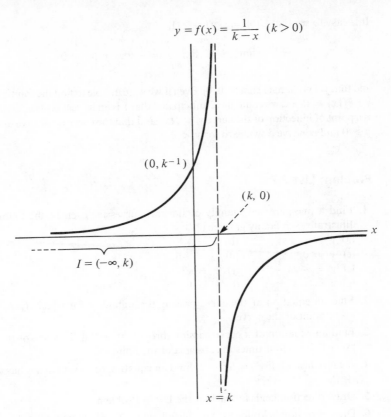

$$y = f(x) = \frac{1}{k-x} \quad (k > 0)$$

$(0, k^{-1})$

$(k, 0)$

x

$I = (-\infty, k)$

$x = k$

Fig. 2.11

confidence in the appropriateness of the DE as part of a mathematical model for describing a physical situation. If we then draw the direction field associated with the DE, we can often obtain useful information about the unique solution that is known to exist. For a solution of $y' = F(x, y)$ given by $y = f(x)$, we know that the solution is increasing when $F(x, y) > 0$, decreasing when $F(x, y) < 0$, and stationary when $F(x, y) = 0$. We can investigate concavity and points of inflection by examining

$$y'' = \frac{d^2y}{dx^2} = \frac{\partial F}{\partial x}\frac{dx}{dx} + \frac{\partial F}{\partial y}\frac{dy}{dx} = F_x + F_yF$$

Even when we are able to solve a DE, an examination of the associated direction field and a geometric analysis are often revealing.

EXAMPLE 11. For the initial-value problem of illustrative Example 9:

$$y' = 1 + y^2, \quad y(0) = 0$$

we find that

$$y'' = 2yy' = 2y(1 + y^2)$$

It is easy to see that $y'(0) = 1$, $y'(x) \geq 1$,

$$\lim_{y \to +\infty} y' = \lim_{y \to -\infty} y' = +\infty, \quad y''(0) = 0$$

and that $y''(x)$ changes sign at $(0, 0)$. Even if we were unable to find the solution given by $y = f(x) = \tan x$, we could have anticipated that f is an increasing function, that $(0, 0)$ is a point of inflection of the curve $y = f(x)$, and that the curve is concave upward for $y > 0$ and concave downward for $y < 0$.

Problem List 2.7

1. Find a one-parameter family of integral curves for each of the following DE. Illustrate each family graphically.
   (a) $y' = 4x$ (b) $y' = 4$
   (c) $y' = 0$ (d) $y' = -6x$
   (e) $y' = -2x + 3$ (f) $y' = x^2$
   (g) $y' = -3x^2$ (h) $y' = x^{-1/2}$

2. Find an equation of the curve passing through $(2, 3)$ for which $dy/dx = x/2$ at every point of the curve.

3. Find an equation of a curve passing through $(0, -3)$ if the slope of the curve at every point is four times the abscissa of the point.

4. A curve has a DE $y'' = 6x - 2$. Find an equation of the curve if it has slope 6 at $(1, 5)$.

5. Draw direction fields for each of the DE in Problem 1.

6. Draw direction fields for each of the following DE. Show several isoclines for each field.

   (a) $y' = y$ (b) $y' = \dfrac{-y}{x}$

   (c) $y' = \dfrac{x + y}{x}$ (d) $y' = xy - 1$

   (e) $y' = y^2$ (f) $y' = x + y$

   (g) $y' = x - y$ (h) $y' = \dfrac{-2y}{1 + x^2}$

7. Draw the direction field determined by the DE $y' = x/2$. Find the integral curve passing through $(3, 4)$.

8. Draw the direction field determined by the DE $y' = y - x$. Find the set of points $P(x, y)$ such that the solution curve through each P is concave upward.

9. (a) Solve the DE $dy/dx = (y - x)/(y + x)$. Write the general solution in polar coordinates. Find the particular solution through $(1, 0)(\theta = 0, r = 1)$.
   (b) Draw the direction field for the DE of part (a). Exhibit the integral curve through $(1, 0)$.

10. Solve the DE $y' = (x + y)/(x - y)$. Write the solution in polar coordinates. See Figure 2.7.

11. Show that the initial-value problem

$$y' = 3y^{2/3}, \quad y(2) = 0$$

of illustrative Example 5 has an infinite number of solutions.

12. Solve the DE $y' = y^{2/3}$ by separating the variables. Find the solution for which $y = 0$ when $x = 1$. Verify that $Y \equiv 0$ provides a second solution. Explain why this illustration does not contradict Peano's Theorem or Picard's Theorem.

13. Solve the initial-value problem

$$y' = y^2, \quad y(0) = 0$$

Explain why the solution is unique. See illustrative Example 10.

14. Solve the initial-value problem:
    (a) $y' = y^2 - 1, y(0) = 0$
    (b) $y' = y^2 - 1, y(0) = 2$
    See Figure 2.6.

15. (a) Find two solutions of the initial-value problem

$$y' = \sqrt{y}, \quad y(0) = 0$$

Explain why Picard's Theorem does not apply.
    (b) Find the unique solution of the initial-value problem

$$y' = \sqrt{y}, \quad y(0) = 1$$

Explain how Picard's Theorem applies.

16. Let P and Q be continuous on $[a, b]$. Prove that the initial-value problem

$$y' + P(x)y = Q(x), \quad y(x_0) = y_0 [a < x_0 < b]$$

has a unique solution on $[a, b]$.

References

2.1 Birkhoff, G., and G.-C. Rota. 1962. *Ordinary Differential Equations*. New York: Ginn.

2.2 Kaplan, Wilfred. 1962. *Ordinary Differential Equations*. Reading, Mass.: Addison-Wesley.

2.3 Kreider, D. L., R. G. Kuller, and D. R. Ostberg. 1968. *Elementary Differential Equations*. Reading, Mass.: Addison-Wesley.

2.4 Murray, F. J., and K. S. Miller. 1954. *Existence Theorems for Ordinary Differential Equations*. New York: New York University Press.

2.5 Simmons, G. F. 1972. *Differential Equations with Applications and Historical Notes*. New York: McGraw-Hill.

3

Applications of First-Order Differential Equations

3.1 Introduction

To study a physical system mathematically, we first identify the significant variables and constants involved. In some instances we restrict our attention to internal processes; in others we also consider certain external forces affecting the system. We then use physical laws governing the behavior of the system. These laws are formulated by making reasonable assumptions, by conducting experiments, and by applying inductive reasoning to data collected from observations. When these laws, which in most important cases involve DE, are translated into mathematical language, they furnish a mathematical model for the physical situation. Such a model does not provide an exact description of the physical process under study; it is instead an approximation that makes it possible to analyze the physical process mathematically. When we solve a DE contained in a model, we are applying logic to obtain mathematical consequences of the model. An appropriate model will enable us to predict future behavior of the physical system, or at least give us additional insight into the physical process. If predictions based on the model do not agree with observed results, we revise the model by changing some of our underlying physical assumptions, or by incorporating new factors previously neglected. We then obtain a new DE, which we hope will result in a more realistic model.

For example, the DE governing the motion of a particle can be obtained without accounting for friction. This DE may be an appropriate model for some motions but may lead to unusable results in others. A more complicated

DE, obtained by assuming some form of frictional force acting on the particle, may be more realistic and hence more useful in studying motions in which friction is significant. A still more complicated DE might be needed for certain ballistic problems, in which factors like the earth's rotation might be relevant.

Historically, DE were first used with outstanding success in the physical sciences. This was primarily because the systems studied were influenced by relatively few factors. As time passed, systems encountered in fields such as economics and biology were also studied with models involving DE. Formulating such models is complicated, since the systems under study are often influenced by many diverse factors. The basic problem in developing such models is to isolate those factors that significantly influence the future behavior of the underlying systems. Computing machines have been important in this extension since they enable investigators to obtain useful approximate solutions to complicated DE.

We are interested in a physical system whose future state can be predicted or approximated by analyzing a mathematical model consisting of a DE and one or more initial conditions. For example, a model of the system might consist of the initial-value problem:

$$y' = F(x, y); \quad y(a) = b$$

To insure that the model is appropriate and useful for studying the system, it is customary to assume that the DE satisfy three requirements:

(1) The DE has a solution. This requirement corresponds to the realization that "something happens" to the system, including the possibility that y remains constant.

(2) The solution of the DE is unique. The physical interpretation of this requirement is that, given the initial state of the system, only one result is possible. This involves the philosophical concept of determinism; we assume that we are dealing with a system whose future is uniquely determined by a given state of the system. We feel that we are dealing with a situation where repetitions of the same experiment under identical conditions lead to the same result.

(3) The solution depends continuously on the initial conditions and on any parameters appearing in the DE. Suppose that the numbers a and b in the initial-value problem are obtained from measurements. Since measurements necessarily involve approximations, we will actually be solving the initial-value problem:

$$Y' = F(x, y); \quad Y(a + h) = b + k$$

where $|h|$ and $|k|$ are small. If the values of the solution given by $Y = Y(x)$ are not close to the values of the solution given by $y = y(x)$, it is clear that the model will be useless for studying the physical situation.

As another example, consider the initial-value problem:

$$y' = ky; \quad y(a) = b$$

where the parameter k is positive. It can be shown that solutions given by $y = y(x, k)$ vary continuously with k. Loosely speaking, this means that solutions of

$$Y' = (k + m)y; \quad Y(a) = b$$

where $|m|$ is sufficiently small, approximate the solution of $y' = ky; y(a) = b$.

Physically, the continuity assumption stipulates that small changes in the initial state of the system or in its physical parameters yield small changes in the system.

An initial-value problem whose DE satisfies requirements 1–3 is termed well-posed, or well-set. For a more complete discussion of these requirements, see Reference 3.10. We assume that we are dealing with regions in which the initial-value problems we consider are well-posed. In more advanced applications, models sometimes employ DE that do not completely satisfy requirements 1–3.

In this chapter we apply many differential equations to diverse fields. Our objective is to give the reader an insight into the way in which DE are formulated and used in real-life situations. We try to restrict ourselves to applications that do not require any deep knowledge of the fields from which the applications are taken. (Additional applications and more detailed models of the systems considered can be found in the references at the end of the chapter.)

In Sections 3.16 through 3.20 we present some geometric applications of DE. One of the basic problems of analytic geometry is that of finding an equation of a curve or graph. This is solved by translating a descriptive property defining a curve C into an algebraic equation $f(x, y) = 0$ that is satisfied by the coordinates of every point on C, and not satisfied by the coordinates of any other points. If the property characterizing the curve C involves not only x and y but also the slope dy/dx of C at $P(x, y)$, the mathematical formulation will involve a DE instead of an algebraic equation. The solution of the DE will generally contain an arbitrary constant, and hence a family of curves having the given property will be obtained. A particular member of the family may be selected by applying a given condition; for example, it may be specified that the required curve C should pass through a fixed point (a, b).

We may also adopt the view (Chapter 2) that the DE defines a direction field. The knowledge that the required curve passes through (a, b) is used to find a specific integral curve of the DE.

3.2 Growth and Decay

Let us denote by Q a quantity that increases in size with time t, and let Q_0 be the value of Q at time $t = 0$. We seek a function f given by $Q = f(t)$ to be the mathematical model for describing and studying the relation between t and Q for $t \geq 0$. If f is to be found by solving a DE, we must use a continuous function f, since otherwise f will not be differentiable. In a physical situation, however, Q at any instant will consist of a specific number of units; that is, an integral number of individuals, molecules, atoms, and the like. For example, if Q is the number of bacteria in a certain culture, Q_0 will be an integer. The value of Q will remain Q_0 until a time $t_1 > t_0 = 0$, when Q will jump from Q_0 to an integral value $Q_1 > Q_0$. Similarly, Q will remain Q_1 until a time $t_2 > t_1$, when Q will assume a new value $Q_2 > Q_1$, the process continuing as the bacteria reproduce by splitting or dividing. We could use a step function f in our model, but such a function would have derivative zero in open intervals such as $t_1 < t < t_2$ and would be discontinuous at values such as t_1 and t_2. A function of this type cannot be used in a model involving a DE.

In many situations of this kind, where the value of Q is large and the intervals $(t_0, t_1), (t_1, t_2), \ldots,$ are short, an appropriate model is formulated by assuming that $Q = f(t)$ defines a continuous function f for $t \geq 0$, and that the time rate of change of Q at any instant is proportional to the value Q at that instant. That is, we assume that

$$\frac{dQ}{dt} = kQ \tag{3.1}$$

where k, known as the *growth constant*, is a positive constant. Equation (3.1) is known as the *DE of organic growth*.

In Section 2.2, we solved (3.1) by separating the variables. For variety, let us write (3.1) in the form $dQ/dt - kQ = 0$, and treat it as a first-order linear DE. Using the integrating factor $e^{\int - k dt} = e^{-kt}$, we obtain

$$Qe^{-kt} = \int 0 \, dt = C$$

or

$$Q = Ce^{kt}$$

Setting $t = 0$ and $Q = Q_0$, we find that $C = Q_0$ and hence

$$Q = Q_0 e^{kt} \tag{3.2}$$

Thus, as shown graphically in Figure 3.1, Q increases exponentially for $t \geq 0$.

EXAMPLE 1. Bacteria in a certain culture increase at a rate proportional to the number present. If the number N increases from 1000 to 2000 in 1 hr, how many are present at the end of 1.5 hr ($t = 1.5$)?

Solution: The DE $dN/dt = kN$ has solution $N = N_0 e^{kt}$. Since $N = 1000$ when $t = 0$, $N_0 = 1000$. Setting $N = 2000$ and $t = 1$ in $N = 1000 e^{kt}$, we obtain

$$2000 = 1000 e^k$$

or

$$e^k = 2$$

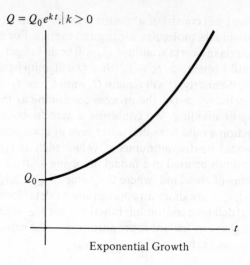

$$Q = Q_0 e^{kt}, \mid k > 0$$

Q_0

t

Exponential Growth

Fig. 3.1

Thus
$$N(t) = 1000(e^k)^t = 1000(2)^t$$
and
$$N(1.5) = 1000(2)^{1.5} = 2000\sqrt{2} \approx 2828$$

EXAMPLE 2. In a culture of yeast, the amount A of active yeast grows at a rate proportional to the amount present. If the original amount A_0 doubles in 2 hr, how long does it take for the original amount to triple?

Solution: The amount A grows exponentially according to $A = A_0 e^{kt}$. Since $A = 2A_0$ when $t = 2$, $2A_0 = A_0 e^{2k}$, and $e^{2k} = 2$.
    Thus, at time t

$$A = A_0(e^{2k})^{t/2} = A_0(2)^{t/2}$$

Setting $A = 3A_0$ and solving for t, we obtain

$$3A_0 = A_0(2)^{t/2}$$

$$\ln 3 = \frac{t}{2} \ln 2$$

and

$$t = \frac{2 \ln 3}{\ln 2} \approx 3.17 \text{ hr}$$

The time required is independent of the original amount A_0.

Nature abounds in quantities that grow at rates proportional to the amount present. The volume of timber in a forest grows at a rate approxi-

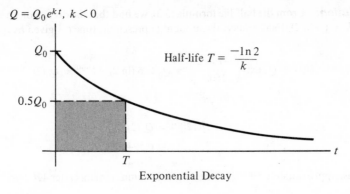

$Q = Q_0 e^{kt},\ k < 0$

Half-life $T = \dfrac{-\ln 2}{k}$

Exponential Decay

Fig. 3.2

mately proportional to the number of cubic feet of wood present at any instant. The amount of a malignant growth increases at a rate proportional to the weight of the growth at any instant, and so on.

If a quantity Q decreases in size with time t in such a way that its rate of change at any instant is proportional to the value of Q at that instant, the process is still governed by DE (3.1). The constant k, which is then negative, is termed the *decay constant*, and (3.1) is known as the *DE of organic decay*.

For a quantity Q undergoing exponential decay, the time required for half of the original amount Q_0 to disintegrate is known as the *half-life* of the decaying substance.

Setting $Q = \frac{1}{2}Q_0$ in $Q = Q_0 e^{kt}$ and solving for t, we obtain

$$\frac{Q_0}{2} = Q_0 e^{kt}$$

$$e^{kt} = \frac{1}{2}$$

$$kt = \ln \frac{1}{2} = -\ln 2$$

and

$$T = t_{\text{hl}} = \frac{-\ln 2}{k} \approx -0.693147 k^{-1} \tag{3.3}$$

We observe from (3.3) that the half-life varies inversely with $|k|$, consistent with the fact that $|k|$ is a measure of the rate at which the substance decays. We note also that the half-life is independent of the original amount Q_0. Exponential decay is illustrated in Figure 3.2.

EXAMPLE 3.   Radium disintegrates at a rate proportional to the amount of radium present at any instant. If the half-life of radium is 1600 yr, what percentage of the original amount Q_0 will remain after 1200 yr?

Solution: From the half-life formula (3.3), we find that the decay constant k is given by $k = (-\ln 2)/1600$. Hence, the amount Q present at time t is given by

$$Q(t) = Q_0 \exp\left(\frac{-t \ln 2}{1600}\right) = Q_0 \exp(\ln 2)^{-t/1600} = Q_0(2)^{-t/1600}$$

and

$$Q(1200) = Q_0(2)^{-0.75}$$
$$\approx 0.59460 Q_0$$

Thus, approximately 59.5% of the original radium remains after 1200 yr.

The science of *radiogeology* applies our knowledge of radioactivity to geology. We know that uranium 238 (atomic weight 238) undergoes radioactive decay with half-life $T = 4.55$ billion years. During decay it becomes radium 226 and eventually ends as nonradioactive lead 206. *Radioactive dating* uses this knowledge to estimate dates of events that took place long ago. One technique uses the ratio of lead to uranium in a rock formation to estimate the time that has elapsed since the lava solidified and formed the mass of rock. The age of the solar system, and hence of the earth, has been estimated by radioactive dating at 4.5 billion years. Other elements such as potassium (half-life 1.3 billion years), thorium (half-life 13.9 billion years), and rubidium (half-life 50 billion years) are used in radioactive dating.

An important breakthrough in radiogeology occurred in 1947 when Willard Frank Libby (b. 1908), an American chemist, discovered *radiocarbon*, a radioactive isotope of carbon designated carbon 14. The relatively short half-life ($T = 5568$ yr) of radiocarbon is used in determining the age of plant and animal remains such as wood, bone, and fossilized pollen. The radiocarbon in people, animals, and plants is continually replaced from the atmosphere so long as they are alive. At death the radiocarbon begins to decay and the ratio of radiocarbon to ordinary carbon (stable carbon 12) decreases with time. This ratio enables scientists to estimate the time of death. For example, scholars have estimated the age of the Dead Sea scrolls by radioactive dating. In 1960, Dr. Libby received the Nobel Prize in chemistry for his discovery of radiocarbon and his contributions to radiogeology and radiochronology.

More extensive accounts of radioactive dating are contained in References 3.4, 3.20, and 3.24. Reference 3.4 contains a particularly interesting description of the use of Pb 210, or white lead (half-life 22 yr), in determining whether a given oil painting is authentic or a forgery.

EXAMPLE 4. It is found that 22% of the original radiocarbon in a wooden archaeological specimen has decomposed. Use the half-life $T = 5568$ yr of radiocarbon to compute the number of years since the specimen was part of a living tree. This should yield a good estimate of the time elapsed since the specimen, a wooden bowl, was used in an ancient civilization.

Solution: Let C denote the amount of radiocarbon present at time t and C_0 the amount present at time $t = 0$. Then

$$C = C_0 e^{kt}$$
$$0.78C_0 = C_0 e^{kt}$$
$$\ln 0.78 = kt$$

and

$$t = \frac{\ln 0.78}{k}$$

From $T = (-\ln 2)/k = 5568$, we obtain

$$k = \frac{-\ln 2}{5568}$$

and therefore

$$t = \frac{5568 \ln 0.78}{-\ln 2} \approx 1996 \text{ yr}$$

At arbitrary time t,

$$C = C_0 \exp\left(\frac{-\ln 2}{5568}\right) t = C_0(2)^{-t/5568}$$

The DE of organic decay governs, at least as an approximation under suitable assumptions, numerous other physical processes and situations. In the troposphere, the rate of decrease of atmospheric pressure with respect to altitude above sea level is proportional to the pressure. The angular velocity of a rotating disk decreases at a rate proportional to the value of the angular velocity. A chemical substance may decompose at a rate proportional to the amount present. Under certain conditions, the rate of dissipation of an electrical charge Q is proportional to the charge Q at any instant. From Lambert's law of absorption, the rate of decrease of radiation intensity of sunlight with respect to the depth of the penetrating medium is proportional to the intensity.

Problem List 3.1

1. Bacteria in a certain culture increase at a rate proportional to the number present. If the number doubles in 1 hr, how long does it take for the original number to quadruple?

2. Bacteria in a certain culture increase at a rate proportional to the number present. If the number doubles in 2 hr, what percentage of the original number will be present at the end of 3 hr?

3. Bacteria increasing at a rate proportional to the number present grow to 10,000 bacteria in 3 hr and 50,000 bacteria in 6 hr. How many were present initially?

4. In a culture of yeast, the time rate of change of the amount of active yeast is proportional to the amount present. If 2 grams are present initially and 4 grams are present after 2 hr ($t = 2$), how many grams are present after 5 hr ($t = 5$)?

5. In a culture of yeast, the amount of active yeast grows at a rate proportional to the amount present. If the amount doubles in 2 hr, what multiple of the original amount of active yeast will be present at the end of 7 hr?

6. Assume that radium disintegrates at a rate proportional to the amount present. If 100 mg reduces to 90 mg in 200 yr, how many milligrams will remain at the end of 1000 yr? Find the half-life T.

7. It takes one year for a radioactive element to lose 90% of its original value y_0. Find the half-life T.

8. Uranium disintegrates at a rate proportional to the amount present at any instant. Given that N_0 grams are present at $t = 0$ and N_1 grams are present at $t = t_1$ years, find a formula for the half-life T of uranium.

9. The amount of a radioactive substance decays at a rate proportional to the amount present. The amount decreases by 50% in 2 hr. Of an original amount of 20 g, how much will be present at the end of 8 hr?

10. If 25% of a radioactive substance disappears in 10 yr, how many years does it take for 60% of the substance to disappear?

11. Charcoal found in the Lascaux cave in southern France has lost 85.5% of its radiocarbon. Using $T = 5568$ yr as the half-life of radiocarbon, find the number of years since the charcoal was living wood. This result gives an estimate of the age of the famous Lascaux prehistoric paintings.

12. The angular velocity ω of a certain wheel decreases at a rate proportional to ω. If $\omega = 200$ rev/sec at $t = 0$, and $\omega = 150$ rev/sec at $t = 10$ sec, how fast is the wheel rotating at $t = 20$ sec?

13. The rate of change of atmospheric pressure p with respect to altitude h is approximately proportional to p. If $p = 14.7$ psi at $h = 0$, and $p = 7.35$ psi at $h = 18{,}000$ ft, what is p at $h = 1000$ ft?

14. A certain chemical substance decomposes at a rate proportional to the amount of the substance present. If 60 g of an original 100 g remain after 1 hr, what expression describes the amount remaining at time t?

15. A certain electrical condenser holding a charge of q_0 coulombs (C) discharges at a rate proportional to the charge q at time t. Find q in terms of t and show that $q \to 0$ as $t \to +\infty$.

3.3 Interest Compounded Continuously

Interest is defined as a charge for borrowed money. It is difficult to give a precise mathematical definition of interest since many different methods of computing interest exist. If a principal of P dollars, invested at interest rate r per annum, grows to $P(1 + r)$ dollars in 1 yr, $P(1 + r)^2$ dollars in 2 yr, and $P(1 + r)^t$ dollars in t yr, r is called the *rate of interest per annum compounded annually*. If the interest rate per annum is r and interest is compounded twice a year, P dollars grows to $P(1 + r/2)$ dollars in 6 mo, $P(1 + r/2)^2$ dollars in 1 yr, $P(1 + r/2)^3$ dollars in 1.5 yr, $P(1 + r/2)^4$ dollars in 2 yr, and $P(1 + r/2)^{2t}$ dollars in t yr.

If P dollars is invested at interest rate r per annum with interest compounded k times per year, the amount a of the original investment at the end of t yr is given by

$$a = P\left(1 + \frac{r}{k}\right)^{kt} = f(t) \tag{3.4}$$

The quantity r/k is the interest rate applied at each compounding, and the exponent kt is the total number of compoundings in t yr.

EXAMPLE 1. One dollar invested at 10% per annum compounded annually ($r = 0.1$) grows to $1(1 + 0.1)^1 = 1.10$ dollars in one year. One dollar invested at 10% per annum compounded twice a year ($k = 2$) grows to $1(1 + 0.05)^2 = 1.1025$ dollars in 1 yr. In the second instance, 10% is called the *nominal interest rate per annum* and 10.25% the *effective interest rate per annum*.

Similarly, if interest is compounded quarterly ($k = 4$), the same dollar becomes $1(1 + 0.025)^4 = 1.1038^+$ dollars in 1 yr and the effective interest rate per annum is $10.38^+\%$.

For fixed values of P, r, and t, the value of a in (3.4) increases as k increases. As $k \to +\infty$, the value of a does not increase without limit, although intuition might suggest this conjecture. Rather,

$$\lim_{k \to +\infty} a = \lim_{k \to +\infty} P\left(1 + \frac{r}{k}\right)^{kt}$$

$$= P \lim_{k \to +\infty} \left[\left(1 + \frac{r}{k}\right)^{k/r}\right]^{rt} = P\left[\lim_{k \to +\infty} \left(1 + \frac{r}{k}\right)^{k/r}\right]^{rt}$$

$$= Pe^{rt}$$

where $e \approx 2.71828$ is the Napierian logarithmic base.

It is reasonable to argue that money should earn interest continuously. But even if interest is compounded every second, the function f in (3.4) is a step function and is not continuous. A continuous model is obtained for continuous compounding by defining the amount A in terms of t by

$$A = Pe^{rt} = F(t) \tag{3.5}$$

This is certainly a reasonable definition since the graphs of f in (3.4) and F in (3.5) are virtually indistinguishable when k is very large. Note that f and F are different functions; a is the value of f at t and A is the value of F at t. Equation (3.5) is more tractable mathematically than (3.4), since da/dt in (3.4) is zero or undefined whereas dA/dt in (3.5) is more useful and meaningful. From (3.5) we obtain

$$\frac{dA}{dt} = (Pe^{rt})r = rA$$

Thus, we see that money invested at compound interest compounded continuously grows according to the DE of organic growth. Conversely, a quantity Q, such as the number of bacteria in a culture, we sometimes say, grows according to the *compound interest law*.

EXAMPLE 2. Given that $10,000 is invested at 6% per annum, find what amount has accumulated after 6 yr if interest is compounded (a) annually; (b) quarterly; (c) continuously.

Solution:

(a) $a = 10,000(1.06)^6 \approx \$14,185.19$

(b) $a = 10,000\left(1 + \dfrac{0.06}{4}\right)^{4(6)} = 10,000(1.015)^{24}$

$\approx \$14,295.03$

(c) $A = 10,000e^{0.36} \approx \$14,333.29$

Although continuous compounding enjoys an advantage over compounding that is done a finite number of times a year, the advantage is not so great as one's intuition might suggest.

EXAMPLE 3. How long does it take for a given amount of money to double at 6% per annum compounded (a) annually? (b) continuously?

Solution:

(a) $2P = P(1.06)^t$

$\ln 2 = t \ln 1.06$

$t = \dfrac{\ln 2}{\ln 1.06} \approx 11.896 \text{ yr}$

(b) $2P = Pe^{0.06t}$

$\ln 2 = 0.06t$

$t = \dfrac{\ln 2}{0.06} \approx 11.552 \text{ yr}$

EXAMPLE 4. A savings and loan association advertises an interest rate per annum of 7.5% compounded continuously. Find the effective interest rate per annum.

Solution: The interest for 1 yr will be $Pe^r - P = P(e^r - 1)$. Thus,

$$e^{0.075} - 1 = 0.0779^-$$

and the effective rate per annum is $7.79^-\%$. Many companies advertise an effective rate of 7.90%. This is obtained by using a 360-day year and multiplying 7.79^- by $\frac{365}{360}$.

This illustrates our statement that there are many different methods of computing interest.

Remark: The nominal rate j, which by continuous compounding is equivalent to an effective rate i, is called the *force of interest*. It is useful in comparing various business propositions.

Problem List 3.2

1. Find the time required for money to double when invested at 7% per annum compounded continuously.
2. Find the amount A at the end of one year if one dollar is invested at 5.25% per annum compounded continuously. What is the effective rate per annum?
3. At what yearly rate of interest compounded continuously does $1000 increase to $3000 in 10 yr?
4. What yearly rate of interest compounded annually is equivalent to 8% per annum compounded continuously?
5. Find the effective interest rate per annum when money is compounded continuously at 100% per annum.
6. Derive a formula for the number T of years required for money to double at $p\%$ per annum compounded continuously.
7. What principal invested at 9% per annum will amount to $300,000 in 30 yr if interest is compounded (a) continuously and (b) semiannually?

3.4 Belt or Cable Friction

Figure 3.3 shows a belt wrapped around a drum of radius r. Assume that tensions T_1 and T_0 are such that $T_1 > T_0$ and the belt is on the point of slipping in the direction from A to B. The tension in the belt varies from $T = T_0$, corresponding to $\theta = 0$, to $T = T_1$, corresponding to $\theta = \angle AOB$, known as the *angle of wrap*. Let T denote the tension at point P corresponding to a fixed but arbitrary value of θ and let $T + \Delta T$ denote the tension at a nearby point corresponding to $\theta + \Delta\theta$. The arc of the belt, of length $\Delta s = r\Delta\theta$, shown to the right of the drum, is in equilibrium under the action of the forces having magnitudes T, $T + \Delta T$, ΔN, and ΔF. The force having magnitude ΔN is the normal force the drum exerts on the section of belt of length Δs, and ΔF is the magnitude of the frictional force opposing slipping. It is easily seen that the force having magnitude $T + \Delta T$ acts at angle $\Delta\theta$ with the tangential direction t along which the force of magnitude T acts. Since the section of belt of length Δs is in equilibrium, the sum of the components of the four external forces acting on it in any direction must be zero. Choosing the tangential direction t and the normal direction n, we obtain

$$T + \Delta F - (T + \Delta T) \cos \Delta\theta = 0 \quad \text{(tangential direction)}$$

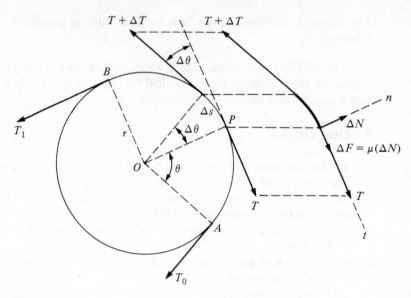

Fig. 3.3

and

$$\Delta N - (T + \Delta T) \sin \Delta\theta = 0 \quad \text{(normal direction)}$$

Due to the assumption that the belt is on the point of slipping, we replace ΔF by $\mu(\Delta N)$, where the constant μ denotes the coefficient of friction between the belt and the drum.

Eliminating ΔN, we obtain

$$T + \mu(T + \Delta T) \sin \Delta\theta - (T + \Delta T) \cos \Delta\theta = 0$$

which, after division by $\Delta\theta$ and rearrangement, becomes

$$T\left(\frac{1 - \cos \Delta\theta}{\Delta\theta}\right) + \mu(T + \Delta T)\frac{\sin \Delta\theta}{\Delta\theta} - \left(\frac{\Delta T}{\Delta\theta}\right)\cos \Delta\theta = 0$$

From our assumption that T is a differentiable function of θ, and the application of some basic calculus, it follows that as $\Delta\theta \to 0$,

$$\frac{1 - \cos \Delta\theta}{\Delta\theta} \to 0, \quad \Delta T \to 0, \quad \frac{\sin \Delta\theta}{\Delta\theta} \to 1, \quad \frac{\Delta T}{\Delta\theta} \to \frac{dT}{d\theta} \quad \text{and} \quad \cos \Delta\theta \to 1$$

Thus,

$$\frac{dT}{d\theta} = \mu T \tag{3.6}$$

and

$$T = T_0 e^{\mu\theta} \tag{3.7}$$

The angle θ must be measured in radians since $\lim_{x \to 0} (\sin x)/x = 1$ only if x is measured in radians.

We introduced this application with two main objectives in mind. First, we wanted to show how we derive a DE by making assumptions about a complicated physical situation. For a discussion of the realism of this model and its extensions to more complicated belt action involving slipping, see Reference 3.19.

Second, we wanted to show that the same mathematical model often governs many seemingly unrelated physical situations. We have discovered that the DE of organic growth governs the way in which belt tension varies, as well as the way in which bacteria increase in number, and the way in which money grows at compound interest compounded continuously. As we proceed, we shall meet other examples in which, from a mathematical point of view, diverse physical situations and processes are indistinguishable, except for the names assigned to their significant elements. This point of view helps many to understand and appreciate the tendency of modern mathematics towards abstraction and unification.

EXAMPLE 1. The coefficient of friction between a cable and a cylinder is 0.25. The minimum tension T_0 in the cable when slipping is impending is 60 lb. Find the maximum tension in the cable if the angle of wrap is (a) $\pi/6$ radians; (b) $\pi/2$ radians; (c) π radians; (d) 2π radians.

Solution:
(a) $T = 60e^{\pi/24} \approx 68.4$ lb   (c) $T = 60e^{\pi/4} \approx 131.6$ lb
(b) $T = 60e^{\pi/8} \approx 88.9$ lb   (d) $T = 60e^{\pi/2} \approx 288.6$ lb

EXAMPLE 2. Given that the coefficient of friction between the fixed horizontal cylinder and the cable in Figure 3.4 is 0.3, find what magnitude of the force F is just sufficient (a) to start the 300-lb body moving upward; (b) to prevent the 300-lb body from moving downward.

Solution: (a) $F = 300e^{0.3\pi} \approx 769.9$ lb
               (b) $300 = Fe^{0.3\pi}$, $F = 300e^{-0.3\pi} \approx 116.9$ lb

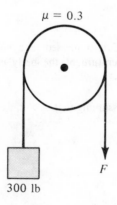

$\mu = 0.3$

F

300 lb

Fig. 3.4

Problem List 3.3

1. A cable is wrapped 1.5 turns around a horizontal drum (angle of wrap $= 3\pi$). Find the coefficient of friction μ between the cable and the drum if a pull of 120 lb just supports a load of 840 lb.

2. A mooring line is wound three times around a post; the coefficient of friction between the line and the post is 0.25. If a ship pulls with a force of 15,000 lb on the line, what force must be exerted at the free end of the rope to prevent the rope from slipping?

3. A cable is wrapped around a post and is on the point of slipping. If the tangential tension T_0 corresponds to $\theta = 0$ and the tangential tension $T_1 > T_0$ corresponds to $\theta = \alpha$, where α is the angle of wrap, what is the tangential tension T_m corresponding to $\theta = \alpha/2$?

4. A rope thrown over a fixed horizontal cylinder (angle of wrap $\alpha = \pi$) requires a pull of 400 lb to start a weight W lb, attached to the other end of the rope, moving upward. For a 100-lb pull just sufficient to prevent the weight from moving downward, find W and the coefficient of friction μ between the rope and the cylinder.

5. The coefficient of friction between a belt and a cylinder is 0.28. Find the angle of wrap if a force of 75 lb withstands a pull of 150 lb.

3.5 Temperature Rate of Change

Under certain conditions, the temperature rate of change of a body is proportional to the difference between the temperature T of the body and the temperature τ of the surrounding medium. This is known as *Newton's law of cooling*. We restrict ourselves to situations in which τ remains constant. We assume also that heat flows rapidly enough that the temperature T of the body is the same at all points of the body at a given time t.

If $T = f(t)$ denotes the temperature of the body at time t, then f satisfies the DE

$$\frac{dT}{dt} = k(T - \tau) \tag{3.8}$$

where $k < 0$.

EXAMPLE 1. A body whose temperature T is initially 200 °C is immersed in a liquid whose temperature τ is constantly 100°C. If the temperature of the body is 150°C at $t = 1$ min, what is its temperature at $t = 2$ min?

Solution: Separating the variables in (3.8) and assuming that $T > \tau$, we obtain

$$\frac{dT}{T - 100} = k\, dt$$

Hence,

$$\ln (T - 100) = kt + c \tag{3.9}$$

From the initial condition, $t = 0$, $T = 200$, we find that $c = \ln 100$.

From the condition $t = 1$, $T = 150$, we find that

$$\ln 50 = k(1) + \ln 100$$

or

$$k = \ln \frac{50}{100} = \ln \frac{1}{2} = -\ln 2$$

Substituting $c = \ln 100$ and $k = -\ln 2$ into (3.9), we obtain

$$\ln (T - 100) = -t \ln 2 + \ln 100 = \ln [100(2)^{-t}],$$

$$T - 100 = 100(2)^{-t}$$

and

$$T = f(t) = 100(1 + 2^{-t})$$

When $t = 2$ min, $T = 100(1 + \frac{1}{4}) = 125\,°C$.

EXAMPLE 2. Integrate the DE $dT/dt = k(T - \tau)$ subject to the condition $T(0) = T_0$.

Solution:

$$\int_{T_0}^{T} \frac{du}{u - \tau} = k \int_0^t dv; \quad k < 0, \quad T > \tau$$

$$\ln (u - \tau)]_{T_0}^{T} = kv]_0^t$$

$$\ln (T - \tau) - \ln (T_0 - \tau) = kt$$

$$\frac{T - \tau}{T_0 - \tau} = e^{kt}$$

$$T = \tau + (T_0 - \tau)e^{kt} \tag{3.10}$$

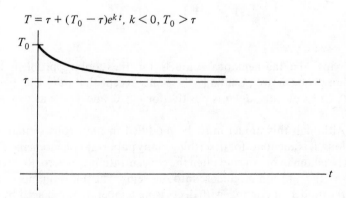

$T = \tau + (T_0 - \tau)e^{kt}, \ k < 0, \ T_0 > \tau$

Fig. 3.5

3.5 Temperature Rate of Change

As $t \rightarrow +\infty$, $T \rightarrow \tau$, and $dT/dt = k(T - \tau) \rightarrow 0$.

That is, as t becomes large, the difference between the temperature of the body and the temperature of the surrounding medium approaches zero, and the rate at which the body cools also approaches zero. See Figure 3.5.

3.6 Diffusion

Let y denote the concentration in milligrams per cubic centimeter of a drug or chemical in a small body, and let y_0 denote the concentration at time $t = 0$. Assume that the body is placed in a container or vat in which the concentration of the drug or chemical is a, where $a > y_0$. The concentration in the small body will increase but a, we assume, remains constant. The assumption is reasonable if the vat is large and the body is small. *Fick's law of diffusion* states that the time rate of movement of a solute across a thin membrane is proportional to the area of the membrane and to the difference in concentration of the solute on the two sides of the membrane. From Fick's law, y satisfies the DE

$$\frac{dy}{dt} = K(a - y) \tag{3.11}$$

where the constant of proportionality K is positive.

If we write (3.11) in the form $dy/dt = (-K)(y - a)$, we observe that (3.11) parallels (3.8),

$$\frac{dT}{dt} = k(T - \tau) \tag{3.8}$$

with y corresponding to T, $-K$ to k, and a to τ. Hence, by (3.10), the solution of (3.8), we obtain

$$\begin{aligned} y &= a + (y_0 - a)e^{(-K)t} \\ &= a - (a - y_0)e^{-Kt} \end{aligned} \tag{3.12}$$

Thus, the mathematical models for the cooling problem and the diffusion problem are essentially the same, except that in (3.12), $a > y_0$ and $K > 0$. Consequently, dy/dt is positive for $t \geq 0$, and $y \rightarrow a$ as $t \rightarrow +\infty$. See Figure 3.6.

Although this model must be modified in numerous similar physical situations, it is adequate for describing many important phenomena. The body might be a human organ, and often the concentration y_0 is zero. Reference 3.21 discusses the infusion of glucose into the veins. The diffusion model is essentially the model for cooling, with decreasing temperature replaced by increasing temperature.

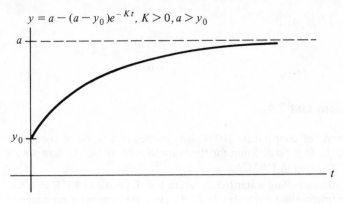

$$y = a - (a - y_0)e^{-Kt}, \; K > 0, \, a > y_0$$

Fig. 3.6

EXAMPLE 1. The concentration of potassium in a kidney is 0.0025 milligram per cubic centimeter. The kidney is placed in a vat in which the potassium concentration is 0.0040 mg/cm^3. In 2 hr the potassium concentration in the kidney is found to be 0.0030 mg/cm^3. What would be the potassium concentration in the kidney 4 hr after it was placed in the vat? How long does it take for the concentration to reach 0.0035 mg/cm^3? Assume that the vat is sufficiently large that the vat concentration $a = 0.004$ mg/cm^3 remains constant.

Solution: Substituting $a = 0.0040$ and $y_0 = 0.0025$ in (3.12), we obtain $y = 0.0040 - 0.0015e^{-Kt}$. Setting $y = 0.0030$ and $t = 2$ yields

$$0.0030 = 0.0040 - 0.0015e^{-2K}$$

$$e^{-2K} = \frac{2}{3}$$

and

$$e^{-K} = \left(\frac{2}{3}\right)^{1/2}$$

Hence, $y = 0.0040 - 0.0015\left(\frac{2}{3}\right)^{t/2}$. When $t = 4$,

$$y = 0.0040 - (0.0015)\left(\frac{4}{9}\right) \approx 0.0033 \text{ mg/cm}^3$$

When $y = 0.0035$,

$$0.0015\left(\frac{2}{3}\right)^{t/2} = 0.0005$$

$$\left(\frac{2}{3}\right)^{t/2} = \frac{1}{3}$$

$$\frac{t}{2}\ln\frac{2}{3} = \ln\frac{1}{3} = -\ln 3$$

and

$$t = \frac{-2 \ln 3}{\ln \dfrac{2}{3}} \approx 5.42 \text{ hr}$$

Problem List 3.4

1. A body of temperature 160 °C is immersed in a liquid of constant temperature 100 °C. If it takes 2 min for the body to cool to 140 °C, how long does it take for it to cool to 110 °C?

2. In 10 min boiling water (temperature 100 °C) cools to 80 °C in a room in which the temperature is constantly 25 °C. How many minutes are required for boiling water to cool to 50 °C under these conditions?

3. A body of temperature 150 °C cools to 140 °C in 1 min when immersed in a liquid of constant temperature 100 °C. How long would it take for the temperature of the body to drop from 120 °C to 110 °C, also a drop of 10 degrees, under the same conditions?

4. Coffee at temperature 180 °F is cooling in a room with constant temperature 75 °F. If it takes 1 min for the coffee temperature to reach 155 °F, what additional time is needed for the coffee to cool to 130 °F?

5. Make the substitution $y = T - \tau$ in $dT/dt = k(T - \tau)$. Find y in terms of t. Note that the difference y between the temperature T of the cooling body and the constant temperature τ of the surrounding medium satisfies the DE of organic decay.

6. Given that T satisfies $dT/dt = k(T - \tau)$, and that $T = T_0$ when $t = 0$, find, in terms of k, the time it takes for T to assume the value $(T_0 + \tau)/2$.

7. The temperature of a body is 40 °C above the constant room temperature τ. In 15 min this temperature difference is 20°. In how many more minutes will the temperature difference be (a) 5° and (b) 4°?

8. The concentration of a drug in a large vat is constantly 0.005 mg/cm³. Two hours after a small drug-free body is placed in the vat, the concentration of the drug in the body is 0.002 mg/cm³. Find the drug concentration in the body 3 hr after it was placed in the vat. How long does it take for the concentration to reach 0.004 mg/cm³?

9. The concentration of a drug in a small body is 0.003 mg/cm³. The body is placed in a large vat in which the concentration of the same drug is 0.006 mg/cm³. In 3 hr the drug concentration of the body rises to 0.004 mg/cm³. At what time is the drug concentration of the body 0.005 mg/cm³? What is the drug concentration of the body 9 hr after it was placed in the vat?

3.7 Mixture Problems; Equation of Continuity

We consider a vat or tank containing Q_0 gallons of a solution in which P_0 pounds of a substance S is dissolved. A second solution flows into the vat at a given rate r_1 gal/min, this solution containing P_1 lb/gal of S. Finally, the mixture in the vat flows out at a given rate r_2 gal/min. We want to find the number of pounds P of S in the vat at time $t \geq 0$.

It is assumed that the mixture in the vat is well-stirred, so that at any given time t_k, $P(t) = P(t_k)$ has the same value at each point in the vat. The rate at which P changes with time is given by

$$\frac{dP}{dt} = (\text{rate } S \text{ flows in}) - (\text{rate } S \text{ flows out}) \qquad (3.13)$$

Equation (3.13) is called the *equation of continuity*. It states that the mass of the quantity S is conserved; that is, no amount of S is created or destroyed in the process.

The rate at which S flows into the vat is given by $P_1 r_1$ lb/min. The rate at which S flows out of the vat is given by $(P/Q(t))r_2$ lb/min, where $(P/Q(t))$ is the concentration of S in the vat at time t; $Q(t)$ is the number of gallons in the vat at time t. If $r_1 = r_2$, $Q(t)$ will have the constant value Q_0. Many interesting cases arise, each leading to a different DE. For example, either r_1 or r_2 may be zero, P_0 or P_1 may be zero, and so on.

EXAMPLE 1. A tank contains 40 gal brine for which the concentration is initially 3 lb salt/gal. A salt solution of 2 lb salt per gallon enters the tank at the rate of 5 gal/min and the well-stirred mixture is drawn off at the same rate. Find the time required for the amount of salt in the tank to be reduced to 100 lb.

Solution: Let P denote the number of pounds of salt in the tank at time t. When $t = 0$, $P = P_0 = 3(40) = 120$ gal. Salt flows into the tank at $2(5) = 10$ lb/min. Since the concentration of salt in the tank at time t is $P/40$ lb/gal, salt flows out of the tank at $(P/40)(5)$ lb/min. From (3.13),

$$\frac{dP}{dt} = 2(5) - \frac{P}{40}(5) \qquad (3.14)$$

If we write (3.14) in the form $dP/dt = \frac{1}{8}(80 - P)$, we observe that the DE governing this process is a special case of the diffusion equation (3.11). Thus we have another example of two different physical situations served by the same mathematical model. Since we have solved the diffusion equation, we leave the solution of (3.14) to Problem 2.

EXAMPLE 2. A tank contains 100 gal brine in which 10 lb salt is dissolved. Brine containing 2 lb salt/gal flows into the tank at 5 gal/min. If the well-stirred mixture is drawn off at 4 gal/min, find (a) the amount of salt in the tank at time t; and (b) the amount of salt in the tank at $t = 10$ min.

Solution: Let $P(t)$ denote the number of pounds of salt in the tank and $Q(t)$ the number of gallons of brine at time t. Then, since 1 gal brine per minute is added to the tank, $Q(t) = 100 + t$. Also, $P(0) = 10$ and $Q(0) = 100$. Since $5(2) = 10$ lb salt is added to the tank per minute, and $[P/(100 + t)](4)$ lb salt per minute is extracted from the tank, P satisfies the DE

$$\frac{dP}{dt} = 10 - \frac{4P}{100 + t} \qquad (3.15)$$

Writing the linear DE (3.15) in the standard form

$$\frac{dP}{dt} + \frac{4P}{100 + t} = 10$$

we find that

$$\exp\left(\int \frac{4\,dt}{100 + t}\right) = \exp\left[4 \ln\left(100 + t\right)\right]$$

$$= \exp\left[\ln\left(100 + t\right)^4\right] = \left(100 + t\right)^4$$

is an integrating factor, and hence

$$P(100 + t)^4 = \int 10(100 + t)^4\,dt = 2(100 + t)^5 + c$$

Setting $t = 0$ and $P = 10$, we obtain

or

$$10(100)^4 = 2(100)^5 + c$$
$$c = -190(100)^4$$

Thus,

$$P(t) = 2(100 + t) - 190(100)^4(100 + t)^{-4}$$

and

$$P(10) = 2(110) - 190(100)^4(110)^{-4} \approx 90.2 \text{ lb}$$

3.8 Rate of Dissolution

A liquid capable of dissolving another substance is known as a solvent, and the dissolved substance is termed the solute. The concentration of the solution at time t is the ratio of solute to solvent at time t; the solution is said to be saturated when the concentration assumes its maximum possible value.

Let Q represent the number of undissolved grams of a solute in a solution at time t. In many important instances, the rate at which Q decreases with respect to the time t is proportional to the amount Q at t and to the difference between the saturation concentration and the concentration at t.

EXAMPLE 1. One hundred grams of a certain solvent is capable of dissolving 50 g of a particular solute. Given that 25 g of the undissolved solute is contained in the solvent at time $t = 0$ and that 10 g dissolves in 2 hr, find the amount Q of undissolved solute (a) at time t, and (b) at $t = 6$.

Solution: The saturation concentration is $50/100$ and the concentration at time t is $(25 - Q)/100$. Hence,

$$\frac{dQ}{dt} = kQ\left(\frac{50}{100} - \frac{25 - Q}{100}\right) = \frac{kQ(25 + Q)}{100}$$

where the constant of proportionality k is negative.

Separating the variables, we obtain

$$\frac{100 \, dQ}{Q(25 + Q)} = k \, dt$$

By resolving $100/[Q(25 + Q)]$ into partial fractions, we find that

$$\frac{100}{Q(25 + Q)} = \frac{4}{Q} - \frac{4}{25 + Q} = 4\left(\frac{1}{Q} - \frac{1}{25 + Q}\right)$$

Hence,

$$\left(\frac{1}{Q} - \frac{1}{25 + Q}\right) dQ = \frac{k \, dt}{4}$$

and

$$\ln Q - \ln (25 + Q) = \frac{kt}{4} - \ln c$$

or

$$\ln \frac{cQ}{25 + Q} = \frac{kt}{4}$$

which can be written

$$\ln \frac{25 + Q}{cQ} = \frac{-kt}{4}$$

Setting $t = 0$ and $Q = 25$ yields

$$\ln \frac{50}{25c} = 0$$

$$\frac{50}{25c} = 1$$

$$c = 2$$

We now set $t = 2$ and $Q = 25 - 10 = 15$ in $(25 + Q)/(2Q) = e^{-kt/4}$ to obtain

$$\frac{40}{30} = e^{-k/2} \quad \text{or} \quad e^{-k/4} = \left(\frac{4}{3}\right)^{1/2}$$

Thus,

$$\frac{25 + Q}{Q} = 2\left(\frac{4}{3}\right)^{t/2} = \frac{25}{Q} + 1$$

$$Q = \frac{25}{2(4/3)^{t/2} - 1} = Q(t)$$

and

$$Q(6) = \frac{25}{2(64/27) - 1} = \frac{675}{101} \approx 6.68 \text{ g}$$

Since this is the number of grams of undissolved solute at $t = 6$, $25 - 6.68 = 18.32$ g dissolved in the first 6 hr. It is interesting to note that $Q \to 0$ as $t \to +\infty$.

Problem List 3.5

1. A 200-gal tank is full of water with 100 lb salt in solution. Pure water enters at 5 gal/min and the well-stirred mixture runs out at the same rate. How much salt is present after 40 min?

2. Complete the solution of illustrative Example 1 of Section 3.7.

3. (a) A tank contains 100 gal pure water. Brine containing 2 lb salt per gallon is added to the tank at the rate of 5 gal/min. How much salt is present after 20 min if the well-stirred mixture runs out at 6 gal/min?
   (b) Find the maximum number of pounds of salt the tank will contain and the number of minutes needed to reach the maximum value.

4. A 10,000-ft^3 room contains 0.1% carbon dioxide. Fresh air containing 0.04% carbon dioxide is admitted to the room at 1000 ft^3/min. Find the percentage of carbon dioxide in the room after 30 min. Assume that the air is well mixed and that air enters and leaves the room at the same rate.

5. A certain solvent weighs 120 g and can dissolve 40 g of a particular solute. Given that 20 g solute is contained in the solvent at time $t = 0$, find the amount Q of undissolved solute at $t = 4$ hr if 5 g solute dissolves in 1 hr. Find also the time required for 10 g solute to dissolve.

3.9 Chemical Reactions; Law of Mass Action

In one type of chemical reaction, the rate at which a chemical A transforms into a second chemical B is proportional to the amount Q of A remaining untransformed at time t. The quantity Q satisfies the DE $dQ/dt = kQ$, $k < 0$.

A chemical reaction of this kind is termed an *internal process*, since the transformation is unaffected by external processes or agencies. It is also referred to as a *first-order reaction*, since dQ/dt is a linear function of Q. We are already familiar with the mathematical model governing this kind of reaction, since Q satisfies the DE of organic decay.

We next consider the situation in which two substances X and Y combine to form a third substance Z. Let us assume that x grams of X react with y grams of Y and that the molecular structures of X and Y are such that a grams of X combine with b grams of Y to form $a + b$ grams of Z. The constants x and y represent initial amounts of X and Y. Let $Q(t)$ denote the number of grams of Z present at time t, with $Q(0)$ equal to zero.

The rate at which Q changes is governed by *the law of mass action*, which asserts that if no temperature change is involved, the rate at which two substances X and Y react to form a third substance is proportional to the product of the amounts of X and Y untransformed at time t.

Since the molecules of X and Y combine in the ratio of a to b, $Q(t)$ consists of $(a/(a + b))Q$ g X and $(b/(a + b))Q$ g Y. Hence, $x - aQ/(a + b)$ g X and $y - bQ/(a + b)$ g Y remain untransformed at time t. By the law of mass action, Q satisfies the DE

$$\frac{dQ}{dt} = k\left(x - \frac{aQ}{a + b}\right)\left(y - \frac{bQ}{a + b}\right) \tag{3.16}$$

where the constant of proportionality k is positive.

This kind of reaction is referred to as a *second-order reaction*, since dQ/dt is a quadratic function of Q. It is also called a *bimolecular reaction*. In solving (3.16), it is customary to rewrite the DE in the form

$$\frac{dQ}{dt} = \frac{kab}{(a + b)^2}\left(\frac{a + b}{a} x - Q\right)\left(\frac{a + b}{b} y - Q\right)$$

or

$$\frac{dQ}{dt} = K(A - Q)(B - Q) \tag{3.17}$$

where

$$K = \frac{kab}{(a + b)^2}, \quad A = \frac{a + b}{a} x, \quad B = \frac{a + b}{b} y$$

If $A = B$, that is, if $x/a = y/b$,

$$(A - Q)^{-2} dQ = K dt, \quad \text{and} \quad (A - Q)^{-1} = Kt + c$$

Setting $t = 0$ and $Q = 0$, we obtain $c = A^{-1}$,

$$\frac{1}{A - Q} = Kt + \frac{1}{A} = \frac{KAt + 1}{A}$$

and

$$Q = A - \frac{A}{KAt + 1}$$

$$= \frac{a + b}{a} x - \frac{a^{-1}(a + b)^2 x}{kbxt + (a + b)}$$

As $t \to +\infty$, $Q \to \frac{a+b}{a} x = A = B$.

If $A \neq B$, substances X and Y can be labeled so that $A > B$. We then write (3.17) in the form

$$\frac{dQ}{(A - Q)(B - Q)} = K \, dt$$

After resolving the left integrand into its partial fractions, we obtain

$$\frac{1}{(A - Q)(B - Q)} = \frac{1}{A - B}\left(\frac{-1}{A - Q} + \frac{1}{B - Q}\right)$$

$$\frac{1}{A - B}[\ln (A - Q) - \ln (B - Q)] = Kt + c_1$$

$$\frac{A - Q}{B - Q} = c \exp [K(A - B)t]$$

where $c = \exp [c_1(A - B)]$.

Multiplying both sides by $B - Q$ and solving for $Q = Q(t)$, we find that

$$Q(t) = \frac{A - Bc \exp [K(A - B)t]}{1 - c \exp [K(A - B)t]}$$

Setting $t = 0$ and $Q = 0$ yields $c = A/B$ and hence

$$Q(t) = \frac{AB\{1 - \exp [K(A - B)t]\}}{B - A \exp [K(A - B)t]}$$

$$= \frac{AB\{1 - \exp [-K(A - B)t]\}}{A - B \exp [-K(A - B)t]}$$

As $t \to +\infty$, $Q \to B = [(a + b)/b]y$, as happened in the case $A = B$. Thus, the limiting value that Q approaches depends on the amount y of substance Y present initially, and on the ratio $b/(a + b)$.

The dissolution DE $dQ/dt = kQ(25 + Q)/100$ of Section 3.8 has the same form as the DE (3.17) when the substitution $u = 25 - Q$ is made. The quantity u, representing the amount of solute dissolved at time t, corresponds to Q, the amount of compound Z formed at time t.

The law of mass action also applies to more complicated reactions involving more than two compounds. An nth-order process involving n compounds is governed by the DE

$$\frac{dQ}{dt} = k(x_1 - r_1Q)(x_2 - r_2Q) \cdots (x_n - r_nQ)$$

where x_i denotes the initial amount of the ith compound, r_i the fraction of $Q(t)$ contributed by the ith compound, and $r_1 + r_2 + \cdots + r_n = 1$.

EXAMPLE 1. Two grams of substance Y combine with 1 g substance X to form 3 g substance Z; that is, the substances combine in the ratio 2:1 by weight. When 100 g Y is thoroughly mixed with 50 g X, it is found that in 10 min 50 g Z has been formed. How many grams of Z can be formed in 20 min? How long does it take to form 60 g Z?

Solution: By (3.16),

$$\frac{dQ}{dt} = k\left(50 - \frac{Q}{3}\right)\left(100 - \frac{2Q}{3}\right) = R(150 - Q)^2$$

where R is a constant. Hence,

$$(150 - Q)^{-2}\, dQ = R\, dt$$

and

$$(150 - Q)^{-1} = Rt + c$$

Setting $t = 0$ and $Q = 0$ yields $c = \frac{1}{150}$.
Setting $t = 10$ and $Q = 50$ yields $\frac{1}{100} = 10R + \frac{1}{150}$, or $R = \frac{1}{3000}$.
Setting $t = 20$ in

$$\frac{1}{150 - Q} = \frac{t}{3000} + \frac{1}{150} \tag{3.18}$$

and solving for Q, we find that $Q(10) = 75$ g.
Setting $Q = 60$ in (3.18) and solving for t, we obtain $t = \frac{40}{3}$ min.

EXAMPLE 2. Three grams of substance Y combine with 2 g substance X whenever 5 g substance Z are formed. When 120 g Y are thoroughly mixed with 90 g X, in 10 min 100 g Z have been formed. How many grams of Z can be formed in 20 min? How long does it take to form 175 g Z?

Solution: By (3.16),

$$\frac{dQ}{dt} = k\left(90 - \frac{2Q}{5}\right)\left(120 - \frac{3Q}{5}\right) = R(225 - Q)(200 - Q)$$

where R is a constant. In $dQ/[(225 - Q)(200 - Q)] = R\, dt$, we resolve $(225 - Q)^{-1}(200 - Q)^{-1}$ into its partial fractions to obtain

$$\frac{1}{25}\frac{-dQ}{225 - Q} + \frac{1}{25}\frac{dQ}{200 - Q} = R\, dt$$

Integrating gives

$$\frac{1}{25}\left[\ln(225 - Q) - \ln(200 - Q)\right] = Rt + c$$

Setting $t = 0$ and $Q = 0$, we get

$$\frac{1}{25}[\ln 225 - \ln 200] = c = \frac{1}{25}\ln\frac{9}{8}$$

Setting $t = 10$ and $Q = 100$ in

$$\ln(225 - Q) - \ln(200 - Q) = 25Rt + \ln 9 - \ln 8$$

we find that

$$25R = \frac{1}{10}\ln\frac{10}{9}$$

Hence,

$$\ln\left[\frac{8(225 - Q)}{9(200 - Q)}\right] = \frac{t}{10}\ln\frac{10}{9}$$

or

$$\frac{8(225 - Q)}{9(200 - Q)} = \left(\frac{10}{9}\right)^{t/10}$$

Setting $t = 20$ and solving for Q, we obtain $Q = \frac{950}{7} \approx 135.7$ g. Setting $Q = 175$ and solving for t, we obtain

$$t = \frac{10\ln(16/9)}{\ln(10/9)} \approx 54.6 \text{ min}$$

We also note that as $t \to +\infty$, $Q \to 200$ g, three-fifths of which comes from Y. Since $\frac{3}{5}(200) = 120$, the amount of Y available approaches zero as $t \to +\infty$.

Another type of reaction is one in which an amount a of one substance A is transformed into a second substance B in such a manner that the rate at which the amount x of B present at time t increases is proportional to x and to the amount $a - x$ of A remaining untransformed at time t. This kind of reaction is termed *autocatalytic* and is governed by the DE

$$\frac{dx}{dt} = kx(a - x), \quad k > 0 \tag{3.19}$$

The solution of (3.19) is left to Problem 3. In Reference 3.13 various other chemical and biochemical reactions are considered.

Problem List 3.6

1. Three grams of substance Y combine with 2 g substance X to form 5 g substance Z. When 90 g Y are thoroughly mixed with 60 g X, it is found that in 20 min, 50 g Z have been formed. How many grams of Z can be formed in 30 min? How long does it take to form 100 g Z?

2. Seven grams of substance Y combine with 3 g substance X whenever 10 g substance Z are formed. When 70 g Y are thoroughly mixed with 60 g X, it is found that in 10 min, 30 g Z are formed. How many grams of Z can be formed in 40 min? How long does it take to form 50 g Z?

3. Solve the DE of an autocatalytic reaction $dx/dt = kx(a - x)$, given that $x = x_0$ when $t = 0$. Evaluate $\lim_{t \to +\infty} x$. Find the value of x for which dx/dt is a maximum, the maximum value of dx/dt, and the value of t for which this maximum is assumed.

3.10 Population Growth

Let $N(t)$ denote the number of individuals in a population at time t. For mathematical convenience, we assume as in Section 3.2 that N is a differentiable function for $t \geq 0$, and that $N(0) = N_0$. In a restricted environment in which the growth of the population is essentially an internal process, it is reasonable to assume that

$$\frac{dN}{dt} = kN$$

where $k = \frac{dN/dt}{N}$, the *growth rate per individual*, is constant. This is the equation of organic growth and decay, yielding $N = $ constant for $k = 0$, and $N = N_0 e^{0.02t}$, and $t = (\ln 2)/0.02 = 50 \ln 2 \approx 34.7$ yr. We note that the interval. It is also customary to write $k = b - d$, where b is the birthrate and d the deathrate.

This model is quite reasonable for an internal growth process over a fairly short time interval, provided that N is not too large. We previously met the model when we considered the growth of bacteria. The model has been applied with limited success to human populations. The world population has been increasing at about 2% per year and hence when $N = 2N_0$, $2N_0 = N_0 e^{0.02t}$, and $t = (\ln 2)/0.02 = 50 \ln 2 \approx 34.7$ yr. We note that the simplistic model agrees well with the facts, since it has been observed that the world's population has been doubling about every 35 years. The model is referred to as the *Malthusian law of population growth*, after Thomas R. Malthus (1766–1834), the English economist who predicted that the world population would increase so rapidly that dire results were in store for the inhabitants of the earth. He predicted geometric growth, which corresponds to exponential growth in the continuous model.

The Malthusian model is unrealistic when N is large. If we take $N = 4$ billion in 1976, the population predicted for 2976 by the model exceeds 1,940,660,000 billion. This is clearly impossible owing to space limitation, since the surface area of the earth is only about 1,860,000 billion square feet.

An improved population model that incorporates a leveling-off effect when N is large was introduced by the Belgian biomathematician P. F. Verhulst in 1837. It employs the DE

$$\frac{dN}{dt} = AN - BN^2 = (A - BN)N \tag{3.20}$$

in which A, the birthrate per individual, is a positive constant, and the deathrate per individual BN is proportional to N where $B > 0$. The birthrate assumes an increase based on cooperation while the deathrate assumes a slowdown from competition, limited resources, limited food supply, limited space, and the like. The term $-BN^2$ has the effect of making the population self-regulating. Equation (3.20) is termed the *logistic equation*, and growth governed by it is called *logistic growth*. The associated model is referred to as the *Verhulst-Pearl model*.

To solve (3.20), we separate the variables to obtain

$$\frac{dN}{N(A - BN)} = dt$$

and resolve $1/[N(A - BN)]$ into its partial fractions $[A^{-1}/N] + [BA^{-1}/(A - BN)]$.

On integrating, we get

$$A^{-1} \ln N - A^{-1} \ln (A - BN) = t + c$$

Setting $t = 0$ and $N = N_0$ gives

$$c = A^{-1} \ln \frac{N_0}{A - BN_0}$$

$$\ln \frac{N(A - BN_0)}{(A - BN)N_0} = At, \quad \ln \frac{(A - BN)N_0}{N(A - BN_0)} = -At$$

$$\frac{A - BN}{N} = \frac{A}{N} - B = \frac{A - BN_0}{N_0} e^{-At}$$

$$\frac{A}{N} = B + (AN_0^{-1} - B)e^{-At}$$

and

$$N = \frac{A}{B + (AN_0^{-1} - B)e^{-At}} \tag{3.21}$$

It is seen by inspection that if $t \to +\infty$, then $N \to AB^{-1}$. This limiting value of N, called the *carrying capacity of the environment*, is independent of N_0. The sigmoid, or **S**-shaped, graph of (3.21) is shown in Figure 3.7. (See Problem 2 for details.) The values of A and B in a particular application can be estimated by using data from observations.

In general, the value of B is small compared with the value of A. About 1850, Verhulst studied the populations of several European countries and the United States. For the United States, he used $A = 0.03134$ and $B =$

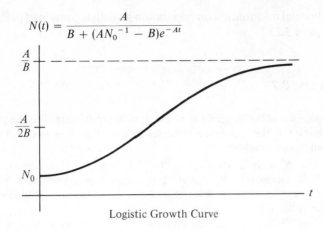

$$N(t) = \frac{A}{B + (AN_0^{-1} - B)e^{-At}}$$

Logistic Growth Curve

Fig. 3.7

$(1.5887)(10)^{-10}$, and assumed $N_0 = N(0) = (3.9)(10)^6$, based on the U.S. population in 1790. These values reduce (3.21) to

$$N \approx \frac{313,400,000}{1.5887 + 78.7703e^{-0.03134t}}$$

This model has predicted the U.S. population up to the present time with remarkable accuracy. For example, $N(10) \approx 5.297$ million compared to the Bureau of the Census figure of 5.308 million for the year 1800. The Verhulst model was also very accurate for predicting population growth in European countries. An exception occurred in Verhulst's home country, Belgium, for which the model was inaccurate owing to the rapid industrialization of that country. This factor, plus the affluence due to the colonization of the Congo, enabled Belgium to support a larger population than the model had predicted.

In 1920, R. Pearl and L. J. Reed used a slight modification of the Verhulst model to study the population of the United States. The model was again highly successful. They chose $N(0) = 98,636,500$, the U.S. population on April 1, 1914. This date marks the turning point when the population rate of growth stopped increasing and started to decrease. It corresponds to the abscissa of the point of inflection in Figure 3.7.

Pearl also applied the Verhulst-Pearl model with outstanding success to populations of other countries and to the growth of the fruit flies (*Drosophila*). Excellent agreement has also been obtained by other investigators in studying the growth of bacterial cultures.

A useful model must be adjusted continually to reflect changes in the birth and death rates. These rates will be influenced by attitudes towards marriage and birth control, and by other sociological trends, and also by technological improvements and changing pollution standards. It is clear that the model a society chooses to use will have critical social and economic implications.

Additional information on population growth is contained in References 3.12, 3.15, and 3.23.

Problem List 3.7

1. The population of a city grows at a rate equal (approximately) to the population at any instant. If the population increases 25% in 10 yr, how long will it take for the population to double?

2. Show that $N = A/[B + (AN_0^{-1} - B)e^{-At}]$ defines a solution of $dN/dt = (A - BN)N$. Given that $N_0 < A/B$, $A > 0$, and $B > 0$, show that N increases with t for $0 \le t < +\infty$, dN/dt increases for $N < A/(2B)$, and dN/dt decreases for $A/(2B) < N < A/B$.

3. In Problem 2 it was shown that dN/dt is a maximum when $N = A/(2B)$. This is the value of N after which the rate of population growth decreases, eventually approaching zero as $t \to +\infty$ and $N \to A/B$. Find the value of t for which $N = A/(2B)$. Find the date of this critical change for the U.S. population, using the values $A = 0.03134$, $B = (1.5887)(10)^{-10}$, and $N_0 = (3.9)(10)^6$ for 1790. These are the values the Verhulst model used.

4. For the logistic DE $dN/dt = (A - BN)N$, find dN/dt for $N = A/(2B)$.

5. Use the Verhulst formula

$$N(t) \approx \frac{313{,}400{,}000}{1.5887 + 78.7703e^{-0.03134t}}$$

to estimate the U.S. population for (a) 1850 $[t = 60]$; (b) 1900 $[t = 110]$; and (c) 1950 $[t = 160]$. Give answers in millions and tenths of a million. Also find the limiting population AB^{-1}.

6. In applying the logistic growth law

$$N = \frac{A}{B + (AN_0^{-1} - B)e^{-At}}$$

to world population, ecologists have estimated $A = 0.029$ and $B = 2.941(10)^{-12}$, with t measured in years. Find the limiting population AB^{-1} and the time required for the population to double if $N_0 = N(0) = 4(10)^9$.

7. For the logistic DE $dn/dt = (A - BN)N$, how does $N \to AB^{-1}$ when $N_0 > AB^{-1}$?

8. The DE $dn/dt = (bN - d)N$, in which the birthrate per individual bN is proportional to N, and the deathrate per individual d is constant, is sometimes used to model the phenomenon known as *population explosion*. Show that if $bN_0 > d$, then there exists a value $t = k$ such that $\lim_{t \to k^-} N(t) = +\infty$.

3.11 Dissemination of Technological Innovations

Let us assume that a region contains R companies, all of which can derive possible advantage by adopting a new technical development. Denote by $N(t)$ the number of these companies that have adopted the innovation at time t, and let $N(0) = N_0$ designate the number of the R companies that have already adopted the innovation at time $t = 0$. We again assume for mathematical convenience that N is differentiable for $t \geq 0$. It seems reasonable to assume that the rate at which $N(t)$ increases is proportional to the number $N(t)$ that have already adopted the innovation, and also to the number $R - N(t)$ of remaining prospective adopters. That is, we assume that

$$\frac{dN}{dt} = kN(R - N) \tag{3.22}$$

where the constant of proportionality k is positive.

If we write (3.22) in the form

$$\frac{dN}{dt} = (kR)N - kN^2$$

and compare with (3.20), we see that kR corresponds to A and k corresponds to B. Thus, N grows logistically and approaches the limiting value $(kR)/k = R$ as $t \to +\infty$. The solution of (3.22) can be obtained by replacing A by kR and B by k in (3.21).

The logistic DE provides a satisfactory model for many diverse physical situations. One interesting example involves the spread of an infectious disease. The constant R denotes the totality of people susceptible to infection, $N(t)$ the number of people infected at time t, and $N(0) = N_0$ the number of people infected at time $t = 0$. The number of people infected at time t grows logistically and approaches R as $t \to +\infty$. When t is small and N is close to N_0, the growth of N resembles exponential growth. During this interval, the term $-kN^2$ has insignificant effect and $dN/dt \approx (kR)N$. Eventually N becomes sufficiently large that the number $R - N$ of those still not infected becomes small, thereby producing a leveling-off effect. As in logistic population growth, the rate of increase of dN/dt increases until $N = R/2$, one-half the limiting value, and decreases thereafter.

A similar logistic DE provides a satisfactory model for studying the spread of a rumor.

Problem List 3.8

1. Of a thousand companies, ten have adopted a new development at time $t = 0$. Given that the number $N(t)$ that have adopted the innovation satisfies the DE $dR/dt = 0.0007N(1000 - N)$, find the number of companies that can be expected to adopt the innovation in 10 yr.

2. The number of people $N(t)$ that have contracted an infectious disease at time t is governed by the DE $dN/dt = kN(R - N)$, where the constant R denotes the total number of people susceptible to infection. Use the substitution $f = N/R$ to find the DE satisfied by the fractional part f of the R people who have been infected at time t.

3.12 Electric Circuits

The current i in an electric circuit is the time rate of change of the quantity of electricity flowing from one element of the circuit to another. At time t sec, the current has value i amperes (A). Figure 3.8 shows a simple RL circuit containing a resistor of resistance R ohms (Ω) and an inductor of inductance L henrys (H) in series with a voltage source of E volts (V). The voltage source, usually a battery or a generator, supplies the electromotive force that produces the current i. It forces the electricity to flow in the direction shown in much the same way that water pressure would generate a flow of water in a circuit. The increase in electrical pressure from $-$ to $+$ can be measured by a voltmeter. The resistor, possibly a light bulb, an electric heater, or a toaster, opposes the current and produces (by Ohm's law) a voltage drop of magnitude $E_R = Ri$. The inductor opposes any change in current and produces a voltage drop of magnitude $E_L = L\, di/dt$.

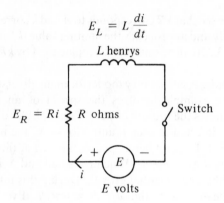

Fig. 3.8

The drops E_R and E_L can also be measured by a voltmeter. We assume that R and L are constant, and that $i = \phi(t)$ and $E = E(t)$ are functions of the time t. The current flows when the switch is closed. According to *Kirchhoff's second law*, the algebraic sum of all the voltage drops around a closed circuit is zero. Thus,

$$E - E_L - E_R = 0 \quad \text{or} \quad E_L + E_R = E$$

Thus, the function ϕ defined by $i = \phi(t)$ satisfies the DE

$$L\frac{di}{dt} + Ri = E(t) \tag{3.23}$$

EXAMPLE 1. In an RL circuit, the inductance is 4 H, the resistance is 20 Ω, and E is constantly equal to 100 V. Given that $i = 0$ when $t = 0$, find the relationship between i and t for $t \geq 0$.

Solution: Equation (3.23) becomes

$$4\frac{di}{dt} + 20i = 100 \quad \text{or} \quad \frac{di}{dt} + 5i = 25$$

This equation has solution given by

$$ie^{5t} = \int 25e^{5t}\,dt = 5e^{5t} + c$$

Setting $i = 0$ and $t = 0$, we obtain $0 = 5 + c$, $c = -5$, and $i = 5 - 5e^{-5t}$.

EXAMPLE 2. Solve Example 1 with $E(t) = 20 \sin 5t$.

Solution: The DE now becomes

$$4\frac{di}{dt} + 20i = 20 \sin 5t \quad \text{or} \quad \frac{di}{dt} + 5t = 5 \sin 5t$$

A general solution is given by

$$ie^{5t} = \int 5e^{5t} \sin 5t\,dt$$

$$= e^{5t}\frac{\sin 5t - \cos 5t}{2} + c$$

Setting $i = 0$ and $t = 0$, we find that $c = \frac{1}{2}$ and hence

$$i = \frac{\sin 5t - \cos 5t}{2} + \frac{e^{-5t}}{2}$$

The term $e^{-5t}/2$ becomes negligible as $t \to +\infty$, and it is called the *transient current*, while the remaining terms $(\sin 5t - \cos 5t)/2$ (without negative exponential) are called the *steady-state* current.

Slightly more complicated circuits will be studied in Section 5.6.

Problem List 3.9

1. An electric circuit consists of an inductance L of 2 H, a resistance R of 20 Ω, and a constant voltage E of 100 V. If the current $i = 0$ A when $t = 0$ sec, what equation will be satisfied by i and t? Draw its graph.

2. In Problem 1, replace the constant voltage $E = 100$ by the variable voltage $E = E(t) = 100 \cos 10t$, and give the information requested.

3. A generator having emf 100 V is connected in series with a 20 Ω resistor and an inductor of 4 H. Given $i = 0$ when $t = 0$, find i in terms of t. Find i for $t = 0.2$ sec.

4. Solve Problem 3, having replaced 100 V by 20 sin 5t V.

5. An inductor of 3 H and a 6 Ω resistor are connected in series with a generator having emf $150e^{-2t} \cos 25t$ V. Find i in terms of t if $i = 20$ when $t = 0$. Find i for $t = 0.5$ sec.

6. Solve $L \, di/dt + Ri = E$ for the conditions $i = 0$ when $t = 0$ and L, R, and E positive constants.

3.13 Flow of Liquid from a Small Orifice

The liquid in the vessel of Figure 3.9 flows out through the small sharp-edged orifice. If there were no loss of energy, the speed of the escaping water would be the same as the speed of a freely falling body, namely, $\sqrt{2gh}$, where h denotes the height in feet of the surface above the orifice at time t. Because of friction and surface tension, the actual speed has been found to be approximately $0.6\sqrt{2gh}$, or $4.8h^{1/2}$ ft/sec when g, the acceleration of gravity, is assumed to be 32 ft/sec². Thus, if the orifice has area A, the fluid leaves the vessel at $4.8Ah^{1/2}$ ft³/sec. Hence, if V denotes the volume of liquid in the vessel at time t,

$$\frac{dV}{dt} = -4.8Ah^{1/2} \tag{3.24}$$

the minus sign indicating that V decreases with time.

We next assume that the shape of the vessel enables us to express V in the form $V = f(h)$. Then $dV/dt = (df/dh)(dh/dt)$, and the resulting DE, in which the variables h and t are separable, is solved to find h in terms of t (or t in terms of h).

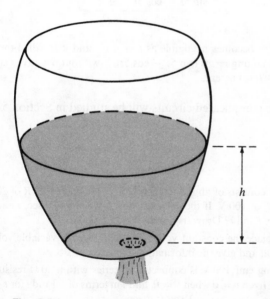

Fig. 3.9

EXAMPLE 1. A tank 4 ft deep has a rectangular cross section 6 ft by 8 ft. The tank is initially filled with water, which runs out through an orifice of radius 1 in. located in the bottom of the tank. Find (a) the time required for the tank to empty; (b) the time required for half the water to drain from the tank; and (c) the height of the water in the tank 20 min after it starts to drain.

Solution: By (3.24)

$$\frac{dV}{dt} = -4.8 \frac{\pi}{(12)^2} h^{1/2}$$

Since $V = (6)(8)h$, $dV/dt = 48\, dh/dt$, and hence

$$48 \frac{dh}{dt} = -4.8 \left(\frac{\pi}{144}\right) h^{1/2}$$

or

$$h^{-1/2}\, dh = \frac{-\pi}{1440}\, dt$$

Thus, $2\sqrt{h} = (-\pi t/1440) + c$. Setting $t = 0$, and $h = 4$, we obtain $c = 4$ and

$$h = \left(2 - \frac{\pi t}{2880}\right)^2 \quad \text{or} \quad t = \frac{2880}{\pi}(2 - \sqrt{h})$$

Setting $h = 0$, we obtain

$$t = \frac{5760}{\pi} \sec = \frac{5760}{60\pi} \min \approx 30.6 \min$$

Setting $h = 2$ gives

$$t = \frac{2880}{\pi}(2 - \sqrt{2}) \sec = \frac{2880}{60\pi}(2 - \sqrt{2}) \min \approx 9.0 \min$$

When $t = 20 \min = 1200 \sec$,

$$h = \left(2 - \frac{1200\pi}{2880}\right)^2 \approx 0.48 \text{ ft}$$

EXAMPLE 2. The conical reservoir in Figure 3.10 has an orifice of 2-in. radius at the vertex of the cone. Find the time required to empty the reservoir if it is filled with water at time $t = 0$.

Solution: Since $r = h/2$ from the law of similar triangles,

$$V = \frac{\pi}{3} r^2 h = \frac{\pi h^3}{12}, \quad \frac{dV}{dt} = \frac{\pi h^2}{4} \frac{dh}{dt}$$

and by (3.24),

$$\frac{\pi h^2}{4} \frac{dh}{dt} = -4.8\pi \left(\frac{2}{12}\right)^2 h^{1/2}$$

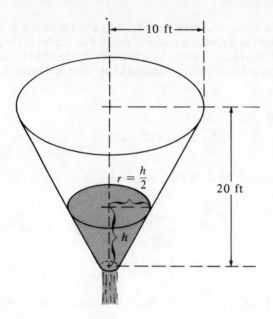

Fig. 3.10

Separating the variables and simplifying, we get

$$h^{3/2}\, dh = -\frac{8}{15}\, dt$$

and hence,

$$\int_{20}^{0} h^{3/2}\, dh = \int_{0}^{t} -\frac{8}{15}\, du$$

$$\frac{2}{5} h^{5/2} \bigg]_{20}^{0} = -\frac{8}{15} t$$

and

$$t = \frac{15}{8}\left(\frac{2}{5}\right) 400\sqrt{20}\ \text{sec}$$

$$= \frac{300\sqrt{20}}{60}\ \text{min} \approx 22.4\ \text{min}$$

This model is suitable for the usual type of orifice. For certain applications, the coefficient of $\sqrt{2gh}$, known as the *coefficient of contraction*, or *discharge coefficient*, may differ from 0.6. A more complicated model is required for other physical assumptions; for example, the air pressure at the liquid level may differ from the air pressure at the level of the orifice.

Problem List 3.10

1. Water 9 ft deep is contained in a tank of rectangular cross section 5 ft by 8 ft. The water runs out through a small orifice of radius 1 in. in the bottom of the tank. Find (a) the time required to empty the tank; (b) the time required to drain half the tank; (c) the height of the water 20 min after it starts to drain.

2. A liquid fills a right circular cylinder of base radius 1 ft and height 5 ft. The liquid runs out through a small orifice of radius 1.2 in. in the bottom of the tank. Find (a) the time required to empty the cylinder; (b) the height of the liquid 1 min after it starts to drain.

3. In illustrative Example 2 of Section 3.13, find the time required for the water level to drop from 10 ft to 5 ft.

4. A hemispherical tank of radius 9 ft is filled with water. How long does it take to drain the water from the tank through an orifice of radius 3 in. located at the bottom of the tank?

3.14 One-Dimensional Heat Flow

We consider heat transfer by conduction in a body whose boundaries are kept at constant temperatures. If no heat is generated or lost internally and if heat transfer by radiation is neglected, the body eventually reaches a *steady state*, in which the temperature T will be a function of the space coordinates x, y, z in the body but independent of the time t. In this section we consider the steady state in which the temperature T depends on a single coordinate x; that is, the steady-state temperature T is given by $T = \phi(x)$. In physical terms, the temperature changes in the y and z directions are negligible compared with the change in the x direction.

Let A denote the area of a surface S in the body perpendicular to the x direction and let **grad** T be the temperature gradient at any point of S. Then the magnitude of **grad** T is dT/dx. The rate Q at which heat flows across S is proportional to A and to dT/dx; that is,

$$Q = -kA\frac{dT}{dx} \qquad (k > 0) \qquad \textbf{(3.25)}$$

where k, the constant of proportionality, is called the *thermal conductivity* of the medium, and where the minus sign means that heat flows in the direction of decreasing temperature. If x is measured in centimeters, A in square centimeters, T in degrees Celsius (absolute temperature), and t in seconds, Q is measured in calories per second.

If S' is any other surface such that the heat flowing across S' is the same as the heat flowing across S, then Q is constant for all such surfaces. This is a form of the equation of continuity and indicates that heat is neither being created nor being lost inside a closed surface.

EXAMPLE 1. The iron bar in Figure 3.11 is 100 cm long, has constant cross-sectional area 4 cm², and is perfectly insulated laterally, so that heat flow takes place only in the x direction. If the left end of the bar is kept at 0 °C and the right end at 60 °C, what is T in terms of x? (The value of k for iron is 0.15.)

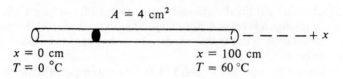

$$A = 4 \text{ cm}^2$$

$x = 0$ cm
$T = 0$ °C

$x = 100$ cm
$T = 60$ °C

Fig. 3.11

Solution: We must solve the two-point boundary-value problem

$$Q = -0.15(4)\frac{dT}{dx}; \quad x = 0, \quad T = 0; \quad x = 100, \quad T = 60$$

From $dT/dx = -5Q/3$, we obtain

$$T = \frac{-5Q}{3}x + c$$

From $x = 0, T = 0$, we find that $c = 0$, and from $x = 100, T = 60$, we find that

$$Q = \frac{-3(60)}{5(100)} = -0.36$$

and hence,

$$T = -\frac{5}{3}(-0.36)x = 0.6x$$

The heat flow is in the direction of the negative x axis at the rate 0.36 cal/sec across every section perpendicular to the x axis.

EXAMPLE 2. A hollow spherical brass ($k = 0.26$) shell has inner radius 4 cm and outer radius 10 cm. If the inner surface temperature is kept at 100 °C and the outer surface temperature at 20 °C, what is the temperature T in terms of r, the radial distance from the center of the shell? What is the temperature on the sphere where $r = 7$ cm? For what value of r is $T = 60$ °C?

Solution: The same amount of heat per second flows across every spherical surface having center at the center of the shell, radius r, and surface area $A = 4\pi r^2$. The flow is in the radial direction, and hence (3.25), with x replaced by r, becomes

$$Q = -(0.26)(4\pi r^2)\frac{dT}{dr}$$

Separating the variables gives

$$dT = \frac{-Qr^{-2}\,dr}{(0.26)(4\pi)} \quad \text{or} \quad dT = Br^{-2}\,dr$$

where $B = -Q/[(0.26)(4\pi)]$. Integrating gives

$$T = \frac{-B}{r} + D$$

Setting $r = 4$, $T = 100$, and then setting $r = 10$, $T = 20$, we obtain $100 = -0.25B + D$, $20 = -0.1B + D$, which yield

$$B = -\frac{1600}{3}, \quad D = -\frac{100}{3}$$

Thus,

$$T = T(r) = \frac{1600}{3r} - \frac{100}{3}$$

When $r = 7$,

$$T = \frac{1600}{21} - \frac{100}{3} = \frac{300}{7} \approx 42.9\,°C$$

When $T = 60$,

$$60 = \frac{1600}{3r} - \frac{100}{3}$$

and

$$r = \frac{40}{7} \approx 5.7 \text{ cm}$$

The value of Q can be found from $Q = (-0.26)(4\pi)B$.

Problem List 3.11

1. A sheet of aluminium ($k = 0.49$) is 10 cm thick. One face is kept at 20 °C and the other face at 80 °C. Assuming that the sheet is sufficiently large for heat flow to be perpendicular to these two faces, find the temperature T in terms of the distance x from the cooler face. What is the amount of heat transmitted per second across a square centimeter of a section parallel to these faces?

2. A cement ($k = 0.0007$) wall is 20 cm thick. The inner surface remains at 20 °C and the outer surface at 5 °C. Find the temperature T in terms of the distance x from the inside surface. What is the heat loss per hour through a square centimeter parallel to the walls? Assume that heat flows perpendicular to the inner and outer walls.

3. A hollow spherical glass ($k = 0.002$) shell has inner radius 6 cm and outer radius 10 cm. If the inner surface temperature is kept at 50 °C and the outer surface temperature at 20 °C, what is the temperature T in terms of r, the radial distance from the center of the shell? Find T for $r = 7, 8$, and 9.

4. A steam pipe is insulated by an asbestos ($k = 0.0002$) covering in the form of a cylindrical shell. The inner radius of the shell (the radius of the pipe) is 3 cm and the outer radius of the shell is 5 cm. If the temperature of the inner surface of the shell

(the temperature of the surface of the pipe) remains at 90 °C and the outer surface of the shell remains at 30 °C, what is the temperature T in terms of the distance r from the axis of the pipe? What is the heat loss per second through the covering per centimeter length of pipe? Assume that the pipe is sufficiently long for heat to flow radially outward from the axis of the pipe.

3.15 Economic Applications

Numerous models have been constructed to describe national economic growth. In a simple model due to E. Domar, the national income Y in dollars per year is expressed as

$$Y = C + I$$

where C is the total dollar value of goods consumed per year and I is the total dollar investment per year. The variables Y, C, and I are assumed to vary with time t.

It is also assumed that I is a linear function of Y; that is

$$I = aY + b$$

where a and b are constants. This yields

$$\frac{dI}{dY} = a, \quad \frac{dY}{dI} = \frac{1}{a}, \quad \text{and} \quad \frac{dY}{dt} = \frac{1}{a}\frac{dI}{dt}$$

The constant a is known as the *marginal propensity to save*. Relating savings to income, it is the fraction of each extra dollar that is saved rather than spent.

Next, let F denote the dollar capacity, or potential output, per year. That is, F is the value Y would have if full employment were realized and maximum use of capital equipment were achieved. The assumption is made that the rate of change of F is proportional to the total investment I; that is,

$$\frac{dF}{dt} = \rho I$$

where ρ is a positive constant.

The situation in which productive capacity is fully realized is known as *economic equilibrium*. This is achieved when aggregate demand equals potential capacity, or

$$Y = F$$

For $Y = F$ to hold for all t, it is necessary that $dY/dt = dF/dt$ for all t.

EXAMPLE 1. Determine total investment I in terms of the time t during which economic equilibrium prevails.

Solution: From $dY/dt = dF/dt$, we obtain

$$\frac{1}{a}\frac{dI}{dt} = \rho I$$

or

$$\frac{dI}{dt} = (a\rho)I$$

This DE has solution $I = I(0)e^{a\rho t}$, where $I(0)$ is the rate of investment at $t = 0$. The conclusion is that in the model, investment must grow exponentially to maintain economic equilibrium.

The Domar model is used in *macroeconomics*, the branch of economics dealing with income and output of whole systems. We now present an application from *microeconomics*, which studies individual areas of economic activity.

Let P denote the price in dollars of a single commodity, Q_d the number of items of the commodity in demand per unit of time, and Q_s the number of items of the commodity in supply per unit of time. We assume that Q_d and Q_s are linear functions of P given by

$$Q_d = a - Pb \qquad Q_s = -c + Pd$$

where $a, b, c,$ and d are positive constants. That is, we assume that Q_d is a linear decreasing function of P, and Q_s a linear increasing function of P. The model is said to be in equilibrium when demand equals supply, that is, when $Q_d = Q_s$.

In equilibrium, the variables in the model have no tendency to change. If the variables Q_d, Q_s, and P vary with time, however, a dynamic analysis is required. If P is given by $P = P(t)$, we are interested in whether or not the model approaches a state of equilibrium as $t \to +\infty$.

We first note that if

$$Q_d = a - Pb = -c + Pd = Q_s$$

then

$$P = \bar{P} = \frac{a + c}{b + d}$$

The quantity $\bar{P}$ is called the *equilibrium price*.

Next, we assume that the rate of change of the price P is proportional to the *excess demand* $Q_d - Q_s$; that is,

$$\frac{dP}{dt} = k(Q_d - Q_s), \quad k > 0$$

Thus,

$$\frac{dP}{dt} = k(a - Pb + c - Pd) = k(a + c) - k(b + d)P$$

or

$$\frac{dP}{dt} + k(b + d)P = k(a + c)$$

Employing the integrating factor

$$\exp\left[\int k(b + d)\, dt\right] = \exp\left[k(b + d)t\right]$$

we obtain

$$P \exp\left[k(b + d)t\right] = k(a + c)\int \exp\left[k(b + d)t\right] dt$$

$$= \frac{a + c}{b + d} \exp\left[k(b + d)t\right] + C$$

Setting $P = P(0) = P_0$ when $t = 0$, we find that

$$C = P_0 - \frac{a + c}{b + d} = P_0 - \bar{P}$$

and

$$P = (P_0 - \bar{P}) \exp\left[-k(b + d)t\right] + \bar{P}$$

If $P_0 = \bar{P}$, then $P = \bar{P} = $ constant. If $P_0 \neq \bar{P}$, then $\lim_{t \to +\infty} P = \bar{P}$ since $k(b + d) > 0$. Since $P(t)$ converges to the equilibrium price $\bar{P}$ as $t \to +\infty$, the equilibrium, it is said, is *dynamically stable*. The situation is depicted graphically in Figure 3.12.

For more complete details, see Reference 3.6.

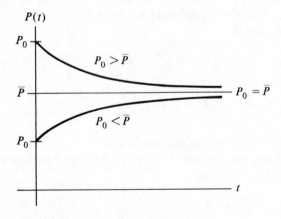

Fig. 3.12

Problem List 3.12

1. Find I in terms of t in the Domar model if $dF/dt = \rho I$ is replaced by
   (a) $dF/dt = \sqrt{t}I$; (b) $dF/dt = e^{kI}$.
2. Assume that $I = I(0)e^{rt}$ in the Domar model, where r, the growth constant, may differ from $a\rho$. Show that this implies that $(dY/dt)/(dF/dt) = r/(a\rho)$. The constant $r/(a\rho)$ is known as the *coefficient of utilization*.
3. In the supply-demand price model, find $Q_s(\bar{P}) = Q_d(\bar{P}) = \bar{Q}$.
4. Solve $dP/dt = k(a + c) - k(b + d)P$, using the substitution $y = P - \bar{P}$.

3.16 Differential Equation of a Family of Functions

Direction fields are determined by DE of the form $dy/dx = F(x, y)$ (Section 2.7). For example, the direction field of the DE $dy/dx = 2x$ consists of the graph of the one-parameter family of functions given by $y = x^2 + c$. (See Reference 3.22 for a discussion of an n-parameter family.) For a DE of order n, the direction field consists of the graph of an n-parameter family of functions. It is natural to ask whether, given an n-parameter family of functions, one can find a DE such that each member of the family is a solution of the DE.

Let us assume that we are given a family of functions defined by

$$y = F(x; c) \tag{3.26}$$

where c is an arbitrary constant. The graph of (3.26) consists of a one-parameter family of curves. For each member of the family (3.26),

$$\frac{dy}{dx} = \frac{\partial}{\partial x} F(x; c) \tag{3.27}$$

If it is possible to eliminate c from (3.26) and (3.27), the result will be a DE of the form

$$G\left(y, \frac{dy}{dx}, x\right) = 0 \tag{3.28}$$

Equation (3.28) will be satisfied by every member of family (3.26) and (3.28) is called a *DE of the family*.

EXAMPLE 1. Find a DE of the family having equation $y = cx^2$.

Solution: Eliminating c from $y = cx^2$ and $dy/dx = 2cx$, we obtain the DE $x\, dy/dx - 2y = 0$. We say that the family is characterized by the DE. The graph of $y = cx^2$ consists of a set of parabolas passing through the origin, plus the x axis. The DE is not defined along the y axis (the line $x = 0$). The origin is a point of every member of the family and is called a *singular point* of the DE.

EXAMPLE 2. Find a DE of the family having equation $y = x^2 + c$.

Solution: The required DE $dy/dx = 2x$ is obtained by differentiating the equation of the family. The constant c did not appear and hence no elimination of c was required since c was eliminated in the differentiation.

If the equation of the family is given implicitly by $g(x, y; c) = 0$, implicit differentiation is used, followed by elimination of c. More generally, to obtain a DE of an n-parameter family having equation $g(x, y; c_1, c_2, \ldots, c_n) = 0$, differentiate n times with respect to x and then eliminate $c_1, c_2, \ldots,$ and c_n from these n equations and the given equation of the family. Any n-parameter family defined by $g(x, y; c_1, c_2, \ldots, c_n) = 0$ is called a *primitive* of the DE. If the family is given explicitly by $y = F(x; c_1, c_2, \ldots, c_n)$, F is a primitive of the DE.

When a primitive contains n essential constants, no more than n differentiations are permitted before the n constants are eliminated. A differential equation of the family cannot have order exceeding n.

EXAMPLE 3. Find a DE of the family having equation $y = cx$.

Solution: From $dy/dx = c$, we obtain $x \, dy/dx = y$ as the required DE. Although every member of $y = cx$ satisfies $d^2y/dx^2 = 0$, this DE is not regarded as a DE of the family since it has order greater than one.

EXAMPLE 4. Find a DE of the family having equation $y = ae^{-3x} + be^{2x}$.

Solution: Two differentiations with respect to x yield

$$y' = -3ae^{-3x} + 2be^{2x}, \quad y'' = 9ae^{-3x} + 4be^{2x}$$

Eliminating a from these two DE yields

$$y'' + 3y' = 10be^{2x} \tag{3.29}$$

Eliminating a between

$$y = ae^{-3x} + be^{2x} \quad \text{and} \quad y' = -3ae^{-3x} + 2be^{2x}$$

gives rise to

$$y' + 3y = 5be^{2x} \tag{3.30}$$

The required DE, obtained by eliminating b from (3.29) and (3.30), is $y'' + y' - 6y = 0$.

A differential equation of a family of functions (curves) is free of parameters and presents in concise form the essential characteristics common to all members of the family. In some investigations it is more fruitful to study a DE of a family than an algebraic equation of the family. The algebraic equation defines, in implicit or explicit form, a general solution of a DE of the family. Such a general solution is not necessarily a complete solution, since singular solutions may exist. Another possibility is that algebraic difficulties may arise in the elimination involved in obtaining the DE. (For more complete details see References 3.1, 3.9, and 3.10.)

Problem List 3.13

1. Find a DE of the family having equation:
   (a) $y = cx^3$
   (b) $x^2 + y^2 = c^2$
   (c) $y = ce^x$
   (d) $y = x + ce^x$
   (e) $x^2 - y^2 = c^2$
   (f) $y^2 = cx$
   (g) $xy = c$
   (h) $y = cx + f(c)$

2. Find a DE of the family (a) of a circles of radius 3 with centers on the x axis; (b) of parabolas with vertices at the origin and foci on the x axis.

3. Find a DE of each of the following two-parameter families.
   (a) $y = ae^x + be^{-x}$
   (b) $y = ax + b$
   (c) $y = x^2 + ax + b$
   (d) $y = a \sin x + b \cos x$
   (e) $y = c \cos (2x + \phi)$

4. Find a DE of all nonvertical straight lines in the xy plane.

5. Find a DE of all circles that are tangent to the coordinate axes and whose centers lie on the line $y = x$.

6. Find a DE of all circles in the xy plane.

7. Find a DE of the folia of Descartes defined by $x^3 + y^3 - 3axy = 0$. See Fig. 2.8.

8. Find a third-order DE satisfied by the functions f, g, and h, given by $f(x) = x^4$, $g(x) = x^2$, and $h(x) = x$.

3.17 Orthogonal Trajectories

Let $y = F(x; c)$ define a one-parameter family of curves in a domain D of the xy plane. Assume that if $P(x, y)$ is an arbitrary point of D, then one and only one member of the family passes through P, and that this curve has a well-defined tangent at P.

If a second family of curves in D has the property that one and only one member passes through P, this curve meeting the member of the first family at right angles at P, the members of the second family are the *orthogonal trajectories* of the first family. The term *trajectory*, meaning solution curve, was originally employed in mechanics.

Orthogonal trajectories are important in applied mathematics. For example, the family of isotherms, or curves of equal temperature, on a weather map is orthogonal to the family of curves representing the direction of heat flow. As a second example, in an electrically charged conducting sheet, the lines (curves) of current flow are orthogonal to the lines (curves) of equal electrical potential.

To find an equation of the orthogonal trajectories of the family $y = F(x; c)$, we find a DE of the family and from it determine the slope $m = m(x, y)$ at $P(x, y)$ in terms of x and y. The expression $m(x, y)$ will not involve the parameter c. Since the slopes of the two families are negative reciprocals at $P(x, y)$, a DE of the required family is given by

$$\frac{dy}{dx} = \frac{-1}{m(x, y)}$$

A primitive of this DE, that is, a general solution, constitutes an equation of the required family.

EXAMPLE 1. Find an equation of the family orthogonal to the family $y = cx$.

Solution: Eliminating c from $y = cx$ and $dy/dx = c$, we obtain $dy/dx = y/x$, a DE of the family. Hence, a DE of the required family is

$$\frac{dy}{dx} = \frac{-1}{y/x} = \frac{-x}{y}$$

From $y\, dy + x\, dx = 0$, we obtain the equation $x^2 + y^2 = k^2$ of the required family, a one-parameter family of circles centered at the origin.

The two families are illustrated in Figure 3.13. The domain D must not include the origin, since infinitely many members of $y = cx$ pass through that point. The circles are perpendicular to the x axis although no one of the circles has a slope at a point at which $y = 0$.

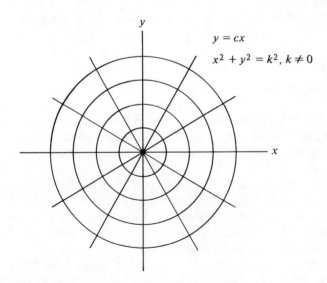

$y = cx$

$x^2 + y^2 = k^2, \, k \neq 0$

Fig. 3.13

EXAMPLE 2. Find the family orthogonal to the family $y = ce^{-x}$ of exponential curves. Determine the member of each family passing through $(0, 4)$.

Solution: From $y' = -ce^{-x}$, we obtain $y' = -y$, a DE of the family of exponential curves. Solving

$$\frac{dy}{dx} = \frac{1}{y} \quad \text{or} \quad y\,dy = dx$$

we obtain the primitive given by $y^2 = 2(x + k)$, a one-parameter family of parabolas. The parabolas are orthogonal to the exponential curves.

Setting $x = 0$ and $y = 4$ in $y = ce^{-x}$ and $y^2 = 2(x + k)$, we find that $c = 4$ and $k = 8$. The required family members through $(0, 4)$ have equations $y = 4e^{-x}$ and $y^2 = 2(x + 8)$. The slopes of these curves at $(0, 4)$ are -4 and $+\frac{1}{4}$, respectively.

In finding the orthogonal trajectories of a given family of curves, it is sometimes simpler to use polar coordinates.

In Figure 3.14, ψ is the angle between the radius vector OP and the tangent to the curve C at $P(r, \theta)$. It is known from the calculus that

$$\tan \psi = \frac{f(\theta)}{f'(\theta)} = \frac{r}{dr/d\theta} = r\frac{d\theta}{dr} = g(\theta)$$

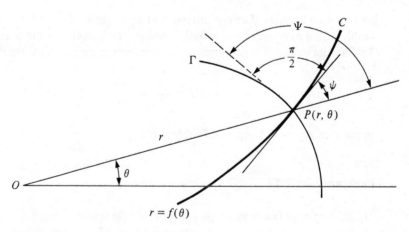

Fig. 3.14

Denoting the corresponding angle for curve Γ by Ψ, and assuming that Γ is perpendicular to C at P, we note that

$$\tan \Psi = \tan\left(\psi + \frac{\pi}{2}\right) = -\cot \psi = \frac{-1}{\tan \psi}$$

Hence, a DE of Γ is given by $r\,d\theta/dr = -1/g(\theta)$, which when solved yields an equation of the family of curves of which Γ is a member.

EXAMPLE 3. Find an equation of the orthogonal trajectories of the family of circles having polar equation $r = f(\theta) = 2a \cos \theta$.

Solution: From

$$\tan \psi = \frac{f(\theta)}{f'(\theta)} = \frac{2a \cos \theta}{-2a \sin \theta} = -\cot \theta = g(\theta)$$

we obtain $\tan \Psi = -1/g(\theta) = \tan \theta$. Hence, $r \, d\theta/dr = \tan \theta$ is a DE of the required family. Separating the variables yields

$$\frac{dr}{r} = \frac{\cos \theta \, d\theta}{\sin \theta},$$

$$\ln |r| = \ln |\sin \theta| + \ln |2c|$$

$$\ln \left| \frac{r}{2c} \right| = \ln |\sin \theta|$$

and

$$r = \pm 2c \sin \theta$$

The required family is a family of circles having centers on the line $\theta = \pi/2$ (y axis).

If two families of curves intersect at an angle α different from $\pi/2$, each family is said to constitute *isogonal*, or *oblique*, trajectories of the other family. To find the isogonal trajectories of a given family, we use the formula

$$\tan \alpha = \frac{m_2 - m_1}{1 + m_2 m_1}$$

in place of $m_2 = -1/m_1$. See Problem 10.

Problem List 3.14

1. Show that the family $xy = c$ is orthogonal to the family $x^2 - y^2 = k$.
2. Find an equation of the family orthogonal to:
   - (a) $y = ax^{-2}$
   - (b) $y = mx + 4$
   - (c) $y = 2x + b$
   - (d) $y = cx^2$
   - (e) $y^2 = cx^3$
   - (f) $x^2 + y^2 + cy = 0$
   - (g) $\dfrac{x^2}{4c^2} + \dfrac{y^2}{c^2} = 1$
   - (h) $2y^2 + x^2 = c^2$
   - (i) $x^2 = 4cy$
   - (j) $x^2 + y^2 = c^2$
   - (k) $x^2 - 2y^2 = k$
   - (l) $y = cx^3$
3. Find an equation of the family orthogonal to the family of circles through the origin with centers on the y axis. Use rectangular coordinates.
4. Find an equation of the family orthogonal to the family of ellipses with vertices at $(\pm 5, 0)$ and centers at the origin.
5. Find the orthogonal trajectories of the family of parabolas $y^2 = cx$ and determine the member of each family that passes through $(1, 2)$.
6. Find an equation of the family orthogonal to the family $y = cx^4$ and depict the two families graphically.

7. A family whose orthogonal trajectories are the members of the same family is said to be *self-orthogonal*. Show that the family $y^2 = 4cx + 4c^2$ is self-orthogonal.

8. Find a polar equation of the family orthogonal to the family having polar equation (a) $r = k\theta,\ \theta > 0$; (b) $r = e^{k\theta}$.

9. Find a polar equation of the family orthogonal to the family having polar equation $r = a(1 + \cos \theta)$. Show that the family of cardioids is self-orthogonal.

10. Find an equation of the family whose members intersect the members of the family $xy = c$ at an angle of $\pi/4$.

3.18 The Tractrix

Let C be a curve for which the distance to the y axis, measured along the tangent to C from an arbitrary point $P(x, y)$ on C, has the constant value a. Let us assume also that C passes through the point $(a, 0)$.

It is seen from Figure 3.15 that the slope of C at $P(x, y)$ is given by

$$\frac{dy}{dx} = \frac{-\sqrt{a^2 - x^2}}{x}$$

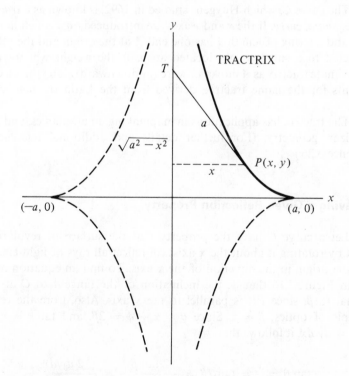

Fig. 3.15

A table of integrals yields

$$y = \int \frac{-\sqrt{a^2 - x^2}}{x} \, dx = a \ln \left(\frac{a + \sqrt{a^2 - x^2}}{x} \right) - \sqrt{a^2 - x^2} + c$$

Setting $x = a$ and $y = 0$, we find that $c = 0$, and hence C has equation

$$y = a \ln \left(\frac{a + \sqrt{a^2 - x^2}}{x} \right) - \sqrt{a^2 - x^2}$$

The second quadrant branch of the curve C in Figure 3.15 has equation

$$y = a \ln \left(\frac{a + \sqrt{a^2 - x^2}}{|x|} \right) - \sqrt{a^2 - x^2}$$

and all four branches are given by

$$y = \pm \left[a \ln \left(\frac{a + \sqrt{a^2 - x^2}}{|x|} \right) - \sqrt{a^2 - x^2} \right]$$

Equations of the third and fourth quadrant branches are obtained from $dy/dx = \sqrt{a^2 - x^2}/x$.

The curve C, which Huygens studied in 1692, is known as a *tractrix*, or *equitangential curve*. If the x and y axes are introduced on a rough horizontal table, and a string of length a has one end A at the origin and the other end connected to a small weight located at $(a, 0)$, the weight will traverse an approximate tractrix as A moves upward (or downward) along the y axis. This accounts for the name tractrix, derived from the Latin *tractum*, meaning "drag."

The tractrix has applications in mapmaking, in mechanics, and in non-Euclidean geometry. (For further details and additional references, see Reference 3.26.)

3.19 Curve Having a Given Reflection Property

Let a curve C have the property that the surface of revolution obtained by rotating it about the x axis will reflect all rays of light emanating from the origin in the direction of the x axis. To find an equation of C, we note in Figure 3.16 that ϕ, the inclination of the tangent to C at $P(x, y)$, is equal to β, since PR is parallel to the x axis. Also, from the reflection principle of optics, $\beta = \alpha$. Since $\theta = \alpha + \phi = 2\beta$, and $\tan \theta = y/x$, and $\tan \beta = dy/dx$, it follows that

$$\tan \theta = \frac{y}{x} = \tan 2\beta = \frac{2 \tan \beta}{1 - \tan^2 \beta} = \frac{2 \, dy/dx}{1 - (dy/dx)^2}$$

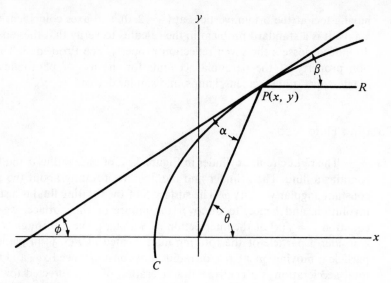

Fig. 3.16

Thus,

$$y - y\left(\frac{dy}{dx}\right)^2 = 2x\frac{dy}{dx}$$

$$y\left(\frac{dy}{dx}\right)^2 + 2x\left(\frac{dy}{dx}\right) - y = 0$$

and

$$\frac{dy}{dx} = \frac{-x \pm \sqrt{x^2 + y^2}}{y}, \quad y \neq 0$$

Writing this DE in the form

$$\mp \frac{1}{2}(x^2 + y^2)^{-1/2}(2x\,dx + 2y\,dy) = dx$$

we obtain

$$\mp(x^2 + y^2)^{1/2} = x + c$$

or

$$x^2 + y^2 = (x + c)^2$$
$$= x^2 + 2cx + c^2$$

Hence, the possible curves having the given reflection property are members of the one-parameter family of parabolas

$$y^2 = 2cx + c^2 = 2c\left(x + \frac{c}{2}\right)$$

having foci at the origin, vertices at $(-c/2, 0)$, and axes coincident with the x axis. It is a standard problem in the calculus to verify that these parabolas do indeed possess the given reflection property. See Problem 3. The reflection property of the parabola accounts for the use of parabolic surfaces in telescopes, radar screens, lamps, and similar devices.

3.20 Rotating Fluid

The right circular cylinder in Figure 3.17, of base radius a and height h, contains a fluid. The cylinder and the fluid are rotating about the y axis at constant angular velocity ω. The surface S of the rotating fluid is a surface of revolution, and hence, to discover the nature of this surface, we seek an equation $y = f(x)$ of the intersection of S and a plane π through the y axis. Consider a particle of the rotating fluid located at $P(x, y)$ in plane π. The particle is moving on a circle of radius x at constant speed $x\omega$, and hence its total acceleration is a centripetal acceleration of $\omega^2 x$ directed towards the center of the circle on which the particle moves. (Assume $x > 0$.) By Newton's second law, the total force acting on the particle has magnitude $m\omega^2 x$, with m denoting the mass of the particle. This force is the resultant of two forces. The first, due to the attractive force the earth exerts on the particle, acts in the direction of the negative y axis, and has magnitude $W = mg$, where W is the weight of the particle, and g the local acceleration of gravity. The second force $\mathbf{F}$, of magnitude F, is due to the action of the surrounding particles on the particle under study. It is easily seen that $\mathbf{F}$ acts in the direction of a line having slope

$$\frac{-mg}{m\omega^2 x} = \frac{-g}{\omega^2 x}$$

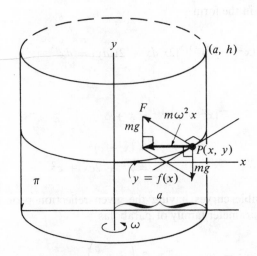

Fig. 3.17

The direction of **F** is also perpendicular to the surface S, since there is no tendency for the particle at P to move over the nearby particles of the fluid. Thus, **F** is perpendicular to the curve $y = f(x)$ at P and hence this curve has slope given by

$$\frac{dy}{dx} = \frac{-1}{-g/(\omega^2 x)} = \frac{\omega^2 x}{g}$$

Integrating gives

$$y = \frac{\omega^2 x^2}{2g} + c$$

For convenience, we assume that $y = 0$ when $x = 0$ and thereby obtain $c = 0$. Hence, the curve $y = f(x)$ is the parabola

$$y = \frac{\omega^2 x^2}{2g}$$

and the surface S is a paraboloid of revolution. (For more details see Reference 3.16.)

EXAMPLE 1. Find the height of the liquid at the axis of the cylinder when the top of the rotating fluid coincides with the top of the cylinder.

Solution: When $x = a$, $y = \omega^2 a^2/(2g)$, and hence the low point of the surface S has height $h - (\omega^2 a^2/(2g))$, where h denotes the height of the cylinder.

Problem List 3.15

1. Show that $\cosh^{-1} u = \ln(u + \sqrt{u^2 - 1})$. Use this result to show that, for $0 < x \le a$,

$$y = a \ln\left(\frac{a + \sqrt{a^2 - x^2}}{x}\right) - \sqrt{a^2 - x^2}$$

can be written in the form

$$y = a \cosh^{-1}\frac{a}{x} - \sqrt{a^2 - x^2}$$

(See Reference 3.22 for definitions of $\cosh u$ and $\cosh^{-1} u$.)

2. Find an equation of the family orthogonal to the family

$$y = \pm\left[\ln\left(\frac{1 + \sqrt{1 - x^2}}{|x|}\right) - \sqrt{1 - x^2}\right] + c$$

3. Given that the curve C in Figure 3.16 is a parabola having equation $y^2 = 2cx + c^2$, and that $\alpha = \beta$, show that PR is parallel to the x axis.

4. If the cylinder in Figure 3.17 is half full of liquid when the liquid and cylinder are at rest, what is the maximum value the angular velocity ω can assume before the liquid begins to overflow?

References

3.1 Agnew, Ralph P. 1962. *Differential Equations*, 2nd ed. New York: McGraw-Hill.

3.2 Allen, Roy G. 1959. *Mathematical Economics*, 2nd ed. New York: St. Martin's Press.

3.3 Bailey, N. T. 1967. *The Mathematical Approach to Biology and Medicine.* New York: John Wiley & Sons.

3.4 Braun, M. 1975. *Differential Equations and Their Applications.* New York: Springer-Verlag.

3.5 Bush, R. R., R. P. Abelson, and R. Hayman. 1956. *Mathematics for Psychologists.* New York: Social Sciences Research Council.

3.6 Chiang, A. C. 1967. *Fundamental Methods of Mathematical Economics.* New York: McGraw-Hill.

3.7 Coleman, James S. 1964. *Introduction to Mathematical Sociology.* New York: Free Press (Macmillan).

3.8 Domar, E. D. 1957. *Essays in the Theory of Economic Growth.* Fair Lawn, N.J.: Oxford University Press.

3.9 Ince, E. L. 1959. *Ordinary Differential Equations.* New York: Dover.

3.10 Kaplan, Wilfred. 1962. *Ordinary Differential Equations.* Reading, Mass.: Addison-Wesley.

3.11 Kooros, A. 1965. *Elements of Mathematical Economics.* Boston: Houghton Mifflin.

3.12 Lotka, A. J. 1956. *Elements of Mathematical Biology.* New York: Dover.

3.13 Margenau, H., and G. M. Murphy. 1956. *The Mathematics of Physics and Chemistry.* New York: Van Nostrand Reinhold.

3.14 Noble, Ben. 1970. *Applications of Undergraduate Mathematics in Engineering.* New York: Macmillan.

3.15 Pearl, R., and L. J. Reed. *Proc. Nat. Acad. Sci.* 6 (1920), p. 275; *Scientific Monthly* (1921), p. 194.

3.16 Polya, G. 1963. *SMSG Studies in Mathematics* 11, Mathematical Methods in Science. Stanford University.

3.17 Rashevsky, N. 1959. *Mathematical Biology of Social Behavior.* Chicago: University of Chicago Press.

3.18 Scarborough, J. B. 1965. *Differential Equations and Applications.* Baltimore: Waverly Press.

3.19 Shames, I. H. 1960. *Engineering Mechanics, Statics, and Dynamics.* Englewood Cliffs, N.J.: Prentice-Hall.

3.20 Simmons, G. F. 1972. *Differential Equations with Applications and Historical Notes.* New York: McGraw-Hill.

3.21 Thrall, R. M., J. A. Mortimer, K. R. Rebman, and R. F. Baum, eds. 1967. *Some Mathematical Models in Biology*, rev. ed. Ann Arbor: University of Michigan.

3.22 Tierney, John A. 1979. *Calculus and Analytic Geometry*, 4th ed. Boston: Allyn and Bacon.

3.23 Verhulst, P. F. Mem. Acad. Roy. Bruxelles, 1844, vol. 18, p. 1; 1846, vol. 20, p. 1.

3.24 *World Book Encyclopedia*. Chicago: Field Enterprises Educational Corporation.

3.25 Wylie, C. R., Jr. 1966. *Advanced Engineering Mathematics*, 3rd ed. New York: McGraw-Hill.

3.26 Yates, Robert C. 1974. *Curves and Their Properties*. Reston, Va.: National Council of Teachers of Mathematics.

Linear Differential Equations

4.1 Introduction; Operator Notation

An ordinary linear differential equation is a DE of the form

$$a_0(x)D_x^n y(x) + a_1(x)D_x^{n-1} y(x) + \cdots + a_{n-1}(x)D_x y(x) + a_n(x)y(x) = f(x)$$

(4.1)

where $a_0, a_1, \ldots, a_n$ and f are continuous functions of x (recall Section 1.1). The symbol D_x is an *operator*, which indicates that the function on which it operates is to be differentiated with respect to x. If L is an operator, L^n is an operator indicating n successive applications of L. Examples are

$$D_x(e^{2x}) = 2e^{2x}$$

$$D_x^2(e^{2x}) = 4e^{2x}$$

$$D_x^3(e^{2x}) = 8e^{2x}$$

$$D_x^3(x^3 + 5x^2 - 1) = 6$$

When it is clear that all indicated differentiations are with respect to x, it is customary to omit the subscript x. We shall follow this practice. Thus,

$$\frac{d}{dx}(\sin x) = D_x(\sin x) = D(\sin x) = \cos x$$

The DE $D^2y(x) + 3Dy(x) + 2y(x) = \sin x$ is often written in the form

$$(D^2 + 3D + 2)y(x) = \sin x \qquad (4.2)$$

The expression $D^2 + 3D + 2$ is called a *differential operator* and (4.2) is read "D squared plus $3D$ plus 2 operating on y of x equals $\sin x$." It is often useful to denote an operator by a single letter. If we let $H = D^2 + 3D + 2$, we write (4.2) in the form

$$Hy(x) = \sin x$$

read "H operating on y of x equals $\sin x$."

If we now let L denote the differential operator

$$a_0(x)D^n + a_1(x)D^{n-1} + \cdots + a_{n-1}(x)D + a_n(x)$$

we can write (4.1) as

$$Ly(x) = f(x) \qquad (4.3)$$

or in the more abbreviated form

$$Ly = f \qquad (4.4)$$

The form (4.4) makes it evident that an operator is a function whose domain and range are sets of functions. We call a function of this type an "operator" rather than a "function."

Definition   Let S be a set of functions and let a be an arbitrary constant. The operator L is called a *linear operator* on S if and only if

$$L(u + v) = Lu + Lv \quad \text{and} \quad L(au) = aLu$$

for all u, v in S.

It is easy to show (see Problem 1) that a linear operator L has the property

$$L(au + bv) = aLu + bLv \qquad (4.5)$$

Property (4.5) states that the image of a linear combination of u and v under L is the same linear combination of the images of u and v under L. The student is asked to show in Problems 2 and 3 that the differentiation operator D and the operator L of (4.4) are linear operators. The theory of linear DE, as contrasted with the theory of nonlinear DE, is based primarily on this result.

Before considering the solution of (4.1), we state the following existence and uniqueness theorem, proved in Reference 4.4.

4.1 Introduction; Operator Notation   **117**

Theorem 4-I Let $a_0, a_1, \ldots, a_n$ and f in (4.1) be continuous on an interval I which may be open, closed, half-open, or infinite. Assume that $a_0(x) \neq 0$ on I and let x_0 be an arbitrary point in I. Then there exists one and only one solution of (4.1), given by $y = \phi(x)$, satisfying the initial conditions

$$\phi(x_0) = y_0, \, \phi'(x_0) = y_0', \ldots, \phi^{(n-1)}(x_0) = y_0^{(n-1)}$$

The solution ϕ satisfies (4.1) at every point of I.

The conditions stated in Theorem 4-I are sufficient but not necessary. It is possible for a unique solution to exist under less stringent conditions.

4.2 Second-Order Linear Differential Equations

Let us, for simplicity, consider (4.1) for $n = 2$. The theory of linear DE of order $n > 2$ is essentially the same as the theory of second-order linear DE. Furthermore, the main applications we wish to consider do not involve DE of order $n > 2$.

We assume, as in Theorem 4-I, that $a_0(x) \neq 0$ on the interval I under consideration, and divide (4.1), with $n = 2$, by $a_0(x)$ to obtain

$$D^2 y(x) + p(x)Dy(x) + q(x)y(x) = \rho(x) \qquad \textbf{(4.6)}$$

where p, q, and ρ are continuous on I. In operator notation, (4.6) takes the form

$$Ly = \rho \qquad \textbf{(4.7)}$$

where $L = \phi(D) = D^2 + pD + q$.

The mathematical theory required for a detailed discussion of (4.6) is quite elaborate. Even a single DE of the form (4.6) may possess an extensive literature. The Bessel equation referred to in Section 1.2 is such an example.

EXAMPLE 1. By Theorem 4-I the DE $2y'' - y' - 3y = 4x + 2$ has a unique solution satisfying the initial conditions

$$y(1) = 5, \quad y'(1) = 2$$

Since $a_0(x) \equiv 2, a_1(x) \equiv -1, a_2(x) \equiv -3, f(x) = 4x + 2$, and a_0, a_1, a_2, and f are continuous on $(-\infty, +\infty)$, the unique solution satisfies the DE for all x.

EXAMPLE 2. The DE $x^2 y'' + xy' + (x^2 - 1)y = 0$ is a special case of Bessel's equation. This DE is of the form (4.1) with $a_0(x) = x^2$, $a_1(x) = x$, $a_2(x) = x^2 - 1$, and $f(x) \equiv 0$. Theorem 4-I applies to this DE when I is any interval that does not contain $x = 0$. Dividing both sides of the DE by x^2, we obtain $y'' + x^{-1}y' + (1 - x^{-2})y = 0$, which is of the form (4.6) with $p(x) = x^{-1}$, $q(x) = 1 - x^{-2}$, and $\rho(x) \equiv 0$. If $x_0 \in I$, Theorem 4-I asserts that the DE has a unique solution, given by $y = \phi(x)$, satisfying the initial conditions

$$\phi(x_0) = y_0, \quad \phi'(x_0) = y_0'$$

Since each term of the left member of (4.6) contains y or one of the derivatives of y whereas the right member $\rho(x)$ depends on x alone, Equation (4.6) is said to be *nonhomogeneous*. The following DE, which is (4.6) with $\rho(x) \equiv 0$, is called the *homogeneous equation* (or *reduced equation*) associated with (4.6):

$$D^2 y(x) + p(x)Dy(x) + q(x)y(x) = 0 \qquad \textbf{(4.8)}$$

The following theorem, referred to as the principle of superposition, states that a linear combination of solutions of a *homogeneous* second-order DE is also a solution. Although the theorem is stated for two functions, it is easily extended to any finite number of functions. A similar theorem holds for higher-order linear DE.

Theorem 4-II If y_1 and y_2 are any two solutions of (4.8), then $y_3 = c_1 y_1 + c_2 y_2$, where c_1 and c_2 are arbitrary constants, is also a solution.

Proof: Since $L = D^2 + pD + q$ in (4.7) is a linear operator,

$$Ly_3(x) = L(c_1 y_1(x) + c_2 y_2(x)) = c_1 Ly_1(x) + c_2 Ly_2(x) = c_1(0) + c_2(0) = 0$$

The coefficients of c_1 and c_2 are zero because y_1 and y_2 are solutions of (4.8).

EXAMPLE 3. The DE $D^2 y(x) + y(x) = 0$ has solutions defined by $y_1(x) = \sin x$ and $y_2(x) = \cos x$. Hence $y_3 = c_1 \sin x + c_2 \cos x$ also defines a solution.

4.3 Linear Independence; The Wronskian

Definition The n functions $y_1, y_2, \ldots, y_n$ are *linearly independent* on an interval I if and only if an identity

$$c_1 y_1(x) + c_2 y_2(x) + \cdots + c_n y_n(x) \equiv 0$$

can hold for every x in I only if the constants $c_1, c_2, \ldots, c_n$ are all zero.

This definition implies that no one function of n linearly independent functions can be expressed as a linear combination of the other $n - 1$ functions. For example, if

$$y_1(x) \equiv k_2 y_2(x) + k_3 y_3(x) + \cdots + k_n y_n(x)$$

on I, then

$$(-1)y_1(x) + k_2 y_2(x) + \cdots + k_n y_n(x) \equiv 0$$

on I, and the coefficient -1 of $y_1(x)$ is not zero.

In the case of two functions y_1 and y_2, linear independence on an interval I implies that neither function is a constant multiple of the other on I. In other words, the ratio of the two functions is not constant on I.

EXAMPLE 1. The functions defined by $y_1 = \sin^2 x$ and $y_2 = 2 - 2\cos^2 x$ are not linearly independent on any interval I, since $y_2 \equiv 2y_1$ for every x in I.

Definitions   Two functions that are not linearly independent on an interval I are said to be *linearly dependent* on I. A set of n functions that are not linearly independent on an interval I are said to be *linearly dependent* on I.

The importance of the concept of linear independence in the theory of linear DE is largely due to the following existence theorem, proved in Reference 4.4.

Theorem 4-III   There exist two linearly independent solutions of the reduced equation $D^2 y(x) + pDy(x) + qy(x) = 0$. These solutions are valid on any interval I on which p and q are continuous.

EXAMPLE 2. The DE $D^2 y(x) + y(x) = 0$ has solutions defined by $y_1 = \sin x$ and $y_2 = \cos x$, valid on any given interval I. These solutions are linearly independent on I since their ratio $y_1/y_2 = \tan x$ is not constant on I.

It is seen by inspection that $y \equiv 0$ defines a solution of the reduced equation $D^2 y(x) + pDy(x) + qy(x) = 0$. This solution is often referred to as the *trivial solution*. The trivial solution and any other solution are easily shown to be linearly dependent. See Problem 7.

A simple test for determining that two solutions y_1 and y_2 of $Ly(x) = 0$ are linearly independent employs the function W of x given by

$$W(y_1, y_2) = \begin{vmatrix} y_1 & y_2 \\ y_1' & y_2' \end{vmatrix} = y_1 y_2' - y_2 y_1' = W(x) \qquad \textbf{(4.9)}$$

The quantity $W(y_1, y_2)$ is called the Wronskian of the functions y_1 and y_2, after the Polish mathematician Hoëné Wronski (1778–1853).

The following theorem, proved in Reference 4.1, provides a practical test for linear independence.

Theorem 4-IV   If $W(y_1, y_2)$ is not identically zero on a given interval I, where y_1 and y_2 are differentiable functions defined on I, then y_1 and y_2 are linearly independent on I.

EXAMPLE 3. The functions defined by $y_1 = \sin x$ and $y_2 = \cos x$ are linearly independent on any given interval I since

$$
\begin{aligned}
W(y_1, y_2) &= \sin x(-\sin x) - \cos x(\cos x) \\
&= -(\sin^2 x + \cos^2 x) \\
&= -1 \neq 0 \quad \text{on } I
\end{aligned}
$$

EXAMPLE 4. The functions defined by $y_1 = e^{-x}$ and $y_2 = e^{2x}$ are linearly independent on any given interval I since

$$
\begin{aligned}
W(y_1, y_2) &= e^{-x}(2e^{2x}) - e^{2x}(-e^{-x}) \\
&= 3e^{-x}e^{2x} = 3e^x > 0 \quad \text{on } I
\end{aligned}
$$

Let p and q in (4.8) be continuous on an interval I, let x_0 be an arbitrary point of I, and let y_1 and y_2 be solutions of (4.8) on I. The following argument shows that on I the Wronskian $W = W(x)$ is everywhere positive, everywhere negative, or identically zero.

$$
\frac{dW}{dx} + pW = (y_1 y_2'' - y_2 y_1'') + p(y_1 y_2' - y_2 y_1')
$$

$$
= y_1(-py_2' - qy_2) - y_2(-py_1' - qy_1) + p(y_1 y_2' - y_2 y_1') = 0
$$

This first-order linear DE has solution

$$
W = W(x) = W(x_0) \exp\left[-\int_{x_0}^{x} p(t)\, dt \right] \tag{4.10}
$$

Thus, since the exponential function is positive, $W(x)$ has the same sign as $W(x_0)$ for all x in I. The relation (4.10) is known as *Abel's identity*, after the Norwegian mathematician Niels Abel (1802–1829).

EXAMPLE 5. The functions defined by $y_1 = e^{2x}$ and $y_2 = e^{-2x}$ may be shown to be solutions of $y'' - 4y = 0$ on $(-\infty, +\infty)$. We find that $W(y_1, y_2) = (e^{2x})(-2e^{-2x}) - (e^{-2x})(2e^{2x}) \equiv -4$ is everywhere negative on $(-\infty, +\infty)$, in agreement with Abel's identity.

EXAMPLE 6. By Theorem 4-IV the functions defined by $y_1 = x$ and $y_2 = x^2$ are linearly independent on $I = (-2, 2)$ since $W(y_1, y_2) = W(x) = x(2x) - x^2(1) \equiv x^2$ is not identically zero on I. Note that $W(0) = 0$ whereas $W(1) = 1 > 0$. It might appear that this contradicts our result based on Abel's identity. No contradiction is involved, however, since y_1 and y_2 do not satisfy a second-order DE of the form $Ly(x) = 0$ on $(-2, 2)$. We note also that we do not allow I to be the degenerate interval $(0, 0)$ consisting of the single point $x = 0$.

The following theorem, also proved in Reference 4.1, enables us to draw a conclusion when $W(y_1, y_2)$ is identically zero on I.

Theorem 4-V If y_1 and y_2 are solutions of $Ly(x) = 0$ on an interval I, and if $W(y_1, y_2) \equiv 0$ on I, then y_1 and y_2 are linearly dependent on I; that is, they are not linearly independent on I.

EXAMPLE 7. The functions defined by $y_1 = \cos 2x$ and $y_2 = 4 \sin^2 x - 2$ are solutions of $y'' + 4y = 0$ on $(-\infty, +\infty)$ since

$$y_1' = -2 \sin 2x, \quad y_2' = 8 \sin x \cos x$$

$$y_1'' = -4 \cos 2x, \quad y_2'' = -8 \sin^2 x + 8 \cos^2 x$$

$$y_1'' + 4y_1 = -4 \cos 2x + 4 \cos 2x \equiv 0,$$

and

$$y_2'' + 4y_2 = -8 \sin^2 x + 8 \cos^2 x + 4(4 \sin^2 x - 2)$$
$$= -8 \sin^2 x + 8 - 8 \sin^2 x + 16 \sin^2 x - 8 \equiv 0$$

Since

$$W(y_1\ y_2) = (\cos 2x)(8 \sin x \cos x) - (4 \sin^2 x - 2)(-2 \sin 2x)$$
$$= (1 - 2 \sin^2 x)(8 \sin x \cos x) - (4 \sin^2 x - 2)(-4 \sin x \cos x)$$
$$= 8 \sin x \cos x \, (1 - 2 \sin^2 x + 2 \sin^2 x - 1) \equiv 0$$

we conclude from Theorem 4-V that y_1 and y_2 are linearly dependent on $(-\infty, +\infty)$. Using the trigonometric identity $\cos 2x \equiv 1 - 2 \sin^2 x$, we see that $y_2 \equiv -2y_1$ on $(-\infty, +\infty)$.

We shall see in the following section that our method for finding a complete solution of $Ly = \rho$ on a given interval I will depend on finding two linearly independent solutions of $Ly(x) = 0$ on I. Theorems 4-IV and 4-V enable us to determine whether or not two solutions of $Ly(x) = 0$ are linearly independent on I. If $W(y_1, y_2) \neq 0$ on I, then y_1 and y_2 are linearly independent on I. If $W(y_1, y_2) \equiv 0$ on I, then y_1 and y_2 are not linearly independent on I.

Thus we are interested in testing solutions of $Ly(x) = 0$ for linear independence. It is interesting to point out, however, that Theorem 4-IV applies to y_1 and y_2 even if these functions do not satisfy a DE of the form $Ly(x) = 0$, whereas Theorem 4-V applies only if y_1 and y_2 do satisfy a DE of the form $Ly(x) = 0$. This distinction is illustrated in Problem 23.

4.4 Complete Solutions of the Homogeneous and Nonhomogeneous Differential Equations

The following theorem, proved in Reference 4.4, concerns the solution of the homogeneous DE $Ly(x) = D^2 y(x) + pDy(x) + qy(x) = 0$.

Theorem 4-VI If y_1 and y_2 are *any* two linearly independent solutions of $Ly(x) = 0$ on a given interval I, then *every* solution on I of $Ly(x) = 0$ can be expressed in the form

$$y_3 = c_1 y_1 + c_2 y_2$$

where c_1 and c_2 are arbitrary constants.

Theorem 4-VI is more complex than Theorem 4-II. Theorem 4-II states merely that a linear combination of any two solutions of $Ly(x) = 0$ is also a solution, whereas Theorem 4-VI states that a linear combination of any two *linearly independent* solutions of $Ly(x) = 0$ yields a *complete* solution.

It is easy to show that a complete solution of $Ly(x) = 0$, given by $y_3 = c_1 y_1 + c_2 y_2$, plus a particular solution of $Ly = \rho$, given by $y_p = \psi(x)$, is a solution of $Ly = \rho$. We merely note that

$$L(y_3 + \psi) = Ly_3 + L\psi = 0 + \rho = \rho$$

The final theorem of this section, also proved in Reference 4.4, states the less obvious fact that $y = y_3 + \psi$ is a complete solution of $Ly = \rho$.

Theorem 4-VII If $y_3 = c_1 y_1 + c_2 y_2$ defines a complete solution of $Ly(x) = 0$ on a given interval I, and $y_p = y_p(x)$ defines *any* particular solution of $Ly = \rho$ on I, then *every* solution of $Ly = \rho$ on I can be expressed in the form

$$y = c_1 y_1 + c_2 y_2 + y_p$$

In Theorem 4-VII, $y_3 = c_1 y_1 + c_2 y_2$ defines a complete solution of the reduced equation $Ly(x) = 0$; y_3 is called the *complementary function* and is often denoted by y_c. Thus, $y = y_c + y_p$.

We now summarize the method of solving $D^2 y + pDy + qy = Ly = \rho$.

To find a complete solution of $Ly = \rho$, first find two linearly independent solutions y_1 and y_2 of $Ly(x) = 0$. Then add to $c_1 y_1 + c_2 y_2$ any particular solution y_p of $Ly = \rho$.

It is not a simple matter to produce even one solution of $Ly(x) = 0$, except for the trivial solution given by $y(x) \equiv 0$. Sometimes a solution can be found by inspection, by guesswork, by trial and error, or by exploiting some feature of an application involving $Ly(x) = 0$. In the following section we present a useful method for finding a solution y_2 of $Ly(x) = 0$ when we know a solution y_1 of $Ly(x) = 0$.

4.5 Reduction of Order

Given a nontrivial solution y_1 of $Ly(x) = 0$, a procedure known as the *method of reduction of order* is often useful for finding a second solution y_2 of $Ly(x) = 0$ such that y_1 and y_2 are linearly independent. The method is due to the French mathematician Jean D'Alembert (1717–1783). Remembering that vy_1 defines a solution when v is constant but that the ratio of y_1 and y_2 is not constant when y_1 and y_2 are linearly independent, it is natural to seek a solution $y_2 = vy_1$ where v is a function of x to be determined.

From

$$y_2' = vy_1' + v'y_1, \quad y_2'' = vy_1'' + 2v'y_1' + y_1v''$$

we obtain

$$
\begin{aligned}
y_2'' + py_2' + qy_2 &= (vy_1'' + 2v'y_1' + y_1v'') + p(vy_1' + v'y_1) + q(vy_1) \\
&= (y_1'' + py_1' + qy_1)v + (2y_1' + py_1)v' + y_1v'' = 0
\end{aligned}
$$

The coefficient of v is identically zero since $Ly_1(x) = 0$. We now seek a solution of the DE

$$y_1 \frac{dv'}{dx} + (2y_1' + py_1)v' = 0$$

which is a first-order linear DE for v'. We have reduced the problem of solving a second-order linear DE to that of solving a first-order linear DE. We now restrict ourselves to an interval I on which $y_1(x)$ is never zero, divide by y_1, and separate variables to obtain

$$\frac{dv'}{v'} + \left(2\frac{y_1'}{y_1} + p\right) dx = 0$$

This yields

$$\ln |v'| + 2 \ln |y_1| + \int p\, dx = \ln |c|$$

It is now convenient to set $c = 1$ since we are merely seeking a particular solution. From

$$\ln |v'y_1^2| = -\int p\, dx$$

we have, assuming for convenience that $v'(x) > 0$,

$$v'y_1^2 = \exp\left(-\int p\, dx\right)$$

or

$$v' = \frac{dv}{dx} = \frac{\exp\left(-\int p\, dx\right)}{y_1^2}$$

A second integration yields

$$v = \int \frac{\exp\left(-\int p\, dx\right)}{y_1^2}\, dx$$

and hence,

$$y_2 = vy_1 = y_1 \int \frac{\exp\left(-\int p\, dx\right)}{y_1^2}\, dx$$

where, again for convenience, we assume that the constant of integration in our second integration is zero.

The solutions y_1 and y_2 are linearly independent. Otherwise their ratio $y_2/y_1 = v$ would be constant and hence v' would be identically zero. This is impossible since

$$v' = \frac{\exp\left(-\int p\, dx\right)}{y_1^2} > 0$$

The linear independence of y_1 and y_2 can also be established by showing that $W(y_1, y_2) = W(y_1, vy_1) \neq 0$ on any given interval I on which $y_1 \neq 0$. See Problem 26.

Although v is undefined at $x = x_0$ when $y_1(x_0) = 0$, the functions y_1 and $y_2 = vy_1$ may still satisfy $Ly(x) = 0$ at $x = x_0$. See Problems 27b, d, e, and f.

By Theorem 4-VI, a complete solution of $Ly(x) = 0$ is given by

$$y = c_1 y_1 + c_2 y_2 = c_1 y_1 + c_2 vy_1 = c_1 y_1 + c_2 y_1 \int \frac{\exp\left(-\int p\, dx\right)}{y_1^2}\, dx$$

The method described also applies to DE of order higher than two. The objective is to reduce the solution of a DE of order n to the solution of a DE of order $n - 1$.

EXAMPLE 1. Find a complete solution of $y'' + 2y' + y = 0$.

Solution: We seek a function y_1 having the property that a linear combination of y_1 and its derivatives is identically zero.

Since $De^{rx} = re^{rx}$ we are prompted to try $y_1 = e^{rx}$. From $r^2 e^{rx} + 2re^{rx} + e^{rx} = 0$ we obtain $r^2 + 2r + 1 = (r + 1)^2 = 0, r = -1$, and $y_1 = e^{-x}$.

Setting $p = 2$ we obtain

$$y_2 = y_1 \int \frac{\exp\left(-\int 2\, dx\right)}{y_1^2}\, dx$$

$$= e^{-x} \int \frac{e^{-2x}\, dx}{e^{-2x}} = xe^{-x}$$

A complete solution, valid on $(-\infty, +\infty)$, is given by

$$y = c_1 e^{-x} + c_2 x e^{-x}$$

EXAMPLE 2. Find a complete solution of $(D^2 + x^{-1}D - 9x^{-2})y(x) = 0$ on $(0, +\infty)$.

Solution: By trial and error, we find that one solution is given by $y_1 = x^3$. We next compute

$$\exp\left(-\int x^{-1}\, dx\right) = \exp(-\ln x) = \exp(\ln x^{-1}) = x^{-1}$$

Since

$$x^3 \int \frac{x^{-1}\, dx}{(x^3)^2} = x^3 \int x^{-7}\, dx = x^3\left(\frac{x^{-6}}{-6}\right) = \frac{x^{-3}}{-6} = y_2$$

a complete solution is given by

$$y = c_1 x^3 + c_2\left(\frac{x^{-3}}{-6}\right) \quad \text{or} \quad y = c_1 x^3 + k x^{-3}$$

We find that $W(y_1, y_2) = -6x^{-1}$. Since $-6x^{-1}$ is never zero on $(0, +\infty)$, the functions given by $y_1 = x^3$ and $y_2 = x^{-3}$ are linearly independent on $(0, +\infty)$.

In the following sections, we shall carry out the program outlined for solving $Ly = \rho$ in the important special case in which the coefficients $p(x)$ and $q(x)$ in (4.6) are constant. Methods for finding a particular solution Y of $Ly = \rho$ will be discussed in Sections 4.8 and 4.9.

Problem List 4.1

1. Show that a linear operator L has the property $L(au + bv) = aLu + bLv$, where a and b are constants and the functions u and v are in the domain of L.

2. Show that the differentiation operator D is a linear operator.

3. Show that the operator

$$L = a_0(x)D^n + a_1(x)D^{n-1} + \cdots + a_{n-1}(x)D + a_n(x)$$

is linear.

4. Given that the equations $y_1 = e^x$ and $y_2 = e^{-2x}$ define solutions of $(D^2 + D - 2)y(x) = 0$, show that $y_3 = c_1 e^x + c_2 e^{-2x}$, where c_1 and c_2 are arbitrary constants, also defines a solution.

5. Find the solution of $D^2 y(x) = 6x$ satisfying $y(0) = 5$ and $y'(0) = 2$.

6. Let $y_1, y_2, \ldots, y_n$ denote solutions of an nth-order linear DE. Show that the linear combination $y = c_1 y_1 + c_2 y_2 + \cdots + c_n y_n$ also defines a solution.

7. Let y_1 be an arbitrary solution of $(D^2 + pD + q)y(x) = 0$. Show that y_1 and y_2, the trivial solution given by $y_2(x) \equiv 0$, are linearly dependent.

8. Show that the functions given by $y_1 = 5x$ and $y_2 \equiv 1$ are linearly independent on any interval I.

9. Evaluate $W(y_1, y_2)$ where $y_1 = x^2$ and $y_2 = x^{-2}$. Show that y_1 and y_2 are linearly independent on any interval that does not include $x = 0$.

10. Given $y_1 = \sin x$ and $y_2 = (1 - \cos^2 x)^{1/2}$, find $W(y_1, y_2)$.

11. The functions given by $y_1 = \sin 2x$ and $y_2 = \cos 2x$ are solutions of the DE $y'' + 4y = 0$. Show that y_1 and y_2 are linearly independent on any interval I. Write a complete solution of $y'' + 4y = 0$.

12. The functions given by $y_1 = e^{3x}$ and $y_2 = e^{-3x}$ are solutions of the DE $y'' - 9y = 0$. Show that these solutions are linearly independent on any interval I. Write a complete solution of $y'' - 9y = 0$.

13. The functions given by $y_1 = x$ and $y_2 = x^{-1}$ are solutions of $y'' + x^{-1}y' - x^{-2}y = 0$ on any interval I not including $x = 0$. Find $W(y_1, y_2)$ and write a complete solution of the DE on I.

14. Find by inspection a particular solution of $y'' + 2y = 6x$.

15. Find by inspection a particular solution of $y'' + 3y = 6$.

16. Given that $y_1 = e^{4x}$ and $y_2 = xe^{4x}$ define solutions of $y'' - 8y' + 16y = 0$, show that y_1 and y_2 are linearly independent on any given interval I. Write a complete solution of $y'' - 8y' + 16y = 0$.

17. Construct a second-order DE for which $y = e^{3x}$ defines a solution.

18. Given that $y_3 = c_1 + c_2e^{-x}$ defines a complete solution of $y'' + y' = 0$ and that $y_p = e^x$ defines a particular solution of $y'' + y' = 2e^x$, find a complete solution of $y'' + y' = 2e^x$.

19. The DE $y'' + 4y = 0$ has a complete solution given by $y_3 = c_1 \sin 2t + c_2 \cos 2t$. If $y_p = \cos 3t$ defines a particular solution of $y'' + 4y = -5 \cos 3t$, what is the solution denoted by $y = y(t)$ of $y'' + 4y = -5 \cos 3t$ satisfying $y(0) = 2$ and $y'(0) = 4$?

20. Show that $y_1 = x + 1$ defines a solution of $y'' - 4y' + y = x - 3$. Show also that $y_2 = 3(x + 1)$ does not define a solution of the same DE. Does this result contradict Theorem 4-II?

21. Show that $y_1 \equiv 4$ and $y_2 = \sqrt{x}$ define solutions of the DE $yy'' + (y')^2 = 0$ on $(0, +\infty)$. Show also that not every linear combination $c_1y_1 + c_2y_2$ defines a solution of the same DE on the same interval. Does this result contradict Theorem 4-II?

22. Show that if the Wronskian of two functions is different from zero at every point of an interval I, then there is no point of I at which both functions are simultaneously zero.

23. Show that the Wronskian of the two functions given by $y_1 = x^3$ and $y_2 = |x|^3$ is identically zero on $-1 < x < 1$ but that y_1 and y_2 are linearly independent on $-1 < x < 1$. Does this result contradict Theorem 4-V?

24. Let y_1 and y_2 denote two solutions of $(D^2 + pD + q)y(x) = 0$ on an interval I. Use Theorem 4-V to show that if $y_1(x) = 0$ and $y_2(x) = 0$ have a common root $x = a$ in I, then y_1 and y_2 are linearly dependent on I. Show also that if $y_1 = y_1(x)$ and $y_2 = y_2(x)$ have maxima or minima at the same point a of I, then y_1 and y_2 are linearly dependent on I.

25. Let y_1 and y_2 define two linearly independent solutions of $(D^2+pD+q)y(x)=0$ on an interval I. Show that if a and b are two consecutive zeros of $y_1(x)$ on I, then there is one and only one number c such that $a < c < b$ and $y_2(c) = 0$. This is the *separation theorem of Sturm* (1803–1855) and is illustrated by $y_1 = \sin x$ and $y_2 = \cos x$, which define linearly independent solutions of $(D^2 + 1)y(x) = 0$ on $(-\infty, +\infty)$. *Hint:* Apply Rolle's theorem to $\phi(x) = y_1(x)/y_2(x)$ on $a \le x \le b$.

26. Let y_1 be a nontrivial solution of $(D^2 + pD + q)y(x) = 0$ and let

$$y_2 = y_1 \int \frac{\exp\left(-\int p\, dx\right)}{y_1^2}\, dx$$

Show that the Wronskian W of y_1 and $y_2 \neq 0$ on an arbitrary interval I on which $y_1(x) \neq 0$.

27. Given the solution y_1, find a complete solution on the specified interval of each of the following DE.

(a) $(D^2 + x^{-1}D - x^{-2})y(x) = 0$, $y_1 = x$, $x > 0$

(b) $(D^2 + 1)y(x) = 0$, $y_1 = \sin x$, $-\infty < x < +8$

(c) $(D^2 - 8D + 16)y(x) = 0$, $y_1 = e^{4x}$, $-\infty < x < +\infty$

(d) $\left[D^2 + x^{-1}D + \left(1 - \frac{1}{4x^2}\right) \right]y(x) = 0$, $y_1 = x^{-1/2}\sin x$, $x > 0$

(e) $(D^2 - 2xD + 2)y(x) = 0$, $y_1 = x$, $-\infty < x < +\infty$

(f) $[D^2 - 2x(1 - x^2)^{-1}D + 2(1 - x^2)^{-1}]y(x) = 0$, $y_1 = x$, $-1 < x < 1$

4.6 Homogeneous Second-Order Linear Differential Equations with Constant Coefficients

There is a method, due to Euler, for completely solving

$$Ly(x) = (D^2 + aD + b)y(x) = 0 \tag{4.11}$$

the special case of (4.8) in which $p(x) \equiv a$ and $q(x) \equiv b$.

Let us see whether (4.11) has a solution of the form $y = e^{rx}$, where r is a constant. This trial solution is suggested by the fact that a first-order linear DE of the form $(D + k)y(x) = 0$ has a solution of the form $y = e^{rx}$. Furthermore, $D^2y + aDy + by$ is a linear combination of D^2y, Dy, and y, so it is conceivable that a function whose derivatives are multiples of the function may satisfy (4.11).

Substituting $y = e^{rx}$ into (4.11), we obtain

$$r^2e^{rx} + are^{rx} + be^{rx} = 0$$

or

$$e^{rx}(r^2 + ar + b) = 0$$

Since e^{rx} is never zero, we conclude that if $y = e^{rx}$ defines a solution of (4.11), then r is a root of the equation

$$r^2 + ar + b = 0 \tag{4.12}$$

The quadratic equation (4.12), obtained by replacing D by r in $D^2 + aD + b$, is termed the *characteristic*, or *auxiliary*, *equation* of the DE (4.11). The roots r_1 and r_2 of (4.12) are given by the quadratic formula

$$r_1, r_2 = \frac{-a \pm \sqrt{a^2 - 4b}}{2}$$

Note that $r_1 + r_2 = -a$ or $a = -(r_1 + r_2)$ and that

$$r_1 r_2 = \frac{(-a)^2 - (a^2 - 4b)}{4} = b$$

It is necessary to consider three separate cases; r_1 and r_2 may be real and unequal, r_1 and r_2 may be real and equal, and r_1 and r_2 may be complex numbers of the form $\alpha \pm i\beta$ ($\beta \neq 0$).

Case I: Roots Real and Unequal

Let us assume that r_1 and r_2 are distinct real roots of (4.12). It is easy to show that $y_1 = e^{r_1 x}$ and $y_2 = e^{r_2 x}$ define solutions of (4.11). For example,

$$(D^2 + aD + b)e^{r_1 x} = e^{r_1 x}(r_1^2 + ar_1 + b) = 0$$

since r_1 is a root of the characteristic equation.

The solutions y_1 and y_2 are linearly independent, since

$$W(y_1, y_2) = e^{r_1 x}(r_2 e^{r_2 x}) - e^{r_2 x}(r_1 e^{r_1 x})$$
$$= e^{(r_1 + r_2)x}(r_2 - r_1)$$

cannot be identically zero on an interval I due to the fact that $e^{(r_1 + r_2)x}$ and $r_2 - r_1$ are both different from zero.

Hence, by Theorem 4-VI, $y = c_1 e^{r_1 x} + c_2 e^{r_2 x}$ defines a complete solution of (4.11), valid on $(-\infty, +\infty)$.

EXAMPLE 1.   Find a complete solution of $(D^2 + D - 6)y(x) = 0$.

Solution: The characteristic equation $r^2 + r - 6 = (r + 3)(r - 2) = 0$ has roots $r_1 = -3$ and $r_2 = 2$ and hence the DE has a complete solution given by

$$y = c_1 e^{-3x} + c_2 e^{2x}$$

EXAMPLE 2.   Find a complete solution of $(D^2 + 5D)y(x) = 0$.

Solution: The characteristic equation $r^2 + 5r = 0$ has roots $r_1 = 0$ and $r_2 = -5$, and hence the DE has a complete solution given by

$$y = c_1 e^{0x} + c_2 e^{-5x} \quad \text{or} \quad y = c_1 + c_2 e^{-5x}$$

EXAMPLE 3.   Find the solution of $(D^2 - 4D + 1)y(x) = 0$ for which $y(0) = 0$ and $y'(0) = \sqrt{3}$.

Solution: The characteristic equation $r^2 - 4r + 1 = 0$ has roots $r_1 = 2 + \sqrt{3}$ and $r_2 = 2 - \sqrt{3}$. A complete solution of the DE is given by

$$y = c_1 \exp(2 + \sqrt{3})x + c_2 \exp(2 - \sqrt{3})x$$

Setting $x = 0$ in this equation and in

$$y' = c_1(2 + \sqrt{3}) \exp(2 + \sqrt{3})x + c_2(2 - \sqrt{3}) \exp(2 - \sqrt{3})x$$

we obtain

$$c_1 + c_2 = 0$$

$$c_1(2 + \sqrt{3}) + c_2(2 - \sqrt{3}) = \sqrt{3}$$

yielding $c_1 = \frac{1}{2}, c_2 = -\frac{1}{2}$.

The required solution is given by

$$y = \frac{1}{2}[\exp(2 + \sqrt{3})x - \exp(2 - \sqrt{3})x]$$

Case II: Roots Real and Equal

When the discriminant $a^2 - 4b$ of the characteristic equation is zero, the roots r_1 and r_2 of the characteristic equation are real and equal. In this case $y_1 = e^{r_1 x}$ defines a solution of (4.11) as in Case I, but we lack a second solution y_2 such that y_1 and y_2 are linearly independent. We note, however, that the DE $D^2 y(x) = 0$ is of this type, since it has characteristic equation $r^2 = 0$, with roots $r_1 = r_2 = 0$. But $D^2 y(x) = 0$ is easily solved by integrating twice with respect to x to obtain the solution given by

$$y = c_1(1) + c_2 x$$

This is a complete solution of $D^2 y(x) = 0$ since the functions defined by $y_1 \equiv 1$ and $y_2 = x$ are linearly independent. (See Problem 5.)

Since $y_2 = x y_1$ in this particular example, it is natural to test to see whether $y_2 = x e^{r_1 x}$ defines a solution of (4.11) when $r_1 = r_2$.

The function y_2 is indeed a solution since

$$y_2' = r_1 x e^{r_1 x} + e^{r_1 x}, \quad a = -(r_1 + r_1) = -2r_1, \quad b = r_1 r_1 = r_1^2,$$

and

$$\begin{aligned} y_2'' + a y_2' + b y_2 &= (r_1^2 x e^{r_1 x} + 2r_1 e^{r_1 x}) \\ &\quad + (-2r_1)(r_1 x e^{r_1 x} + e^{r_1 x}) + r_1^2 x e^{r_1 x} \\ &= e^{r_1 x}(r_1^2 x + 2r_1 - 2r_1^2 x - 2r_1 + r_1^2 x) \\ &= 0 \end{aligned}$$

The functions y_1 and y_2 are linearly independent on an arbitrary interval I since their ratio $y_2/y_1 = x$ is not constant on I. See Problem 6 for a proof of linear independence using the Wronskian $W(e^{r_1 x}, x e^{r_1 x})$. Hence when $r_1 = r_2$, the DE (4.11) has a complete solution given by $y = c_1 e^{r_1 x} + c_2 x e^{r_1 x}$.

The expression for y_2 may also be found from the formula

$$y_2 = y_1 \int \frac{\exp\left(-\int p\,dx\right)}{y_1^2}\,dx$$

of Section 4.5.

With $y_1 = e^{r_1 x}$ and $p = a$, this formula yields

$$y_2 = e^{r_1 x} \int \frac{\exp\left(-ax\right)}{e^{2r_1 x}}\,dx$$

Since $a = -(r_1 + r_2) = -2r_1$,

$$y_2 = e^{r_1 x} \int e^{-2r_1 x} e^{-ax}\,dx = e^{r_1 x} \int e^{ax} e^{-ax}\,dx$$

$$= e^{r_1 x} \int dx = x e^{r_1 x}.$$

EXAMPLE 4. Find a complete solution of $(D^2 - 8D + 16)y(x) = 0$.

Solution: The characteristic equation $r^2 - 8r + 16 = 0$ has $r = 4$ as a repeated root. Hence the DE has a complete solution given by $y = c_1 e^{4x} + c_2 x e^{4x}$.

EXAMPLE 5. Solve the initial-value problem

$$(D^2 + 4D + 4)y(x) = 0; \quad y(0) = 4, \quad y'(0) = -5$$

Solution: The characteristic equation $r^2 + 4r + 4 = (r + 2)^2 = 0$ has $r = -2$ as a repeated root. Setting $x = 0$ in

$$y = c_1 e^{-2x} + c_2 x e^{-2x}$$

and

$$y' = -2c_1 e^{-2x} - 2c_2 x e^{-2x} + c_2 e^{-2x}$$

we obtain

$$y(0) = c_1 = 4 \quad \text{and} \quad y'(0) = -2(4) + c_2 = -5$$

From $c_1 = 4$ and $c_2 = 3$ we obtain the required solution

$$y = 4e^{-2x} + 3x e^{-2x} = e^{-2x}(4 + 3x)$$

Case III: Roots Not Real

If the roots of the characteristic equation $r^2 + ar + b = 0$ are not real, then they must be *complex conjugates*, since a and b are real numbers. This situation prevails when the discriminant $a^2 - 4b$ of the characteristic equation is negative. Let these complex roots be denoted by $r_1 = \alpha + i\beta$ and $r_2 = \alpha - i\beta$, where α and β are real, $\beta \neq 0$, and $i = \sqrt{-1}$ is the *complex imaginary unit* having the property that $i^2 = -1$.

Since $a = -(r_1 + r_2) = -2\alpha$ and $b = r_1 r_2 = \alpha^2 - (i\beta)^2 = \alpha^2 + \beta^2$, the characteristic equation can be written in the form

$$r^2 - 2\alpha r + (\alpha^2 + \beta^2) = 0$$

If $\beta = 0$, then $r = \alpha$ is a double root of the characteristic equation $r^2 - 2\alpha r + \alpha^2 = 0$, and the DE reduces to $(D^2 - 2\alpha D + \alpha^2)y(x) = 0$. We know that this DE has a solution given by $y = e^{\alpha x}$.

If $\alpha = 0$, then $r^2 + \beta^2 = 0$ is the characteristic equation, and the DE reduces to $(D^2 + \beta^2)y(x) = 0$.

But it is seen by inspection that this DE has solutions given by $y_1 = \sin \beta x$ and $y_2 = \cos \beta x$.

These two special DE suggest that when α and β are both different from zero, we try f_1 and f_2 defined by $y_1 = f_1(x) = e^{\alpha x} \sin \beta x$ and $y_2 = f_2(x) = e^{\alpha x} \cos \beta x$ as solutions of (4.11). (See Problem 7.)

The following computation, with $a = -2\alpha$ and $b = \alpha^2 + \beta^2$, verifies that f_1 is a solution of (4.11).

$$
\begin{aligned}
y_1'' + ay_1' + by_1 &= y_1'' - 2\alpha y_1' + (\alpha^2 + \beta^2)y_1 \\
&= (-\beta^2 e^{\alpha x} \sin \beta x + \alpha\beta e^{\alpha x} \cos \beta x + \alpha\beta e^{\alpha x} \cos \beta x \\
&\quad + \alpha^2 e^{\alpha x} \sin \beta x) - 2\alpha(\beta e^{\alpha x} \cos \beta x + \alpha e^{\alpha x} \sin \beta x) \\
&\quad + (\alpha^2 + \beta^2)e^{\alpha x} \sin \beta x \\
&= e^{\alpha x}(-\beta^2 \sin \beta x + 2\alpha\beta \cos \beta x + \alpha^2 \sin \beta x \\
&\quad - 2\alpha\beta \cos \beta x - 2\alpha^2 \sin \beta x \\
&\quad + \alpha^2 \sin \beta x + \beta^2 \sin \beta x) = 0
\end{aligned}
$$

For a verification that f_2 is also a solution of (4.11), see Problem 8.

The functions y_1 and y_2 are linearly independent on an arbitrary interval I since their ratio $y_2/y_1 = \cot \beta x$ is not constant on I. See Problem 8 for a proof of linear independence using the Wronskian $W(e^{\alpha x} \sin \beta x, e^{\alpha x} \cos \beta x)$.

Hence, when $\beta \neq 0$, it follows from Theorem 4-VI that

$$
\begin{aligned}
y &= c_1 e^{\alpha x} \sin \beta x + c_2 e^{\alpha x} \cos \beta x \\
&= e^{\alpha x}(c_1 \sin \beta x + c_2 \cos \beta x)
\end{aligned}
\tag{4.13}
$$

defines a complete solution of (4.11).

Another way of obtaining (4.13) is to write

$$y_1 = e^{r_1 x} \quad \text{and} \quad y_2 = e^{r_2 x}$$

as

$$y_1 = e^{(\alpha + i\beta)x} \quad \text{and} \quad y_2 = e^{(\alpha - i\beta)x}$$

We then proceed formally, without justifying our use of e^z when z is complex, and write

$$y_1 = e^{\alpha x} e^{i\beta x} \quad \text{and} \quad y_2 = e^{\alpha x} e^{-i\beta x}$$

This step is suggested by the familiar law of exponents $e^{u+v} = e^u e^v$, known to hold when u and v are real numbers.

To assign a reasonable meaning to $e^{i\beta x}$ and $e^{-i\beta x}$, we recall that

$$e^z = \exp z = 1 + \frac{z}{1!} + \frac{z^2}{2!} + \frac{z^3}{3!} + \cdots + \frac{z^{n-1}}{(n-1)!} + \cdots$$

for all real values of z. This is known as the Maclaurin series for e^z. We next set $z = i\theta$ where θ is real and $i = \sqrt{-1}$. This yields

$$e^{i\theta} = 1 + \frac{(i\theta)}{1!} + \frac{(i\theta)^2}{2!} + \frac{(i\theta)^3}{3!} + \cdots + \frac{(i\theta)^{n-1}}{(n-1)!} + \cdots$$

Replacement of i^2 by -1, $i^3 = i^2 i$ by $-i$, $i^4 = (i^2)^2$ by $+1, \ldots,$ and $i^{4m+k} = i^{4m} i^k = (i^4)^m i^k$ by i^k, where $k = 0, 1, 2,$ or 3, we obtain

$$e^{i\theta} = 1 + \frac{i\theta}{1!} - \frac{\theta^2}{2!} - \frac{i\theta^3}{3!} + \frac{\theta^4}{4!} + \frac{i\theta^5}{5!} - \frac{\theta^6}{6!} - \frac{i\theta^7}{7!} + \cdots$$

$$= \left(1 - \frac{\theta^2}{2!} + \frac{\theta^4}{4!} - \frac{\theta^6}{6!} + \cdots\right) + i\left(\frac{\theta}{1!} - \frac{\theta^3}{3!} + \frac{\theta^5}{5!} - \frac{\theta^7}{7!} + \cdots\right)$$

$$= \cos \theta + i \sin \theta$$

This is the famous *Euler identity*. Setting $\theta = \beta x$ and $\theta = -\beta x$, we obtain

$$e^{i\beta x} = \cos \beta x + i \sin \beta x$$

and

$$e^{-i\beta x} = \cos (-\beta x) + i \sin (-\beta x)$$
$$= \cos \beta x - i \sin \beta x$$

These formulas yield

$$y_1 = e^{\alpha x}(\cos \beta x + i \sin \beta x) \quad \text{and} \quad y_2 = e^{\alpha x}(\cos \beta x - i \sin \beta x)$$

The linear combinations

$$y_3 = \frac{1}{2i} y_1 - \frac{1}{2i} y_2 \quad \text{and} \quad y_4 = \frac{1}{2} y_1 + \frac{1}{2} y_2$$

yield

$$y_3 = e^{\alpha x} \sin \beta x \quad \text{and} \quad y_4 = e^{\alpha x} \cos \beta x$$

and the linear combination $y = c_1 y_3 + c_2 y_4$ yields (4.13).

For a justification of the use of complex numbers in this development, see any standard reference on the theory of functions of a complex variable, for example, Reference 4.5. From a practical point of view, the logical justification of a method by which a solution of a DE is obtained is not very important. If a solution obtained by any method whatsoever can be verified by substituting the solution into the DE, the solution has the same standing as one produced by a completely logical development.

EXAMPLE 6. Find a complete solution of $(D^2 - 4D + 13)y(x) = 0$

Solution: By the quadratic formula, the characteristic equation $r^2 - 4r + 13 = 0$ has roots $r_1 = 2 + 3i$ and $r_2 = 2 - 3i$.

$$y = e^{2x}(c_1 \sin 3x + c_2 \cos 3x)$$

EXAMPLE 7. Find the solution of $(D^2 + 4)y(x) = 0$ for which $y(0) = 0$ and $y'(0) = 6$.

Solution: The characteristic equation $r^2 + 4 = 0$ has roots $r_1 = 0 + 2i$ and $r_2 = 0 - 2i$. A complete solution of $(D^2 + 4)y(x) = 0$ is given by

$$y = e^{0x}(c_1 \sin 2x + c_2 \cos 2x)$$

or

$$y = c_1 \sin 2x + c_2 \cos 2x$$

Substituting $x = 0$ and $y = 0$, we find that $c_2 = 0$. Hence,

$$y = c_1 \sin 2x \quad \text{and} \quad y' = 2c_1 \cos 2x$$

Substituting $x = 0$ and $y' = 6$, we find that $c_1 = 3$. Hence the required particular solution is given by $y = 3 \sin 2x$.

We shall find (Chapter 5) that, when the behavior of a physical system is governed by a linear, constant coefficient DE, the nature of the roots of the associated characteristic equation determines the mode of behavior of the system.

Problem List 4.2

1. Find a complete solution of the following.
   (a) $(D^2 - 7D + 12)y(x) = 0$ (b) $(D^2 + D - 6)y(x) = 0$
   (c) $(D^2 + 7D + 6)y(x) = 0$ (d) $(D^2 - 4)y(x) = 0$
   (e) $(D^2 - 9D)y(x) = 0$ (f) $(D^2 + 5D)y(x) = 0$
   (g) $(D^2 - 4D + 1)y(x) = 0$ (h) $16y'' - 8y' - 3y = 0$
   (i) $y'' - 6y' + 9y = 0$ (j) $y'' + 4y' + 4y = 0$
   (k) $y'' + 25y = 0$ (l) $y'' - 6y' + 25y = 0$
   (m) $y'' + 10y' + 29y = 0$ (n) $y'' + 2y' + 2y = 0$

2. For each of the following DE, find the particular solution that satisfies the given conditions.

(a) $D^2y(x) = 0$; $y = 4$ and $y' = 3$ when $x = 0$

(b) $(D^2 - 1)y(x) = 0$; $y = 1$ and $y' = 3$ when $x = 0$

(c) $(D^2 + D)y(x) = 0$; $y = 5$ and $y' = 2$ when $x = 0$

(d) $(D^2 - 4D + 4)y(x) = 0$; $y = 1$ when $x = 0$; $y = 3e^2$ when $x = 1$

(e) $y'' - y' - 2y = 0$; $y = 1$ and $y' = 8$ when $x = 0$

(f) $y'' + 9y = 0$; $y = 5$ and $y' = 6$ when $x = 0$

(g) $y'' + 2y' + 2y = 0$; $y = 3$ and $y' = -3$ when $x = 0$

3. Solve the DE $(D + 3)y(x) = 0$ by considering the characteristic equation $r + 3 = 0$.

4. Solve the DE $(D - 5)y(x) = 0$ by considering the characteristic equation $r - 5 = 0$.

5. Show that the solutions of $D^2y(x) = 0$ given by $y_1 \equiv 1$ and $y_2 = x$ are linearly independent.

6. Employ the Wronskian to show that the functions defined by $y_1 = e^{r_1 x}$ and $y_2 = xe^{r_1 x}$ are linearly independent on an arbitrary interval I.

7. Evaluate:

(a) $\lim_{\alpha \to 0} e^{\alpha x}(c_1 \sin \beta x + c_2 \cos \beta x)$

(b) $\lim_{\beta \to 0} e^{\alpha x}(c_1 \sin \beta x + c_2 \cos \beta x)$

8. Show that $y_2 = e^{\alpha x} \cos \beta x$ defines a solution of $(D^2 + aD + b)y(x) = [D^2 - 2\alpha D + (\alpha^2 + \beta^2)]y(x) = 0$. Also show that y_1 and y_2 are linearly independent on an arbitrary interval I, given that y_1 is defined by $y_1 = e^{\alpha x} \sin \beta x$ and $\beta \neq 0$.

9. Write a complete solution of the second-order homogeneous DE whose characteristic equation has roots:

(a) $2, -5$ (b) $3, 3$

(c) $0, -4$ (d) $3, \pm 5i$

(e) $\pm 2i$ (f) $3 + \sqrt{5}, 3 - \sqrt{5}$

10. Write a DE of the form $(D^2 + aD + b)y(x) = 0$ having a complete solution given by

(a) $y = c_1 e^{5x} + c_2 e^{-3x}$

(b) $y = c_1 e^x + c_2 xe^x$

(c) $y = c_1 \sin x + c_2 \cos x$

(d) $y = e^{-2x}(c_1 \sin 4x + c_2 \cos 4x)$

11. Solve the DE $y'(x) = y(-x)$.

12. Solve the DE $(D^2 + \omega^2)y(x) = 0$ by letting $v = y' = dy/dx$ and writing

$$y'' = \frac{d^2y}{dx^2} = \frac{dv}{dx} \quad \text{in the form} \quad v\frac{dv}{dy}$$

13. Show that a complete solution of the DE $(D^2 + \omega^2)y(x) = 0$ can be denoted by $y = A \sin(\omega x + \alpha)$ or by $y = B \cos(\omega x + \theta)$, where A, B, α, and θ denote arbitrary constants.

14. The functions defined by $y_1 = e^{r_1 x}$ and $y_2 = e^{r_2 x}$ are solutions of the DE $(D^2 - [r_1 + r_2]D + r_1 r_2)y(x) = 0$. Show that the function defined by

$$y_3 = \lim_{r_1 \to r_2} \frac{e^{r_2 x} - e^{r_1 x}}{r_2 - r_1}$$

is a solution of the DE $(D^2 - 2r_2 D + r_2^2)y(x) = 0$.

15. Let y_1 and y_2 define linearly independent solutions of the constant coefficient DE $(D^2 + aD + b)y(x) = 0$. Show that the Wronskian W of y_1 and y_2 is constant if and only if $a = 0$.

16. Find a complete solution of the DE $(D^2 - k^2)y(x) = 0$. Show that the result can be written in the form $y = A \cosh kx + B \sinh kx$.

4.7 The Nonhomogeneous Equation; Method of Undetermined Coefficients

A particular solution y_p of $Ly = (D^2 + aD + b)y = \rho$ given by $y_p = \psi(x)$ can always be found whenever a complete solution of $Ly = (D^2 + aD + b)y = 0$ is known (Reference 4.4). In many simple cases, y_p can be found by a procedure known as the *method of undetermined coefficients*.

The method applies when $\rho(x)$ is a linear combination of terms of the form $x^m e^{\alpha x} \sin \beta x$ or $x^m e^{\alpha x} \cos \beta x$, where α and β are real and m is a nonnegative integer. For example, $\rho(x)$ might be x^4, xe^{-x}, $x \cos \sqrt{2}x$, $xe^x \sin x$, or $e^{3x} \cos x$; but not $\sec x$, $\ln x$, x^{-3}, or $x^{1/2} \sin x$. Essentially, ρ must be a solution of a linear, homogeneous, constant coefficient DE. See Reference 4.4 for details.

We assume that ρ and its successive derivatives constitute a finite set of linearly independent functions. A trial solution $y_p = \psi(x)$ which is a linear combination of these linearly independent functions is assumed. The conjecture that there exists a specific linear combination of this type such that $(D^2 + aD + b)y = \rho$ seems reasonable. The method is best illustrated by examples.

EXAMPLE 1. Find a particular solution of $(D^2 - D - 2)y(x) = 8e^{3x}$.

Solution: Since all derivatives of $8e^{3x}$ are of the form Ae^{3x}, we try $y_p = Ae^{3x}$. Substitution into the DE yields

$$9Ae^{3x} - 3Ae^{3x} - 2Ae^{3x} = 8Ae^{3x}$$

Dividing by e^{3x}, we obtain $4A = 8$, or $A = 2$. This proves that if the DE has a particular solution of the form $y_p = Ae^{3x}$, then A must be 2. It is easily verified that $y_p = 2e^{3x}$ does indeed define a particular solution of $y'' - y' - 2y = 8e^{3x}$.

EXAMPLE 2. Find a particular solution of

$$(D^2 - 7D + 12)y(x) = 12x^2 + 10x - 11$$

Solution: Since differentiation reduces the degree of a polynomial, it is natural to assume that $y_p(x)$ is a polynomial of degree n whenever $\rho(x)$ is a polynomial of degree n.
Trying $y_p = Ax^2 + Bx + C$ yields

$$(2A) - 7(2Ax + B) + 12(Ax^2 + Bx + C) = 12x^2 + 10x - 11$$

Equating the coefficients of like powers of x, we obtain the equations

$$2A - 7B + 12C = -11$$

$$-14A + 12B = 10$$

$$12A = 12$$

From these equations we find that $A = 1$, $B = 2$, and $C = \frac{1}{12}$. It is easily shown that $y_p = x^2 + 2x + \frac{1}{12}$ defines a particular solution.

EXAMPLE 3. Find a particular solution of $(D^2 + 2D)y(x) = 80 \sin 4x$.

Solution: Since all derivatives of $80 \sin 4x$ are of the form $A \sin 4x$ or $B \cos 4x$, we try $y_p = A \sin 4x + B \cos 4x$.
Subsitution into the DE yields

$$(-16A \sin 4x - 16B \cos 4x) + 2(4A \cos 4x - 4B \sin 4x) = 80 \sin 4x$$

Setting $4x = 0$ and then $4x = \pi/2$, we obtain the equations

$$-16B + 8A = 0$$

$$-16A - 8B = 80$$

from which we obtain $A = -4$ and $B = -2$.
It is easily shown that $y_p = -4 \sin 4x - 2 \cos 4x$ defines a particular solution of the given DE.

If the suggested trial solution is part of the complete solution of the reduced equation, the usual trial solution will not yield a particular solution of $Ly = \rho$, since it will produce zero rather than $\rho(x)$ when substituted into the left member. That is, $Ly_p(x) = 0$ whereas we want $Ly_p(x) = \rho(x)$.

Suppose that we seek, for example, a particular solution of $Dy(x) = 5$. The usual trial solution $y_p = A$, where A is a constant, cannot succeed since $Dy(x) = 0$ has a complete solution given by $y = c$ and hence $DA = 0$. We note that by simple integration with respect to x, however, $Dy(x) = 5$ has a solution given by $y = 5x$. Since $5x$ is of the form xA, this simple example suggests that we try x times the suggested trial solution whenever the usual trial solution is part of a complete solution of $Ly(x) = 0$. The next example illustrates this situation.

EXAMPLE 4. Find a particular solution of $(D^2 - 3D)y(x) = 2e^{3x}$.

Solution: Since $y_p = Ae^{3x}$ is part of the complete solution of $(D^2 - 3D)y(x) = 0$ given by $y = c_1 + c_2e^{3x}$, the given DE cannot have a particular solution of the form $y_p = Ae^{3x}$. Hence we try $y_p = Axe^{3x}$. Then

$$y_p' = 3Axe^{3x} + Ae^{3x} \quad \text{and} \quad y_p'' = 9Axe^{3x} + 3Ae^{3x} + 3Ae^{3x}$$

Substituting into $(D^2 - 3D)y(x) = 2e^{3x}$, we obtain

$$(9Axe^{3x} + 6Ae^{3x}) - 3(3Axe^{3x} + Ae^{3x}) \equiv 2e^{3x}$$

and we find that $A = \frac{2}{3}$. We then verify that $y_p = \frac{2}{3}xe^{3x}$ defines a particular solution of $(D^2 - 3D)y(x) = 2e^{3x}$.

More generally, if the characteristic equation of $Ly(x) = 0$ has a root $r = \alpha + i\beta$ of multiplicity k, and $Y = \psi(x)$ defines the usual trial solution resulting from r, we try to find a particular solution of $Ly = \rho$ of the form $y_p = x^k\psi(x)$.

EXAMPLE 5. Find a particular solution of $(D^2 - 3D + 2)y(x) = 12$.

Solution: Instead of trying $y_p = A$, we merely note *by inspection* that $y_p \equiv 6$ defines a particular solution.

EXAMPLE 6. Find a complete solution of $(D^2 - D - 2)y(x) = 8e^{3x}$.

Solution: From $r^2 - r - 2 = 0$, we obtain $r_1 = 2$ and $r_2 = -1$. Hence, a complete solution of the reduced equation is given by $y_c = c_1e^{2x} + c_2e^{-x}$, often termed the *complementary function*. In Example 1, the given DE was found to possess the particular solution given by $y_p = 2e^{3x}$. Hence, by Theorem 4-VII, $y = c_1e^{2x} + c_2e^{-x} + 2e^{3x}$ defines a complete solution. That is, *every* solution of $(D^2 - D - 2)y(x) = 8e^{3x}$ can be expressed in the form

$$y = c_1e^{2x} + c_2e^{-x} + 2e^{3x}$$

EXAMPLE 7. Find the solution of $(D^2 - 3D)y(x) = 2e^{3x}$ satisfying the conditions $y = 1$ and $y' = 3$ when $x = 0$.

Solution: It follows from Example 4 that the given DE has a complete solution given by $y = c_1 + c_2e^{3x} + \frac{2}{3}xe^{3x}$. Differentiating yields

$$y' = 3c_2e^{3x} + 2xe^{3x} + \frac{2}{3}e^{3x}$$

Using the given conditions, we obtain the equations $1 = c_1 + c_2$ and $3 = 3c_2 + \frac{2}{3}$, and hence $c_2 = \frac{7}{9}$ and $c_1 = \frac{2}{9}$. Thus the required particular solution is given by $y = \frac{2}{9} + \frac{7}{9}e^{3x} + \frac{2}{3}xe^{3x}$.

EXAMPLE 8. Find a complete solution of $(D^2 - 1)y(x) = 5e^x \sin x$.

Solution: From $r^2 - 1 = 0$, we find $r = \pm 1$, and hence the complementary function is given by $y_c(x) = c_1e^x + c_2e^{-x}$.

The derivatives of $5e^x \sin x$ involve $e^x \sin x$ and $e^x \cos x$. Hence we try $y_p(x) = \psi(x)Ae^x \sin x + Be^x \cos x$. Substituting

$$Dy_p(x) = Ae^x \cos x + Ae^x \sin x - Be^x \sin x + Be^x \cos x$$

and

$$D^2y_p(x) = -Ae^x \sin x + Ae^x \cos x + Ae^x \cos x + Ae^x \sin x$$
$$- Be^x \cos x - Be^x \sin x + Be^x \cos x - Be^x \sin x$$

into the given DE, we obtain

$$(2Ae^x \cos x - 2Be^x \sin x) - (Ae^x \sin x + Be^x \cos x) = 5e^x \sin x$$

Setting $x = 0$ and $x = \pi/2$ yields the equations

$$-2A - B = 0 \quad \text{and} \quad e^{\pi/2}(-2B - A) = 5e^{\pi/2}$$

from which we find that $A = -1$ and $B = -2$. Thus $y_p(x) = -e^x \sin x - 2e^x \cos x$. It is a simple exercise to verify that

$$y(x) = y_c(x) + y_p(x) = c_1 e^x + c_2 e^{-x} - e^x \sin x - 2e^x \cos x$$

defines a complete solution.

   Although the method of undetermined coefficients applies to only a limited class of DE, it is usually simple to apply and is effective in solving many DE arising in applications. The method does not apply to a DE such as $Ly(x) = \tan x$, for which the successive derivatives of $\rho(x) = \tan x$ do not form a finite set. In the following section we consider an alternative method of finding a particular solution of $Ly(x) = \rho(x)$.

Problem List 4.3

1. Find a particular solution of:
   (a) $(D^2 + 5D - 4)y(x) = 6e^x$
   (b) $(D^2 + 4D + 2)y(x) = 5e^{-x}$
   (c) $(D^2 - D + 3)y(x) = 8e^{2x}$
   (d) $(D^2 + 3D + 4)y(x) = 12x^2 - 6x - 8$
   (e) $(D^2 - D + 2)y(x) = 8$
   (f) $y'' - 6y' + y = 24 \sin x$
   (g) $y'' + 2y' + 10y$
   $\quad = 2 \sin 2x + 10 \cos 2x$
   (h) $y'' + y' + 4y = 12e^x + 4x^2 + 2x + 2$
   (i) $y'' - 2y' = 5e^{2x}$
   (j) $y'' + y = 3 \sin x$

2. Find a complete solution of:
   (a) $(D^2 - D - 2)y(x) = 8e^x$
   (b) $(D^2 + 4)y(x) = 26e^{3x}$
   (c) $(D^2 - 9)y(x) = 6x^2 + 4x$
   (d) $(D^2 + 1)y(x) = 9 \sin 2x$
   (e) $y'' + 4y' + 13y$
   $\quad = -3 \sin 4x + 16 \cos 4x$
   (f) $y'' - y = e^{-x}$
   (g) $y'' - y = 6e^x$
   (h) $y'' + y = 4 \sin x$

3. For each DE find the solution satisfying the given conditions.
   (a) $(D^2 - 2D)y(x) = e^{3x}$; $y = \frac{4}{3}$ and $y' = 5$ when $x = 0$
   (b) $(D^2 + 1)y(x) = 20e^{-3x}$; $y = 2$ and $y' = -2$ when $x = 0$
   (c) $(D^2 + D)y(x) = x^2 + 2x + 1$; $y = 1$ when $x = 0$; $y = 12 + e^{-3}$ when $x = 3$
   (d) $y'' - 4y' + 3y = e^{2x}$; $y = 0$ and $y' = 0$ when $x = 0$
   (e) $y'' + y = 2x$; $y = 1$ and $y' = 2$ when $x = 0$
   (f) $y'' + 16y = 32 \sin 4x$; $y = 0$ and $y' = 0$ when $x = 0$

4. Given that ψ_1 is a particular solution of $Ly = (D^2 + pD + q)y = \rho_1$ and that ψ_2 is a particular solution of $Ly = \rho_2$, show that $y_p = \psi_1 + \psi_2$ is a particular

solution of $Ly = \rho_1 + \rho_2$. That is, show that a particular solution of $Ly = \rho_1 + \rho_2$ can be obtained by superposition of particular solutions of $Ly = \rho_1$ and $Ly = \rho_2$.

5. Show that $y = xe^{kx}$ defines a particular solution of $(D^2 - 2kD + k^2)y(x) = 0$.

6. Find a second-order DE having a complete solution defined by $y = c_1 e^x + c_2 e^{-2x}$.

7. Find a second-order DE having a complete solution defined by $y = c_1 \sin x + c_2 \cos x + e^x$.

8. The number k is a repeated root of the characteristic equation of $y'' - 2ky' + k^2 y = 0$, and hence $y = e^{kx}$ defines a solution. Find a second solution of the form $y = v(x)e^{kx}$.

9. Solve the DE $xy'' - 4y' = x^6$ by letting $u = y'$.

10. Find a complete solution of:
    (a) $y'' - 2y' + y = 12xe^x$
    (b) $y'' - 4y' + 4y = 6e^{2x}$
    (c) $y'' + \omega^2 y = E \sin \beta x, \beta \neq \omega$
    (d) $(D^2 + 1)y(x) = 4 \cosh x$
    (e) $(D^2 - 1)y(x) = 4x \sin x$

11. Find a particular solution of $(D^2 + 1)y(x) = e^x \sin x$.

12. Find a particular solution of $(D^2 + 2D + 2)y(x) = e^{-x} \cos x$ by letting $y(x) = v(x)e^{-x}$.

4.8 Variation of Parameters

The method of undetermined coefficients for finding a particular solution of the nonhomogeneous equation $Ly = \rho$ has the advantage of simplicity. We now present an alternative method, known as the *variation of parameters* (or *variation of constants*), devised by Euler in 1739 and refined by Lagrange in 1774. This method is more general in application than the method of undetermined coefficients, and it also applies to linear DE with nonconstant coefficients.

We begin by assuming that the functions p, q, and ρ are continuous on an interval I, and that we know two linearly independent solutions y_1 and y_2 of

$$Ly(x) = (D^2 + pD + q)y(x) = 0 \qquad (4.14)$$

on I.

To find a particular solution ψ of

$$Ly = (D^2 + pD + q)y = \rho \qquad (4.15)$$

on I, we might try $y_p = \psi_1 y_1$ or $y_p = \psi_2 y_2$, where ψ_1 and ψ_2 are functions of x to be determined. Another possibility is to try the linear combination $y_p = k_1(\psi_1 y_1) + k_2(\psi_2 y_2)$. It turns out that this trial solution is fruitful, so let us replace $k_1 \psi_1$ by v_1 and $k_2 \psi_2$ by v_2, and assume that (4.15) has a particular solution of the form

$$y_p = v_1 y_1 + v_2 y_2 \qquad (4.16)$$

where v_1 and v_2 are functions of x to be determined. We observe that (4.16) is the same as the complete solution $y = c_1 y_1 + c_2 y_2$ of (4.14) with the constants c_1 and c_2 replaced by the functions v_1 and v_2. This accounts for the name variation of parameters (or constants).

We next note that

$$y_p' = v_1 y_1' + v_2 y_2' + v_1' y_1 + v_2' y_2$$

Before computing y_p'', we observe that y_p'' will not involve v_1'' or v_2'' if we assume that

$$v_1' y_1 + v_2' y_2 = 0 \qquad (4.17)$$

on I. We may be able to afford this assumption since we have two functions v_1 and v_2 at our disposal and our only remaining requirement is that $Ly_p = \rho$.

We now find that y_p'', under assumption (4.17), is given by

$$y_p'' = v_1 y_1'' + v_2 y_2'' + v_1' y_1' + v_2' y_2'$$

Substituting y_p, y_p', and y_p'' into (4.15) yields

$$(v_1 y_1'' + v_2 y_2'' + v_1' y_1' + v_2' y_2') + p(v_1 y_1' + v_2 y_2') + q(v_1 y_1 + v_2 y_2) = \rho$$

which on rearrangement becomes

$$v_1(y_1'' + py_1' + qy_1) + v_2(y_2'' + py_2' + qy_2) + v_1' y_1' + v_2' y_2' = \rho$$

or

$$v_1(Ly_1) + v_2(Ly_2) + v_1' y_1' + v_2' y_2' = \rho$$

and since $Ly_1(x) = Ly_2(x) = 0$,

$$v_1' y_1' + v_2' y_2' = \rho \qquad (4.18)$$

Solving (4.17) and (4.18) for v_1' and v_2', we obtain

$$v_1' = \frac{-y_2 \rho}{y_1 y_2' - y_2 y_1'} \qquad v_2' = \frac{y_1 \rho}{y_1 y_2' - y_2 y_1'} \qquad (4.19)$$

This unique solution is always possible since each denominator in (4.19) is the Wronskian $W(y_1, y_2)$, and W is different from zero for every x in I. This follows from a form of the contrapositive of Theorem 4-V, which states that if y_1 and y_2 are linearly independent solutions of $Ly(x) = 0$, then W is different from zero for at least one x_0 in I. By Abel's identity, W is therefore different from zero for every x in I.

It is always possible to find v_1 and v_2 from v_1' and v_2', since the right members of (4.19) are continuous on I. Thus,

$$v_1 = \int \frac{-y_2 \rho}{W} dx \qquad v_2 = \int \frac{y_1 \rho}{W} dx \qquad (4.20)$$

It is customary to assign each constant of integration in (4.20) the value zero. We now have

$$y_p = v_1 y_1 + v_2 y_2 = y_1 \int \frac{-y_2 \rho}{W} \, dx + y_2 \int \frac{y_1 \rho}{W} \, dx \qquad \textbf{(4.21)}$$

We have proved that if $Ly = \rho$ has a particular solution of the form $y_p = v_1 y_1 + v_2 y_2$, where y_1 and y_2 are linearly independent solutions of $Ly(x) = 0$ on I, and if (4.17) holds on I, and if the constants of integration c_1 and c_2 in (4.20) are both zero, then y_p can be found and must be given by (4.21) and will be free of arbitrary constants. It is a simple matter to verify that the function y_p given by (4.21) is indeed a particular solution of $Ly = \rho$. This verification involves a reversal of the steps by which (4.21) was obtained. See Problem 9.

Two final points are worth mentioning. First, it may be difficult to express v_1 and v_2 except by integrals. Second, if we write (4.20) as $v_1 = V_1 + c_1, v_2 = V_2 + c_2$, then

$$y = v_1 y_1 + v_2 y_2$$
$$= c_1 y_1 + c_2 y_2 + V_1 y_1 + V_2 y_2$$

defines a complete solution of $Ly = \rho$.

The method of variation of parameters is readily extended to DE of order greater than 2. See Reference 4.1.

EXAMPLE 1. Find a particular solution y_p of $(D^2 + 1)y(x) = \tan x$ on $(-\pi/2, \pi/2)$.

Solution: From $r^2 + 1 = 0$, $r = \pm i$, and $y_c = c_1 \sin x + c_2 \cos x$, we find that $W(\sin x, \cos x) = (\sin x)(-\sin x) - (\cos x)(\cos x) \equiv -1$.

Substituting $W = -1$, $y_1 = \sin x$, $y_2 = \cos x$, and $\rho(x) = \tan x$ into (4.20), we obtain

$$v_1 = \int \frac{-\cos x \tan x}{-1} \, dx = \int \sin x \, dx = -\cos x$$

and

$$v_2 = \int \frac{\sin x \tan x}{-1} \, dx = -\int \frac{\sin^2 x \, dx}{\cos x} = -\int \frac{(1 - \cos^2 x) \, dx}{\cos x}$$

$$= \int (\cos x - \sec x) \, dx$$

$$= \sin x - \ln (\sec x + \tan x)$$

Hence, by (4.21), a particular solution of $(D^2 + 1)y(x) = \tan x$ is given by

$$y_p = [-\cos x][\sin x] + [\sin x - \ln (\sec x + \tan x)][\cos x]$$
$$= -\cos x \ln (\sec x + \tan x)$$

A complete solution of $(D^2 + 1)y(x) = \tan x$ on $(-\pi/2, \pi/2)$ is given by

$$y = c_1 \sin x + c_2 \cos x - \cos x \ln (\sec x + \tan x)$$

Note that $\sec x + \tan x = (1 + \sin x)/\cos x > 0$ on $(-\pi/2, \pi/2)$.

EXAMPLE 2. Find a complete solution of $(D^2 - D - 2)y(x) = 8e^{3x}$.

Solution: From $(r^2 - r - 2) = (r - 2)(r + 1) = 0$, $r_1 = 2$, $r_2 = -1$, and $y_c = c_1 e^{2x} + c_2 e^{-x}$, we find that

$$W(e^{2x}, e^{-x}) = (e^{2x})(-e^{-x}) - (e^{-x})(2e^{2x}) = -3e^x$$

Substitution of $W = -3e^x$, $y_1 = e^{2x}$, $y_2 = e^{-x}$, and $\rho(x) = 8e^{3x}$ into (4.20) yields

$$v_1 = \int \frac{-e^{-x}(8e^{3x})\, dx}{-3e^x} = \frac{8}{3} \int e^x\, dx = \frac{8}{3} e^x$$

and

$$v_2 = \int \frac{e^{2x}(8e^{3x})\, dx}{-3e^x} = \frac{-8}{3} \int e^{4x}\, dx = \frac{-2}{3} e^{4x}$$

Hence, by (4.21),

$$y_p = \left(\frac{8}{3} e^x \right)(e^{2x}) + \left(\frac{-2}{3} e^{4x} \right)(e^{-x}) = 2e^{3x}$$

defines a particular solution of the given DE. A complete solution is given by

$$y = c_1 e^{2x} + c_2 e^{-x} + 2e^{3x}$$

valid on $(-\infty, +\infty)$.

Remark: The same particular solution of this DE was found by the method of undetermined coefficients in illustrative Example 1 of Section 4.7.

EXAMPLE 3. Given that the equation $(D^2 + x^{-1}D - x^{-2})y(x) = 0$ has a complete solution given by $y = c_1 x + c_2 x^{-1}$, find the solution of

$$(D^2 + x^{-1}D - x^{-2})y(x) = 4x^{-1}$$

for which $y = 1$ and $y' = 1$ when $x = 1$.

Solution: Substituting $y_1 = x$, $y_2 = x^{-1}$, $W(x, x^{-1}) = (x)(-x^{-2}) - (x^{-1})(1) = -2x^{-1}$, and $\rho = 4x^{-1}$ into (4.20), we obtain

$$v_1 = \int \frac{-x^{-1}(4x^{-1})\, dx}{-2x^{-1}} = 2 \ln x$$

and

$$v_2 = \int \frac{x(4x^{-1})\, dx}{-2x^{-1}} = -x^2$$

By (4.21), $Ly(x) = 4x^{-1}$ has a particular solution given by $y_p = 2x \ln x - x$ and a complete solution given by

$$y = c_1 x + c_2 x^{-1} + 2x \ln x - x$$

Differentiating, we get

$$y' = c_1 - c_2 x^{-2} + 2 \ln x + 2 - 1$$

Substituting $x = 1$, $y = 1$, and $x = 1$, $y' = 1$, we obtain the equations

$$1 = c_1 + c_2 - 1, \quad 1 = c_1 - c_2 + 1$$

which yield $c_1 = c_2 = 1$.

The required solution of $Ly(x) = 4x^{-1}$ is given by

$$y = x + x^{-1} + 2x \ln x - x$$
$$= x^{-1} + 2x \ln x$$

valid on $(0, +\infty)$.

Problem List 4.4

1. Find a particular solution of $(D^2 + 1)y(x) = \rho(x)$ by variation of parameters if $\rho(x)$ is:

   (a) $\cot x, 0 < x < \dfrac{\pi}{2}$ (b) $\sec x, \dfrac{-\pi}{2} < x < \dfrac{\pi}{2}$

   (c) $\sec^3 x, \dfrac{-\pi}{2} < x < \dfrac{\pi}{2}$ (d) $6e^x$

   (e) $\csc x, 0 < x < \pi$ (f) $4 \sin x$

2. Find a particular solution of $(D^2 - 1)y(x) = \rho(x)$ by variation of parameters if $\rho(x)$ is (a) e^x; (b) $8xe^{-x}$.

3. Given that $(D^2 - x^{-1}D + x^{-2})y(x) = 0$ has a complete solution given by $y = c_1 x + c_2 x \ln x$, find the solution of

$$(D^2 - x^{-1}D + x^{-2})y(x) = x^2$$

on $(0, +\infty)$ for which $y = 0$ and $y' = 0$ when $x = 1$.

4. Find a complete solution of $(D^2 + 2D + 1)y(x) = x^{-1}e^{-x}$ on $(0, +\infty)$.

5. Solve $(D + p)y = q$ by variation of parameters.

6. Find a particular solution of $(D^2 - 1)y(x) = x^{-1}$ on $(0, +\infty)$.

7. Given that $Ly(x) = [D^2 + x^{-1}D + (1 - 0.25x^{-2})]y(x) = 0$ has linearly independent solutions given by $y_1 = x^{-1/2} \sin x$ and $y_2 = x^{-1/2} \cos x$, find a particular solution of $Ly(x) = x^{-1/2}$ on $(0, +\infty)$.

8. Apply variation of parameters to $(D^2 + \omega^2)y(x) = \rho(x)$ to obtain the particular solution given by

$$y_p = \omega^{-1} \int_0^x \rho(t) \sin \omega(x - t) \, dt$$

9. Show that the function given by (4.21) satisfies DE (4.15), $(D^2 + pD + q)y(x) = \rho(x)$.

10. Find the solution of $(D^2 + 1)y(x) = \cos^2 x$ for which $y(0) = y'(0) = 0$.

11. (a) Let y_1 be a solution of $Ly(x) = (D^2 + pD + q)y(x) = 0$. Apply the method of *reduction of order* of Section 4.5 to find a particular solution of $Ly = (D^2 + pD + q)y = \rho$. That is, try a particular solution of the form $y_p = vy_1$ where v is a function of x to be determined. Show that

$$v = \int \left\{ \left[y_1^{-2} \exp\left(-\int p \, dx \right) \right] \left[\int y_1 \rho \exp\left(\int p \, dx \right) dx \right] \right\} dx$$

(b) Use the method of part (a) to find a particular solution of $(D^2 + 1)y(x) = \tan x$ on $(-\pi/2, \pi/2)$.

(c) Use the method of part (a) to find a particular solution of $(D^2 + 1)y(x) = \sec x$ on $(-\pi/2, \pi/2)$.

(d) Use the method of part (a) to find the solution of $(D^2 + 2D + 1)y(x) = x^{-1}e^{-x}$ on $(0, +\infty)$ for which $y(1) = y'(1) = 0$.

12. (a) Let $r_1 = \frac{1}{2}(-a + \sqrt{a^2 - 4b})$ and $r_2 = \frac{1}{2}(-a - \sqrt{a^2 - 4b})$ denote the roots of the equation $r^2 + ar + b = 0$. Solve $(D^2 + aD + b)y(x) = \rho(x)$ by letting $z = dy/dx - r_1 y$.

(b) Find a complete solution of each DE by the method of part (a):
   (i) $(D^2 - D - 2)y(x) = 8e^{3x}, \; -\infty < x < +\infty$
   (ii) $(D^2 - 1)y(x) = 5e^x \sin x, \; -\infty < x < +\infty$
   (iii) $(D^2 - 4D + 4)y(x) = 6e^{2x}, \; -\infty < x < +\infty$
   (iv) $(D^2 - 1)y(x) = 8xe^{-x}, \; -\infty < x < +\infty$
   (v) $(D^2 + 2D + 1)y(x) = x^{-1}e^{-x}, \; 0 < x < +\infty$

4.9 Operator Methods of Finding a Particular Solution of $Ly(x) = 0$

To construct an algebra of differential operators of the form (Section 4.1)

$$L = a_0 D^n + a_1 D^{n-1} + \cdots + a_{n-1}D + a_n$$

meaning a set of rules by which these operators can be combined to form new operators, we begin by defining simple operations on these operators. We are here restricting ourselves to operators in which the coefficients a_0, $a_1, \ldots, a_n$ are constants. The algebra of operators then furnishes a method of solving DE.

Addition of the differential operators L_1 and L_2 is defined by

$$(L_1 + L_2)y = L_1 y + L_2 y \tag{4.22}$$

EXAMPLE 1. $[(D^2 + D) + (D - 1)]y = (D^2 + D)y + (D - 1)y$
$$= D^2 y + Dy + Dy - y$$
$$= D^2 y + 2Dy - y$$
$$= (D^2 + 2D - 1)y$$

The operator ϕL, where ϕ is a function of x and L is a differential operator, is defined by

$$[\phi(x)L]y(x) = \phi(x)[Ly(x)] \tag{4.23}$$

If $\phi(x)$ is a constant c, then (4.23) reduces to

$$[cL]y(x) = c[Ly(x)] \tag{4.24}$$

EXAMPLE 2. $[(x^2 - 1)(D^2 + 2D)]y(x) = (x^2 - 1)[(D^2 + 2D)y(x)]$

EXAMPLE 3. $[5(D^2 - 3)]y(x) = 5[(D^2 - 3)y(x)]$

Multiplication is defined by

$$(L_1 L_2)y = L_1(L_2 y) \tag{4.25}$$

EXAMPLE 4. $[(D^2 + D)(D + 3)]y = (D^2 + D)[(D + 3)y]$

Using these definitions, we now can show that differential operators can be combined by the usual laws of algebra; that is, as if they were polynomials in D.

This result is obtained by noting that $De^{rx} = re^{rx}$, $D^n e^{rx} = r^n e^{rx}$, and $cD^n e^{rx} = cr^n e^{rx}$. We then let

$$L = a_0 D^n + a_1 D^{n-1} + \cdots + a_{n-1}D + a_n$$
$$= p(D)$$

and call $p(r)$, the polynomial in r obtained by replacing D by r in L, the *characteristic polynomial* of L. It is then easy to see that

$$L(e^{rx}) = p(r)e^{rx} \tag{4.26}$$

Hence, L is completely specified by its characteristic polynomial; in other words, if we know $p(r)$, we can obtain $L = p(D)$ by replacing r by D in $p(r)$. Suppose, for example, that we are given $L_1 = p_1(D)$ and $L_2 = p_2(D)$ having characteristic polynomials $p_1(r)$ and $p_2(r)$. To find $L_3 = L_1 L_2 = p_3(D)$, we note that

$$L_3(e^{rx}) = (L_1 L_2)e^{rx} = L_1(L_2 e^{rx}) = L_1[p_2(r)e^{rx}]$$
$$= p_2(r)L_1(e^{rx}) = p_2(r)[p_1(r)e^{rx}]$$
$$= [p_2(r)p_1(r)]e^{rx}$$

Thus, $p_3(r)$, the characteristic polynomial of L_3, is given by $p_3(r) = p_2(r)p_1(r)$. Hence $L_3 = p_3(D) = p_2(D)p_1(D)$. That is, L_3 is obtained by multiplying the polynomials $p_2(D)$ and $p_1(D)$.

EXAMPLE 5. $[(D + 3)(D - 1)]y = [D^2 + 2D - 3]y$

EXAMPLE 6. $[(D - 1)(D + 3)]y = [D^2 + 2D - 3]y = [(D + 3)(D - 1)]y$

Multiplication of operators is commutative; that is,

$$(L_1 L_2)y = (L_2 L_1)y$$

since $p_2(r)p_1(r) = p_1(r)p_2(r)$. In other words, the multiplication of their characteristic polynomials is commutative.

The algebra of differential operators with *nonconstant* coefficients is more complicated. To verify that multiplication in this algebra is noncommutative, see Problem 4.

Definition If $Ly(x) = f(x)$, then the inverse of L, denoted by $L^{-1} = 1/L$, is an operator satisfying

$$L^{-1}f(x) = \frac{1}{L} f(x) = y(x)$$

Thus, $[LL^{-1}]f(x) = L[L^{-1}f(x)] = Ly(x) = f(x)$.

EXAMPLE 7. Since $Dx^3 = 3x^2, D^{-1}(3x^2) = (1/D)(3x^2) = x^3 + c$.

This example shows that an inverse operator applied to a function produces not a single function but an infinite set of functions, one corresponding to each choice of c. Since our objective is to find a particular solution of $Ly(x) = \rho(x)$, we select for $L^{-1}f$ the function obtained by equating to zero any constants of integration arising during the determination of $L^{-1}f$. Thus, we assume that $D^{-1}(3x^2) = x^3$.

This example also shows us that the inverse operator D^{-1} is an "integrating" operator and could be denoted by $D^{-1} = \int$.

EXAMPLE 8. Since $(D + 1) \sin x = \cos x + \sin x$,

$$\frac{1}{D + 1}(\cos x + \sin x) = \sin x$$

Let us now write (4.26) in the form $L(e^{rx})/p(r) = e^{rx}$, $p(r) \neq 0$. Then

$$L^{-1}(e^{rx}) = L^{-1}\left[\frac{L(e^{rx})}{p(r)}\right] = \frac{L^{-1}[L(e^{rx})]}{p(r)} = \frac{(L^{-1}L)e^{rx}}{p(r)} = \frac{e^{rx}}{p(r)}, \quad p(r) \neq 0$$

$$(4.27)$$

EXAMPLE 9. Find a particular solution of $(D^2 - D - 6)y(x) = e^{5x}$.

Solution: By (4.27),

$$y(x) = \frac{1}{D^2 - D - 6} e^{5x} = \frac{e^{5x}}{(5)^2 - 5 - 6} = \frac{e^{5x}}{14}$$

EXAMPLE 10. Find a particular solution of

$$Ly(x) = (D^2 + 3D + 5)y(x) = 4e^{-2x} + 10$$

Solution: We see by inspection that $Ly(x) = 10$ has a solution given by $(y_p)_1 \equiv 2$. From

$$\frac{1}{D^2 + 3D + 5}(4e^{-2x}) = 4\frac{1}{D^2 + 3D + 5}(e^{-2x})$$

$$= 4\frac{e^{-2x}}{(-2)^2 + 3(-2) + 5}$$

$$= \frac{4e^{-2x}}{3}$$

we find that $Ly(x) = 4e^{-2x}$ has a solution $(y_p)_2$ given by $(y_p)_2 = \frac{4}{3}e^{-2x}$. Hence $Ly(x) = 4e^{-2x} + 10$ has a particular solution y_p given by $y_p = (y_p)_1 + (y_p)_2 = 2 + \frac{4}{3}e^{-2x}$.

Let us now assume that $L = p(D)$ is a quadratic polynomial in D having linear factors $(D - b)$ and $(D - a)$. To find a particular solution of

$$Ly(x) = [(D - a)(D - b)]y(x)$$
$$= (D - a)[(D - b)y(x)] = \rho(x) \qquad (4.28)$$

let us introduce a new variable u by letting $(D - b)y(x) = u(x)$. Then

$$(D - a)u(x) = \rho(x) \qquad (4.29)$$

Since (4.29) is a first-order linear DE, it has an integrating factor $e^{\int p\,dx} = e^{\int -a\,dx} = e^{-ax}$ and solutions defined by $u(x)e^{-ax} = \int e^{-ax}\rho(x)\,dx + c$. Hence

$$u(x) = e^{ax}\int e^{-ax}\rho(x)\,dx + ce^{ax}$$

and

$$(D - b)y(x) = e^{ax}\int e^{-ax}\rho(x)\,dx + ce^{ax} \qquad (4.30)$$

To enable us to operate on both sides of (4.28) by the operator $1/(D - a)$, and to write $(D - b)y(x) = [1/(D - a)]\rho(x)$, we are motivated by (4.28) and (4.30) to adopt the following definition of the inverse operator $1/(D - a)$.

Definition

$$\frac{1}{D-a}\phi(x) = e^{ax}\int e^{-ax}\phi(x)\,dx + ce^{ax} \qquad (4.31)$$

Since we intend to use (4.31) to find a particular solution of $Ly(x) = \rho(x)$, we assign the arbitrary constant c the value 0. Thus

$$\frac{1}{D-a}\phi(x) = e^{ax}\int e^{-ax}\phi(x)\,dx \qquad (4.32)$$

The definition we have adopted for $1/(D-a)$ is also appropriate when $a = 0$, since (4.32) then reduces to $(1/D)\phi(x) = \int \phi(x)\,dx$.

EXAMPLE 11. Find a particular solution of $(D^2 - 2D)y(x) = 8\cos x$.

Solution:

$$(D-2)\,Dy(x) = 8\cos x$$

$$(D-2)y(x) = \frac{1}{D}(8\cos x) = 8\sin x$$

$$y(x) = \frac{1}{D-2}(8\sin x)$$

$$= 8e^{2x}\int e^{-2x}\sin x\,dx$$

$$= 8e^{2x}\left[\frac{e^{-2x}}{5}(-2\sin x - \cos x)\right]$$

$$= \frac{8(-2\sin x - \cos x)}{5}$$

EXAMPLE 12. Find a particular solution of $(D^2 + D - 2)y(x) = 8x^2 - 24x$.

Solution:

$$(D+2)(D-1)y(x) = 8x^2 - 24x$$

$$(D+2)y(x) = \frac{1}{D-1}(8x^2 - 24x)$$

$$= e^x\int e^{-x}(8x^2 - 24x)\,dx$$

$$= 8e^x\int x^2 e^{-x}\,dx - 24e^x\int xe^{-x}\,dx$$

$$= 8e^x[-e^{-x}(x^2 + 2x + 2)] - 24e^x[e^{-x}(-x - 1)]$$
$$= 8(-x^2 - 2x - 2 + 3x + 3)$$
$$= -8(x^2 - x - 1)$$

$$y(x) = \frac{1}{D+2}[-8(x^2 - x - 1)]$$

$$= -8e^{-2x}\int e^{2x}(x^2 - x - 1)\, dx$$

$$= -8e^{-2x}\int x^2 e^{2x}\, dx + 8e^{-2x}\int xe^{2x}\, dx + 8e^{-2x}\int e^{2x}\, dx$$

$$= -8e^{-2x}\left[\frac{e^{2x}}{8}(4x^2 - 4x + 2)\right]$$

$$+ 8e^{-2x}\left[\frac{e^{2x}}{4}(2x - 1)\right] + 8e^{-2x}\left[\frac{e^{2x}}{2}\right]$$

$$= (-4x^2 + 4x - 2) + (4x - 2) + (4)$$

$$= -4x^2 + 8x$$

EXAMPLE 13. Find a particular solution of $(D + 1)^2 y(x) = x^3 e^{-x}$.

Solution:

$$(D + 1)y(x) = \frac{1}{D+1}(x^3 e^{-x})$$

$$= e^{-x}\int e^x(x^3 e^{-x})\, dx = \frac{x^4 e^{-x}}{4}$$

$$y(x) = \frac{1}{D+1}\left(\frac{x^4 e^{-x}}{4}\right)$$

$$= e^{-x}\int e^x\left(\frac{x^4 e^{-x}}{4}\right) dx = \frac{x^5 e^{-x}}{20}$$

Our purpose here has been to introduce operational methods. These methods were developed by the electrical engineer Oliver Heaviside (1850–1925), who was employed by the Great Northern Telegraph Company. He developed the Heaviside operational calculus as a practical method of solving linear DE with constant coefficients, a type of DE he frequently encountered in his investigations of telegraph transmission lines and cables. Many mathematicians found fault with Heaviside's work because he could not justify many of his procedures. Later developments placed his methods on a sound rigorous basis.

A more extensive treatment of operational and symbolic methods will be deferred to Chapter 7, on the Laplace transform. This transform is an operator whose development was a refinement of Heaviside's approach.

Problem List 4.5

1. Verify that $(D + 4)(D - 2)y(x) = (D - 2)(D + 4)y(x)$.
2. Use an operator method to find a particular solution:
   (a) $(D^2 + 4D + 5)y(x) = 4e^{3x}$

(b) $(D^2 - 1)y(x) = e^{-x}$
(c) $(D^2 + 4D)y(x) = 9 \sin x$
(d) $(D^2 + D - 2)y(x) = xe^x$
(e) $(D^2 - 1)y(x) = e^x \cos x$
(f) $(D + 1)^2 y(x) = x^{1/2}e^{-x}$

3. Given that L_1, L_2, and L_3 are linear differential operators, prove that $L_1 + (L_2 + L_3) = (L_1 + L_2) + L_3$.

4. Show that $(D - e^x)(xD + 1) \neq (xD + 1)(D - e^x)$.

5. Show that $(xD + 2)(xD - 2) \neq x^2 D^2 - 4$.

6. (a) Prove by mathematical induction that

$$D^n[e^{ax}\phi(x)] = e^{ax}(D + a)^n \phi(x)$$

(b) Prove that $p(D)e^{ax}\phi(x) = e^{ax}p(D + a)\phi(x)$.

(c) Use the formula of part (b) to find a particular solution of $(D^2 + 3D + 2)y(x) = e^{-x}\cos x$.

4.10 Higher-Order Linear Equations

The results we have outlined for second-order linear DE are readily extended to linear DE of order greater than two. For proofs of these extensions, see References 4.1 and 4.4.

By Theorem 4-I the DE

$$Ly(x) = [D^n + p_1(x)D^{n-1} + p_2(x)D^{n-2} + \cdots$$
$$+ p_{n-1}(x)D + p_n(x)]y(x) = \rho(x) \tag{4.33}$$

where $p_1, p_2, \ldots, p_n$, and ρ are continuous on an interval I, and $x_0 \in I$, has a unique solution on I. This solution, given by $y = \phi(x)$, satisfies the initial conditions

$$\phi(x_0) = y_0, \quad \phi'(x_0) = y_0', \ldots, \quad \phi^{(n-1)}(x_0) = y_0^{(n-1)}$$

EXAMPLE 1. It is easily verified that the function defined by $y = x^3 - 3$ is a solution on $(0, +\infty)$ of the initial-value problem

$$(D^3 - x^{-1}D^2 + xD - 3)y(x) = 9; \quad y(1) = -2, \quad y'(1) = 3, \quad y''(1) = 6$$

By Theorem 4-I, this solution is unique.

The principle of superposition applies almost immediately to the reduced equation $Ly(x) = 0$ of DE (4.33) from the fact that L is a linear operator.

Theorem 4-VIII If $y_1, y_2, \ldots, y_n$ are any n solutions of $Ly(x) = 0$, then $y = c_1 y_1 + c_2 y_2 + \cdots + c_n y_n$, where $c_1, c_2, \ldots, c_n$ are arbitrary constants, is also a solution.

Example 2. The DE $(D^3 - 2D^2 - D + 2)y(x) = 0$ has solutions defined by $y_1 = e^x$, $y_2 = e^{-x}$, and $y_3 = e^{2x}$. Hence $y_4 = c_1 e^x + c_2 e^{-x} + c_3 e^{2x}$ also defines a solution.

Linear dependence was defined in Section 4.3 as follows:

Definition A set of n functions $y_1, y_2, \ldots, y_n$ that are not linearly independent on an interval I are said to be *linearly dependent* on I.

Theorem 4-III generalizes to the following:

Theorem 4-IX There exist n linearly independent solutions of the reduced equation $Ly(x) = 0$. These solutions are valid on any interval I on which $p_1, p_2, \ldots, p_n$ are continuous.

The Wronskian W of n functions $y_1, y_2, \ldots, y_n$, each of which can be differentiated $(n - 1)$ times on an interval I, is defined on I by the determinant

$$W = W(y_1, y_2, \ldots, y_n) = \begin{vmatrix} y_1 & y_2 & y_3 & \cdots & y_n \\ y_1' & y_2' & y_3' & \cdots & y_n' \\ y_1'' & y_2'' & y_3'' & \cdots & y_n'' \\ \vdots & \vdots & \vdots & & \vdots \\ y_1^{(n-1)} & y_2^{(n-1)} & y_3^{(n-1)} & \cdots & y_n^{(n-1)} \end{vmatrix} \tag{4.34}$$

For n linearly independent solutions $y_1, y_2, \ldots, y_n$ of $Ly(x) = 0$ on an interval I, Abel's identity generalizes to

$$W = W(x) = W(x_0) \exp\left[-\int_{x_0}^{x} p_1(t)\, dt \right] \tag{4.35}$$

As in the case $n = 2$, x_0 is an arbitrary point of I and W is everywhere positive, everywhere negative, or identically zero on I.

The Wronskian W can be obtained by employing any of the standard methods for evaluating a determinant. When $n = 3$,

$$W = \begin{vmatrix} y_1 & y_2 & y_3 \\ y_1' & y_2' & y_3' \\ y_1'' & y_2'' & y_3'' \end{vmatrix} = y_1 \begin{vmatrix} y_2' & y_3' \\ y_2'' & y_3'' \end{vmatrix} - y_1' \begin{vmatrix} y_2 & y_3 \\ y_2'' & y_3'' \end{vmatrix} + y_1'' \begin{vmatrix} y_2 & y_3 \\ y_2' & y_3' \end{vmatrix}$$

$$= y_1(y_2' y_3'' - y_2'' y_3') - y_1'(y_2 y_3'' - y_2'' y_3) + y_1''(y_2 y_3' - y_2' y_3)$$

$$= y_1 y_2' y_3'' + y_2 y_3' y_1'' + y_3 y_1' y_2'' - y_3 y_2' y_1'' - y_2 y_1' y_3'' - y_1 y_3' y_2''$$

Example 3. Find the Wronskian of the functions defined by $y_1 \equiv 1$, $y_2 = x$, and $y_3 = x^2$.

$$W(x) = \begin{vmatrix} 1 & x & x^2 \\ 0 & 1 & 2x \\ 0 & 0 & 2 \end{vmatrix} = 1 \begin{vmatrix} 1 & 2x \\ 0 & 2 \end{vmatrix} - (0) \begin{vmatrix} x & x^2 \\ 0 & 2 \end{vmatrix} + (0) \begin{vmatrix} x & x^2 \\ 1 & 2x \end{vmatrix}$$

$$\equiv 1(2 - 0) = 2$$

Theorem 4-IV, used to test linear independence when $n = 2$, has the following generalization.

Theorem 4-X If each of $y_1, y_2, \ldots, y_n$ can be differentiated $(n - 1)$ times on an interval I, and $W(y_1, y_2, \ldots, y_n)$ is not identically zero on I, then $y_1, y_2, \ldots, y_n$ are linearly independent on I.

EXAMPLE 4. The functions of Example 3, given by $y_1 \equiv 1$, $y_2 = x$, and $y_3 = x^2$, are linearly independent on an arbitrary interval I since $W \equiv 2$ is not identically zero on I.

EXAMPLE 5. The functions given by $y_1 = e^x$, $y_2 = \sin x$, and $y_3 = \cos x$ are linearly independent on an arbitrary interval I since

$$W(x) = \begin{vmatrix} e^x & \sin x & \cos x \\ e^x & \cos x & -\sin x \\ e^x & -\sin x & -\cos x \end{vmatrix}$$

$$= e^x(-\cos^2 x - \sin^2 x) - e^x(-\sin x \cos x + \sin x \cos x)$$
$$+ e^x(-\sin^2 x - \cos^2 x) = -2e^x < 0$$

is not identically zero on I.

Theorem 4-V generalizes to the following.

Theorem 4-XI If $y_1, y_2, \ldots, y_n$ are solutions of $Ly(x) = 0$ on a given interval I, and if $W(y_1, y_2, \ldots, y_n) \equiv 0$ on I, then $y_1, y_2, \ldots, y_n$ are linearly dependent on I; that is, they are not linearly independent on I.

Theorems 4-VI and 4-VII become the following.

Theorem 4-XII If $y_1, y_2, \ldots, y_n$ are *any* n linearly independent solutions of $Ly(x) = 0$ on a given interval I, then *every* solution on I of $Ly(x) = 0$ can be expressed in the form $y = c_1 y_1 + c_2 y_2 + \cdots + c_n y_n$ where $c_1, c_2, \ldots, c_n$ are arbitrary constants.

Theorem 4-XIII If $y = c_1 y_1 + c_2 y_2 + \cdots + c_n y_n$ defines a complete solution of $Ly(x) = 0$ on a given interval I, and $y_p = \psi(x)$ defines *any* particular solution of $Ly = \rho$ on I, then *every* solution of $Ly = \rho$ on I can be expressed in the form $y = c_1 y_1 + c_2 y_2 + \cdots + c_n y_n + y_p$.

EXAMPLE 6. It is seen by inspection that the functions defined by $y_1 \equiv 1$, $y_2 = x$, and $y_3 = x^2$ satisfy $D^3 y(x) = 0$ on $I = (-\infty, +\infty)$. In Example 4 we found that y_1, y_2, and y_3 are linearly independent on I. By Theorem 4-XII $y = c_1(1) + c_2 x + c_3 x^2$ defines a complete solution of $D^3 y(x) = 0$ on I.

It is also seen by inspection that the function defined by $y_p = e^x$ is a solution of $D^3 y(x) = e^x$ on I. By Theorem 4-XIII $y = c_1(1) + c_2(x) + c_3(x^2) + e^x$ defines a complete solution of $D^3 y(x) = e^x$ on I.

In Section 4.5 we presented a method for reducing the order of a homogeneous DE. The method applied when one solution of the DE was known. The process reduced the solution of a second-order DE to the solution of a first-order DE and was very effective since we were able to solve first-order DE. Theoretically, the method of reduction of order can be extended to DE of higher order. The solution of a fourth-order DE can be made to depend on the solution of a third-order DE. Applying the method a second time, the solution depends on being able to solve a second-order DE. A third application reduces the problem to that of solving a first-order DE. Practically, the method is seldom used since its application for $n > 2$ is generally quite complicated.

The method of solving second-order linear DE with constant coefficients extends readily to DE of the form $Ly = (D^n + a_1 D^{n-1} + a_2 D^{n-2} + \cdots + a_{n-1} D + a_n)y = \rho$ where $a_1, a_2, \ldots, a_n$ are real constants. We find that $y = e^{rx}$ defines a solution of $Ly(x) = 0$ if r is a root of the characteristic equation

$$g(r) = r^n + a_1 r^{n-1} + \cdots + a_{n-1} r + a_n = 0 \qquad (4.36)$$

For $n = 2$, the two roots (not necessarily distinct) are easily found from the quadratic formula. For larger n, we need some facts concerning the solutions of algebraic equations. It is known that the equation $g(r) = 0$ has exactly n roots $r_1, r_2, \ldots, r_n$. Some of these roots may be repeated and some may not be real numbers.

EXAMPLE 7. The equation $r^3 - r = r(r - 1)(r + 1) = 0$ has roots 0, 1, and -1.

EXAMPLE 8. The equation $r^3 - r^2 - 8r + 12 = (r - 2)^2(r + 3) = 0$ has roots 2, 2, and -3.

EXAMPLE 9. The equation $r^4 - 1 = (r^2 - 1)(r^2 + 1) = 0$ has roots 1, -1, i, and $-i$, where $i = \sqrt{-1}$.

It is also known that if $\alpha + i\beta$ is a root of $g(r) = 0$, where α and β are real and $i = \sqrt{-1}$, then the conjugate $\alpha - i\beta$ is also a root of $g(r) = 0$. This result depends on the fact that the coefficients $a_1, a_2, \ldots, a_n$ are real.

If r_1 is a root of $g(r) = 0$, then $x - r_1$ is a factor of $g(r)$. If $r_2 = (\alpha + i\beta)$ is a root of $g(r) = 0$, then $x - r_2 = x - \alpha - i\beta$ is a factor of $g(r)$. Since $g(\alpha - i\beta) = 0$, $x - r_3 = x - (\alpha - i\beta) = x - \alpha + i\beta$ is also a factor of $g(r)$. But

$$(x - r_2)(x - r_3) = [(x - \alpha) - i\beta][(x - \alpha) + i\beta] = x^2 - 2\alpha + \alpha^2 + \beta^2$$

Thus, it is sometimes possible to solve $g(r) = 0$ by factoring $g(r)$ into linear and quadratic factors that involve real numbers. When the roots of $g(r) = 0$ cannot be found by factoring $g(r)$, approximate roots, real or complex of the form $i\beta$ ($\beta \neq 0$), must be found. These in turn lead to approximate solutions of $Ly(x) = 0$.

If the roots $r_1, r_2, \ldots, r_n$ of $g(r) = 0$ are distinct, the functions given by $y_1 = e^{r_1 x}$, $y_2 = e^{r_2 x}, \ldots, y_n = e^{r_n x}$ are linearly independent and Theorem 4-XII provides a complete solution of $Ly(x) = 0$. If $r_1 = r_2$ is a double real root of $g(r) = 0$, then a set of n linearly independent solutions is obtained by replacing $y_2 = e^{r_2 x}$ by $y_2 = xe^{r_1 x}$. If $r_1 = r_2 = r_3$ is a triple real root of $g(r) = 0$, a case not encountered when $n = 2$, it seems reasonable to try $y_3 = x^2 e^{r_1 x}$. It develops that y_3 is a solution of $Ly(x) = 0$ and the n functions given by $y_1 = e^{r_1 x}$, $y_2 = xe^{r_1 x}$, $y_3 = x^2 e^{r_1 x}$, $y_4 = e^{r_4 x}, \ldots, y_n = e^{r_n x}$ are linearly independent, assuming that $r_4, r_5, \ldots, r_n$ are distinct.

In general, if r_1 is a real root of multiplicity k, we incorporate the k functions given by $y_1 = e^{r_1 x}$, $y_2 = xe^{r_1 x}$, $y_3 = x^2 e^{r_1 x}$, $y_4 = x^3 e^{r_1 x}, \ldots$, $y_k = x^{k-1} e^{r_1 x}$ in a complete solution of $Ly(x) = 0$.

It can be shown that each member of the set $(y_1, y_2, \ldots, y_k, y_{k+1}, \ldots, y_n)$ is a solution of $Ly(x) = 0$ and that the n functions are linearly independent.

If $\alpha + i\beta$ is a repeated root ($\beta \neq 0$), then $\alpha - i\beta$ is also a repeated root, and we employ the functions given by $e^{\alpha x} \sin \beta x$, $xe^{\alpha x} \sin \beta x$, $e^{\alpha x} \cos \beta x$, and $xe^{\alpha x} \cos \beta x$.

If $\alpha + i\beta$ is a triple root of $g(x) = 0$, we employ the functions given by $e^{\alpha x} \sin \beta x$, $xe^{\alpha x} \sin \beta x$, $x^2 e^{\alpha x} \sin \beta x$, $e^{\alpha x} \cos \beta x$, $xe^{\alpha x} \cos \beta x$, and $x^2 e^{\alpha x} \cos \beta x$. The procedure extends readily to cases in which $\alpha + i\beta$ is a multiple root of order higher than three. This method yields a set of n linearly independent solutions of $Ly(x) = 0$.

The method of undetermined coefficients is often useful in finding a particular solution y_p of the nth-order constant coefficient DE $Ly = \rho$. The method of variation of parameters is also applicable and is more general in application but is usually more complicated to apply. After a particular solution y_p of $Ly = \rho$ is obtained, a complete solution of $Ly = \rho$ is provided by Theorem 4-XIII.

Algebraic difficulties may arise in solving the characteristic equation $r^n + a_1 r^{n-1} + \cdots + a_{n-1} r + a_n = 0$. In each of the following examples, the characteristic equation is simple and is solved by factoring.

EXAMPLE 10. Find a complete solution of $(D^3 - 7D + 6)y(x) = 0$.

Solution: From $r^3 - 7r + 6 = (r - 1)(r - 2)(r + 3) = 0$, we obtain $r_1 = 1$, $r_2 = 2, r_3 = -3$. A complete solution of the DE is given by

$$y = c_1 e^x + c_2 e^{2x} + c_3 e^{-3x}$$

EXAMPLE 11. Find a complete solution of $(D^3 - 3D - 2)y(x) = 0$.

Solution: From $r^3 - 3r - 2 = (r + 1)^2(r - 2) = 0$, we obtain $r_1 = r_2 = -1$ and $r_3 = 2$. A complete solution of the DE is given by

$$y = c_1 e^{-x} + c_2 x e^{-x} + c_3 e^{2x}$$

EXAMPLE 12. Find a complete solution of $(D^3 + D)y(x) = 0$.

Solution: From $r^3 + r = r(r^2 + 1) = 0$, we obtain $r_1 = 0$, $r_2 = 0 + i$, and $r_3 = 0 - i$. A complete solution of the DE is given by

$$y = c_1 + c_2 \sin x + c_3 \cos x$$

EXAMPLE 13. Find a complete solution of $(D^4 + 8D^2 + 16)y(x) = 0$.

Solution: From $r^4 + 8r^2 + 16 = (r^2 + 4)^2 = 0$, we obtain $r_1 = r_2 = 2i$ and $r_3 = r_4 = -2i$. A complete solution of the DE is given by

$$y = c_1 \sin 2x + c_2 \cos 2x + c_3 x \sin 2x + c_4 x \cos 2x$$

EXAMPLE 14. Find a complete solution of $[(D + 3)(D^2 + 2D + 5)^3]y(x) = 0$.

Solution: From $(r + 3)(r^2 + 2r + 5)^3 = 0$, we obtain $r_1 = -3$, $r_2 = r_3 = r_4 = -1 + 2i$, and $r_5 = r_6 = r_7 = -1 - 2i$. A complete solution is given by $y = c_1 e^{-3x} + e^{-x}(c_2 \sin 2x + c_3 \cos 2x) + x e^{-x}(c_4 \sin 2x + c_5 \cos 2x) + x^2 e^{-x}(c_6 \sin 2x + c_7 \cos 2x)$.

EXAMPLE 15. Solve the initial-value problem:

$$Ly(x) = (D^3 + D)y(x) = 20e^{-2x}, \quad y(0) = y'(0) = y''(0) = 0$$

Solution: From $r(r^2 + 1) = 0$, we obtain $r_1 = 0$, $r_2 = i$, and $r_3 = -i$. A complete solution of $Ly(x) = 0$ is given by $y = c_1 + c_2 \sin x + c_3 \cos x$. Trying $y_p = Ae^{-2x}$ in $Ly(x) = 20e^{-2x}$, we obtain $y'_p = -2Ae^{-2x}$, $y''_p = 4Ae^{-2x}$, $y'''_p = -8e^{-2x}$, $-8Ae^{-2x} - 2Ae^{-2x} = 20e^{-2x}$, and $A = -2$. A complete solution of $Ly(x) = 20e^{-2x}$ is given by $y = c_1 + c_2 \sin x + c_3 \cos x - 2e^{-2x}$.

From

$$y' = c_2 \cos x - c_3 \sin x + 4e^{-2x}$$

$$y'' = -c_2 \sin x - c_3 \cos x - 8e^{-2x}$$

and the initial conditions, we obtain

$$y(0) = c_1 + c_3 - 2 = 0, \quad y'(0) = c_2 + 4 = 0, \quad y''(0) = -c_3 - 8 = 0$$

These three equations yield $c_1 = 10$, $c_2 = -4$, and $c_3 = -8$. The initial-value problem has the unique solution given by $y = 10 - 4 \sin x - 8 \cos x - 2e^{-2x}$.

EXAMPLE 16. Find a particular solution of $(D + 1)^3 y(x) = x^3 e^{-x}$.

Solution:

$$(D + 1)^2 y(x) = \frac{1}{D + 1}(x^3 e^{-x})$$

$$= e^{-x} \int e^x (x^3 e^{-x})\, dx = \frac{x^4 e^{-x}}{4}$$

$$(D + 1)y(x) = \frac{1}{D + 1}\left(\frac{x^4 e^{-x}}{4}\right)$$

$$= e^{-x} \int e^x \left(\frac{x^4 e^{-x}}{4}\right) dx = \frac{x^5 e^{-x}}{20}$$

$$y(x) = \frac{1}{D + 1}\left(\frac{x^5 e^{-x}}{20}\right) = e^{-x} \int e^x \left(\frac{x^5 e^{-x}}{20}\right) dx = \frac{x^6 e^{-x}}{120}$$

Problem List 4.6

1. Find the Wronskian of the functions given by:
   (a) $1, x, x^3$
   (b) e^x, e^{-2x}, e^{3x}
   (c) $e^{r_1 x}, e^{r_2 x}, e^{r_3 x}$
   (d) $e^{-x}, \sin x, \cos x$
   $(r_1, r_2, r_3$ distinct)
   (e) $1, x, x^2, x^3$
   (f) $\sin^2 x, \cos^2 x, \cos 2x$

2. Show that the functions defined by $y_1 = e^{-2x}$, $y_2 = \sin x$, and $y_3 = \cos x$ are linearly independent on $(-\infty, +\infty)$.

3. Show that the functions defined by $y_1 \equiv 1$, $y_2 = x$, $y_3 = x^2$, and $y_4 = \ln x$ are linearly independent on $(0, +\infty)$.

4. Show that the functions defined by $y_1 \equiv 1$, $y_2 = x$, $y_3 = xe^x$, and $y_4 = x^2 e^x$ are linearly independent on $(-\infty, +\infty)$.

5. Show that the functions defined by $y_1 = e^x$, $y_2 = e^{-x}$, and $y_3 = \sinh x$ are not linearly independent on $(-\infty, +\infty)$.

6. Prove, without employing the Wronskian, that the $n + 1$ functions defined by $y_1 \equiv 1$, $y_2 = x$, $y_3 = x^2, \ldots, y_n = x^{n-1}$, $y_{n+1} = x^n$, where n is a positive integer, are linearly independent on $(-\infty, +\infty)$.

7. Find a complete solution of:
   (a) $(D^3 - D)y(x) = 0$
   (b) $(D^3 - 2D^2 - 5D + 6)y(x) = 0$

8. Find a complete solution of:
   (a) $(D^3 - D^2 - 9D + 9)y(x) = 0$
   (b) $y^{(iv)} - 5y'' + 4y = 0$
   (c) $(D^4 + D)y(x) = 0$
   (d) $(D^2 - 4)(D^2 + 6D + 13)y(x) = 0$
   (e) $D^4 y(x) = 0$
   (f) $y''' - 3y'' = 0$
   (g) $(D^4 + 8D + 16)y(x) = 0$
   (h) $(D + 4)(D + 2)^2(D - 3)^3 y(x) = 0$

9. A third-order linear constant coefficient DE $Ly(x) = 0$ has a complete solution given by $y = c_1 e^{2x} + c_2 \sin 3x + c_3 \cos 3x$. Write an expression in terms of undetermined coefficients that should be used as a trial solution to find a

particular solution of each of the following DE. Do not evaluate the unknown coefficients.

(a) $Ly(x) = 6x + 3$ (b) $Ly(x) = 7e^{2x}$

(c) $Ly(x) = 7e^{-3x}$ (d) $Ly(x) = 4\cos 3x$

(e) $Ly(x) = 3\cos 4x$ (f) $Ly(x) = 2\sin x$

(g) $Ly(x) = 3\sin 4x - \cos 4x$ (h) $Ly(x) = 4\sin x \cos x$

10. Find a complete solution of:

(a) $(D^3 - 3D^2)y(x) = 60e^{-2x}$

(b) $(D^3 - D)y(x) = 4\sin x$

(c) $D''y(x) = 5e^x$

(d) $(D^3 + 4D)y(x) = 24\sin 2x$

(e) $D^4 y(x) = 288 \ln x$

11. Find the unique solution of the initial-value problem:

(a) $D^3 y(x) = e^{-x}$, $y(0) = y'(0) = y''(0) = 0$

(b) $(D^3 + D)y(x) = 12\sin 2x$, $y(0) = y'(0) = y''(0) = 0$

(c) $D^3 y(x) = x^{-2}$, $y(1) = y'(1) = y''(1) = 0$

(d) $D^2(D - 1)^2 y(x) = 6$, $y(0) = y'(0) = y''(0) = y'''(0) = 0$

12. Show that $W(e^{r_1 x}, e^{r_2 x}, e^{r_3 x})$ is never zero if and only if r_1, r_2, and r_3 are distinct.

13. Use an operator method to find a particular solution of:

(a) $(D^3 + 2D + 1)y(x) = e^{-x}$ (b) $(D + 1)^3 y(x) = 120xe^{-x}$

(c) $(D^3 - D^2)y(x) = \cos x$ (d) $(D^4 - 2D^2 + 1)y(x) = 16e^{-x}$

14. For the DE $(D - 1)^2(D + 1)y(x) = e^{2x} + 4$:

(a) Find a particular solution.

(b) Find a complete solution.

15. Use (4.30) to find a complete solution of $(D - a)^n y = 0$ where n is a positive integer.

4.11 The Euler Equation

An ordinary linear DE

$$Ly(x) = [a_0(x)D^n + a_1(x)D^{n-1} + \cdots + a_{n-1}(x)D + a_n(x)]y(x) = f(x)$$

is generally quite difficult to solve. Finding even one solution of $Ly(x) = 0$ can present a formidable challenge. In Chapter 8 we discuss solutions of $Ly(x) = 0$ that are defined by infinite series.

It is sometimes possible to reduce a variable coefficient DE to a DE with constant coefficients by employing a suitable change of variables. An effective substitution with this property is often difficult to find. A simple illustration is afforded by the Euler DE

$$Ly(x) = (x^n D_x^n + a_1 x^{n-1} D_x^{n-1} + \cdots + a_{n-1} x D_x + a_n)y(x) = \rho(x) \quad (4.37)$$

where x and D_x have the same exponent in each term. Restricting ourselves to an interval on which $x > 0$, we eliminate the variable x by the substitution

$$x = e^t \quad \text{or} \quad t = \ln x$$

Denoting d/dt by D and noting that $dt/dx = 1/x$, we obtain

$$\frac{dy}{dx} = \frac{dy}{dt}\frac{dt}{dx} = \frac{dy}{dt}\frac{1}{x} = \frac{1}{x}Dy$$

$$\frac{d^2y}{dx^2} = \frac{d}{dx}\left(\frac{dy}{dt}\frac{1}{x}\right) = \frac{dy}{dt}\left(-\frac{1}{x^2}\right) + \frac{1}{x}\frac{d^2y}{dt^2}\frac{dt}{dx}$$

$$= \frac{1}{x^2}\frac{d^2y}{dt^2} - \frac{1}{x^2}\frac{dy}{dt} = \frac{1}{x^2}D(D-1)y$$

$$\frac{d^3y}{dx^3} = \frac{d}{dx}\left[\frac{1}{x^2}\left(\frac{d^2y}{dt^2} - \frac{dy}{dt}\right)\right]$$

$$= \frac{1}{x^2}\left(\frac{d^3y}{dt^3} - \frac{d^2y}{dt^2}\right)\frac{dt}{dx} + \left(\frac{d^2y}{dt^2} - \frac{dy}{dt}\right)\left(\frac{-2}{x^3}\right)$$

$$= \frac{1}{x^3}\left(\frac{d^3y}{dt^3} - 3\frac{d^2y}{dt^2} + 2\frac{dy}{dt}\right) = \frac{1}{x^3}D(D-1)(D-2)y$$

$$\vdots$$

$$\frac{d^ny}{dx^n} = \frac{1}{x^n}D(D-1)(D-2)\cdots(D-n+1)y \qquad (4.38)$$

A proof of (4.38) may be constructed by mathematical induction. See Problem 7.

Substitution from (4.38) into the Euler DE (4.37) yields a constant coefficient DE involving y and t that is solved by the methods of Section 4.10. Replacing t by $\ln x$, we obtain a solution of the Euler DE.

On an interval on which $x < 0$, the substitution $x = -e^t$ or $t = \ln(-x)$ is used.

EXAMPLE 1. Find a complete solution of $x^2y'' + xy' = 0$ on $(0, +\infty)$.

Solution: Substituting for y' and y'' from (4.38), we obtain

$$x^2x^{-2}D(D-1)y + xx^{-1}Dy = 0$$

from which

$$[D(D-1) + D]y = D^2y = 0$$

from which $y = c_1 + c_2t$.
A complete solution is given by $y(x) = c_1 + c_2\ln x$.

EXAMPLE 2. Find a complete solution of

$$x^3y''' + 5x^2y'' + 2xy' - 2y = 0 \quad \text{on } (0, +\infty)$$

Solution: Substituting for y', y'', and y''' from (4.38), we obtain

$$x^3x^{-3}D(D-1)(D-2)y + 5x^2x^{-2}D(D-1)y + 2xx^{-1}Dy - 2y = 0$$

Simplifying,

$0^2 - 20 \not{5} D + 2$

$$(D - 1)[D(D - 2) + 5D + 2]y = (D - 1)(D + 1)(D + 2)y = 0$$

from which

$$(r - 1)(r + 1)(r + 2) = 0; \quad r = 1, -1, \text{ and } -2$$

and

$$y = c_1 e^t + c_2 e^{-t} + c_3 e^{-2t}$$

Replacement of t by $\ln x$ yields

$$y(x) = c_1 \exp(\ln x) + c_2 \exp(-\ln x) + c_3 \exp(-2 \ln x)$$

or

$$y(x) = c_1 x + c_2 x^{-1} + c_3 x^{-2}$$

This defines a complete solution since $W(x, x^{-1}, x^{-2}) \neq 0$ on $(0, +\infty)$. See Problem 2.

The Euler DE was discovered by Euler; it is sometimes attributed to Cauchy and it is often referred to as the Cauchy DE or the Euler–Cauchy DE. Although the DE is somewhat special, it plays an important role in the advanced theory of DE.

Problem List 4.7

1. Find a complete solution of $x^2 y'' + xy' = 0$, the DE of Example 1, on $(-\infty, 0)$.

2. Show that the solution in Example 2 is a complete solution; that is, show that $W(x, x^{-1}, x^{-2}) \neq 0$ on $(0, +\infty)$.

3. Find a complete solution on $(0, +\infty)$ of:
   (a) $x^2 y'' - 2y = 0$
   (b) $xy' + 4y = 0$
   (c) $x^2 y'' + \frac{1}{4} y = 0$
   (d) $x^2 y'' + xy' + y = 0$
   (e) $x^3 y''' + 4x^2 y'' = 0$
   (f) $x^3 y''' + 6x^2 y'' + 5xy' + 3y = 0$

4. Find a complete solution on $(0, +\infty)$ of the DE

$$x^2 y'' + xy' - 9y = 8x$$

5. Show that if $y = x^r$ defines a solution of $x^2 y'' + \alpha xy' + \beta y = 0$, $x > 0$, then r is a root of

$$r^2 + (\alpha - 1)r + \beta = 0$$

6. Employ the substitution $u = x + 3$ to find a complete solution on $(-3, +\infty)$ of $(x + 3)^2 y''(x) + 5(x + 3)y'(x) + 4y(x) = 0$.

7. Prove (4.38) by mathematical induction.

160   *Linear Differential Equations*

8. Find the solution on $(0, +\infty)$ of the initial-value problem

$$x^3 y'' + 2x^2 = (xy' - y)^2; \quad y(1) = 0, \quad y'(1) = 2$$

Hint: Let $y = -x \ln u$ where $u > 0$.

References

4.1 Greenspan, Donald. 1960. *Theory and Solution of Ordinary Differential Equations.* New York: Macmillan.

4.2 Heaviside, Oliver. 1950. *Electromagnetic Theory.* New York: Dover.

4.3 Ince, E. L. 1959. *Ordinary Differential Equations.* New York: Dover.

4.4 Kaplan, Wilfred. 1962. *Ordinary Differential Equations.* Reading, Mass.: Addison-Wesley.

4.5 Nehari, Z. 1968. *Introduction to Complex Analysis.* Boston: Allyn and Bacon.

5

Applications of
Second-Order
Equations

Rectilinear Motion; The Harmonic Oscillator

The motion of a particle moving in a straight line is governed by a DE of the form

$$G\left(t, x, \frac{dx}{dt}, \frac{d^2x}{dt^2}\right) = 0 \tag{1.9}$$

This follows from Newton's second law. (See Section 1.3.) We have solved problems in which (1.9) had the particularly simple form $d^2x/dt^2 = f(t)$. We now consider problems whose solutions involve various other forms of (1.9).

In Figure 5.1, a small block of mass m is attached to one end of a spring, and the other end of the unstretched spring is attached to a fixed wall. The block is pulled to the right, stretching the spring x_0 feet, and is then released from rest (with initial velocity zero). The block moves on a smooth (frictionless) plane.

According to *Hooke's law*, named after the English experimental scientist Robert Hooke (1635–1703), the force in pounds that stretches or compresses a spring is proportional to the change in length of the spring. This law applies approximately to all elastic materials, provided that the extension (or compression) does not exceed the so-called elastic limit of the material. Let us denote the magnitude $|\mathbf{F}|$ of this force by $k|x|$; $k > 0$ is known as the *spring constant*. The force $\mathbf{F} = (kx)\mathbf{i}$ is called a *restoring force*, since it always acts towards the origin and tends to restore the block

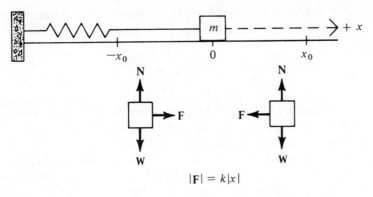

$$|\mathbf{F}| = k|x|$$

Fig. 5.1

to its original (or *equilibrium*) position at $x = 0$. Note that $kx > 0$ when $x > 0$, $kx = 0$ when $x = 0$, and $kx < 0$ when $x < 0$.

The force system acting on the block when the block is to the right (and also to the left) of the equilibrium position is shown in Figure 5.1. The vector $\mathbf{W}$ represents the force the earth exerts on the block and the vector $\mathbf{N}$ represents the force the plane exerts on the block.

Each of the vectors $\mathbf{W}$ and $\mathbf{N}$ has magnitude w, the weight of the block in pounds.

By Newton's second law, the unbalanced force acting on the block in the x direction equals the mass of the block times its acceleration in the x direction. Since the unbalanced force in the x direction has value $-kx$,

$$m\frac{d^2x}{dt^2} = -kx \tag{5.1}$$

whether x is positive, negative, or zero.

The mass m is measured in slugs and is obtained by dividing the weight w of the block in pounds by g, the acceleration of gravity in feet per second squared. That is, $m = w/g$. In numerical work we shall use the approximation $g \approx 32 \text{ ft/sec}^2$. The DE governing the motion of the block can now be written

$$\frac{d^2x}{dt^2} + \frac{kg}{w}x = 0 \tag{5.2}$$

This is a linear DE with constant coefficients. The characteristic equation $r^2 + (kg/w) = 0$ has roots $\pm i\sqrt{kg/w}$ and (5.2) has a complete solution given by

$$x = c_1 \sin\sqrt{kg/w}\,t + c_2 \cos\sqrt{kg/w}\,t \tag{5.3}$$

Substituting $t = 0$ and $x = x_0$, we obtain $c_2 = x_0$.
Differentiating (5.3) with respect to t yields

$$v = \frac{dx}{dt} = c_1\sqrt{\frac{kg}{w}}\cos\sqrt{\frac{kg}{w}}\,t - c_2\sqrt{\frac{kg}{w}}\sin\sqrt{\frac{kg}{w}}\,t \tag{5.4}$$

Substituting $v = 0$ and $t = 0$, we obtain $c_1 = 0$. Hence the particular solution of (5.2) satisfying the initial conditions we have imposed is given by

$$x = x_0 \cos \sqrt{\frac{kg}{w}} \, t \qquad (5.5)$$

Equation (5.5) reveals that the block moves forever back and forth between the points $x = x_0$ and $x = -x_0$. The motion is said to be *simply harmonic*, Equation (5.2) is termed the *DE of a simple harmonic motion*, and the block is called a *harmonic oscillator*. The time required for the block to go from $x = x_0$ to $x = -x_0$ and back again is called the *period* of the motion. It is equal to the fundamental period of the periodic function given by

$$\psi(t) = \cos \sqrt{\frac{kg}{w}} \, t$$

and has the value

$$T = \frac{2\pi}{\sqrt{kg/w}} = 2\pi \sqrt{\frac{w}{kg}} \sec \qquad (5.6)$$

The reciprocal of T denotes the number of complete oscillations or cycles of the block per second and is called the *frequency* of the motion.

The graph of $x = x_0 \cos \sqrt{kg/w} \, t$ is shown in Figure 5.2. The maximum displacement x_0 of the block from the equilibrium position $x = 0$ is called the *amplitude* of the motion.

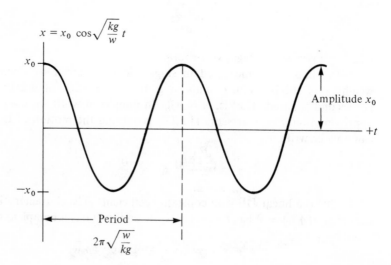

Fig. 5.2

EXAMPLE 1. Assume that the spring of Figure 5.1 is stretched 1 in. by a force of 12 lb. For a block weighing 2 lb and released from rest at $x = 3$ in., find the amplitude and period of the resulting motion.

Solution: Letting $f = 12$ and $s = \frac{1}{12}$ in $f = ks$ (Hooke's law), we find that $k = 144$ lb/ft. The DE (5.2) becomes

$$\frac{d^2x}{dt^2} + \frac{(144)(32)}{2}x = 0$$

Since $x = \frac{1}{4}$ ft when $t = 0$, the required particular solution is given by (5.5), which reduces to $x = \frac{1}{4}\cos 48t$.

The motion has amplitude $\frac{1}{4}$ ft and period $2\pi/48 \approx 0.13$ sec. The graph in Figure 5.3 shows how the displacement x varies with the time t during one period of the motion.

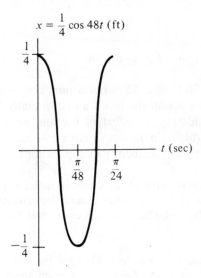

$x = \dfrac{1}{4}\cos 48t$ (ft)

Fig. 5.3

The solution of DE (5.2) can also be obtained by the method of Section 2.5.

Letting $\omega^2 = kg/w$ and writing the acceleration in the form

$$\frac{d^2x}{dt^2} = \frac{dv}{dt} = \frac{dv}{dx}\cdot\frac{dx}{dt} = v\frac{dv}{dx}$$

we obtain $v(dv/dx) + \omega^2 x = 0$.

Separating the variables and integrating yields

$$v\,dv = -\omega^2 x\,dx, \qquad \int_0^v z\,dz = -\omega^2 \int_{x_0}^x u\,du, \qquad \frac{v^2}{2} = \frac{x_0^2\omega^2}{2} - \frac{\omega^2 x^2}{2}$$

and

$$v = \frac{dx}{dt} = \pm\omega\sqrt{x_0^2 - x^2}$$

This is an intermediate integral of $d^2x/dt^2 + \omega^2 x = 0$. Again separating the variables and integrating, we obtain

$$\frac{dx}{\sqrt{x_0^2 - x^2}} = \pm \omega \, dt$$

$$\sin^{-1} \frac{x}{x_0} = \pm \omega t + \phi$$

and

$$x = x_0 \sin(\pm \omega t + \phi)$$
$$= x_0 \cos \phi \sin(\pm \omega t) + x_0 \sin \phi \cos(\pm \omega t)$$

which is of the form (5.3),

$$x = c_1 \sin \omega t + c_2 \cos \omega t$$

The motion of the block in Figure 5.1 is not a pure simple harmonic motion since some friction must act on the block and eventually the block will come to rest. Before considering the effect of friction, we present an example showing one way in which a pure simple harmonic motion can be induced. This kind of motion appears frequently in applications.

EXAMPLE 2. In Figure 5.4, the point $P(x, y)$ moves around the circle of radius r at constant speed $|\mathbf{V}|$. Assuming that P starts at P_0 and moves counterclockwise around the circle, show that the projection $P(x, 0)$ of $P(x, y)$ undergoes a simple harmonic motion on the x axis.

Solution: Let s denote the arc length from P_0 to P. Then $s = r(\theta - \phi) = r\theta - r\phi$, and $ds/dt = |\mathbf{V}| = r(d\theta/dt) = $ constant. Thus $d\theta/dt = \omega$ is constant and $\theta = \omega t + \phi$

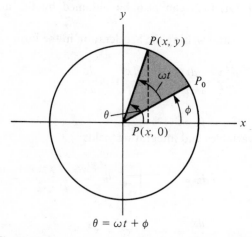

$$\theta = \omega t + \phi$$

Fig. 5.4

since $\theta = \phi$ when $t = 0$. We now have

$$x = r \cos(\omega t + \phi)$$

$$v = \frac{dx}{dt} = -r\omega \sin(\omega t + \phi)$$

and

$$a = \frac{dv}{dt} = \frac{d^2x}{dt^2} = -r\omega^2 \cos(\omega t + \phi)$$

Since $d^2x/dt^2 = -\omega^2 x$,

$$x = c_1 \sin \omega t + c_2 \cos \omega t$$

$$v = \frac{dx}{dt} = c_1 \omega \cos \omega t - c_2 \omega \sin \omega t$$

and $P(x, 0)$ undergoes a simple harmonic motion. If $x = r$ and $v = 0$ when $t = 0$, then $c_2 = r$, $c_1 = 0$, and $x = r \cos \omega t$. This situation would prevail when $\phi = 0$.

We next assume that the block in Figure 5.1 is acted on by a force of friction that opposes the motion. It has been determined experimentally that the magnitude of frictional resistance is proportional to $|dx/dt|^\alpha$ where $1 \leq \alpha \leq 2$. See Reference 5.5. For simplicity we shall use $\alpha = 1$, a reasonable assumption when $|dx/dt|$ is small.

Thus, the resistance in the positive x direction is assumed to be given by $-c(dx/dt)$, where c, called the *resistance coefficient*, is positive. Note that when the block moves to the right $(dx/dt > 0)$, the resistance acts to the left; and when the block moves to the left $(dx/dt < 0)$, the resistance acts to the right. The resistance is zero at any instant when $dx/dt = 0$.

Applying Newton's second law, we obtain, instead of (5.1),

$$m\frac{d^2x}{dt^2} = -kx - c\frac{dx}{dt} \tag{5.7}$$

or

$$\frac{d^2x}{dt^2} + \frac{cg}{w}\frac{dx}{dt} + \frac{kg}{w}x = 0 \tag{5.8}$$

as the DE governing the motion. The character of the motion is determined by the roots of the characteristic equation of (5.8). If these roots are real and unequal, the motion is *overdamped*. The resistance is enough to prevent the block from oscillating. If the roots are real and equal, the motion is *critically damped*. Again no oscillation takes place. Finally, if the roots are conjugate complex numbers $\alpha \pm i\beta$ with $\beta \neq 0$, the motion is *underdamped*. In this case the block oscillates, but the amplitude of the motion tends to zero as $t \to +\infty$. The different possibilities will be illustrated by examples.

EXAMPLE 3. The block in Figure 5.1 weighs 32 lb, the spring constant $k = 36$ lb/ft, and the resistance coefficient $c = 13$. Determine the resulting motion if $x = \frac{1}{2}$ ft and $v = 0$ ft/sec when $t = 0$ sec.

Solution: Equation (5.8) becomes

$$\frac{d^2x}{dt^2} + 13\frac{dx}{dt} + 36x = 0$$

The characteristic equation $r^2 + 13r + 36 = 0$ has roots -4 and -9, and hence the DE has a complete solution given by

$$x = c_1 e^{-4t} + c_2 e^{-9t} \tag{5.9}$$

The velocity $v = dx/dt$ is given by

$$v = -4c_1 e^{-4t} - 9c_2 e^{-9t} \tag{5.10}$$

Substituting $x = \frac{1}{2}$ and $t = 0$ into (5.9), and $v = 0$ and $t = 0$ into (5.10), we find that $c_1 = \frac{9}{10}$ and $c_2 = -\frac{4}{10}$. Thus, the displacement x is given by

$$x = \frac{9e^{-4t} - 4e^{-9t}}{10}$$

$$= \frac{e^{-4t}}{10}(9 - 4e^{-5t})$$

It is easily seen that x is never zero and that $x \to 0$ as $t \to +\infty$. The motion is overdamped and the displacement-time curve is shown in Figure 5.5.

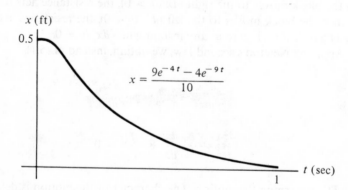

Fig. 5.5

EXAMPLE 4. Solve Example 3 for $c = 12$ instead of $c = 13$.

Solution: The characteristic equation $r^2 + 12r + 36 = 0$ has a repeated root $r = -6$. The DE has a complete solution given by $x = e^{-6t}(c_1 + c_2 t)$. Setting $x = \frac{1}{2}$ and $t = 0$, we find that $c_1 = \frac{1}{2}$.

Substituting $v = 0$ and $t = 0$ into

$$v = \frac{dx}{dt} = e^{-6t}(c_2) - 6e^{-6t}\left(\frac{1}{2} + c_2 t\right)$$

$$= e^{-6t}(c_2 - 3 - 6c_2 t)$$

we find that $c_2 = 3$. The displacement x is given by

$$x = e^{-6t}\left(\frac{1}{2} + 3t\right)$$

and the displacement-time curve resembles that of Figure 5.5 except that x approaches zero more rapidly. The motion is critically damped.

EXAMPLE 5. Solve Example 3 for $c = 8$ instead of $c = 13$.

Solution: The characteristic equation $r^2 + 8r + 36 = 0$ has roots $-4 \pm 2\sqrt{5}\,i$. The DE has a complete solution given by

$$x = e^{-4t}(c_1 \sin 2\sqrt{5}\,t + c_2 \cos 2\sqrt{5}\,t)$$

Setting $x = \frac{1}{2}$ and $t = 0$, we find that $c_2 = \frac{1}{2}$.
Substituting $v = 0$ and $t = 0$ into

$$v = \frac{dx}{dt} = e^{-4t}(2\sqrt{5}\,c_1 \cos 2\sqrt{5}\,t - \sqrt{5} \sin 2\sqrt{5}\,t)$$

$$- 4e^{-4t}\left(c_1 \sin 2\sqrt{5}\,t + \frac{1}{2}\cos 2\sqrt{5}\,t\right)$$

we find that $c_1 = 1/\sqrt{5}$. The displacement x is given by

$$x = e^{-4t}\left(\frac{1}{\sqrt{5}}\sin 2\sqrt{5}\,t + \frac{1}{2}\cos 2\sqrt{5}\,t\right)$$

The motion is underdamped; the factor e^{-4t} is termed the *damping factor* and the factor $(1/\sqrt{5})\sin 2\sqrt{5}\,t + \frac{1}{2}\cos 2\sqrt{5}\,t$ is termed the *harmonic factor*. Figure 5.6 shows the displacement-time curve, which is drawn by first sketching the curves $f(t) = \pm e^{-4t}$, and observing that the height of the displacement curve is numerically equal to the value of the damping factor when the harmonic factor is numerically equal to 1, and zero when the harmonic factor has the value zero. It is also helpful to write

$$x = e^{-4t}\left(\frac{1}{\sqrt{5}}\sin 2\sqrt{5}\,t + \frac{1}{2}\cos 2\sqrt{5}\,t\right)$$

in the form

$$x = 0.3\sqrt{5}\,e^{-4t}\sin(2\sqrt{5}\,t + \phi)$$

where $\cos\phi = \frac{2}{3}$ and $\sin\phi = \sqrt{5}/3$. Although the motion is not periodic, the quantity $2\pi/(2\sqrt{5}) = \pi/\sqrt{5}$ is termed the *quasi period*.

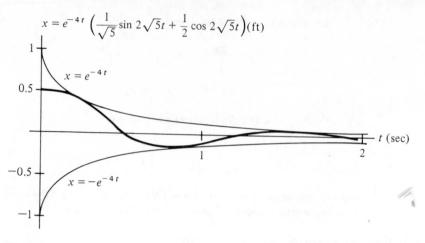

$$x = e^{-4t}\left(\frac{1}{\sqrt{5}}\sin 2\sqrt{5}t + \frac{1}{2}\cos 2\sqrt{5}t\right)(\text{ft})$$

Fig. 5.6

We now assume that the block of Figure 5.1 is acted on by an external force in the x direction, which we denote by $F(t)$. In many applications, $F(t)$ is periodic and has the form $F_0 \sin \omega t$ or $F_0 \cos \omega t$, where F_0 and ω are constant. This could be achieved by causing the wall in Figure 5.1 to undergo a simple harmonic motion. The function F is called the *forcing function* and $F(t)$ is called the *driving*, or *impressed*, *force*. The DE governing the motion becomes

$$\frac{w}{g}\frac{d^2x}{dt^2} + c\frac{dx}{dt} + kx = F(t) \qquad (5.11)$$

EXAMPLE 6. Solve Example 5 for an impressed force $F(t) = 72 \cos 6t$ acting on the block.

Solution: The motion is governed by the DE

$$\frac{d^2x}{dt^2} + 8\frac{dx}{dt} + 36x = 72 \cos 6t$$

We try a particular solution given by $x_p = A \sin 6t + B \cos 6t$, and find by the method of undetermined coefficients that $B = 0$ and $A = \frac{3}{2}$. The displacement x is given by

$$x = e^{-4t}(c_1 \sin 2\sqrt{5}t + c_2 \cos 2\sqrt{5}t) + \frac{3}{2}\sin 6t$$

Differentiating to find $v = dx/dt$, and then using the initial conditions, we find that $c_1 = -7/(2\sqrt{5})$ and $c_2 = \frac{1}{2}$.

The terms in the solution that involve e^{-4t} have an appreciable effect on the motion only for small values of t, since they contribute very little to the value of x when t is large. These terms are called the *transient solution*. The remaining term $\frac{3}{2}\sin 6t$ is called the *steady-state solution*, since it gives the displacement (approximate) for large values of t. The steady-state solution is periodic and has the same period as the impressed external force. The motion is approximately a simple harmonic motion of period $\pi/3$ and amplitude $\frac{3}{2}$.

There is another way of viewing DE (5.11). In this approach, $F(t)$ is regarded as an input to the physical system governed by the DE, and $x(t)$, the solution of the DE, is regarded as the output, or the response of the system to the input $F(t)$. In the problems we consider, for most inputs $F(t)$ the output $x(t)$ will be determined uniquely except for the transient solution. The transient solution depends on the initial conditions.

The input-output point of view is very fruitful in many applications. It is often used, for example, in analyzing economic systems. (For more details on input-output analysis, see Reference 5.4.)

EXAMPLE 7. The block in Figure 5.7 weighs w pounds. When it is held at rest by the spring, its position is the equilibrium position. Find its displacement from the equilibrium position if it is pulled x_0 feet below the equilibrium position and released from rest.

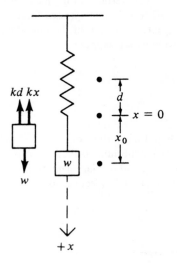

Fig. 5.7

Solution: Let x denote the displacement of the block, measured positive downward from the equilibrium position. If d denotes the stretch in the spring when the block is in the equilibrium position ($x = 0$) and k denotes the spring constant, then from Hooke's law, $w = kd$.

When the block is x feet from the equilibrium position, the spring exerts a force of $-k(d + x) = -kd - kx = -w - kx$ pounds in the x direction. Since the earth pulls on the block with a force of w pounds, from Newton's second law we have

$$\frac{w}{g}\frac{d^2x}{dt^2} = w + (-w - kx)$$

or

$$\frac{d^2x}{dt^2} + \frac{kg}{w}x = 0 \qquad (5.12)$$

Using the initial conditions $x = x_0$ and $v = dx/dt = 0$ when $t = 0$, we find that the displacement x is given by $x = x_0 \cos \sqrt{kg/w}\, t$.

The block undergoes a simple harmonic motion about the point $x = 0$. The DE is simple since the displacement is measured from the equilibrium position. If an impressed force $F(t)$ is applied to the block, DE (5.12) becomes

$$\frac{d^2x}{dt^2} + \frac{kg}{w} x = \frac{g}{w} F(t)$$

and we proceed as in Example 6.

If a resistance force is also acting, the DE becomes

$$\frac{d^2x}{dt^2} + \frac{cg}{w}\frac{dx}{dt} + \frac{kg}{w} x = \frac{g}{w} F(t)$$

From the input-output point of view, a vertical spring-mass system can be regarded as a horizontal spring-mass system with an impressed force equal to the weight of the moving block. If we translate the origin in the horizontal system d units to the right, the term in the DE involving w disappears.

Problem List 5.1

For these problems, use the value $g = 32$.

1. The block of Figure 5.8 weighs 3 lb and the spring constant $k = 150$ lb/ft. The block is pulled 4 in. to the right of the equilibrium position ($x = 0$) and released from rest. Find the acceleration, velocity, and displacement at time t and determine the period and amplitude of the motion. Neglect friction.

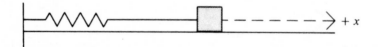

Fig. 5.8

2. The block in Figure 5.8 weighs 2 lb and the spring constant $k = 16$ lb/ft. The block is given an initial velocity of 48 ft/sec at the equilibrium position ($x = 0$). Find the displacement and velocity at time t, the velocity at $t = \pi/48$ sec, and the amplitude and frequency of the motion. Neglect friction.

3. The block in Figure 5.8 weighs 2 lb and the spring constant $k = 144$ lb/ft. After being pulled 6 in. to the right, the block is given initial velocity -12 ft/sec. Find the displacement and velocity at time t, the maximum speed, and the period and amplitude of the motion. Neglect friction.

4. The spring in Figure 5.8 will be stretched 6 in. by a force of 81 lb. If the block weighs 1 lb and is given initial velocity 12 ft/sec at $x = 0$, how far to the right will the block move? Neglect friction.

5. A particle moves around the circle $x^2 + y^2 = r^2$ at constant speed. Prove that the projection of the particle on the y axis undergoes a simple harmonic motion.

6. The motion of a body moving in a resisting medium is governed by the DE

$$\frac{d^2x}{dt^2} + \frac{1}{2}\frac{dx}{dt} = 0$$

If $dx/dt = 50$ ft/sec and $x = 0$ ft when $t = 0$ sec, what is the displacement x in terms of t, $\lim_{t \to +\infty} x$, and $\lim_{t \to +\infty} v$?

7. An 8-lb body is dropped from rest, subject to air resistance of $v/20$ lb, where $v = dx/dt$ is the velocity of the body in ft/sec. Find the speed of the body t sec after release.

8. A body of weight 32 lb is thrown vertically downward from the top of a high building with initial speed 40 ft/sec. The air resistance is $2v$ lb, where $v = dx/dt$ is the velocity in ft/sec. Find the displacement x and the velocity v in terms of the time t.

9. A 64-lb body falls from rest under the influence of gravity. It is also acted on by air resistance, given by $8v$ lb; v is the velocity in ft/sec. Find v in terms of t and find the limiting velocity.

10. A 96-lb boat is being towed against water resistance of $6v$ lb, where v is the velocity of the boat in ft/sec. If the force in the tow rope is 36 lb constantly, and $v = 0$ when $t = 0$, what is v for $t = 1$ sec? What is the limiting velocity?

11. The block in Figure 5.8 weighs 16 lb, the spring constant $k = 36$ lb/ft, and the resistance coefficient $c = 9$. If the block is pulled 4 in. to the right and released from rest, what are the displacement and the velocity at time t?

12. Solve Problem 11 for $k = 32$ and $c = 8$.

13. Solve Problem 11 for $c = 6$.

14. The motion of a particle is governed by the DE

$$\frac{d^2x}{dt^2} + \frac{cg}{w}\frac{dx}{dt} + \frac{kg}{w}x = 0$$

Show that if the motion is either overdamped or critically damped, the particle will not be located at the origin ($x = 0$) for more than one value of t.

15. The block in Figure 5.8 weighs 32 lb, the spring constant $k = 16$ lb/ft, the resistance coefficient $c = 2$, and the block is acted on by an external impressed force $F(t) = 96 \cos 4t$. For $x = \frac{1}{2}$ ft and $v = 0$ ft/sec when $t = 0$ sec, what are the displacement and velocity at time t?

16. A particle moves along the x axis. At time $t = 0$ the particle is at the origin and has velocity $dx/dt = 3$ ft/sec. Find the displacement and velocity at $t = \pi/6$ if the motion is governed by the DE

$$\frac{d^2x}{dt^2} + 2\frac{dx}{dt} + 10x = 0$$

17. The block in Figure 5.9 weighs 32 lb and the spring constant $k = 25$ lb/ft. The block is pulled down 3 in. below the equilibrium position ($x = 0$) and released from rest. Find the displacement and velocity at time t and determine the period of the motion. Neglect friction.

+x

Fig. 5.9

18. Solve Problem 17 for a block that is given initial velocity -5 ft/sec instead of being released from rest. Find the amplitude of the motion and the maximum speed of the block.

19. A 4-lb weight suspended from a spring stretches the spring 2 in. The weight is then pulled 3 in. below the equilibrium position and released from rest. Find the displacement x and velocity v in terms of t. State the amplitude and period of the motion.

20. An 8-lb weight is attached to a spring with spring constant $k = 12$ lb/ft. The weight is raised 5 in. above the equilibrium position ($x = 0$) and dropped from rest. Find x in terms of t and state the frequency of the motion.

21. A rectilinear motion is governed by the DE

$$\frac{d^2x}{dt^2} + \beta \frac{dx}{dt} + 64x = 0$$

where $\beta > 0$. For what value of β is the motion critically damped?

22. The differential equation

$$\frac{d^2x}{dt^2} + 8\frac{dx}{dt} + 64x = 128 \cos 8t$$

has transient solution $x = e^{-4t}(3 \cos 4\sqrt{3}\,t - 11\sqrt{3} \sin 4\sqrt{3}\,t)$. Find the steady-state solution.

23. A body attached to the lower end of a vertical spring has acceleration $a = d^2x/dt^2 = -16x$ ft/sec^2, where x is the distance of the body from the equilibrium position at time t. If the body passes through its equilibrium position at $t = 0$ with $v = 12$ ft/sec, what is x in terms of t?

24. A 64-lb weight is attached to the end of a vertical spring having spring constant $k = 50$ lb/ft. The weight is given a downward velocity of 20 ft/sec from the equilibrium position ($x = 0$). Find the displacement x in terms of the time t if the

motion is resisted by a damping force numerically equal to $12\,|v|$ lb. State the quasi period and classify the motion as damped oscillatory, critically damped, or overdamped.

25. Gravity at an internal point varies with the distance x of the point from the center of the earth. We assume that the earth is a sphere of radius 4000 miles. Show that a particle dropped from rest into a hole extending diametrically through the earth will undergo a simple harmonic motion. Find (a) the period in hours and (b) the speed in mph for $x = 0$.

26. A cylindrical buoy of diameter 1 ft and weight 200 lb floats vertically in the water. Show that the buoy undergoes a simple harmonic motion when it is depressed slightly and released. Neglect friction and assume that water weighs 62.4 lb/ft^3. By Archimedes' principle, the buoy is buoyed up by the weight of the displaced water. Find the period of the motion.

27. A paratrooper and parachute have total weight 200 lb. At the instant the parachute opens, the paratrooper is plunging vertically downward at 40 ft/sec. If air resistance in pounds varies as $|v|^{5/4}$ where $|v|$ is the speed, and if air resistance is 128 lb when $|v| = 16$ ft/sec, how long will it take for the speed to reach 30 ft/sec? Express the answer as a definite integral and use the trapezoidal rule

$$\int_a^b f(x)\,dx \approx h\left(\frac{1}{2}y_0 + y_1 + y_2 + \cdots + y_{n-1} + \frac{1}{2}y_n\right)$$

with $n = 10$ to approximate the answer.

5.2 Resonance

Let us consider the motion of an oscillating mechanical system governed by the following DE, which is (5.11) with $w/g = m$.

$$m\frac{d^2x}{dt^2} + c\frac{dx}{dt} + kx = F(t)$$

We assume that the input is also oscillatory and seek the nature of the output x corresponding to a periodic input of the form $F(t) = F_0 \cos \omega t$. For simplicity, we begin by assuming that the resistance coefficient c equals zero. Although some friction is always present, an analysis of the theoretical case $c = 0$ is instructive and suggests what we might expect for small values of c. Letting

$$\omega_0 = \sqrt{\frac{k}{m}} = \sqrt{\frac{kg}{w}} \tag{5.12}$$

we first note that the DE

$$\frac{d^2x}{dt^2} + \omega_0^2 x = \frac{F_0}{m}\cos \omega t \tag{5.13}$$

has complementary function given by

$$x_c = c_1 \cos \omega_0 t + c_2 \sin \omega_0 t$$

If the frequency $\omega/(2\pi)$ of the input is the same as the natural frequency $\omega_0/(2\pi)$ of the system, we try to find a particular solution X of (5.13) of the form $X = t(c_1 \cos \omega_0 t + c_2 \sin \omega_0 t)$. We find (see Problem 3) that $X = [F_0/(2m\omega_0)]t \sin \omega_0 t$. Therefore, (5.13) has a complete solution given by

$$x = c_1 \cos \omega_0 t + c_2 \sin \omega_0 t + \frac{F_0}{2m\omega_0} t \sin \omega_0 t$$

Whereas the input had constant amplitude F_0, the values of the output x are unbounded. This is seen by observing that as $t \to +\infty$ along a sequence for which $\sin \omega_0 t = \pm 1$,

$$\left| \frac{F_0}{2m\omega_0} t \sin \omega_0 t \right| \to +\infty$$

The absolute value of $c_1 \cos \omega_0 t + c_2 \sin \omega_0 t$ does not exceed $\sqrt{c_1^2 + c_2^2}$. See Problem 4. The graph of $X = [F_0/(2m\omega_0)]t \sin \omega_0 t$ is sketched in Figure 5.10.

The phenomenon we have described is known as *pure, or undamped, resonance* and is also termed *sympathetic vibration*. It is approximated when

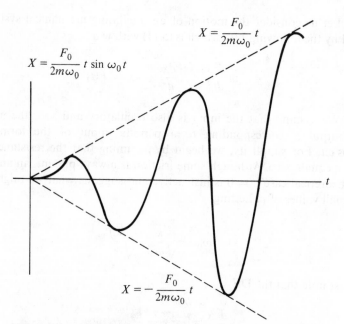

$$X = \frac{F_0}{2m\omega_0} t \sin \omega_0 t$$

$$X = \frac{F_0}{2m\omega_0} t$$

$$X = -\frac{F_0}{2m\omega_0} t$$

Fig. 5.10

c is small and the input and natural frequencies coincide. The forcing function is said to be in resonance with the system. Resonance can be both harmful and useful and must be considered in designing systems. Bridges, cars, ships, planes, and the like are vibrating systems and can be affected adversely by resonance. A body oscillating on a spring will move farther and farther from the equilibrium position when resonance occurs. The spring may eventually break or Hooke's law will no longer apply. Soldiers crossing a bridge often break step so that they will not generate a frequency equal to the natural frequency of the bridge. When a tree is uprooted by applying a periodic force to produce swaying, a resonant effect is being used to advantage. A rocking motion applied to a car often succeeds in getting the car out of a hole or ditch. Desirable amplification effects are often produced in electric circuits by resonant effects.

EXAMPLE 1. A body oscillating about its equilibrium position ($x = 0$) moves according to the DE

$$\frac{d^2x}{dt^2} + 16x = 24 \cos 4t$$

If $x = 0$ and $dx/dt = 0$ when $t = 0$, what is x in terms of t? Compute x for

$$t = \frac{\pi}{8}, \frac{2\pi}{8}, \frac{3\pi}{8}, \frac{8\pi}{8}, \frac{9\pi}{8}, \text{ and } \frac{21\pi}{8}$$

Solution: $r^2 + 16 = 0, r = \pm 4i$

$$x_c = c_1 \cos 4t + c_2 \sin 4t$$

The particular solution $x_p = 3t \sin 4t$ is readily found by the method of undetermined coefficients. Thus,

$$x = c_1 \cos 4t + c_2 \sin 4t + 3t \sin 4t$$

and

$$v = -4c_1 \sin 4t + 4c_2 \cos 4t + 12t \cos 4t + 3 \sin 4t$$

Using the initial conditions, we find that $c_1 = c_2 = 0$ and hence

$$x = 3t \sin 4t$$

The following table is easily constructed:

t	0	$\dfrac{\pi}{8}$	$\dfrac{2\pi}{8}$	$\dfrac{3\pi}{8}$	$\dfrac{8\pi}{8}$	$\dfrac{9\pi}{8}$	$\dfrac{21\pi}{8}$
x	0	$\dfrac{3\pi}{8}$	0	$\dfrac{-9\pi}{8}$	0	$\dfrac{27\pi}{8}$	$\dfrac{63\pi}{8}$

The displacement-time curve resembles Figure 5.10.

An interesting oscillation takes place when ω and ω_0 are unequal but differ only slightly; that is, when the input and natural frequencies are nearly equal.

If $x = 0$ and $dx/dt = 0$ when $t = 0$, the DE (5.13) has the solution given by (see Problem 6)

$$x = \frac{F_0}{m(\omega_0^2 - \omega^2)}(\cos \omega t - \cos \omega_0 t) \qquad (5.14)$$

Applying the trigonometric identity

$$\cos \alpha - \cos \beta = -2 \sin \frac{\alpha + \beta}{2} \sin \frac{\alpha - \beta}{2}$$

we write (5.14) in the form

$$x = \left[\frac{2F_0}{m(\omega_0^2 - \omega^2)} \sin\left(\frac{1}{2}(\omega_0 + \omega)t\right) \right] \sin \frac{1}{2}\left((\omega_0 - \omega)t\right) \qquad (5.15)$$

Equation (5.14) gives the displacement x as the superposition of two sinusoidal waves whose periods differ slightly. Since the period of $\sin[\frac{1}{2}(\omega_0 - \omega)t]$ is much larger than the period of $\sin[\frac{1}{2}(\omega_0 + \omega)t]$, Equation (5.15) makes it evident that the superposition in (5.14) results in a displacement-time curve of the type displaced in Figure 5.11.

This phenomenon is known as the occurrence of *beats*. It manifests itself when two adjacent piano keys are struck simultaneously. The variations in loudness produce the beats. Two strings are tuned to the same frequency by tightening one of them until they vibrate together without producing beats. The phenomenon of beats is also important in electrical theory.

Now let us consider the situation in which the resistance coefficient c in (5.11) is a small positive constant. If $c^2 - 4mk < 0$, the characteristic

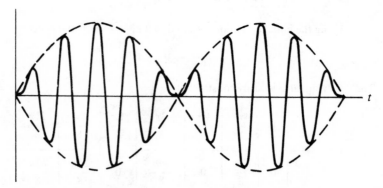

Fig. 5.11 Displacement-time curve. Beats occur when the input and natural frequencies differ by a small amount.

equation of (5.11) will have complex roots $\alpha \pm i\beta$ where $\alpha < 0$, $\beta > 0$, and the transient solution will have the form

$$x_c = e^{\alpha t}(c_1 \cos \beta t + c_2 \sin \beta t)$$

Assuming a particular solution of the form

$$X = A \cos \omega t + B \sin \omega t \tag{5.16}$$

we find (see Problem 7) by the method of undetermined coefficients

$$x = \frac{F_0}{m(\omega_0^2 - \omega^2)} (\cos \omega t - \cos \omega_0 t)$$

that

$$= \frac{2F_0}{m(\omega_0^2 - \omega^2)} \left(\sin \frac{\omega_0 + \omega}{2} t \right) \left(\sin \frac{\omega_0 - \omega}{2} t \right)$$

$$A = \frac{(k - \omega^2 m)F_0}{(k - \omega^2 m)^2 + \omega^2 c^2}$$

$$B = \frac{\omega c F_0}{(k - \omega^2 m)^2 + \omega^2 c^2}$$

Letting $\tan \phi = \omega c/(k - \omega^2 m)$, we then write the steady-state solution (5.16) in the form

$$X = \frac{F_0}{\sqrt{(k - \omega^2 m)^2 + \omega^2 c^2}} \cos (\omega t - \phi)$$

Since $x_c \to 0$ as $t \to +\infty$, the output $x = x_c + X$ will approximate a harmonic oscillation whose frequency $\omega/(2\pi)$ is the same as the frequency of the input $F(t) = F_0 \cos \omega t$. The amplitude of the output x is approximately equal to the coefficient

$$\frac{F_0}{\sqrt{(k - \omega^2 m)^2 + \omega^2 c^2}} = G(\omega) \tag{5.17}$$

since the contribution of the transient solution x_c to the amplitude is negligible except for small t. Thus, if we regard F_0, k, m, and c as fixed, the amplitude in (5.17) depends on ω, whereas the amplitude of the input had fixed value F_0. It is easy to show (see Problem 8) that $G(\omega)$ takes on its maximum value when

$$\omega = \sqrt{\frac{k}{m} - \frac{c^2}{2m^2}}$$

When ω has this value, the system is said to be in *damped* (or *practical*) *resonance*. In designing systems it is extremely important to know when the steady-state amplitude is large. Harmful and beneficial effects are possible, and even though $G(\omega)$ does not become infinite with t, the situation is analogous to pure resonance in an undamped system.

EXAMPLE 2. In Example 6 of Section 5.1 we found that the steady-state solution of

$$\frac{32}{32}\frac{d^2x}{dt^2} + 8\frac{dx}{dt} + 36x = 72 \cos 6t$$

is given by $x_p = \frac{3}{2} \sin 6t$. The amplitude $\frac{3}{2}$ can be obtained by setting $F_0 = 72$, $k = 36$, $\omega = 6$, $m = 1$, and $c = 8$ in

$$\frac{F_0}{\sqrt{(k - \omega^2 m)^2 + \omega^2 c^2}}$$

This yields

$$\frac{72}{\sqrt{(36 - 36[1])^2 + 36(64)}} = \frac{72}{6(8)} = \frac{3}{2}$$

To obtain the maximum amplitude corresponding to $F(t) = 72 \cos \omega t$, we replace $\omega = 6$ by

$$\omega = \sqrt{\frac{k}{m} - \frac{c^2}{2m^2}} = \sqrt{\frac{36}{1} - \frac{(8)^2}{2(1)^2}} = \sqrt{36 - 32} = 2$$

This yields the maximum amplitude

$$\frac{F_0}{\sqrt{(k - \omega^2 m)^2 + \omega^2 c^2}} = \frac{72}{\sqrt{[36 - (2)^2(1)]^2 + (2)^2(8)^2}}$$

$$= \frac{72}{\sqrt{(32)^2 + (16)^2}}$$

$$= \frac{72}{16\sqrt{5}} = \frac{9\sqrt{5}}{10} \approx 2.01$$

Problem List 5.2

1. The motion of a body moving on a spring is governed by the DE $d^2x/dt^2 + 81x = 16 \sin \omega t$. For what value of ω does resonance occur?

2. A 64-lb body is attached to the lower end of a vertical spring whose spring constant k is 32 lb/ft. A force given by $F(t) = 6 \cos \omega t$ lb is applied to the body. For what value of ω will resonance occur? Neglect friction.

3. Find a particular solution of the DE

$$\frac{d^2x}{dt^2} + \omega_0^2 x = \frac{F_0}{m} \cos \omega_0 t$$

of the form $X = t(c_1 \cos \omega_0 t + c_2 \sin \omega_0 t)$.

4. Show that the absolute value of $c_1 \cos \omega_0 t + c_2 \sin \omega_0 t$ does not exceed $\sqrt{c_1^2 + c_2^2}$.

5. A body oscillating about its equilibrium position ($x = 0$) moves according to the DE $d^2x/dt^2 + 4x = 8 \cos 2t$. If $x = 0$ and $dx/dt = 0$ when $t = 0$, what is x for

$$t = \frac{\pi}{4}, \frac{3\pi}{4}, \frac{5\pi}{4}, \frac{6\pi}{4}, \frac{10\pi}{4}, \frac{11\pi}{4}, \frac{12\pi}{4}, \text{ and } \frac{13\pi}{4}$$

6. Show that if $x = 0$ and $dx/dt = 0$ when $t = 0$, and $\omega \neq \omega_0$, then the DE

$$\frac{d^2x}{dt^2} + \omega_0^2 x = \frac{F_0}{m} \cos \omega t$$

has a solution given by

$$x = \frac{F_0}{m(\omega_0^2 - \omega^2)} (\cos \omega t - \cos \omega_0 t)$$

7. By the method of undetermined coefficients, find a particular solution of the DE

$$m \frac{d^2x}{dt^2} + c \frac{dx}{dt} + kx = F_0 \cos \omega t$$

8. Given that $\omega > 0$, show that

$$G(\omega) = \frac{F_0}{\sqrt{(k - \omega^2 m)^2 + \omega^2 c^2}}$$

takes on its maximum value when $\omega = \sqrt{(k/m) - c^2/(2m^2)}$.

9. Given that $m = 1$, $k = 36$, and $F_0 = 72$, find the maximum amplitude of the steady-state solution of

$$m \frac{d^2x}{dt^2} + c \frac{dx}{dt} + kx = F_0 \cos \omega t$$

for (a) $c = 1$ and (b) $c = 0.1$.

5.3 Work and Kinetic Energy

Newton's second law, which we have applied to rectilinear motions, also applies to the motion of a particle moving along a curve C in three dimensions. The law then takes the form

$$\mathbf{F} = m\mathbf{A} = m \frac{d\mathbf{V}}{dt} \tag{5.18}$$

where $\mathbf{F} = F_x \mathbf{i} + F_y \mathbf{j} + F_z \mathbf{k}$ is a vector representing the resultant of the forces acting on the particle,

$$\mathbf{A} = \frac{d^2x}{dt^2} \mathbf{i} + \frac{d^2y}{dt^2} \mathbf{j} + \frac{d^2z}{dt^2} \mathbf{k}$$

is the vector acceleration of the particle and $\mathbf{V} = v_x\mathbf{i} + v_y\mathbf{j} + v_z\mathbf{k}$ is the vector velocity of the particle. Equation (5.18) is a second-order vector DE; the integration of this DE is the central objective of the branch of mechanics known as dynamics. In this section we obtain an intermediate integral of (5.18) that is often very useful in applications.

The vector DE (5.18) is equivalent to the three scalar DE

$$F_x = m\frac{d^2x}{dt^2}, \quad F_y = m\frac{d^2y}{dt^2}, \quad \text{and} \quad F_z = m\frac{d^2z}{dt^2}$$

Writing $F_x = m(d^2x/dt^2)$ in the form

$$F_x = m\frac{dv_x}{dt} = m\frac{dv_x}{dx}\frac{dx}{dt} = mv_x\frac{dv_x}{dx}$$

we separate the variables to obtain

$$F_x\, dx = mv_x\, dv_x \tag{5.19}$$

Let us assume that the particle moves along the curve from $P_1(x_1, y_1, z_1)$ to $P_2(x_2, y_2, z_2)$. Integrating (5.19) yields

$$\int_{x_1}^{x_2} F_x\, dx = m\int_{(v_x)_1}^{(v_x)_2} v_x\, dv_x = \frac{m}{2}v_x^2\Big]_{(v_x)_1}^{(v_x)_2} = \frac{m}{2}(v_x)_2^2 - \frac{m}{2}(v_x)_1^2 \tag{5.20}$$

The quantity $\int_{x_1}^{x_2} F_x\, dx$ is the *work* done on the particle by the x component F_x of $\mathbf{F}$ as the particle moves from P_1 to P_2. The quantity $(m/2)v^2$, where $v = ds/dt = \sqrt{v_x^2 + v_y^2 + v_z^2}$ is the speed of the particle, is called the *kinetic energy* of the particle, and the right member of (5.20) is the change in kinetic energy associated with the x component of the velocity. Similarly, we obtain the equations

$$\int_{y_1}^{y_2} F_y\, dy = \frac{m}{2}(v_y)_2^2 - \frac{m}{2}(v_y)_1^2 \tag{5.21}$$

and

$$\int_{z_1}^{z_2} F_z\, dz = \frac{m}{2}(v_z)_2^2 - \frac{m}{2}(v_z)_1^2 \tag{5.22}$$

Adding (5.20), (5.21), and (5.22), we obtain

$$\int_{x_1}^{x_2} F_x\, dx + \int_{y_1}^{y_2} F_y\, dy + \int_{z_1}^{z_2} F_z\, dz$$

$$= \frac{m}{2}[(v_x)_2^2 + (v_y)_2^2 + (v_z)_2^2] - \frac{m}{2}[(v_x)_1^2 + (v_y)_1^2 + (v_z)_1^2]$$

$$= \frac{m}{2}v_2^2 - \frac{m}{2}v_1^2 \tag{5.23}$$

Equation (5.23) is called the *energy integral* of Newton's second law and expresses symbolically the *work-energy* principle. In words, this principle states that the work done on the particle equals the change in its kinetic energy.

NOTE: Instead of defining the work done by **F** as the sum of the work done by the three components of **F**, we could define work by the line integral $\int_C F_t \, ds$, where $F_t = \mathbf{F} \cdot \mathbf{T}$ is the tangential component of **F** at an arbitrary point P of C. Then

$$\int_C F_t \, ds = \int_C \mathbf{F} \cdot \mathbf{T} \, ds$$

$$= \int_C \left(F_x \frac{dx}{ds} + F_y \frac{dy}{ds} + F_z \frac{dz}{ds} \right) ds$$

$$= \int_C F_x \, dx + F_y \, dy + F_z \, dz$$

$$= \int_C F_x \, dx + \int_C F_y \, dy + \int_C F_z \, dz$$

Example.   A particle of weight w moves along a curve C from $P_1(x_1, y_1, z_1)$ to $P_2(x_2, y_2, z_2)$ and is acted on by the single force of gravity. Find the work done.

Solution: Let the force field through which the particle moves be given by $\mathbf{F} = -w\mathbf{k}$. The work done by gravity is given by $\int_{z_1}^{z_2} (-w) \, dz = -w(z_2 - z_1)$, and is numerically equal to the weight w multiplied by the numerical value of the vertical displacement $|z_2 - z_1|$. The work done is positive when $z_2 < z_1$, negative when $z_2 > z_1$, and zero when $z_2 = z_1$.

The work done is independent of the path C along which the particle moves. A force field such as $\mathbf{F} = -w\mathbf{k}$ having this property is said to be *conservative*. It is shown in Reference 5.8 that a force field

$$\mathbf{F} = F_x(x, y, z)\mathbf{i} + F_y(x, y, z)\mathbf{j} + F_z(x, y, z)\mathbf{k}$$

is conservative in a domain D in three-space whenever there exists a function ϕ of x, y, and z such that

$$\mathbf{F} = \operatorname{grad} \phi$$

in D. For $\mathbf{F} = -w\mathbf{k}$, it is easy to see that $\mathbf{F} = \operatorname{grad} \phi$, where $\phi(x, y, z) = -wz$.

Let v be the speed of a particle of weight w at an arbitrary point $P(x, y, z)$ on a curve C and let v_1 be its speed at a fixed point $P_1(x_1, y_1, z_1)$ of C. Then

$$-w(z - z_1) = \frac{m}{2} v^2 - \frac{m}{2} v_1^2$$

or

$$\frac{m}{2} v^2 + wz = \frac{m}{2} v_1^2 + wz_1 = \text{constant} \tag{5.24}$$

The term $(m/2)v^2$ represents the kinetic energy of the particle; the second term wz is called the *potential energy* of the particle. The sum of the kinetic energy and the potential energy remains constant throughout the motion. Equation (5.24) is called the *conservation law of mechanical energy*. It holds for a particle moving through an arbitrary conservative force field.

5.4 Impulse and Momentum; Motion of a Rocket

We again consider the motion of a particle in three-dimensional space. The particle moves in a force field $\mathscr{F}$ that may represent a single force acting on the particle or the resultant of several forces acting on the particle. In Section 5.3, a displacement integral of $\mathscr{F}$ led to the work-energy equation. We now consider a time integral of $\mathscr{F}$. Integrating Newton's second law, $\mathbf{F} = m\mathbf{A} = md\mathbf{V}/dt$, over the time interval $t_2 - t_1$, we obtain

$$\int_{t_1}^{t_2} \mathbf{F}\, dt = \int_{t_1}^{t_2} m\frac{d\mathbf{V}}{dt}\, dt = m\mathbf{V}_2 - m\mathbf{V}_1 \tag{5.25}$$

The vector $m\mathbf{V}$ is called the *linear momentum* of the particle and the integral $\int_{t_1}^{t_2} \mathbf{F}\, dt$ is called the *impulse* of the force $\mathbf{F}$ during the interval $t_2 - t_1$. Equation (5.25), known as the *impulse-momentum equation*, states that the impulse of a force $\mathbf{F}$ acting on a particle over a time interval equals the change in linear momentum of the particle during that time interval. Since (5.25) is a vector equation, it can be applied in any direction, using vector components. It is equivalent to the three scalar equations

$$\int_{t_1}^{t_2} F_x\, dt = m(v_x)_2 - m(v_x)_1 \qquad \int_{t_1}^{t_2} F_y\, dt = m(v_y)_2 - m(v_y)_1$$

$$\int_{t_1}^{t_2} F_z\, dt = m(v_z)_2 - m(v_z)_1$$

The impulse-momentum principle is particularly useful in situations for which very little is known about force patterns acting over very short intervals. The nature of these forces can be studied by observing changes in linear momentum occasioned by the forces under study.

EXAMPLE 1. A 6.4-lb body A moving 80 ft/sec in the direction of the negative x axis is struck by an object B. After impact, body A moves in the opposite direction at 100 ft/sec. Given that the time of impact was 0.01 sec, find the average force $\bar{\mathbf{F}}$ acting on body A during the period of impact.

Solution: Applying (5.25), we obtain

$$\int_0^{0.01} \mathbf{F}\, dt = \frac{6.4}{32}(100\mathbf{i}) - \frac{6.4}{32}(-80\mathbf{i}) = 36\mathbf{i}$$

From $\bar{\mathbf{F}}(0.01) = 36\mathbf{i}$, we obtain $\bar{\mathbf{F}} = 3600\mathbf{i}$.

For a system of n particles, it is shown in Reference 5.8, by applying Newton's second law to the individual particles, that the resultant force $\mathbf{F}$ acting on the system satisfies

$$\mathbf{F} = \sum_{i=1}^{n} m_i \frac{d\mathbf{V}_i}{dt} \qquad (5.26)$$

Integrating (5.26) between t_1 and t_2, we obtain

$$\int_{t_1}^{t_2} \mathbf{F}\, dt = \left(\sum_{i=1}^{n} m_i \mathbf{V}_i \right)_2 - \left(\sum_{i=1}^{n} m_i \mathbf{V}_i \right)_1 \qquad (5.27)$$

The left member of (5.27) is the impulse of the resultant force $\mathbf{F}$ on the system of particles during the interval $t_2 - t_1$, and the right member is called the change in the linear momentum of the system during the interval $t_2 - t_1$. That is, the *momentum of the system* is defined to be $\sum_{i=1}^{n} m_i \mathbf{V}_i$. The left member is also the impulse of the resultant of the external forces acting on the system, since by Newton's third law, the internal forces act in opposite pairs of equal magnitude and hence their impulses have zero sum.

If no external forces act on the system, the left member of (5.27) is zero. Thus, if no external forces act on a system during an arbitrary time interval, there is no change in the linear momentum of the system during that interval. This is the important *law of conservation of linear momentum*.

To illustrate, consider a rocket moving in the direction of the x axis acted on by the internal forces owing to the discharge of the rocket fuel and a resultant external force $\mathbf{F} = F_x \mathbf{i}$. Let w denote the weight of the rocket (including unburned fuel) and v the velocity of the rocket at time t. Assume that fuel is burned at the constant rate $k > 0$ and that $w = w_0$ when $t = 0$. Then $w = w_0 - kt$, and the linear momentum at time t is given by $[(w_0 - kt)/g]v$. At time $t + \Delta t$, the remaining mass will be given by $(w_0 - kt - k\,\Delta t)/g$, and we denote its velocity by $v + \Delta v$. We assume that fuel is discharged at the constant speed c, where $c > 0$, and we regard the fuel discharged during the interval Δt as a particle of mass $k\,\Delta t/g$, having velocity $v - c$ at time $t + \Delta t$. Applying (5.27) in the x direction yields

$$\left[\frac{w_0 - kt - k\,\Delta t}{g}(v + \Delta v) + \frac{k\,\Delta t}{g}(v - c) \right] - \left[\frac{w_0 - kt}{g} v \right]$$
$$= \int_{t}^{t + \Delta t} F_x\, dt = F_{av}\, \Delta t \qquad (5.28)$$

where F_{av} is the average external force acting in the x direction during the interval Δt. Writing (5.28) in the form

$$\frac{w_0 - kt}{g} \frac{\Delta v}{\Delta t} - \frac{k}{g} \Delta v = \frac{kc}{g} + F_{av}$$

and letting $\Delta t \to 0$, we obtain

$$\frac{w_0 - kt}{g} \frac{dv}{dt} = \frac{kc}{g} + F_x \tag{5.29}$$

as the DE governing the motion of the rocket. If no external forces act on the rocket, $F_x = 0$ and (5.29) reduces to

$$(w_0 - kt)\frac{dv}{dt} = kc \tag{5.30}$$

If $v = v_0$ when $t = 0$, Equation (5.30) can be written in the form $dv = kc\, dt/(w_0 - kt)$ and integrated to give

$$\int_{v_0}^{v} du = -c \int_{0}^{t} \frac{-k\, dz}{w_0 - kz},$$

$$v - v_0 = -c \ln (w_0 - kz)]_0^t = -c \ln \left(\frac{w_0 - kt}{w_0}\right),$$

and

$$v = \frac{dx}{dt} = v_0 + c \ln \left(\frac{w_0}{w_0 - kt}\right) \tag{5.31}$$

We note that in (5.31) $w_0 - kt > 0$ and hence $t < w_0/k$; that is, the weight kt of the burned fuel must be less than the original total weight w_0.

Assuming that $x = x_0$ when $t = 0$, a second integration yields (see Problem 4)

$$x = x_0 + ct - \frac{c}{k}(w_0 - kt) \ln \frac{w_0}{w_0 - kt} + v_0 t \tag{5.32}$$

Example 2. A rocket, 80 percent fuel and 20 percent structure, discharges fuel at a velocity relative to the rocket of 9000 ft/sec. The constant rate of burning of the fuel is 10 percent of the original fuel per second. Assuming that no external forces are acting, find the velocity and distance traveled at the end of (a) 2 sec, (b) 5 sec, and (c) 10 sec, if $x = 0$ and $v = 0$ when $t = 0$.

Solution: Since the original amount of fuel is $0.8w_0$, $k = (0.1)(0.8)w_0 = 0.08w_0$. Setting $k = 0.08w_0$, $v = 0$, $c = 9000$, and $x_0 = 0$ in (5.31) and (5.32), we obtain

$$v(t) = 9000 \ln \frac{w_0}{w_0 - 0.08w_0 t} = 9000 \ln \frac{1}{1 - 0.08t}$$

and

$$x(t) = 9000t - \frac{9000}{0.08w_0}(w_0 - 0.08w_0 t) \ln \frac{w_0}{w_0 - 0.08w_0 t}$$

$$= 9000t - 112{,}500(1 - 0.08t) \ln \frac{1}{1 - 0.08t}$$

These equations yield

$$v(2) \approx 1569 \text{ ft/sec} \quad v(5) \approx 4597 \text{ ft/sec} \quad v(10) \approx 14{,}485 \text{ ft/sec}$$

$$x(2) \approx 1524 \text{ ft} \quad\quad x(5) \approx 10{,}519 \text{ ft} \quad\quad x(10) \approx 53{,}788 \text{ ft}$$

It is interesting to note that $v(t)$ and $x(t)$ are independent of w_0.

5.5 The Simple Pendulum

A *simple pendulum* consists of a particle of weight w supported by a straight rod or piece of string of length l. The particle is free to oscillate in a vertical plane; the mass of the particle is assumed to be concentrated at a point; and the weight of the rod is assumed to be negligible. See Figure 5.12(a).

The particle moves under the action of the forces represented in Figure 5.12(a) by the vectors $\mathbf{B}$ and $\mathbf{W}$, where $|\mathbf{B}| = b$ denotes the tension in the rod and $|\mathbf{W}| = w$ denotes the gravitational attraction the earth exerts on the particle. In Figure 5.12(b), the vector $\mathbf{W}$ is resolved into its two components tangent and normal to the circle on which the particle moves.

Let s denote the distance the particle moves on the circle, with $s = 0$ and $\theta = 0$ at the lowest point of the circle, and with s measured positively as shown in Figure 5.12(a).

From $s = l\theta$, we obtain

$$v = \frac{ds}{dt} = l\frac{d\theta}{dt} = l\omega \quad \text{and} \quad a_t = \frac{dv}{dt} = l\frac{d^2\theta}{dt^2} = l\frac{d\omega}{dt}$$

where a_t is the tangential component of the acceleration. Since the motion is circular, the normal component of the acceleration is

$$a_n = \frac{v^2}{l} = \frac{l^2(d\theta/dt)^2}{l} = l\left(\frac{d\theta}{dt}\right)^2$$

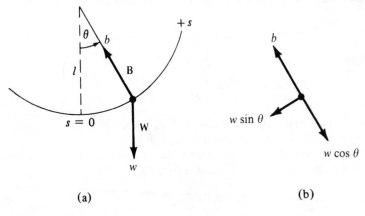

(a)

(b)

Fig. 5.12

We apply Newton's second law in the normal direction to obtain

$$b - w \cos \theta = \frac{w}{g} l \left(\frac{d\theta}{dt} \right)^2 \tag{5.33}$$

and in the tangential direction to obtain

$$-w \sin \theta = \frac{w}{g} l \frac{d^2\theta}{dt^2} \tag{5.34}$$

In (5.33), the normal direction is chosen as positive towards the center of the circle and in (5.34) the tangential direction is chosen as positive in the direction of increasing arc length s, corresponding to the direction of increasing angular displacement θ.

We now write the DE (5.34) in the form

$$\frac{d^2\theta}{dt^2} + \frac{g}{l} \sin \theta = 0 \tag{5.35}$$

Equation (5.35) is nonlinear and cannot be solved by the methods of Chapter 4. However, we know that for small values of θ (in the interval $-\pi/36 < \theta < \pi/36$, for example), θ and $\sin \theta$ have approximately the same value. This suggests that we replace $\sin \theta$ by θ in (5.35), to obtain

$$\frac{d^2\theta}{dt^2} + \frac{g}{l} \theta = 0 \tag{5.36}$$

We assume that θ is measured in radians, t in seconds, g in feet per second squared, and l in feet.

EXAMPLE.   A pendulum is displaced so that the rod makes an angle θ_0 radians with the vertical when released from rest. Find θ in terms of t.

Solution: Since (5.36) is the DE of a simple harmonic motion,

$$\theta = c_1 \sin \sqrt{\frac{g}{l}} t + c_2 \cos \sqrt{\frac{g}{l}} t$$

and

$$\omega = \frac{d\theta}{dt} = c_1 \sqrt{\frac{g}{l}} \cos \sqrt{\frac{g}{l}} t - c_2 \sqrt{\frac{g}{l}} \sin \sqrt{\frac{g}{l}} t$$

Using the initial conditions $\theta = \theta_0$ and $d\theta/dt = 0$ when $t = 0$, we find that $c_2 = \theta_0$ and $c_1 = 0$. Thus $\theta = \theta_0 \cos \sqrt{g/l} t$.

The *period* T of the motion is the time in seconds required for θ to change from θ_0 to $-\theta_0$ and back to θ_0. It is given by

$$T = \frac{2\pi}{\sqrt{g/l}} = 2\pi \sqrt{\frac{l}{g}} \tag{5.37}$$

This formula is approximately true for small oscillations and is frequently used in elementary physics. It can be used to approximate g from observed values of T and l. It is interesting to note that in this approximation applying to small oscillations, the period is independent of the initial conditions.

It is possible to obtain an intermediate or first integral of DE (5.35) by elementary methods. Since

$$\frac{d^2\theta}{dt^2} = \frac{d\omega}{dt} = \frac{d\theta}{dt}\frac{d\omega}{d\theta} = \omega\frac{d\omega}{d\theta}, \quad \text{then} \quad \omega\frac{d\omega}{d\theta} = -\frac{g}{l}\sin\theta$$

Separating variables and integrating gives

$$\frac{\omega^2}{2} = \frac{g}{l}\cos\theta + C$$

Setting $\omega = 0$ when $\theta = \theta_0$, we find that $C = -(g/l)\cos\theta_0$, and hence

$$\frac{\omega^2}{2} = \frac{g}{l}(\cos\theta - \cos\theta_0)$$

or

$$\omega = \frac{d\theta}{dt} = \pm\left(\frac{2g}{l}\right)^{1/2}(\cos\theta - \cos\theta_0)^{1/2} \tag{5.38}$$

From (5.38),

$$dt = \pm\left(\frac{l}{2g}\right)^{1/2}\frac{d\theta}{(\cos\theta - \cos\theta_0)^{1/2}} \tag{5.39}$$

It is not possible to integrate the right member of (5.39) in terms of elementary functions. Since $\cos(-\theta) \equiv \cos\theta$, the period τ of the motion is given by

$$\tau = 4\left(\frac{l}{2g}\right)^{1/2}\int_0^{\theta_0}\frac{d\theta}{(\cos\theta - \cos\theta_0)^{1/2}} \tag{5.40}$$

The integral on the right is known as an *elliptic integral of the first kind*. If $\cos\theta - \cos\theta_0$ is replaced by $2(\sin^2(\theta_0/2) - \sin^2(\theta/2))$, and if the substitution $\sin(\theta/2) = k\sin\phi$ is used with $k = \sin(\theta_0/2)$, Equation (5.40) becomes

$$\tau = 4\left(\frac{l}{g}\right)^{1/2}\int_0^{\pi/2}\frac{d\phi}{(1 - k^2\sin^2\phi)^{1/2}} = 4\sqrt{\frac{l}{g}}K \tag{5.41}$$

where

$$K = \int_0^{\pi/2}\frac{d\phi}{(1 - k^2\sin^2\phi)^{1/2}}$$

Values of K in terms of $\theta_0/2 = \sin^{-1}k$ have been extensively tabulated. See Reference 5.11.

5.5 The Simple Pendulum

If the integrand $(1 - k^2 \sin^2 \phi)^{-1/2}$ in (5.41) is expanded by the binomial theorem in an infinite series in $k^2 \sin^2 \phi$, the formula for τ becomes

$$\tau = 4\left(\frac{l}{g}\right)^{1/2} \int_0^{\pi/2} \left(1 + \frac{k^2}{2} \sin^2 \phi + \frac{1 \cdot 3}{2 \cdot 4} k^4 \sin^4 \phi + \cdots \right) d\phi$$

Term-by-term integration using one of Wallis's formulas (see Reference 5.12) yields

$$\tau = 2\pi\left(\frac{l}{g}\right)^{1/2} \left[1 + \left(\frac{1}{2}\right)^2 k^2 + \left(\frac{1 \cdot 3}{2 \cdot 4}\right)^2 k^4 + \cdots \right]$$

As $k \to 0$, then $\theta_0 \to 0$, and $\tau \to 2\pi\sqrt{l/g}$, in agreement with the period formula (5.37) for small oscillations. As θ_0 increases, k increases, and hence τ increases with increasing amplitude θ_0.

The replacement of $\sin \theta$ by θ in the DE $(d^2\theta/dt^2) + (g/l) \sin \theta = 0$, referred to as the *linearization* of the DE, converts a nonlinear DE to a linear DE. Although $\theta \approx \sin \theta$ for small θ, it is far from obvious that solutions of the linearized DE will approximate solutions of the nonlinear DE even for small θ. The linearization process amounts to replacement of $\sin \theta$ by its Maclaurin series

$$\theta - \frac{\theta^3}{3!} + \frac{\theta^5}{5!} - \cdots$$

and retention of the first term θ of the series.

Problem List 5.3

1. A hockey puck of weight w slides along the ice at 50 ft/sec. How far will it slide if the coefficient of friction μ equals 0.05? (The frictional force is μN, where $N = w$ is the normal force.) Use $g = 32$.

2. A baseball is thrown at an angle θ with the horizontal at a speed of 90 ft/sec. Find the speed of the ball when it is 50 ft above the ground. Use $g = 32$.

3. A 64-lb body A is moving in the positive x direction at 30 ft/sec. After being struck by a second body B, body A continues in the same direction at 50 ft/sec. Find the average force $\bar{F}$ acting on body A during impact if impact lasted 0.01 sec. Use $g = 32$.

4. Integrate DE (5.31) to obtain DE (5.32).

5. A rocket, 90 percent fuel and 10 percent structure, discharges fuel at a velocity relative to the rocket of 10,000 ft/sec. The constant rate of discharge is 5 percent of the original fuel per second. Assuming no action by external forces, find the velocity and the distance traveled at the end of (a) 10 sec, (b) 20 sec if $x = 0$ and $v = 0$ when $t = 0$.

6. Let the original fuel portion of a rocket be denoted by fw_0, where w_0 is the original fuel plus structure, and $0 < f < 1$. Assume that the rocket is fired vertically upward from the surface of the earth with $x = 0$ and $v = 0$ when $t = 0$. In Equation (5.29), set $F_x = -(w_0 - kt)$ and assume that g is constant. Find the *burnout velocity* $v (fw_0/k)$ and the *burnout height* $x (fw_0/k)$.

7. Let $m = m(v)$ denote the mass of a particle moving with velocity v at time
   According to Einstein's special theory of relativity, $m(v) = m_0/\sqrt{1 - (v^2/c^2)}$,
   where $m_0 = m(0)$ is the *rest mass* and $c \approx 186{,}000$ miles/sec is the speed of light.
   Assume that the particle moves in a straight line under the action of a constant
   force f. For $v = 0$ and $x = 0$ when $t = 0$, find (a) $v(t)$; (b) $\lim_{t \to +\infty} v(t)$; and
   (c) $x(t)$.

8. Find the period of a simple pendulum whose rod has length (a) $l = 4$ in.;
   (b) $l = 16$ in. Use $g = 32.17$ and assume that the oscillations are small.

9. A simple pendulum undergoing small oscillations has a period T of (a) 1 sec;
   (b) 2 sec. Find the length l of the pendulum in each case. Use $g = 32.17$.

10. A pendulum of length 1 ft undergoing small oscillations at a certain location
    has period 1.110 sec. Find the acceleration of gravity g at the same location.

11. Use Equations (5.33) and (5.38) to find the tension b in the rod of a simple
    pendulum. Express b in terms of w, θ, and θ_0.

12. The tension in the rod of a simple pendulum does no work on the pendulum
    since it acts perpendicular to the path of the pendulum. Derive Equation (5.38)
    using the work-energy equation.

13. A simple pendulum whose rod has length $l = 2$ ft is released from rest with ini-
    tial angular displacement $\theta_0 = \pi/3$ radians. Find the speed of the pendulum
    when it passes its lowest position ($\theta = 0$). Use $g = 32.17$.

14. Find the period of a simple pendulum undergoing small oscillations in a me-
    dium that resists the motion with a force proportional to the velocity of the
    pendulum. Use the approximation $\sin \theta \approx \theta$.

5.6 Electric Circuits

We have found (Section 3.12) that in the electric circuit of Figure 3.8 the
current i in amperes satisfies the DE

$$L\frac{di}{dt} + Ri = E(t) \tag{3.23}$$

Figure 5.13 contains an additional element known as a capacitor. This
type of element stores electrical energy in the circuit. The voltage drop across a

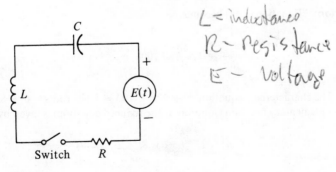

$L =$ inductance
$R \sim$ resistance
$E \sim$ voltage

Fig. 5.13

capacitor is proportional to the charge q in coulombs on the capacitor and is given by $C^{-1}q$, where C^{-1} is the constant of proportionality. The constant C is called the *coefficient of capacitance*, or simply the *capacitance*. Applying Kirchhoff's law to the circuit in Figure 5.13 yields the DE

$$L\frac{di}{dt} + Ri + \frac{1}{C}q = E(t) \qquad (5.42)$$

The current i in amperes equals the time rate of change of q; that is,

$$i(t) = \frac{dq(t)}{dt} \qquad (5.43)$$

Using (5.43) in (5.42), we get the DE

$$L\frac{d^2q}{dt^2} + R\frac{dq}{dt} + \frac{1}{C}q = E(t) \qquad (5.44)$$

satisfied by the charge q.

Differentiating both sides of (5.42) with respect to t and replacing dq/dt by i, we get the DE

$$L\frac{d^2i}{dt^2} + R\frac{di}{dt} + \frac{1}{C}i = \frac{d}{dt}E(t) \qquad (5.45)$$

satisfied by the current i.

It is assumed that L, R, and C are constants, with the inductance L given in henrys, the resistance R in ohms, and the capacitance C in farads. The impressed voltage $E(t)$ is given in volts and the time t in seconds.

Example. A circuit of the kind shown in Figure 5.13 consists of an inductor of 1 henry, a resistor of 12 ohms, a capacitor of 0.01 farad, and a generator having voltage given by $E(t) = 24 \sin 10t$. Find the charge q and the current i at time t if $q = 0$ and $i = 0$ when $t = 0$.

Solution: Equation (5.44) becomes

$$\frac{d^2q}{dt^2} + 12\frac{dq}{dt} + 100q = 24 \sin 10t \qquad (5.46)$$

The characteristic equation $r^2 + 12r + 100 = 0$ has roots $-6 \pm 8i$ and hence the complementary function (solution of the reduced equation) is given by

$$q_c = e^{-6t}(c_1 \cos 8t + c_2 \sin 8t)$$

Trying a particular solution of (5.46) of the form $q_p = A \cos 10t + B \sin 10t$, we

determine that $A = -\frac{1}{5}$ and $B = 0$. A complete solution of (5.46) is given by

$$q = e^{-6t}(c_1 \cos 8t + c_2 \sin 8t) - \frac{1}{5} \cos 10t$$

Differentiation with respect to t yields

$$\frac{dq}{dt} = i = e^{-6t}(-8c_1 \sin 8t + 8c_2 \cos 8t)$$

$$-6e^{-6t}(c_1 \cos 8t + c_2 \sin 8t) + 2 \sin 10t$$

Substituting $t = 0$ and $q = 0$ in the first equation, and $t = 0$ and $i = 0$ in the second equation, we find that $c_1 = \frac{1}{5}$ and $c_2 = \frac{3}{20}$. Hence the charge q is given by

$$q(t) = \frac{e^{-6t}}{20}(4 \cos 8t + 3 \sin 8t) - \frac{1}{5} \cos 10t$$

and the current $i = dq/dt$ by

$$i(t) = \frac{-5e^{-6t}}{2} \sin 8t + 2 \sin 10t$$

The *transient current*, given by $i_t = -\frac{5}{2}e^{-6t} \sin 8t$, is determined by the initial conditions and becomes negligible soon after $t = 0$. The *steady-state current*, given by $i_s = 2 \sin 10t$, approximates the actual current when the transient current becomes negligible.

Comparing Equation (5.44) with the equation

$$\frac{w}{g} \frac{d^2x}{dt^2} + c \frac{dx}{dt} + kx = F(t) \qquad (5.11)$$

we note the correspondences

mass $m = w/g \leftrightarrow$ inductance L

friction constant $c \leftrightarrow$ resistance R

spring constant $k \leftrightarrow$ inverse capacitance $1/C$

impressed force $F \leftrightarrow$ impressed voltage E

displacement $x \leftrightarrow$ charge q

velocity $v = dx/dt \leftrightarrow$ current $i = dq/dt$

This analogy also prevails in more complicated circuits, and consequently makes it possible to analyze certain mechanical systems by building and studying their electrical analogs. An *analog computer* is a device that simulates mechanical and physical systems by replacing the DE governing them by their electrical analogs, and then solving approximately the resulting DE for the electric circuits. Thus we have another manifestation of the manner in which the same mathematical model or theory can be used to describe more than one concrete application.

Problem List 5.4

1. Write a DE involving the charge q on the capacitor and the time t. Do not solve the DE.
   (a) A resistor of 50 ohms (Ω), an inductor of 2 henrys (H), and a capacitor of 0.05 farad (F) are in series with an emf of 40 volts (V).
   (b) An inductor of 2 H, a resistor of 40 Ω, and a capacitor of 0.001 F are in series with a generator having an emf of 220 cos 10t V.
   (c) An emf of $E(t) = 24 \sin 10t$ V is in series with a 12-Ω resistor, a 2-H inductor, and a 0.01-F capacitor.
   (d) An emf $E(t) = 200\sqrt{t}\,e^{-4t}$ is connected in series with a 10-Ω resistor and a 0.01-F capacitor.

2. The circuit in Figure 5.13 consists of an inductor of L H, a resistor of R Ω, a capacitor of C F, and a generator whose voltage at time t sec is given by $E(t)$. Find the charge q (in coulombs) and the current i (in amperes) at time t sec. In each case assume that $q = 0$ and $i = 0$ at $t = 0$.
   (a) $L = 1, R = 6, C = 0.01, E(t) = 100$
   (b) $L = 0.5, R = 5, C = 0.005, E(t) = 50 \sin 20t$
   (c) $L = 2, R = 8, C = \frac{1}{1250}, E(t) = 200 \cos 25t$

3. An inductor of 0.5 H is in series with a constant 50-V emf and a capacitor of 0.02 F. At $t = 0$, the charge q on the capacitor is 2 coulombs (C) and the current i is zero amperes (A). Find q and i in terms of t.

4. A circuit consists of an inductor of 10 H, a capacitor of 0.1 F, and an emf of 120 cos t V at t sec after the switch is closed. If the resistance R is assumed to be zero, and if the initial charge on the capacitor is 1 C, find the charge q on the capacitor at time t.

5. Show that the charge q and the current i satisfying DE (5.44) and (5.45) will be oscillatory if $R < 2\sqrt{L/C}$.

6. Given that the charge q satisfies

$$L\frac{d^2q}{dt^2} + R\frac{dq}{dt} + \frac{1}{C}q = 0$$

and that $q = q_0$ and $i = dq/dt = 0$ at $t = 0$, find q and i at time t. Assume $R < 2\sqrt{L/C}$.

5.7 A Curve of Pursuit

   Many interesting differential equations arise in determining the path one object or one individual follows in pursuing a second object or individual. The tractrix studied in Section 3.18 is a path or a curve of this type.

EXAMPLE 1. Ship A, located at $(x_0, 0)$, where $x_0 > 0$, sights ship B at the origin. Ship A pursues ship B at constant speed a as ship B sails in the positive y direction at constant speed ka. Where and when does A intercept B if A always sails directly towards B and $0 < k < 1$?

Solution: Figure 5.14 shows ship A at $P(x, y)$ and ship B at $Q(0, kat)$ at time t. To determine the path of ship A, we note that

$$\frac{dy}{dx} = \frac{y - kat}{x - 0} \quad \text{or} \quad x\frac{dy}{dx} - y = -kat$$

Differentiating both sides with respect to x, we get

$$x\frac{d^2y}{dx^2} + \frac{dy}{dx}(1) - \frac{dy}{dx} = x\frac{d^2y}{dx^2} = -ka\frac{dt}{dx}$$

Remembering that $ds/dx < 0$, we substitute

$$\frac{dt}{dx} = \frac{dt}{ds}\frac{ds}{dx} = \frac{ds/dx}{ds/dt} = -\frac{\sqrt{1 + (dy/dx)^2}}{a}$$

to obtain the following DE of the path of A:

$$x\frac{d^2y}{dx^2} = k\sqrt{1 + \left(\frac{dy}{dx}\right)^2}$$

Letting $p = dy/dx$ and separating the variables gives

$$x\frac{dp}{dx} = k\sqrt{1 + p^2} \quad \text{and} \quad \frac{dp}{\sqrt{1 + p^2}} = k\frac{dx}{x}$$

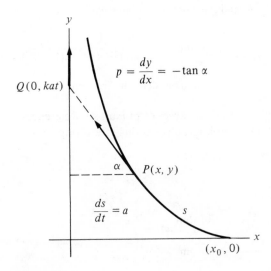

$$p = \frac{dy}{dx} = -\tan \alpha$$

$Q(0, kat)$

$$\frac{ds}{dt} = a$$

$P(x, y)$

s

$(x_0, 0)$

Fig. 5.14

Integrating gives

$$\sinh^{-1} p = k \ln x + c$$

and, from setting $p = 0$ and $x = x_0$,

$$0 = k \ln x_0 + c, \quad c = -k \ln x_0$$

and

$$\sinh^{-1} p = k \ln x - k \ln x_0 = k \ln \left(\frac{x}{x_0}\right) = \ln \left(\frac{x}{x_0}\right)^k$$

Hence,

$$p = \sinh \left[\ln \left(\frac{x}{x_0}\right)^k\right] = \frac{1}{2}\left\{\exp \ln \left(\frac{x}{x_0}\right)^k - \exp\left[-\ln \left(\frac{x}{x_0}\right)^k\right]\right\}$$

$$= \frac{1}{2}\left\{\exp \ln \left(\frac{x}{x_0}\right)^k - \exp \ln \left(\frac{x_0}{x}\right)^k\right\}$$

and

$$p = \frac{dy}{dx} = \frac{1}{2}\left[\left(\frac{x}{x_0}\right)^k - \left(\frac{x_0}{x}\right)^k\right] = \frac{1}{2}\left[\left(\frac{x}{x_0}\right)^k - \left(\frac{x}{x_0}\right)^{-k}\right]$$

Integrating both sides with respect to x gives

$$y = \frac{x_0}{2}\left[\frac{(x/x_0)^{1+k}}{1+k} - \frac{(x/x_0)^{1-k}}{1-k}\right] + C$$

Setting $x = x_0$ and $y = 0$, we get

$$0 = \frac{x_0}{2}\left[\frac{1}{1+k} - \frac{1}{1-k}\right] + C \quad \text{and hence} \quad C = \frac{x_0 k}{1 - k^2}$$

When $x = 0$, then $y = C$, and therefore ship A intercepts ship B

$$\text{at} \quad \left(0, \frac{x_0 k}{1 - k^2}\right) \quad \text{at time} \quad t = \frac{1}{ka}\frac{x_0 k}{1 - k^2} = \frac{x_0}{a(1 - k^2)}$$

The cases $k = 1$ and $k > 1$ are considered in Problems 2 and 3.

For additional material on curves of pursuit see References 5.1 and 5.13, each of which contains a bibliography on this interesting subject.

Problem List 5.5

1. Find: (a) $\lim\limits_{k \to 0^+} \dfrac{x_0 k}{1 - k^2}$ (b) $\lim\limits_{k \to 0^+} \dfrac{x_0}{a(1 - k^2)}$

   (c) $\lim\limits_{k \to 1^-} \dfrac{x_0 k}{1 - k^2}$ (d) $\lim\limits_{k \to 1^-} \dfrac{x_0}{a(1 - k^2)}$

   Explain the significance of these results in Example 1.

2. In Example 1, find y in terms of x when $k = 1$. Let l = distance PQ in Figure 5.14. Show that when $k = 1, l > x_0/2$, and that $\lim_{x \to 0^+} l = x_0/2$.

3. In Example 1, find y in terms of x and x_0 when $k = 2$. Also find y in terms of x_0 when $x = x_0/2$.

4. The formula $\int (dp/\sqrt{1 + p^2}) = \sinh^{-1} p + C$, used in Example 1, is often written

$$\int \frac{dp}{\sqrt{1 + p^2}} = \ln(p + \sqrt{1 + p^2}) + C$$

Show that $\ln(p + \sqrt{1 + p^2}) = \sinh^{-1} p$.

5. Ship A, located at $(0, 0)$, sights ship B at $(1, 0)$. Ship A pursues ship B at constant speed a as ship B sails along the line $x = 1$ in the direction of increasing y. Find an equation of the path of ship A if A always sails directly toward B and B also sails at constant speed a. Show that the distance l between A and B is given by $l = [(1 - x)^2 + 1]/2$, where x is the abscissa of A.

5.8 Cables; The Catenary

In Figure 5.15, a flexible, inextensible cable is suspended at rest from two points, A and B, that do not lie in the same vertical line. We assume that the cable has the shape of a curve, with equation $y = f(x)$, where f is differentiable and $f'(0) = 0$. That is, we assume that the y axis is perpendicular to the cable at its low point $C(0, c)$. Arc length s along the cable is measured positively from $C(0, c)$ as shown, and $P(x, y)$ denotes an arbitrary point on the cable. The section of cable from C to P is held in equilibrium under the action of the three forces represented in Figure 5.15(b) by vectors $\mathbf{T}$, $\mathbf{T_0}$, and $\mathbf{W}$ having magnitudes t, t_0, and w. Since the cable is flexible, the vectors $\mathbf{T}$ and $\mathbf{T_0}$, representing tensions at C and P, act tangent to the cable.

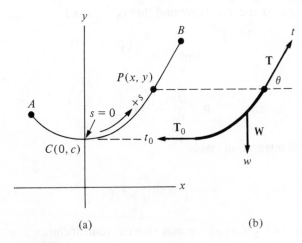

(a)   (b)

Fig. 5.15

The vector **W** acts downward and represents the resultant of the gravitational attraction of the earth on the portion of cable between C and P. It follows from the equilibrium of the three acting forces that

$$t \sin \theta = w \quad \text{and} \quad t \cos \theta = t_0 \tag{5.47}$$

where θ denotes the inclination of the tangent to the cable at P.

Eliminating t by dividing the first equation by the second, we obtain

$$\tan \theta = \frac{dy}{dx} = \frac{w}{t_0} \tag{5.48}$$

Replacing w by ks, where k denotes the weight per unit length of the cable, we obtain

$$\frac{dy}{dx} = \frac{k}{t_0} s \tag{5.49}$$

Differentiation of both sides with respect to x yields

$$\frac{d^2 y}{dx^2} = \frac{k}{t_0} \frac{ds}{dx} = \frac{k}{t_0} \sqrt{1 + \left(\frac{dy}{dx}\right)^2}$$

as the DE of the curve assumed by the cable.

Letting $p = dy/dx$ and separating variables, we obtain

$$\frac{dp}{dx} = \frac{k}{t_0} \sqrt{1 + p^2} \quad \text{and} \quad \frac{dp}{\sqrt{1 + p^2}} = \frac{k}{t_0} dx$$

Integrating gives $\sinh^{-1} p = (kx/t_0) + c_1$.

Setting $p = 0$ and $x = 0$, we find that $c_1 = 0$ and

$$\sinh^{-1} p = \frac{kx}{t_0}$$

or

$$p = \frac{dy}{dx} = \sinh \frac{kx}{t_0} \tag{5.50}$$

A second integration yields

$$y = \frac{t_0}{k} \cosh \frac{kx}{t_0} + c_2$$

The constant c_2 is zero if we make the convenient choice $y = t_0/k$ when $x = 0$. Thus, we choose the point $C(0, c)$ in Figure 5.15 as the point $C(0, t_0/k)$

and obtain

$$y = \frac{t_0}{k} \cosh \frac{kx}{t_0} = \frac{t_0}{2k} (e^{kx/t_0} + e^{-kx/t_0}) \tag{5.51}$$

as an equation of the cable. A curve having an equation of this form is known as a *catenary*, after the Latin *catena*, meaning "chain." Many transmission lines, telegraph cables, and cables of suspension bridges hang in the form of catenaries.

Writing (5.51) in the form

$$y = c \cosh \frac{x}{c} = \frac{c}{2} (e^{x/c} + e^{-x/c}) \tag{5.52}$$

and noting that $t_0 = kc$, we can regard (5.52) as an equation of a one-parameter family of catenaries, where the parameter c denotes the length of a portion of cable of weight t_0. Each member of the family is symmetric with respect to the y axis.

From (5.49) and (5.52), we have

$$s = \frac{t_0}{k} \frac{dy}{dx} = c \sinh \frac{x}{c} = \frac{c}{2} (e^{x/c} - e^{-x/c}) \tag{5.53}$$

and

$$w = ks = kc \sinh \frac{x}{c} = t_0 \sinh \frac{x}{c}$$

From (5.47) and (5.53), we obtain

$$t^2 \sin^2 \theta + t^2 \cos^2 \theta = t^2$$

$$t^2 = w^2 + t_0^2 = k^2 s^2 + k^2 c^2$$

$$= k^2 c^2 \sinh^2 \frac{x}{c} + k^2 c^2$$

$$= k^2 c^2 \cosh^2 \frac{x}{c}$$

and

$$t = kc \cosh \frac{x}{c}$$

$$= t_0 \cosh \frac{x}{c} = ky$$

Cables hang in a great variety of curves. The catenary was derived by assuming a particular form for w in

$$\frac{dy}{dx} = \frac{w}{t_0} \tag{5.48}$$

Another curve is obtained by taking $w = kx$ instead of $w = ks$ in (5.48). This assumes that the cable carries a uniform load of k pounds per horizontal foot. Cables of suspension bridges are often loaded in approximately this fashion.

With $w = kx$ in (5.48), we have $dy/dx = kx/t_0$.

Integrating with respect to x, we obtain $y = [kx^2/(2t_0)] + M$.

Letting $y = 0$ when $x = 0$, we get $M = 0$ and

$$y = \frac{kx^2}{2t_0} = \frac{x^2}{2c} \tag{5.54}$$

Equation (5.54) represents a one-parameter family of parabolas with the parameter c denoting the length of the horizontal projection of a portion of the loaded cable of weight t_0.

If the points A and B of Figure 5.15 from which the cable is suspended have the same ordinate $y_A = y_B$, the distance between them is called the *span* of the cable. The vertical distance $y_A - c = y_B - c$ between A (or B) and the lowest point $C(0, c)$ of the cable is called the *dip*, or *sag*, of the cable.

Problem List 5.6

1. Find the height of the cable $y = 200 \cosh (x/200)$ at its endpoints if it is fixed at $x = -100$ ft and at $x = 200$ ft.

2. For the cable $y = c \cosh (x/c)$, show that $y^2 - s^2 = c^2$.

3. Show that if t_1 is the tension at $P_1(x_1, y_1)$ and t_2 is the tension at $P_2(x_2, y_2)$, where P_1 and P_2 are points of the cable $y = c \cosh x/c$, then

$$t_2 - t_1 = k(y_2 - y_1)$$

k denoting the weight per unit length of the cable.

4. The cable $y = 100 \cosh (x/100)$ is fixed at $x = -100$ ft and at $x = +100$ ft. Given that the cable weighs 2 lb/ft, find the span, the sag, the tension at the low point, and the tension at the supports. Determine the length of the cable.

5. A 90-ft cable weighing 2 lb/ft is suspended from points A and B on the same level. Find an equation of the cable, the span, and the sag if the tension t_0 at the low point of the cable is 120 lb.

6. A parabolic cable of a suspension bridge carries a horizontal load of 800 lb per ft. Find the tension t_0 at the low point and the tension t_B at one of the supports if the cable has span 600 ft and sag 40 ft.

5.9 Deflection of Beams; the Elastic Curve

The weightless rectangular beam in Figure 5.16(a) has edges AB and DC horizontal and edges AD and BC vertical. The x and y axes pass through the centroid O of rectangle $ABCD$, with the x axis perpendicular to the plane of rectangle $ABCD$, and the y axis parallel to edge AD. Since the beam is assumed

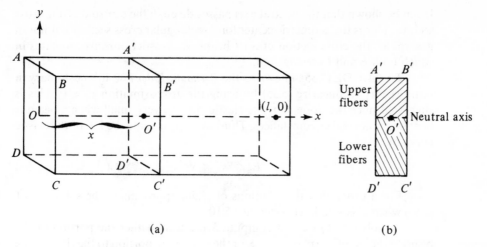

(a)   (b)

Fig. 5.16

weightless, the x axis passes through the centroid of every cross section of the beam parallel to rectangle $ABCD$. For example, the centroid O' of section $A'B'C'D'$, located x units from O, is on the x axis. The same is true of the centroid $(l, 0)$ of the right end of the beam, l denoting the length of the beam.

The centroids of the cross sections lie on the curve having equation $y \equiv 0$, and this curve describes the shape of the beam. When the weight of the beam is considered, and when other vertical forces act on the beam, the centroids of the cross sections lie on a curve having equation $y = f(x)$. This curve is known as the *elastic curve* of the beam.

In Figure 5.16(b), the line through O' parallel to $A'B'$ is known as the *neutral axis* of the cross section. It divides the cross section into two regions, one consisting of the upper fibers and the other of the lower fibers. The forces tending to bend the beam place one of these sets of fibers in tension and the other in compression. The fibers along the neutral axis are in neither tension nor compression.

The algebraic sum of the moments of the external forces and couples acting on one side of the beam, taken about the neutral axis, is called the *bending moment* at x and is denoted by $M(x)$. We adopt the convention that the moment of any force or couple that tends to compress the upper fibers is positive and the moment of any force or couple that tends to place the upper fibers in tension is negative.

It has been shown (see Reference 5.9) that a DE of the elastic curve is given by

$$\frac{y''}{[1 + (y')^2]^{3/2}} = \frac{M(x)}{EI} \tag{5.55}$$

where E is Young's modulus of elasticity and I is the moment of inertia of a cross section about the neutral axis. The value of E depends on the material from which the beam is made and I depends on the shape of the cross section.

It can be shown that the neutral axis passes through the centroid of the cross section. This is the geometric center for a rectangular cross section but not in general for the cross section of an *I* beam. We restrict ourselves to cases in which both *E* and *I* are constant.

To solve DE (5.55), known as the *Bernoulli-Euler law*, it is customary in beam theory to linearize (5.55) by using the approximation $y' \approx 0$. This is suggested since the slope y' of the elastic curve is very small when a beam is bent under normal circumstances. Thus, we approximate the DE of the elastic curve by

$$EIy'' = M(x) \tag{5.56}$$

For a proof that the solutions of (5.56) approximate the solutions of (5.55) when y' is small, see Reference 5.10.

To solve (5.56) we first compute $M(x)$, using either the portion of the beam to the left of the plane $x = x$ or the remaining portion to the right of the plane $x = x$. Two integrations of $M(x)$ with respect to x yield a solution of (5.56) containing two arbitrary constants. These constants are determined using two sets of initial conditions. A particular solution given by $y = f(x)$ is an equation of the elastic curve into which the beam is bent by the applied loads, and y measures the vertical displacement or *deflection* of the beam from its theoretical position as displayed in Figure 5.16(a).

EXAMPLE 1. Find an equation of the elastic curve for a beam *l* ft long carrying a uniformly distributed load of *w* lb/ft. Assume that the beam is *simply supported*, that is, that an upward force of $wl/2$ lb acts at each end. See Figure 5.17.

Solution: We shall use the forces acting on the portion of the beam to the left of $x = x$ to find $M(x)$. The moment due to the left supporting force is $(wl/2)(x)$ and is

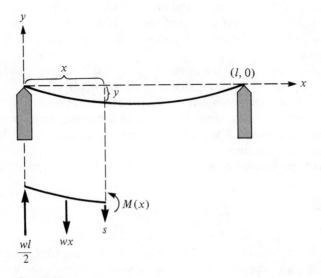

Fig. 5.17

positive since this force tends to compress the upper fibers at $x = x$. To find the moment of the distributed force due to the weight of the left portion of the beam, we use the resultant of wx lb acting downward at $x/2$ ft from the origin. The moment of this force about $x = x$ is $-(wx)(x/2)$ and is negative since this force tends to place the upper fibers at $x = x$ in tension. Thus, $M(x) = -wx^2/2 + wlx/2$. The left portion is held in equilibrium by the two forces used in finding $M(x)$, in addition to an equal and opposite couple of moment $|M(x)|$ and a vertical shearing force $\mathbf{S}$ of magnitude s at $x = x$ supplied by the right portion of the beam.

Substituting $M(x)$ into (5.56), we obtain

$$EIy'' = \frac{wx^2}{2} + \frac{wlx}{2}$$

Two integrations with respect to x yield

$$EIy' = -\frac{wx^3}{6} + \frac{wlx^2}{4} + c_1$$

and

$$EIy = -\frac{wx^4}{24} + \frac{wlx^3}{12} + c_1 x + c_2$$

Using the initial conditions $x = 0$, $y = 0$, and $x = l$, $y = 0$, we obtain

$$c_2 = 0 \quad \text{and} \quad c_1 = \frac{wl^3}{24} - \frac{wl^3}{12} = \frac{-wl^3}{24}$$

Thus, the elastic curve has equation

$$EIy = \frac{-wx^4}{24} + \frac{wlx^3}{12} - \frac{wl^3 x}{24}$$

or

$$y = \frac{-w}{24EI}(x^4 - 2lx^3 + l^3 x)$$

EXAMPLE 2. Find an equation of the elastic curve of an *l*-ft beam, which carries a uniformly distributed load of w lb/ft, and which is built in horizontally at the left end and unsupported at the right end. This type of beam, resembling a diving board, is called a *cantilever beam*.

Solution: In Figure 5.18 it is convenient to use the portion of the beam to the right of $x = x$ to find $M(x)$. The only force acting on the right portion, other than a vertical shearing force $\mathbf{S}$ of magnitude s at $x = x$, is the resultant of the distributed weight $w(l - x)$ acting $(l - x)/2$ ft to the right of $x = x$. Thus, $M(x) = -w(l - x)^2/2$ and is negative since the distributed weight tends to place the upper fibers in tension. The advantage of using the right portion of the beam is that it is not necessary to compute the unknown couple and the upward supporting force exerted by the wall on the left end of the beam. Integrating $EIy'' = -w(l - x)^2/2$, we obtain

$$EIy' = \frac{w(l - x)^3}{6} + c_1$$

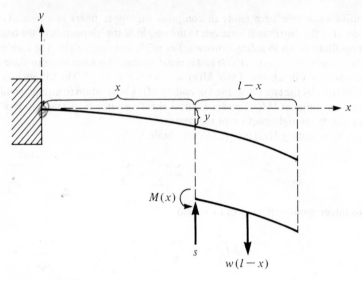

Fig. 5.18

Since $y' = 0$ when $x = 0$, $c_1 = -wl^3/6$. A second integration yields

$$Ely = \frac{-w(l - x)^4}{24} - \frac{wl^3x}{6} + c_2$$

Since $y = 0$ when $x = 0$

$$c_2 = \frac{wl^4}{24} \quad \text{and} \quad Ely = \frac{-w(l - x)^4}{24} - \frac{wl^3x}{6} + \frac{wl^4}{24}$$

Expanding $(l - x)^4$, simplifying, and solving for y, we obtain

$$y = \frac{-w}{24EI}(x^4 - 4lx^3 + 6l^2x^2)$$

The maximum deflection d of a beam is the maximum value assumed by $|y|$ for $0 \leq x \leq l$. Maximum deflections are important when one is designing beams. Since $|y|$ varies inversely with the constant EI, known as the *flexural rigidity*, the quantity EI provides a measure of the tendency of a beam to resist bending. The value of E can be increased by choosing a less elastic material; the value of I can be increased by changing the distribution of the cross-sectional area relative to the neutral axis. An I beam provides great resistance to bending because of its large moment of inertia about its neutral axis.

In Example 1, it is fairly obvious that the maximum deflection occurs at $x = l/2$. Similarly for Example 2, it is intuitively clear that the maximum deflection occurs at $x = l$. In more complicated problems, the equation $y' = 0$ can be solved to determine where the maximum deflection occurs. See Problem 8.

EXAMPLE 3. In Example 1 find the maximum deflection d in terms of w, l, E, and I. Find d in inches if $w = 30$ lb per ft, $l = 10$ ft, $E = 30(10)^6$ lb/in.^{2}, and $I = 32$ in.^{4}.

Solution: When $x = l/2$,

$$|y| = d = \frac{w}{24EI}\left|\frac{l^4}{16} - \frac{2l^4}{8} + \frac{l^4}{2}\right| = \frac{5\,wl^4}{384EI}$$

For the given data,

$$d = \frac{(5)(30)(10)[(10)(12)]^3}{(384)(30)(10)^6(32)} = \frac{45}{6400} \approx 0.007 \text{ in.}$$

Problem List 5.7

1. Show in Example 1 that $y' = 0$ when $x = 1/2$.
2. In Example 2, find the maximum deflection d in terms of w, l, E, and I.
3. A 10-ft steel cantilever beam $[E = 30(10)^6 \text{ lb/in.}^2]$ carries a uniformly distributed load of 1500 lb/ft. Given that $I = 45$ in.^{4}, use the result in Problem 2 to find the maximum deflection d.
4. An l-ft cantilever beam, built in horizontally at the left end, has weight W lb suspended from its right end. Find an equation of the elastic curve if the beam is assumed weightless.
5. An 8-ft oak beam $[E = 15(10)^5 \text{ lb/in.}^2]$ carrying a uniformly distributed load of 150 lb/ft, is simply supported at both ends. Given that $I = 57$ in.^{4}, find (a) the deflection at $x = 4$ ft, and (b) the maximum deflection d.
6. A simply supported beam of length l ft carries a concentrated load of W lb at $x = l/2$. Find the maximum deflection d if the weight of the beam is neglected.
7. An l-ft beam, built in horizontally at the left end and simply supported at the right end, carries a uniformly distributed load of w lb per ft. Find an equation of the elastic curve.
8. Find the value of x in Problem 7 at which the maximum deflection occurs.
9. An l-ft beam, built in horizontally at both ends, carries a uniformly distributed load of w lb/ft. Find an equation of the elastic curve and find the maximum deflection d in terms of w, l, E, and I. *Hint:* Assume a negative bending moment $-M(0)$ at the left end of the beam. Use the conditions $x = 0$, $y = 0$; $x = 0$, $y' = 0$ and $x = l/2$, $y' = 0$ to determine M_0 and the two constants of integration.

5.10 Concentration of a Substance Inside and Outside a Living Cell

Let I denote the region interior to a living cell and let c_i denote the concentration in grams per cubic centimeter of a substance (solute) dissolved in the cell liquid (solvent). Carbon dioxide and lactic acid provide examples of such solutes. Assuming that the solute is produced inside the cell at Qg cm^{-3} sec^{-1}, and that a steady state prevails in which $\partial c_i/\partial t \equiv 0$, we know (from

Reference 5.7) that c_i satisfies the partial DE

$$D_i\left(\frac{\partial^2 c_i}{\partial x^2} + \frac{\partial^2 c_i}{\partial y^2} + \frac{\partial^2 c_i}{\partial z^2}\right) + Q = 0 \qquad (5.57)$$

where D_i is the coefficient of diffusion for the medium interior to the cell. Let us consider the case in which D_i and Q are both constant, c_i depending on x, y, and z.

Let E denote the region exterior to I and consider the case in which I is spherical. It is convenient to introduce spherical coordinates ρ, θ, ϕ satisfying

$$x = \rho \sin \phi \cos \theta \quad y = \rho \sin \phi \sin \theta \quad z = \rho \cos \phi$$

with the origin at the center of the cell whose radius we denote by ρ_0.

In spherical coordinates the expression

$$\frac{\partial^2 c_i}{\partial x^2} + \frac{\partial^2 c_i}{\partial y^2} + \frac{\partial^2 c_i}{\partial z^2}$$

known as the *Laplacian* of c_i, becomes

$$\frac{\partial^2 c_i}{\partial \rho^2} + \frac{2}{\rho}\frac{\partial c_i}{\partial \rho} + \frac{1}{\rho^2}\frac{\partial^2 c_i}{\partial \phi^2} + \frac{\cos \phi}{\rho^2 \sin \phi}\frac{\partial c_i}{\partial \phi} + \frac{1}{\rho^2 \sin^2 \phi}\frac{\partial^2 c_i}{\partial \theta^2}$$

(See Problem 6.)

Because of the spherical symmetry of the physical situation, the partial derivatives of c_i with respect to ϕ and θ vanish, c_i depends on the radial distance ρ alone, and (5.57) reduces to the ordinary DE

$$D_i\left(\frac{d^2 c_i}{d\rho^2} + \frac{2}{\rho}\frac{dc_i}{d\rho}\right) + Q = 0 \qquad (5.58)$$

The concentration c_e outside the cell satisfies

$$\frac{d^2 c_e}{d\rho^2} + \frac{2}{\rho}\frac{dc_e}{d\rho} = 0 \qquad (5.59)$$

in which no term corresponding to Q appears, since we assume that no solute is generated chemically outside the cell.

Although (5.58), after division by D_i, has variable coefficient $2(D_i \rho)^{-1}$, it is readily solved by the substitution $u = \rho c_i$. From

$$\frac{du}{d\rho} = \rho \frac{dc_i}{d\rho} + c_i \quad \text{and} \quad \frac{d^2 u}{d\rho^2} = \rho \frac{d^2 c_i}{d\rho^2} + 2 \frac{dc_i}{d\rho}$$

and (5.58), we obtain

$$\frac{d^2u}{d\rho^2} = \frac{-Q}{D_i} \rho$$

Two integrations with respect to ρ yield

$$u = \frac{-Q}{6D_i} \rho^3 + A\rho + B$$

and hence

$$c_i = -\frac{Q}{6D_i} \rho^2 + A + \frac{B}{\rho}, \quad 0 \le \rho \le \rho_0 \tag{5.60}$$

Similarly the substitution $v = \rho c_e$ yields (see Problem 1)

$$c_e = C + \frac{K}{\rho}, \quad \rho_0 \le \rho \tag{5.61}$$

The constants B and C in (5.60) and (5.61) are determined by making two physical assumptions. The first is that c_i has a finite value at the origin. For this to hold, B must be zero; otherwise, c_i would be undefined at the origin. We next assume that $\lim_{\rho \to +\infty} c_e = k$, where k is the concentration that prevailed at every point exterior to the cell before the diffusion from the cell began. It is reasonable to assume that after the steady state sets in, the value of c_e will still be close to k at points where ρ is very large. This would not hold if any solute were being created outside the cell. Thus, C has the constant nonnegative value k. The value $k = 0$ would prevail if there were no solute outside the cell before the outward diffusion process began.

The constants A and K in

$$c_i = -\frac{Q}{6D_i} \rho^2 + A \tag{5.62}$$

and

$$c_e = k + \frac{K}{\rho} \tag{5.63}$$

are determined from the following boundary conditions prevailing at $\rho = \rho_0$:

$$\frac{dc_i}{d\rho} = \frac{D_e}{D_i} \frac{dc_e}{d\rho} \tag{5.64}$$

and

$$\frac{dc_i}{d\rho} = \frac{-h}{D_i}(c_i - c_e) \tag{5.65}$$

where D_e is the coefficient of diffusion for the medium exterior to the cell and h is the *permeability* of the cellular membrane.

The manner in which these boundary conditions, based on Fick's law and the equation of continuity, are obtained is discussed in Reference 5.7.

From

$$\frac{dc_i}{d\rho} = \frac{-Q\rho}{3D_i} \quad \text{and} \quad \frac{dc_e}{d\rho} = \frac{-K}{\rho^2} \tag{5.66}$$

and (5.64), with $\rho = \rho_0$, we obtain

$$\frac{-Q\rho_0}{3D_i} = \frac{D_e}{D_i}\left(\frac{-K}{\rho_0^2}\right)$$

which yields $K = Q\rho_0^3/(3D_e)$. Thus,

$$c_e = k + \frac{Q\rho_0^3 \rho^{-1}}{3D_e} \tag{5.67}$$

A slightly more complicated computation (see Problem 2) yields

$$A = k + \frac{Q\rho_0}{3h} + \frac{Q\rho_0^2}{6D_i} + \frac{Q\rho_0^2}{3D_e} \tag{5.68}$$

After substituting A into (5.62) and rearranging terms, we find that

$$c_i = k + \frac{Q\rho_0}{3h} + \frac{Q}{6D_i}(\rho_0^2 - \rho^2) + \frac{Q\rho_0^2}{3D_e} \tag{5.69}$$

Since we assumed that the solute was produced inside the cell, Q was positive in the development given. If the solute is consumed inside the cell, the same model is appropriate, with $Q < 0$. Oxygen and sugar are examples of substances consumed inside living cells.

Problem 5 treats briefly the situation in which the solute is produced inside the cell at a rate proportional to the concentration c_i. Instead of treating Q as a constant, we assume that $Q = ac_i$, where a is a positive constant. This kind of reaction is termed *autocatalytic*, reflecting that the solute catalyzes its own creation.

Problem List 5.8

1. Use the substitution $v = \rho c_e$ in (5.59) to obtain (5.61).
2. Derive (5.68) and (5.69) from (5.62), (5.65), (5.66), and (5.67).
3. The concentration of the solute is discontinuous at the boundary of the spherical shell. Compute $c_i - c_e$ for $\rho = \rho_0$.
4. Find the value of c_i at the center of the cell.

5. A substance is produced inside a spherical living cell of radius ρ at a rate proportional to the concentration c_i of the dissolved substance. Use (5.58) to show that

$$c_i = \frac{A \sin\left(\sqrt{a/D_i}\,\rho\right)}{\rho} + \frac{B \cos\left(\sqrt{a/D_i}\,\rho\right)}{\rho}$$

where $a > 0$ is the constant of proportionality and A and B are arbitrary constants.

6. Show that in spherical coordinates the expression

$$\frac{\partial^2 f}{\partial x^2} + \frac{\partial^2 f}{\partial y^2} + \frac{\partial^2 f}{\partial z^2}$$

known as the Laplacian of f, becomes

$$\frac{\partial^2 f}{\partial \rho^2} + \frac{2}{\rho}\frac{\partial f}{\partial \rho} + \frac{1}{\rho^2}\frac{\partial^2 f}{\partial \phi^2} + \frac{\cos\phi}{\rho^2 \sin\phi}\frac{\partial f}{\partial \phi} + \frac{1}{\rho^2 \sin^2\phi}\frac{\partial^2 f}{\partial \theta^2}$$

5.11 A Microeconomic Market Model

We have previously considered (Section 3.15) a simple market model in which Q_d, the number of items of a commodity in demand per unit of time, and Q_s, the number of items of the commodity in supply per unit of time, are linear functions of the price P of a single item. There is a more sophisticated model, in which Q_d and Q_s depend linearly not only on P but also on dP/dt and d^2P/dt^2. This model incorporates the effect of trends in prices on demand and supply. Terms involving $dP/dt = P'$ account for the effect of rising or falling prices, and terms involving $d^2P/dt^2 = P''$ account for the effect of increasing or decreasing price rates of change. Let

$$Q_d = a - Pb + AP' + BP'' \quad \text{and} \quad Q_s = -c + Pd + CP' + DP''$$

where a, b, c, and d are positive constants as in the simpler model, and the constants A, B, C, D may be positive, negative, or zero.

For economic equilibrium to prevail, Q_d must equal Q_s. This leads to the following DE for P:

$$(B - D)P'' + (A - C)P' - (b + d)P = -(a + c)$$

We consider the case $B \neq D$, which leads to the second-order DE

$$P'' + \frac{A - C}{B - D}P' - \frac{b + d}{B - D}P = \frac{-(a + c)}{B - D} \tag{5.70}$$

We see by inspection that

$$P_p = (a + c)/(b + d) = \bar{P}$$

defines a particular solution of (5.70). The positive constant $\bar{P} = (a + c)/(b + d)$ is called the *equilibrium price*. The equilibrium is said to be *dynamically stable* if and only if $\lim_{t \to +\infty} P(t) = \bar{P}$. The importance of the concept of a dynamically stable equilibrium in a model is discussed in Reference 5.2. To determine whether or not a dynamically stable equilibrium prevails, we examine the roots r_1 and r_2 of the characteristic equation

$$r^2 + \frac{A - C}{B - D} r - \frac{b + d}{B - D} = 0$$

of DE (5.70). We first note that since $b + d > 0$, r_1 and r_2 are both nonzero. If r_1 and r_2 are real and unequal, a complete solution of (5.70) is given by

$$P = P(t) = c_1 e^{r_1 t} + c_2 e^{r_2 t} + \bar{P}$$

The arbitrary constants c_1 and c_2 can be determined from the initial values $P(0)$ and $P'(0)$. Since c_1 (or c_2) will be zero only for very special conditions, we require for dynamic stability that $\lim_{t \to +\infty} P(t) = \bar{P}$ when c_1 and c_2 are both different from zero. It is easily seen that $\lim_{t \to +\infty} P(t) = \bar{P}$ if and only if $r_1 < 0$ and $r_2 < 0$. (If $r > 0$, $\lim_{t \to +\infty} e^{rt} = +\infty$.)

If $r_1 = r_2$, a complete solution of (5.70) is given by

$$P = P(t) = c_1 e^{r_1 t} + c_2 t e^{r_1 t} + \bar{P}$$

In this case $\lim_{t \to +\infty} P(t) = \bar{P}$ if and only if $r_1 < 0$. (See Problem 1.)

Finally, if $r_1 = \alpha + i\beta$ and $r_2 = \alpha - i\beta$ where $\beta \neq 0$ and $i^2 = -1$, a complete solution of (5.70) is given by

$$P = P(t) = e^{\alpha t}(c_1 \sin \beta t + c_2 \cos \beta t) + \bar{P}$$

In this case of complex, nonreal characteristic roots, it is easy to see that $\lim_{t \to +\infty} P(t) = \bar{P}$ if and only if $\alpha < 0$. (The absolute value of the harmonic factor $c_1 \sin \beta t + c_2 \cos \beta t$ cannot exceed $|c_1| + |c_2|$.)

Thus, it is possible to determine, without obtaining a complete solution, whether the solution of DE (5.70) does or does not converge to $\bar{P}$.

EXAMPLE 1. Given

$$Q_d = 8 - 2P - 3P' + 3P'' \quad \text{and} \quad Q_s = -6 + 4P + 2P' + 4P''$$

determine whether or not equilibrium is dynamically stable.

Solution: Setting $Q_d = Q_s$, we obtain $P'' + 5P' + 6P = 14$.

The characteristic equation $r^2 + 5r + 6 = (r + 3)(r + 2) = 0$ has roots $r_1 = -3$ and $r_2 = -2$. Since r_1 and r_2 are both negative, the equilibrium is dynamically stable and

$$\lim_{t \to +\infty} P(t) = \bar{P} = \frac{a + c}{b + d} = \frac{8 + 6}{2 + 4} = \frac{7}{3}$$

EXAMPLE 2. Given $Q_d = 10 - P - P' + 2P''$ and $Q_s = -5 + 4P + P' + 3P''$, find the price P in terms of t if $P(0) = 5$ and $P'(0) = 1$. Determine whether or not economic equilibrium is dynamically stable.

Solution: Setting $Q_d = Q_s$, we obtain

$$P'' + 2P' + 5P = 15 \tag{5.71}$$

The characteristic equation $r^2 + 2r + 5 = 0$ has roots $r_1 = -1 + 2i$ and $r_2 = -1 - 2i$. A particular solution of (5.71) is defined by $P_p = 3 = \bar{P}$ and a complete solution of (5.71) is given by

$$P = P(t) = e^{-t}(c_1 \sin 2t + c_2 \cos 2t) + 3$$

Differentiation with respect to time yields

$$P'(t) = e^{-t}(2c_1 \cos 2t - 2c_2 \sin 2t)$$
$$-e^{-t}(c_1 \sin 2t + c_2 \cos 2t)$$

From $P(0) = 5$ and $P'(0) = 1$, we obtain $5 = c_2 + 3$ and $1 = 2c_1 - c_2$, from which $c_2 = 2$ and $c_1 = \frac{3}{2}$. Hence,

$$P = P(t) = e^{-t}\left(\frac{3}{2} \sin 2t + 2 \cos 2t\right) + 3$$

Since $\alpha = -1 < 0$, the equilibrium is dynamically stable and

$$\bar{P} = \lim_{t \to +\infty} P(t) = 3$$

Problem List 5.9

1. Show that if $r < 0$, $\lim_{t \to +\infty} te^{rt} = 0$.
2. Without solving a DE for P, determine whether or not equilibrium is dynamically stable.
   (a) $Q_d = 8 - 2P + 3P' + 2P''$
       $Q_s = -5 + P - 2P' - P''$
   (b) $Q_d = 9 + 3P + 2P' + 3P''$
       $Q_s = -7 - 2P + P' + 2P''$
   (c) $Q_d = 10 + P + 2P' + 2P''$
       $Q_s = -6 - 3P - 2P' + P''$
3. Find $\bar{P}$, P in terms of t, and determine whether or not equilibrium is dynamically stable.

(a) $Q_d = 7 - P + P' + P''$
$Q_s = -9 + P + 4P' + 2P''$
$P(0) = 2, P'(0) = -1$

(b) $Q_d = 6 - 5P + 2P' - 4P''$
$Q_s = -4 + 5P - 2P' - 2P''$
$P(0) = 3, P'(0) = -1$

References

5.1 Bellman, R., and K. L. Cooke. 1971. *Modern Elementary Differential Equations*, 2nd ed. Reading, Mass.: Addison-Wesley.

5.2 Chiang, A. C. 1967. *Fundamental Methods of Mathematical Economics*. New York: McGraw-Hill.

5.3 Ince, E. L. 1959. *Ordinary Differential Equations*. New York: Dover.

5.4 Kaplan, Wilfred. 1962. *Ordinary Differential Equations*. Reading, Mass.: Addison-Wesley.

5.5 Polya, G. 1963. *SMSG Studies in Mathematics* 11, Mathematical Methods in Science. Stanford University.

5.6 Rashevsky, N. 1959. *Mathematical Biology of Social Behavior*. Chicago: University of Chicago Press.

5.7 Rashevsky, N. 1948. *Mathematical Biophysics*. Chicago: University of Chicago Press.

5.8 Shames, I. H. 1960. *Engineering Mechanics, Statics and Dynamics*. Englewood Cliffs, N.J.: Prentice-Hall.

5.9 Sokolnikoff, I. S. 1956. *Mathematical Theory of Elasticity*, 2nd ed. New York: McGraw-Hill.

5.10 Sokolnikoff, I. S., and R. M. Redheffer, 1966. *Mathematics of Physics and Modern Engineering*. New York: McGraw-Hill.

5.11 Standard Mathematical Tables, 20th ed. 1972. Cleveland: Chemical Rubber Co.

5.12 Tierney, John A. 1979. *Calculus and Analytic Geometry*, 4th ed. Boston: Allyn and Bacon.

5.13 Yates, Robert C. 1974. *Curves and Their Properties*. Reston, Va.: National Council of Teachers of Mathematics.

Systems of
Differential
Equations

6.1 Solving a System of *n* Simultaneous Differential Equations

By a solution of the DE

$$\frac{dx}{dt} - 2\frac{dy}{dt} + y = 0 \qquad (6.1)$$

we mean a pair of two real functions, denoted by $x = f(t)$ and $y = g(t)$, each defined on a set S of real numbers, where S is the union of non-overlapping intervals on the t axis, such that

$$f'(t) - 2g'(t) + g(t) \equiv 0$$

for all t in S.

A solution of (6.1) is very general, since we can let g be any differentiable function of t and then find $x = f(t)$ by integrating $2g'(t) - g(t)$. By a solution of the system of two DE consisting of (6.1) and

$$\frac{dy}{dt} - x = 0 \qquad (6.2)$$

we mean a pair of functions f and g of t such that *both* (6.1) and (6.2) are satisfied for all t in the common domain of f and g, or in some subset of this common domain. In this case it is no longer possible to let g be an arbitrary differentiable function of t, since such a function would generally not yield the same function f on being substituted into (6.1) and (6.2).

The situation is somewhat analogous to that presented by an algebraic equation in two unknowns such as $2x + 3y - 18 = 0$. Solutions can be obtained at will by assigning y arbitrarily and finding the corresponding value of x. However, the system of two equations

$$2x + 3y - 18 = 0, \quad 3x - 4y + 7 = 0$$

has the unique solution pair $x = 3$ and $y = 4$. One of the most straightforward methods for finding this solution is to eliminate y (or x) from the two equations and then solve the resulting equation for x (or y). Elimination is also effective in solving (6.1) and (6.2). It is somewhat more complicated, since we must eliminate not only y (or x), but also any derivatives of y (or x) appearing. From (6.2) we obtain $dx/dt = d^2y/dt^2$.

Substituting d^2y/dt^2 for dx/dt in (6.1) yields

$$\frac{d^2y}{dt^2} - 2\frac{dy}{dt} + y = 0$$

Using the repeated root $r = 1$ of the characteristic equation $r^2 - 2r + 1 = 0$, we obtain

$$y = c_1e^t + c_2te^t \tag{6.3}$$

From (6.2), we find that

$$x = \frac{dy}{dt} = c_1e^t + c_2e^t + c_2te^t \tag{6.4}$$

It is easy to verify (see Problem 1) that (6.3) and (6.4) define a solution of the system consisting of (6.1) and (6.2). The common domain of the solution functions is $-\infty < t < +\infty$.

In solving a system of DE by elimination, we often differentiate both sides of a DE with respect to t. This sometimes introduces extraneous solutions by introducing additional constants. The solution functions should be substituted into the original DE of the system to determine whether the constants in the solution are independent or whether one or more relations hold among these constants. To illustrate, the DE

$$\frac{dx}{dt} = 2t \tag{6.5}$$

has a complete solution given by

$$x = t^2 + c_1 \tag{6.6}$$

Differentiating both sides of (6.5) with respect to t, we obtain

$$\frac{d^2x}{dt^2} = 2 \tag{6.7}$$

It is easy to show that a complete solution of (6.7) is given by

$$x = t^2 + c_2 t + c_3 \tag{6.8}$$

Every solution of $dx/dt = 2t$ is included in $x = t^2 + c_2 t + c_3$. This is seen by setting $c_2 = 0$ and comparing with (6.6). However, a particular function given by $x = t^2 + c_2 t + c_3$ is a solution of $dx/dt = 2t$ only if $c_2 = 0$. The function given by $x = t^2 + 2t + c_3$, for example, is a solution of $d^2x/dt^2 = 2$ but is not a solution of $dx/dt = 2t$. Extraneous solutions were introduced when we differentiated both sides of $dx/dt = 2t$ with respect to t to obtain $d^2x/dt^2 = 2$.

The method of elimination is often complicated by cumbersome algebraic manipulations. A more systematic approach is achieved by writing a system of DE involving two unknown functions in the *normal form*,

$$P_1(D)x + P_2(D)y = \phi(t) \tag{6.9}$$

$$P_3(D)x + P_4(D)y = \psi(t) \tag{6.10}$$

where $D = d/dt$ denotes the differentiation operator, and the coefficients of x and y are polynomials in D. The system given by (6.9) and (6.10) is then solved by treating P_1, P_2, P_3, and P_4 as ordinary algebraic coefficients. For example, (6.1) and (6.2) written in normal form become

$$Dx + (-2D + 1)y = 0 \tag{6.11}$$

and

$$-x + Dy = 0 \tag{6.12}$$

Applying the operator D to both sides of (6.12), we obtain

$$-Dx + D^2y = 0 \tag{6.13}$$

Adding (6.11) and (6.13) yields

$$(D^2 - 2D + 1)y = 0$$

from which we get

$$y = c_1 e^t + c_2 t e^t$$

In this simple example we then find x from $x = Dy$.

The basis of the operational method is that if $x = f(t)$ and $y = g(t)$ satisfy (6.9) and (6.10), they also satisfy

$$P_5(D)[P_1(D)x + P_2(D)y] = P_5(D)\phi(t)$$

and

$$P_6(D)[P_3(D)x + P_4(D)y] = P_6(D)\psi(t)$$

where $P_5(D)$ and $P_6(D)$ are arbitrary polynomials in D. It is possible, as indicated earlier, that extraneous solutions may be introduced in the operational method. Hence, possible solutions should be substituted into the original DE of the system to determine whether any relations exist among the arbitrary constants.

The number of essential arbitrary constants in a solution of the system given by (6.9) and (6.10) does not exceed the sum of the orders of the two DE. More specifically, it is shown in Reference 6.5 that the number of essential arbitrary constants equals the degree of the determinant

$$\begin{vmatrix} P_1(D) & P_2(D) \\ P_3(D) & P_4(D) \end{vmatrix} = P_1(D)P_4(D) - P_2(D)P_3(D)$$

For the system given by (6.11) and (6.12),

$$\begin{vmatrix} D & -2D + 1 \\ -1 & D \end{vmatrix} = D^2 - 2D + 1$$

has degree two and the solution given by (6.3) and (6.4) contains two essential arbitrary constants.

EXAMPLE 1. Solve the system

$$\frac{dx}{dt} + \frac{dy}{dt} = 4 - x + y$$

$$\frac{dx}{dt} + \frac{dy}{dt} = e^{2t} + x - y$$

Solution: Writing the system in the normal form, we obtain

$$(D + 1)x + (D - 1)y = 4 \tag{6.14}$$

$$(D - 1)x + (D + 1)y = e^{2t} \tag{6.15}$$

To eliminate x, we apply $D - 1$ to (6.14), apply $D + 1$ to (6.15), and subtract, to obtain

$$[(D - 1)^2 - (D + 1)^2]y = -4 - (2e^{2t} + e^{2t})$$

or

$$-4Dy = -4 - 3e^{2t}$$

Hence,

$$Dy = 1 + \frac{3}{4} e^{2t} \quad \text{and} \quad y = t + \frac{3}{8} e^{2t} + c$$

Applying $D + 1$ to (6.14), applying $D - 1$ to (6.15), and subtracting to eliminate y, we obtain

$$[(D + 1)^2 - (D - 1)^2]x = 4 - (2e^{2t} - e^{2t})$$

or

$$4Dx = 4 - e^{2t}$$

Hence,

$$Dx = 1 - \frac{1}{4} e^{2t} \quad \text{and} \quad x = t - \frac{1}{8} e^{2t} + k$$

In this case c and k are not independent. Substituting the expressions for x and y into (6.14) yields

$$\left(1 - \frac{1}{4} e^{2t}\right) + \left(t - \frac{1}{8} e^{2t} + k\right) + \left(1 + \frac{3}{4} e^{2t}\right) - \left(t + \frac{3}{8} e^{2t} + c\right) = 4$$

which implies that $k = c + 2$.
Thus (6.14) is satisfied by

$$x = f(t) = t - \frac{1}{8} e^{2t} + c + 2, \quad y = g(t) = t + \frac{3}{8} e^{2t} + c$$

It is a simple matter to verify that f and g also satisfy (6.15).
Solutions should be checked by substituting into the original DE. Even though functions satisfy the DE of the normal form of a system, it is possible that an error may have been made in writing the original system in normal form. See Problems 1, 2, and 3.
In this example

$$\begin{vmatrix} D + 1 & D - 1 \\ D - 1 & D + 1 \end{vmatrix} = (D + 1)^2 - (D - 1)^2 = 4D$$

has degree one and hence the solution contains one essential arbitrary constant.

EXAMPLE 2. Find the solution of

$$x - \frac{dy}{dt} = 2t - 1 \qquad \frac{dx}{dt} + y = t - t^2$$

for which $x(0) = 0$ and $y(0) = -1$.

Solution: Substituting $dy/dt = x - 2t + 1$ into

$$\frac{d^2 x}{dt^2} + \frac{dy}{dt} = 1 - 2t$$

obtained by differentiating the second DE with respect to t, yields

$$\frac{d^2x}{dt^2} + x = 0$$

Hence,

$$x = A \cos t + B \sin t, \quad A = 0$$

$$x = B \sin t \quad \text{and} \quad \frac{dx}{dt} = B \cos t$$

Thus, $y = t - t^2 - B \cos t$, $-1 = -B$, $B = 1$, and

$$x = \sin t, \quad y = t - t^2 - \cos t$$

Verification is left to Problem 3.

The problem presented in Example 2 is known as an initial-value problem.

In general, elimination applies to systems of n DE in n unknown functions of a single variable t. For $n > 2$, the unknown functions are often denoted by $x_1, x_2, \ldots, x_n$. The letter t is frequently used to denote an element of the domain of the solution functions since t denotes time in many important applications.

The following system of DE, known as a *first-order system in normal form*, is significant in the theoretical development of systems of DE.

$$\frac{dx_1}{dt} = F_1(t, x_1, x_2, \ldots, x_n)$$

$$\frac{dx_2}{dt} = F_2(t, x_1, x_2, \ldots, x_n)$$
$$\vdots$$
$$\frac{dx_n}{dt} = F_n(t, x_1, x_2, \ldots, x_n)$$

(6.16)

The following existence and uniqueness theorem is proved in Reference 6.6.

Theorem 6-I   Let $\bar{t}, \bar{x}_1, \bar{x}_2, \ldots, \bar{x}_n$ be a set of $n + 1$ real numbers. If there exists a positive number h such that the functions $F_1, F_2, \ldots, F_n$ are continuous and possess continuous first partial derivatives with respect to $x_1, x_2, \ldots, x_n$ for $|t - \bar{t}| < h$, $|x_1 - \bar{x}_1| < h$, $|x_2 - \bar{x}_2| < h, \ldots, |x_n - \bar{x}_n| < h$, then the system (6.16) has a unique solution given by

$$x_1 = x_1(t), \quad x_2 = x_2(t), \ldots, \quad x_n = x_n(t)$$

for which

$$\bar{x}_1 = x_1(\bar{t}), \quad \bar{x}_2 = x_2(\bar{t}), \ldots, \quad \bar{x}_n = x_n(\bar{t})$$

The solution given by $x = \sin t$, $y = t - t^2 - \cos t$, found for the system of Example 2, is unique by Theorem 6-I.

One reason for the importance of system (6.16) is that an nth order DE of the form

$$\frac{d^n x}{dt^n} = F\left(t, x, \frac{dx}{dt}, \frac{d^2 x}{dt^2}, \ldots, \frac{d^{n-1} x}{dt^{n-1}}\right) \tag{6.17}$$

can be replaced by a system of type (6.16). This is accomplished by letting

$$x_1 = x, \quad x_2 = \frac{dx}{dt}, \quad x_3 = \frac{d^2 x}{dt^2}, \ldots, \quad x_n = \frac{d^{n-1} x}{dt^{n-1}}$$

Then the system

$$\frac{dx_1}{dt} = x_2$$

$$\frac{dx_2}{dt} = x_3$$

$$\vdots \tag{6.18}$$

$$\frac{dx_{n-1}}{dt} = x_n$$

$$\frac{dx_n}{dt} = F(t, x_1, x_2, x_3, \ldots, x_n)$$

is of the form (6.16). It is in many ways simpler to analyze system (6.18) than the single nth-order DE (6.17). It is often advantageous to replace an initial-value problem for an nth order DE by an initial-value problem for a system of type (6.18).

The DE (6.17) and the system of DE (6.18) are equivalent in the following sense. If $x_1, x_2, \ldots, x_n$ constitute a solution of (6.18), then $x = x_1$ is a solution of (6.17). Conversely, if x is a solution of (6.17), then $x_1 = x$, $x_2 = dx/dt$, $x_3 = d^2 x/dt^2, \ldots, x_n = d^{n-1} x/dt^{n-1}$ constitute a solution of (6.18).

When we obtained

$$\frac{d^2 y}{dt^2} - 2\frac{dy}{dt} + y = 0$$

from (6.1) and (6.2), we were obtaining a second-order DE from a system of the form (6.18). Thus, either type of replacement may be advantageous.

Another advantage of replacing an nth-order DE by a system of type (6.16) is that the system approach is more tractable when numerical methods are used to find approximate solutions. This is because it is simpler to approximate a first derivative numerically than a higher-order derivative.

Several special cases of system (6.16) are of particular interest.
If F_i, $i = 1, 2, \ldots, n$, is of the form

$$a_{i1}(t)x_1 + a_{i2}(t)x_2 + \cdots + a_{in}(t)x_n + \rho_i(t)$$

the system (6.16) is called *linear*. In this case, the derivatives of the unknown
functions are linear functions of $x_1, x_2, \ldots, x_n$, and the coefficients of
$x_1, x_2, \ldots, x_n$ depend on t. If these coefficients are constants, F_i has the
form

$$a_{i1}x_1 + a_{i2}x_2 + \cdots + a_{in}x_n + \rho_i(t)$$

If $\rho_i(t) \equiv 0$ for $i = 1, 2, \ldots, n$, the system is *homogeneous*; otherwise
it is *nonhomogeneous*. The theory of nonhomogeneous linear systems, developed in References 6.5 and 6.6, is analogous to the theory presented in
Chapter 4 of nonhomogeneous second-order linear DE. For $n = 2$, a nonhomogeneous linear system with constant coefficients has the form (with
$x_1 = x$ and $x_2 = y$)

$$\frac{dx}{dt} = ax + by + \phi(t) \qquad \frac{dy}{dt} = cx + dy + \psi(t)$$

The corresponding homogeneous system is given by

$$\frac{dx}{dt} = ax + by \qquad \frac{dy}{dt} = cx + dy$$

If $F_i(t, x_1, x_2, \ldots, x_n)$ in (6.16) has the form $G_i(x_1, x_2, \ldots, x_n)$ for
$i = 1, 2, \ldots, n$, then the system is said to be *autonomous*. The rates of change
of the unknown functions are independent of t. If t denotes time, the system
is time-invariant and the reaction of a physical system governed by an
autonomous system of DE is the same for different time applications of a given
input. If $x_1 = x_1(t)$, $x_2 = x_2(t), \ldots, x_n = x_n(t)$ constitute a solution of an
autonomous system, and t_0 is a constant, then $x_1(t - t_0), x_2(t - t_0), \ldots,$
$x_n(t - t_0)$ also define a solution.

As an application of an autonomous system, we set $n = 3$ in (6.16),
denote x_1 by x, x_2 by y, x_3 by z, and assume that dx/dt, dy/dt, and dz/dt are
independent of t. Next we assume that $P(x, y, z)$ is a point in three-dimensional
space and that t denotes time. We then interpret system (6.16) as governing the
flow of a fluid in three-space. System (6.16) defines a vector field with the vector
$\mathbf{V} = (dx/dt, dy/dt, dz/dt)$ being associated with the point $P(x, y, z)$. The
components of $\mathbf{V}$ are the velocities in the x, y, and z directions of the fluid
particle located at P. A velocity field is similar to a direction field except that
we associate with each point P at which $\mathbf{V} \neq (0, 0, 0)$ not only a specific
direction but also a magnitude, the magnitude being the speed $v = |\mathbf{V}|$ of the
fluid particle at P. The solution curves, or integral curves, of (6.16) are

traversed as t varies over a parametric interval. In the general case these integral curves are termed *orbits, trajectories,* or *paths* of the system of DE. The orbits are oriented, with the positive direction defined as the direction corresponding to the direction of increasing t. In the application to fluid flow the orbits are called *streamlines* and are the paths along which the fluid particles move. The positive direction on a streamline is the actual direction of flow of the fluid particles. In an initial-value problem the point $P_0(x_0, y_0, z_0)$ corresponds to the number $t = t_0$ such that $x(t_0) = x_0$, $y(t_0) = y_0$, and $z(t_0) = z_0$. The solution provides a parametric representation of the streamline through P_0. This type flow is called *steady* since the velocity at a point $P(x, y, z)$ does not change with time. In a more general type fluid flow, the velocity vector field changes with time as t varies over a parametric interval.

Problem List 6.1

1. Verify that $x = c_1 e^t + c_2 e^t + c_2 t e^t$, $y = c_1 e^t + c_2 t e^t$, constitute a solution on $-\infty < t < +\infty$ of the system

$$\frac{dx}{dt} - 2\frac{dy}{dt} + y = 0, \quad \frac{dy}{dt} - x = 0$$

2. Verify that $x = t - \frac{1}{8} e^{2t} + c + 2$, $y = t + \frac{3}{8} e^{2t} + c$, constitute a solution on $-\infty < t < +\infty$ of the system

$$(D + 1)x + (D - 1)y = 4, \quad (D - 1)x + (D + 1)y = e^{2t}$$

Also verify that the original DE of Example 1

$$\frac{dx}{dt} + \frac{dy}{dt} = 4 - x + y \quad \text{and} \quad \frac{dx}{dt} + \frac{dy}{dt} = e^{2t} + x - y$$

are satisfied.

3. Verify that $x = \sin t$, $y = t - t^2 - \cos t$, constitute a solution on $-\infty < t < +\infty$, of the initial-value problem

$$x - \frac{dy}{dt} = 2t - 1, \quad \frac{dx}{dt} + y = t - t^2; \quad x(0) = 0, \quad y(0) = -1$$

4. Rewrite in normal form, using the operator $D = d/dt$:

(a) $\dfrac{dx}{dt} - 3\dfrac{dy}{dt} + 2x - 4y + t^2 = 0$

$2\dfrac{dx}{dt} + \dfrac{dy}{dt} - x + 2y - t = 0$

(b) $\dfrac{d^2x}{dt^2} + 2\dfrac{dx}{dt} - 3\dfrac{dy}{dt} + 2x - 2y + \sin t = 0$

$\dfrac{d^2y}{dt^2} + 3\dfrac{dx}{dt} + \dfrac{dy}{dt} - 2x + 4y - \cos t = 0$

5. Reduce to a first-order system in normal form:

(a) $\dfrac{d^2y}{dt^2} + 4\dfrac{dy}{dt} + 3ty - e^{-t} = 0$

(b) $\dfrac{d^3x}{dt^3} - t^2\dfrac{d^2x}{dt^2} - \dfrac{dx}{dt} + x^2 = 0$

6. Solve simultaneously:

(a) $\dfrac{dx}{dt} + y = x, \quad \dfrac{dy}{dt} - 3y = 0$

(b) $\dfrac{dx}{dt} - y = 4, \quad \dfrac{dx}{dt} + \dfrac{dy}{dt} = 2x$

(c) $\dfrac{dx}{dt} + \dfrac{dy}{dt} = 2y, \quad \dfrac{dy}{dt} = x$

(d) $\dfrac{dx}{dt} + 2y = 4, \quad \dfrac{dy}{dt} - 2x = 0$

(e) $\dfrac{dx}{dt} + \dfrac{dy}{dt} = -2x, \quad 2\dfrac{dx}{dt} + 3\dfrac{dy}{dt} = -5y$

(f) $\dfrac{dy}{dt} = 2x - 1, \quad \dfrac{dx}{dt} = 1 + 2y$

7. Given that $x(0) = 3$ and $y(0) = 1$, solve

$$\dfrac{dy}{dt} + x + y = 0, \quad \dfrac{dx}{dt} + 3x + 3y = 0$$

8. Given $x(0) = 1$ and $y(0) = 0$, solve

$$\dfrac{dx}{dt} + y = 2, \quad \dfrac{dx}{dt} + 2x + \dfrac{dy}{dt} = 0$$

9. Given $x(0) = 0$ and $y(0) = 5$, solve

$$\dfrac{dx}{dt} + t\dfrac{dy}{dt} = 2t, \quad t\dfrac{dx}{dt} - \dfrac{dy}{dt} = -x$$

10. Solve simultaneously:

(a) $(D - 1)x + Dy = 2t + 1, \quad (2D + 1)x + 2Dy = t$

(b) $(D - 1)x + (2D + 1)y = 2, \quad (D + 1)x + (2D + 3)y = 0$

(c) $(D + 2)x + Dy = 0, \quad 2Dx + (3D + 5)y = 0$

(d) $(D^2 - 4)x + Dy = 0, \quad -4Dx + (D^2 + 2)y = 0$

11. Given $y(0) = -1$ and $y'(0) = 2$, solve simultaneously
$$(D - 4)x + (D - 8)y = 0, \quad (D + 1)x + (2D + 1)y = e^{-t}$$

6.2 Plane Autonomous Systems; Critical Points; Stability

When $n = 2$ and system (6.16) is autonomous, it is customary to denote x_1 by x, x_2 by y, F_1 by f, and F_2 by g. The system is then given by the DE $dx/dt = f(x, y); dy/dt = g(x, y)$.

A vector field $\mathbf{V} = (dx/dt, dy/dt)$ is determined in the xy plane, known as the *phase plane*. Solutions of the system, denoted by $x = x(t)$ and $y = y(t)$, can be interpreted as parametric equations of curves in the phase plane with t serving as the parameter. These curves are the orbits (or trajectories, or paths) of the system and are solution curves of the first-order DE

$$\frac{dy}{dx} = \frac{f(x, y)}{g(x, y)}$$

The orbits of a system are often simpler to find than the solutions given by $x = x(t)$ and $y = y(t)$. Furthermore, information regarding the solutions can often be obtained by studying the orbits.

Definition A point (x, y) at which $f(x, y) = 0$ and $g(x, y) = 0$ is called a *critical point* or *equilibrium point* of the autonomous system

$$\frac{dx}{dt} = f(x, y); \quad \frac{dy}{dt} = g(x, y)$$

At a critical point the vector $\mathbf{V}$ is the zero vector $\mathbf{V} = (0, 0)$ and the vector field has no direction at such a point. Both the slope $dy/dx = f(x, y)/g(x, y)$ and $dx/dy = g(x, y)/f(x, y)$ are undefined at a critical point.

A velocity vector field $\mathbf{V} = (dx/dt, dy/dt, 0)$, in which $dz/dt \equiv 0$, reduces to the two-dimensional type vector field we are considering. The flow pattern in the plane $z = k$, where k is an arbitrary constant, is identical to the flow pattern in the plane $z = 0$, namely, in the xy or phase plane. A fluid flow described by this type vector field is termed a *plane flow*. In all types of fluid flow a critical point is called a *stagnation point*.

There is no loss of generality in assuming that a critical point is located at $(0, 0)$ instead of at (x_0, y_0). The substitutions $u = x - x_0, v = y - y_0$ yield a system of DE for u and v with a critical point at $(0, 0)$.

EXAMPLE 1. In Section 5.5 we found that the motion of a simple pendulum is governed by the DE

$$\frac{d^2\theta}{dt^2} + \frac{g}{l} \sin \theta = 0$$

The corresponding autonomous system of DE, with $x = \theta$, consists of

$$\frac{dx}{dt} = y, \quad \frac{dy}{dt} = -\frac{g}{l} \sin x$$

This system has infinitely many critical points, located in the xy plane at $(\pm n\pi, 0), n = 0, 1, 2, \ldots$.

It is often possible to obtain information concerning solutions of a system of DE by studying and classifying the critical points of the system. It is assumed that the conditions of the existence and uniqueness Theorem 6-I are satisfied. Let $P_0(x_0, y_0)$ be an isolated critical point. That is, assume that P_0 is the only critical point is some circle centered at P_0. The constant solution given by $x \equiv x_0$, $y \equiv y_0$ corresponds to P_0; that is, the orbit of the constant solution consists of the single point P_0. Now suppose that $t = \tau$ corresponds to a point $P(x, y)$ near $P_0(x_0, y_0)$ and think of P as a particle moving in the xy plane as t increases. The particle moves on the same orbit regardless of the value of τ. It may return to P if the motion is periodic but it can never cross another orbit. The crucial question is whether the particle moves toward P_0, away from P_0, or around P_0 in some fashion. If to every $\varepsilon > 0$, there corresponds a $\delta > 0$ such that for an *arbitrary* point $P(x, y)$ within distance δ of P_0, where $x = x(\tau)$ and $y = y(\tau)$, the entire orbit through $P(x, y)$ remains within distance ε of P_0 for all $t \geq \tau$, then P_0 is called a *stable critical point*. The point P may or may not approach P_0 as $t \to +\infty$.

A critical point that is not stable is said to be *unstable*.

If P_0 is stable and if P does approach P_0 as $t \to +\infty$, that is, if $\lim_{t \to +\infty} x(t) = x_0$ and $\lim_{t \to \infty} y(t) = y_0$, then P_0 is called an *asymptotically stable critical point*. See Figure 6.1.

EXAMPLE 2. Solve and find the orbits of the system

$$\frac{dx}{dt} = x, \quad \frac{dy}{dt} = 2y$$

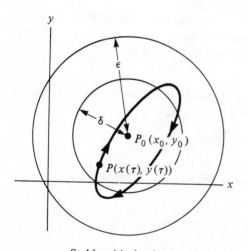

Stable critical point P_0

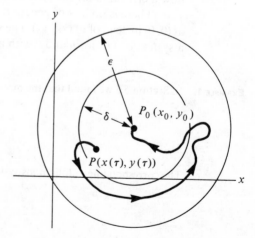

Asymptotically stable critical point P_0

Fig. 6.1

Solution: The origin $(0, 0)$ is the only critical point. From $dy/dx = 2y/x$ we obtain

$$x\, dy - 2y\, dx = 0, \quad \frac{x^2\, dy - 2xy\, dx}{x^4} = d\left(\frac{y}{x^2}\right) = 0, \quad \frac{y}{x^2} = k, \quad \text{and} \quad y = kx^2$$

The orbits consist of the parabolic branches emanating from the origin in Figure 6.2, plus the positive and negative x and y axes. The origin $(0, 0)$ corresponds to the solution given by $x \equiv 0$, $y \equiv 0$. The vector $\mathbf{V} = (x, 2y)$ points away from the origin at every point in the xy plane except at the origin itself, where $\mathbf{V}$ has no direction. As t increases, points move along the orbits away from the origin, which is an unstable critical point.

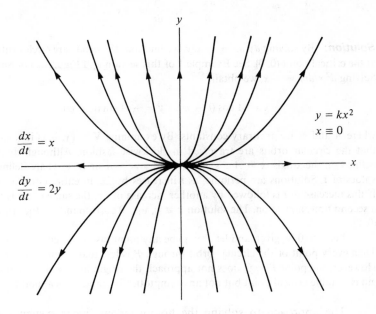

Fig. 6.2

Solutions of this simple autonomous system are given by

$$x = c_1 e^t, \quad y = c_2 e^{2t}$$

Since $\lim_{t \to +\infty} e^t = \lim_{t \to +\infty} e^{2t} = +\infty$, it is easy to verify that orbits are traversed in directions away from the origin.

EXAMPLE 3. Solve and find the orbits of the system

$$\frac{dx}{dt} = -x, \quad \frac{dy}{dt} = -2y$$

Solution: The DE involving x and y is the same as in Example 2. Hence the orbits are the same as those of Example 2 except for orientation. The vector $\mathbf{V} = (-x, -2y)$ has, except at the critical point $(0, 0)$, direction opposite to that of $\mathbf{V} = (x, 2y)$. Thus as t increases, orbits are traversed in the direction of the origin, which is an asymptotically stable critical point.

Inspecting the solutions given by

$$x = c_1 e^{-t}, \quad y = c_2 e^{-2t}$$

and noting that $\lim_{t \to +\infty} e^{-t} = \lim_{t \to +\infty} e^{-2t} = 0$, we see that $x \to 0$ and $y \to 0$ as $t \to +\infty$.

EXAMPLE 4. Solve and find the orbits of the system

$$\frac{dx}{dt} = y, \quad \frac{dy}{dt} = -x$$

Solution: By solving $dy/dx = -x/y$, we find that the orbits are circles with centers at the critical point $(0, 0)$. See Example 1 of this section and Figure 2.3 of Section 2.7. Solving $d^2x/dt^2 = -x$, we obtain

$$x = A \cos (t + \phi), \quad y = -A \sin (t + \phi)$$

where A and ϕ are arbitrary constants. By examining $\mathbf{V} = (y, -x)$ it is easy to see that the circular orbits are traversed in clockwise fashion. Although dy/dx is not defined on the x axis, $x = A \cos (t + \phi)$ and $y = -A \sin (t + \phi)$ are defined for all values of t. Solutions are periodic; as t increases by 2π an entire circle is traversed. If this increase in t is followed by another increase of 2π, the same circle is traversed a second time, and so on. The solution $x \equiv 0$, $y \equiv 0$ corresponds to the critical point $(0, 0)$.

Let $\varepsilon > 0$ be given and let $P(x, y)$ be any point inside the circle $x^2 + y^2 = \varepsilon^2$. Then every point of the circular orbit through P is less than ε units from the origin. However, the point $P(x, y)$ does not approach the origin as $t \to +\infty$. Thus the origin is a stable critical point but not an asymptotically stable critical point.

The approach to solving the homogeneous linear system with constant coefficients

$$\frac{dx}{dt} = ax + by, \quad \frac{dy}{dt} = cx + dy$$

is similar to that used in solving a homogeneous linear DE with constant coefficients. Solutions of the form

$$x = Ae^{rt}, \quad y = Be^{rt}$$

are sought. This yields

$$Are^{rt} = aAe^{rt} + bBe^{rt}, \quad Bre^{rt} = cAe^{rt} + dBe^{rt}$$

or

$$(a - r)A + bB = 0$$

$$cA + (d - r)B = 0$$

In order for this system of two equations in A and B to have a solution other than the trivial solution $A = B = 0$, the determinant of the coefficients

$$\begin{vmatrix} a - r & b \\ c & d - r \end{vmatrix}$$

must equal zero and hence r must be a root of

$$(a - r)(d - r) - bc = 0$$

This equation, usually written as

$$r^2 - (a + d)r + (ad - bc) = 0$$

is known as the *characteristic equation* of the system. The nature of the critical point $(0, 0)$ is determined by the character of the roots of the characteristic equation.

In Example 2, $dx/dt = x$, $dy/dt = 2y$, $a = 1$, $b = 0$, $c = 0$, $d = 2$ and the characteristic equation $r^2 - 3r + 2 = (r - 1)(r - 2) = 0$ has roots 1 and 2.

In Example 3, $dx/dt = -x$, $dy/dt = -2y$, $a = -1$, $b = 0$, $c = 0$, $d = -2$ and the characteristic equation $r^2 + 3r + 2 = (r + 1)(r + 2) = 0$ has roots -1 and -2.

In Example 4, $dx/dt = y$, $dy/dt = -x$, $a = 0$, $b = 1$, $c = -1$, $d = 0$, and the characteristic equation $r^2 + 1 = 0$ has roots $0 \pm i$.

A detailed treatment of the topics introduced in this section is contained in Reference 6.6.

Problem List 6.2

1. Find all critical points of the system
   (a) $dx/dt = x - y$, $dy/dt = 3y$
   (b) $dx/dt = 2y + 1$, $dy/dt = 2x - 1$
2. Solve and find the orbits of each system. Determine whether the critical point $(0, 0)$ is stable, asymptotically stable, or unstable.
   (a) $dx/dt = 2x$, $dy/dt = y$
   (b) $dx/dt = 2y + 1$, $dy/dt = 2x - 1$
   (c) $dx/dt = -y$, $dy/dt = x$
   (d) $dx/dt = x$, $dy/dt = -y$
   (e) $dx/dt = x$, $dy/dt = y$
3. The motion of a point moving in a straight line is governed by the DE

$$\frac{d^2x}{dt^2} + \frac{cg}{w}\frac{dx}{dt} + \frac{kg}{w}x = 0$$

Determine an equivalent system of DE. Describe the critical points of the system in terms of the velocity dx/dt and the acceleration d^2x/dt^2.

4. The motion of a pendulum is governed by the system (see Example 1)

$$dx/dt = y, \quad dy/dt = -\frac{g}{1}\sin x$$

The angular velocity $y = dx/dt$ and the angular acceleration $d^2x/dt^2 = dy/dt$ are both zero at the critical points $(0, 0)$ and $(\pi, 0)$. Discuss stability at each of these points from the viewpoint of physical considerations. Find equations of the orbits.

5. Linearize the system of Problem 4 governing the motion of a pendulum. Solve the system and find equations of the orbits. Show that the origin is a stable critical point but not an asymptotically stable critical point.

6. Solve the initial-value problem

$$dx/dt = -x - y, \quad dy/dt = x - y; \quad x(0) = 0, \quad y(0) = 1$$

Describe the orbit through $(0, 1)$ and show that a point $P(x, y)$ on this orbit approaches the critical point $(0, 0)$ as $t \to +\infty$.

7. Solve the initial-value problem

$$dx/dt = y(x + y), \quad dy/dt = x(x + y); \quad t = 0, \quad x = 0, \quad y = 1$$

Find an equation of the orbit through $(0, 1)$.

8. Given that $x = x_1(t), y = y_1(t)$, and $x = x_2(t), y = y_2(t)$ both constitute solutions of the system

$$\frac{dx}{dt} = a_{11}(t)x + a_{12}(t)y, \quad \frac{dy}{dt} = a_{21}(t)x + a_{22}(t)y$$

on an interval $I = [t_0, t_1]$, show that

$$x = c_1x_1(t) + c_2x_2(t), \quad y = c_1y_1(t) + c_2y_2(t)$$

is also a solution of the system on I, where c_1 and c_2 are arbitrary constants.

9. (a–e) Find and solve the characteristic equation of each of the systems in Problem 2.

6.3 System Governing the Motion of a Projectile

To obtain a mathematical model for the flight of the projectile shown in Figure 6.3, we assume that the projectile is acted on only by the downward gravitational force exerted by the earth. Let the vector $\mathbf{F}$ represent this force and let the vectors $\mathbf{R}$, $\mathbf{V}$, and $\mathbf{A}$ represent the position, velocity, and acceleration of the projectile at time t. By Newton's second law, $\mathbf{F} = m\mathbf{A} = m(d^2\mathbf{R}/dt^2)$. Also, $|\mathbf{F}| = w$, where w is the weight of the projectile, and $|\mathbf{A}| = g$, where $g \approx 32.2$ ft/sec^2 is the acceleration of gravity. Thus, $w = mg$.

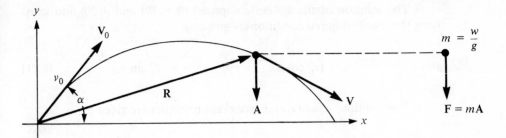

Fig. 6.3

Taking scalar components in the x and y directions, we get

$$0 = m\frac{d^2x}{dt^2} \quad \text{and} \quad -w = -mg = m\frac{d^2y}{dt^2}$$

In this simple model the motion of the projectile is governed by the system

$$\frac{d^2x}{dt^2} = 0 \tag{6.19}$$

$$\frac{d^2y}{dt^2} = -g \tag{6.20}$$

of two DE in two unknown functions.

We assume that at time $t = 0$, the projectile has position vector $\mathbf{R_0} = 0\mathbf{i} + 0\mathbf{j}$ and velocity vector $\mathbf{V_0} = (v_0 \cos \alpha)\mathbf{i} + (v_0 \sin \alpha)\mathbf{j}$. The angle α is called the *angle of elevation*; $\mathbf{V_0}$, the *initial, or muzzle, velocity*; and v_0, the *initial speed*. In scalar form, the initial conditions are $t = 0$, $x = 0$, $y = 0$, $dx/dt = v_0 \cos \alpha$, and $dy/dt = v_0 \sin \alpha$.

Integrating (6.19) with respect to t gives

$$\frac{dx}{dt} = c_1 = \frac{dx}{dt}\bigg]_{t=0} = v_0 \cos \alpha$$

Integrating again gives $x = (v_0 \cos \alpha)t + c_2$.
Since $x = 0$ when $t = 0$, it follows that $c_2 = 0$.
Integrating (6.20) with respect to t, we get $dy/dt = -gt + c_3$.
Since $dy/dt = v_0 \sin \alpha$ when $t = 0$, it follows that $c_3 = v_0 \sin \alpha$.
Integrating again gives

$$y = \frac{-gt^2}{2} + (v_0 \sin \alpha)t + c_4$$

Since $y = 0$ when $t = 0$, it follows, that $c_4 = 0$.

6.3 System Governing the Motion of a Projectile   **229**

The solution of the system comprised of (6.19) and (6.20) and satis-fying the specified initial conditions is given by

$$x = (v_0 \cos \alpha)t \qquad y = -\frac{gt^2}{2} + (v_0 \sin \alpha)t \qquad (6.21)$$

The position, velocity, and acceleration vectors are given by

$$\mathbf{R} = [(v_0 \cos \alpha)t]\mathbf{i} + \left[-\frac{gt^2}{2} + (v_0 \sin \alpha)t\right]\mathbf{j}$$

$$\mathbf{V} = (v_0 \cos \alpha)\mathbf{i} + (-gt + v_0 \sin \alpha)\mathbf{j}$$

$$\mathbf{A} = 0\mathbf{i} - g\mathbf{j}$$

By eliminating t in Equations (6.21), we can easily show (see Prob-lem 1) that the path of the projectile is the parabola having Cartesian equation

$$y = x \tan \alpha - \frac{gx^2}{2v_0^2} \sec^2 \alpha \qquad (6.22)$$

When additional forces acting on the projectile, such as frictional forces, are considered, the analysis is similar, except that a more compli-cated system of DE is obtained. Numerical methods of approximation are often used to integrate approximately the DE occurring in models arising in ballistics. Reference 6.2 discusses the effect of the Coriolis force due to the earth's rotation.

Problem List 6.3

1. Eliminate t from Equations (6.21) to obtain (6.22).
   In the following problems a projectile has position vector

$$\mathbf{R} = [(v_0 \cos \alpha)t]\mathbf{i} + \left[-\frac{gt^2}{2} + (v_0 \sin \alpha)t\right]\mathbf{j}$$

When $t = 0$,

$$\mathbf{R} = \mathbf{R}_0 = 0\mathbf{i} + 0\mathbf{j}$$

and

$$\mathbf{V} = \mathbf{V}_0 = (v_0 \cos \alpha)\mathbf{i} + (v_0 \sin \alpha)\mathbf{j}$$

2. (a) Find the value of t (except for $t = 0$), called the *time of flight*, at which the y component of $\mathbf{R}$ is zero.
   (b) Find the x component of $\mathbf{R}$, called the *range*, corresponding to this value of t.

(c) Find the maximum height assumed by the projectile. How is the time required to assume this height related to the time of flight?

3. A projectile has acceleration $\mathbf{A} = -32\mathbf{j}$. For $\mathbf{R}_0 = 0\mathbf{i} + 0\mathbf{j}$, and $\mathbf{V}_0 = 240\mathbf{i} + 320\mathbf{j}$, find the position, velocity, and acceleration (a) at $t = 2$ sec; (b) when the projectile is at its highest point.

4. A projectile has acceleration $\mathbf{A} = -32\mathbf{j}$. For $\mathbf{R}_0 = 0\mathbf{i} + 0\mathbf{j}$, and $\mathbf{V}_0 = 160\mathbf{i} + 160\mathbf{j}$, find:
    (a) the height of the projectile at $t = 3$ sec
    (b) the maximum height of the projectile
    (c) the time of flight
    (d) the range

5. Find the position vector of the vertex of the parabolic path of a projectile if $v_0 = 512$ ft/sec, $\alpha = 45$ deg, and $g = 32$ ft/sec^2.

6. What angle of elevation is required to yield a maximum altitude of 1600 ft if the initial speed is 640 ft per sec? Use $g = 32$ ft/sec^2.

7. For what angle of elevation will a gun have maximum range?

8. A ball is thrown from the top of a 160-ft cliff at an angle of elevation of 30 deg with initial speed 96 ft/sec. How far horizontally from the base of the cliff does the ball strike the ground? Use $g = 32$ ft/sec^2.

9. Find $v = |\mathbf{V}|$ for a projectile. For what value of t does v assume its minimum value? Find this minimum value and state where it occurs on the parabolic path of the projectile. (Use the result of Problem 2(c)).

10. The range for a projectile has value $(v_0^2/g) \sin 2\alpha \le v_0^2/g$. Show that if $k < v_0^2/g$, the range will be k

$$\text{for } \alpha = \frac{1}{2}\sin^{-1}\frac{gk}{v_0^2} \quad \text{and for} \quad \alpha = \frac{\pi}{2} - \frac{1}{2}\sin^{-1}\frac{gk}{v_0^2}$$

Note that $\alpha = \pi/4$ when $k = v_0^2/g$.

6.4 Central Force System; Newton's Law of Universal Gravitation; Kepler's Laws; Planetary Motion

In Figure 6.4, the particle of constant mass m has position vector $\mathbf{R}$ and is acted upon by force $\mathbf{F}$. By Newton's second law

$$\mathbf{F} = m\mathbf{A} = m\frac{d\mathbf{V}}{dt} = \frac{d}{dt}(m\mathbf{V})$$

where $\mathbf{V}$ and $\mathbf{A}$ denote the velocity and acceleration of the particle at time t. Hence,

$$\mathbf{R} \times \mathbf{F} = \mathbf{R} \times \frac{d}{dt}(m\mathbf{V})$$

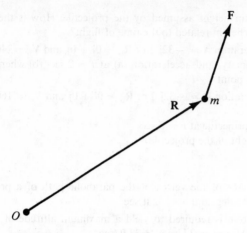

Fig. 6.4

Since (see Reference 6.2)

$$\frac{d}{dt}(\mathbf{R} \times m\mathbf{V}) = \mathbf{R} \times \frac{d}{dt}(m\mathbf{V})\frac{d\mathbf{R}}{dt} \times m\mathbf{V}$$

$$= \mathbf{R} \times \frac{d}{dt}(m\mathbf{V}) + \mathbf{V} \times m\mathbf{V} = \mathbf{R} \times \frac{d}{dt}(m\mathbf{V})$$

it follows that

$$\mathbf{R} \times \mathbf{F} = \frac{d}{dt}(\mathbf{R} \times m\mathbf{V}) \tag{6.23}$$

We now consider the special case in which the force **F** always acts towards (or away from) the origin *O*. See Figure 6.5. A force system of this type is called a *central force system*.

Since $\mathbf{R} \times \mathbf{F} \equiv \mathbf{0}$, it follows that

$$\frac{d}{dt}(\mathbf{R} \times m\mathbf{V}) = \mathbf{0}$$

by (6.23) and hence

$$\mathbf{R} \times m\mathbf{V} = \text{constant} = m\mathbf{H}$$

The vector $m\mathbf{V}$ is the *linear momentum* of the particle; the vector $m\mathbf{H}$, the *moment of momentum* of the particle about the origin. If $\mathbf{H} = \mathbf{0}$, then **R**

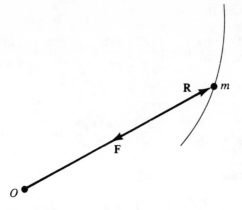

Fig. 6.5

and **V** are parallel and the path of the particle passes through the origin O. Otherwise, since **R** and **V** are then perpendicular to **H**, the path or orbit of the particle lies in a plane perpendicular to **H**.

In studying central force systems it is advantageous to let Γ be the $r\theta$ polar coordinate plane with the pole at the origin O. See Figure 6.6.

To find the velocity **V** and the acceleration **A** in polar coordinates, we write the position vector **R** in the form $r\mathbf{L}$ where $\mathbf{L} = (\cos\theta)\mathbf{i} + (\sin\theta)\mathbf{j}$ is the *radial unit vector*. Then $d\mathbf{L}/d\theta = (-\sin\theta)\mathbf{i} + (\cos\theta)\mathbf{j} = \mathbf{M}$ is the *transverse unit vector* and

$$\mathbf{V} = \frac{d\mathbf{R}}{dt} = \frac{dr}{dt}\mathbf{L} + r\frac{d\mathbf{L}}{dt} = \frac{dr}{dt}\mathbf{L} + r\frac{d\mathbf{L}}{d\theta}\frac{d\theta}{dt}$$

$$= \frac{dr}{dt}\mathbf{L} + r\frac{d\theta}{dt}\mathbf{M} \tag{6.24}$$

The scalar dr/dt is the *radial component* of **V**, and the scalar $r(d\theta/dt)$ is the *transverse component* of **V**.

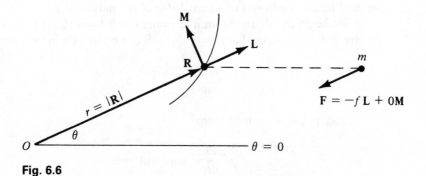

Fig. 6.6

Differentiating (6.24) yields

$$\mathbf{A} = \frac{d\mathbf{V}}{dt} = \frac{dr}{dt}\frac{d\mathbf{L}}{dt} + \frac{d^2r}{dt^2}\mathbf{L} + \left(r\frac{d\theta}{dt}\right)\frac{d\mathbf{M}}{dt} + \left(r\frac{d^2\theta}{dt^2} + \frac{dr}{dt}\frac{d\theta}{dt}\right)\mathbf{M}$$

Replacing

$$\frac{d\mathbf{L}}{dt} = \frac{d\mathbf{L}}{d\theta}\frac{d\theta}{dt} \quad \text{by} \quad \frac{d\theta}{dt}\mathbf{M} \qquad \frac{d\mathbf{M}}{dt} = \frac{d\mathbf{M}}{d\theta}\frac{d\theta}{dt} \quad \text{by} \quad -\frac{d\theta}{dt}\mathbf{L}$$

and simplifying, we obtain

$$\mathbf{A} = \left[\frac{d^2r}{dt^2} - r\left(\frac{d\theta}{dt}\right)^2\right]\mathbf{L} + \left[r\frac{d^2\theta}{dt^2} + 2\frac{dr}{dt}\frac{d\theta}{dt}\right]\mathbf{M} \qquad \textbf{(6.25)}$$

In (6.25) the scalar coefficient of **L** is the *radial component of* **A**, and the scalar coefficient of **M** is the *transverse component* of **A**.

Let us assume that the force **F** acting on the particle has magnitude f and acts toward O. Then

$$\mathbf{F} = -f\mathbf{L} + 0\mathbf{M}$$

and Newton's second law $\mathbf{F} = m\mathbf{A}$ applied in the radial and transverse directions yields

$$\frac{d^2r}{dt^2} - r\left(\frac{d\theta}{dt}\right)^2 = -fm^{-1} \qquad \textbf{(6.26)}$$

and

$$r\frac{d^2\theta}{dt^2} + 2\frac{dr}{dt}\frac{d\theta}{dt} = 0m^{-1} = 0 \qquad \textbf{(6.27)}$$

The basic problem is to find a solution for the system of DE given by (6.26) and (6.27) involving two unknown functions ϕ and ψ where $r = \phi(t)$ and $\theta = \psi(t)$. These functions yield parametric equations of the path or orbit of the moving particle. Another objective is to find r in terms of θ, that is, to find a polar equation $r = \beta(\theta)$ of the orbit. To carry out this program, we need initial conditions and knowledge of the nature of f in (6.26).

We begin by obtaining an interesting result from (6.27) that does not involve f. If we assume that $r \neq 0$, this DE can be written in the form

$$\frac{1}{r}\frac{d}{dt}\left(r^2\frac{d\theta}{dt}\right) = 0 \qquad \textbf{(6.28)}$$

From (6.28), we conclude that

$$r^2\frac{d\theta}{dt} = \text{constant} = h \qquad \textbf{(6.29)}$$

We use the polar coordinate area formula to observe in Figure 6.7 that the shaded area swept out as the particle moves from (r_1, θ_1) at time t_1 to (r, θ) at time t is given by

$$A = \frac{1}{2} \int_{\theta_1}^{\theta} [\beta(u)]^2 \, du$$

Then

$$\frac{dA}{dt} = \frac{dA}{d\theta}\frac{d\theta}{dt} = \frac{1}{2}[\beta(\theta)]^2 \frac{d\theta}{dt} = \frac{1}{2} r^2 \frac{d\theta}{dt}$$

and by (6.29),

$$\frac{dA}{dt} = \frac{h}{2} \tag{6.30}$$

The quantity dA/dt is called the *areal velocity*. The area swept out as the particle moves from (r_1, θ_1) at t_1 to (r_2, θ_2) at t_2 is given by

$$A]_{t_1}^{t_2} = \int_{t_1}^{t_2} \frac{h}{2} \, dt = \frac{h}{2}(t_2 - t_1) \tag{6.31}$$

which is the same for any interval of duration $t_2 - t_1$. This proves the following theorem, known as the *law of areas* for a central force system.

Theorem 6-II For a particle moving in a central force system, the radius vector from the center of the force system to the moving particle sweeps out equal areas in equal time intervals.

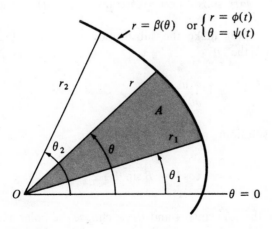

Fig. 6.7

We now apply (6.26) and (6.27) to the specific central force system in which the particle in motion is a planet moving in an orbit under the attractive force exerted by the sun at the pole. Newton's law of universal gravitation asserts that two particles attract each other with a force whose magnitude is directly proportional to the product of their masses and inversely proportional to the distance between them. It can be shown that this law generalizes to forces exerted by homogeneous spheres if we assume the masses to be concentrated at the centers of the spheres. If M denotes the mass of the sun and m the mass of the planet under study, then f in (6.26) is given by

$$f = G \frac{Mm}{r^2}$$

where G is the *universal gravitational constant*. If we let $k = GM$ for convenience, (6.26) becomes

$$\frac{d^2r}{dt^2} - r\left(\frac{d\theta}{dt}\right)^2 = \frac{-k}{r^2} \tag{6.32}$$

Replacing $d\theta/dt$ by h/r^2 from (6.29), we obtain

$$\frac{d^2r}{dt^2} = \frac{h^2}{r^3} - \frac{k}{r^2} \tag{6.33}$$

To obtain $r = \phi(t)$, let $dr/dt = p$. Then

$$\frac{d^2r}{dt^2} = \frac{dp}{dt} = \frac{dr}{dt}\frac{dp}{dr} = p\frac{dp}{dr}$$

We omit this somewhat complicated computation of $r = \phi(t)$.

To obtain $\theta = \psi(t)$, substitute $r = \phi(t)$ into $r^2(d\theta/dt) = h$ and separate the variables.

To obtain $r = \beta(\theta)$, make the substitution $r = 1/u$ in (6.33). This yields (see Problem 2) the DE

$$\frac{d^2u}{d\theta^2} + u = \frac{k}{h^2} \tag{6.34}$$

having a complete solution

$$u = A \cos \theta + B \sin \theta + \frac{k}{h^2}$$

To determine the constants A and B, we choose the polar axis so that the moving planet is closest to the origin when $\theta = 0$.

Since $u = r^{-1}$ will assume its maximum when r assumes its minimum, $du/d\theta$ will be zero and $d^2u/d\theta^2$ will be negative when $\theta = 0$. Thus,

$$\frac{du}{d\theta} = -A \sin \theta + B \cos \theta$$

$$-A(0) + B(1) = 0$$

and

$$B = 0$$

From $d^2u/d\theta^2 = -A \cos \theta$ and $-A(1) < 0$, we conclude that $A > 0$. Hence,

$$r = \frac{1}{u} = \frac{1}{(k/h^2) + A \cos \theta}$$

or

$$r = \frac{h^2/k}{1 + (Ah^2/k) \cos \theta} = \beta(\theta) \tag{6.35}$$

To determine A, we let $r = r_0$ when $\theta = 0$. This yields $A = r_0^{-1} - kh^{-2}$. The constant h is determined by assuming that $d\theta/dt = (d\theta/dt)_0$ when $r = r_0$. Then, by (6.29), $h = r_0^2(d\theta/dt)_0 = r_0 v_0$, where $v_0 = r_0(d\theta/dt)_0$.

Noting that (6.35) has the form

$$r = \frac{ep}{1 + e \cos \theta} \tag{6.36}$$

where e and p are positive, we conclude (see Reference 6.7) that the orbit of the planet is a conic having eccentricity $e = Ah^2/k$. A comet may have a parabolic path ($e = 1$), a hyperbolic path ($e > 1$), or an elliptical path ($e < 1$). The planets, however, are known to move in closed paths and hence their orbits must be ellipses with the sun at one focus. This is *Kepler's first law*.

Kepler's second law states that as a planet moves around its orbit, the radius vector from the sun to the planet sweeps out equal areas in equal time intervals. This is a direct application of Theorem 6-II to the solar system.

Kepler's third law states that for each planet, T^2 is proportional to a^3, where the period T is the time required for the planet to make one complete orbit of the sun, and a is the mean or average of the maximum and minimum distances of the planet from the sun. It is easily seen that a is the semimajor axis of the elliptical orbit.

To prove Kepler's third law, we first note from (6.35) that r has minimum value $h^2k^{-1}(1 + e)^{-1}$ and maximum value $h^2k^{-1}(1 - e)^{-1}$. Hence,

$$a = \frac{1}{2}\left[\frac{h^2}{k(1 + e)} + \frac{h^2}{k(1 - e)}\right] = \frac{h^2}{k(1 - e^2)} = \frac{h^2a^2}{kb^2}$$

where b is the semiminor axis of the ellipse ($a^2 = b^2 + c^2$ and $e = c/a$). Hence

$$b^2 = \frac{h^2 a}{k} \tag{6.37}$$

Using the formula $A = \pi a b$ for the area of an ellipse, we set $t_2 - t_1 = T$ in (6.31), to obtain

$$\pi a b = \frac{hT}{2} \quad \text{or} \quad T^2 = \frac{4\pi^2 a^2 b^2}{h^2}$$

Eliminating b^2 by (6.37), we have

$$T^2 = \left(\frac{4\pi^2}{k}\right) a^3 \tag{6.38}$$

where $4\pi^2/k$ is the constant of proportionality in Kepler's third law.

Kepler's laws of planetary motion suggested the law of universal gravitation to Newton. See Problem 6. Newton then derived Kepler's laws, as we have done, from the law of gravitation. This involves a relatively simple application of deductive mathematical analysis, whereas Kepler discovered his laws inductively from observed data during a lifetime of study. Newton's approach is also much more general, since it provides a model not only for planetary motion but also for satellite motion, motion of atomic particles, and so forth.

In (6.29) we assumed that $h > 0$ and that $r > 0$. This amounts to the assumption that $d\theta/dt > 0$, or that θ increases with t, thus making the orbit an oriented path. An example in which $h = 0$ would be provided by a meteor crashing into the sun.

6.5 Motion of a Particle in the Gravitational Field of the Earth; Satellite Motion

To apply Equations (6.26) and (6.27) to the motion of a particle moving under the attraction of the earth, locate the earth at the pole and let M denote the mass of the earth and m the mass of the moving particle of weight w. To find f in (6.26), we let $R \approx 3960$ miles denote the mean radius of the earth, and we note that when $r = R$, then $f = w = mg$. Substituting into $f = G(Mm/r^2)$, we obtain

$$mg = \frac{GMm}{R^2}$$

$$GM = gR^2$$

and

$$f = mg\left(\frac{R}{r}\right)^2, \quad \text{where } r \geq R$$

With this value of f substituted into (6.26), the system consisting of (6.26) and (6.27) provides a model for studying the motion of the moon, a meteorite, a rocket, a satellite, and the like.

Let us consider the question whether or not a satellite launched at a distance $r = r_0$ from the center of the earth will go into orbit. It is shown in Reference 6.7 that the result does not depend on the angle at which the satellite is launched. Hence, for simplicity, we assume that when $t = 0$, then $\theta = 0, r = r_0$, and

$$\mathbf{V} = \mathbf{V}_0 = r_0\left(\frac{d\theta}{dt}\right)_0 \mathbf{M} = v_0\mathbf{M}$$

See Figure 6.8.

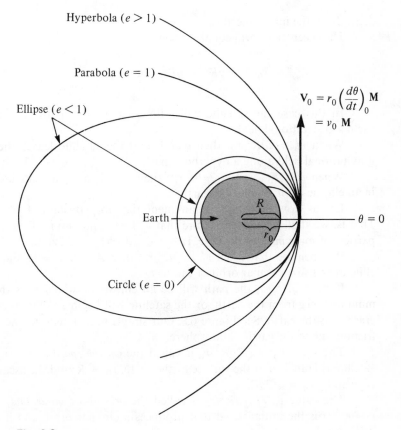

Fig. 6.8

To find an equation of the orbit of the satellite, we set

$$k = gR^2, \quad h = r_0^2\left(\frac{d\theta}{dt}\right)_0 = r_0 v_0$$

and

$$A = r_0^{-1} - kh^{-2}$$

$$= \frac{1}{r_0} + \frac{gR^2}{r_0^2 v_0^2} = \frac{r_0 v_0^2 - gR^2}{r_0^2 v_0^2}$$

in (6.35) to obtain

$$r = \frac{r_0^2 v_0^2/(gR^2)}{1 + [(r_0 v_0^2 - gR^2)/(gR^2)]\cos\theta} = \beta(\theta) \qquad \textbf{(6.39)}$$

The value of the eccentricity

$$e = \frac{r_0 v_0^2 - gR^2}{gR^2} = \frac{r_0 v_0^2}{gR^2} - 1$$

depends on the initial speed v_0.

The eccentricity will equal 1 when

$$r_0 v_0^2 - gR^2 = gR^2 \quad \text{or} \quad v_0^2 = \frac{2gR^2}{r_0}$$

In this case the satellite will follow a parabolic path and leave the earth's gravitational field.

When $v_0^2 > 2gR^2/r_0$, then $e > 1$, and the satellite leaves the earth's gravitational field along a hyperbolic path.

When $gR^2/r_0 < v_0^2 < 2gR^2/r_0$, then $e < 1$, and the satellite travels in an elliptic orbit around the earth.

Let us now relax our requirement that the constant A be positive. That is, we shall no longer require that r_0 be the minimum distance of the particle of mass m from the pole. However, $(dr/dt)_0$ is still zero.

We then see that when $v_0^2 = gR^2/r_0$, then $A = 0$, $e = 0$, and the satellite moves in a circular orbit of radius r_0.

If $v_0^2 < gR^2/r_0$, the path will be an ellipse for which r_0 is the maximum distance from the earth, or the satellite will begin an elliptic orbit and crash into the earth. See Figure 6.8. Our simple model neglects the effect of friction due to the earth's atmosphere.

The value $v_0 = \sqrt{2gR^2/r_0}$ is called the *escape speed* at $r = r_0$. If the satellite is launched at the surface of the earth, $r_0 = R$, and the escape speed is given by $v_0 = \sqrt{2gR}$.

The value $v_0 = \sqrt{gR^2/r_0}$ is called the *orbital*, or *go-around*, *speed* at $r = r_0$. It is the initial speed that produces a circular orbit, and if $r_0 = R$, it is given by $v_0 = \sqrt{gR}$.

EXAMPLE 1. Find the orbital speed and the escape speed at the earth's surface if $R = 3960$ miles and $g = 32.3$ ft/sec^2.

Solution:

$$v_0 = \sqrt{\frac{32.2}{5280}} (3960)$$

$$\approx 4.914 \text{ miles/sec}$$

$$\approx 17,691 \text{ miles/hr}$$

The escape speed is approximately $\sqrt{2}(4.914) \approx 6.950$ miles/sec $\approx$ 25,019 miles/hr.

By (6.38), the orbital time for a satellite is given by

$$T = \frac{2\pi a^{3/2}}{k^{1/2}} = \frac{2\pi a^{3/2}}{g^{1/2}R}$$

where a denotes the semimajor axis of the elliptical orbit.

From (6.39), we obtain

$$a = \frac{1}{2} [\beta(0) + \beta(\pi)]$$

$$= \frac{1}{2} \left[r_0 + \frac{r_0^2 v_0^2}{2gR^2 - r_0 v_0^2} \right] = \frac{gR^2}{2gr_0^{-1}R^2 - v_0^2}$$

Hence

$$T = \frac{2\pi}{g^{1/2}R} \left(\frac{gR^2}{2gr_0^{-1}R^2 - v_0^2} \right)^{3/2}$$

or

$$T = \frac{2\pi gR^2}{(2gr_0^{-1}R^2 - v_0^2)^{3/2}} \tag{6.40}$$

EXAMPLE 2. Find the orbital time for a satellite in circular orbit at the surface of the earth.

Solution: Setting $r_0 = R \approx 3960$ miles, $g \approx 32.2/5280$ miles/sec^2, and $v_0 = \sqrt{gR}$ in (6.40), we obtain

$$T = \frac{2\pi gR^2}{(2gR - gR)^{3/2}} = \frac{2\pi R^{1/2}}{g^{1/2}}$$

$$\approx 2\pi \left[\frac{(3960)(5280)}{32.2} \right]^{1/2}$$

$$\approx 5063 \text{ sec} \approx 84.4 \text{ min}$$

Problem List 6.4

1. Show that $d\mathbf{M}/d\theta = -\mathbf{L}$.
2. Let $r = u^{-1}$ in (6.33) to obtain (6.34).
3. Show that the speed of a planet is greatest when the planet is closest to the sun.

4. What value of k in (6.32) would produce the circular orbit $r = a$? Assume that $\theta = 0$, $r = a$, $dr/dt = 0$, and $\mathbf{V} = a(d\theta/dt)_0\mathbf{M}$ when $t = 0$. Find $\theta = \psi(t)$ and show that T^2 is proportional to a^3 where T is the period.

5. Find f in (6.26) if the orbit of the particle moving in the central force system has polar equation $r = e^\theta$.

6. Find f in (6.26) if the orbit of the particle moving in the central force system has polar equation $r = ep/(1 + e \cos \theta)$.

In Problems 7–13 use $R \approx 3960$ miles and $g \approx 32.2/5280$ miles/sec^2.

7. A satellite is given an initial speed of 20,000 miles/hr tangent to the surface of the earth. Show that its orbit is an ellipse. Find its maximum distance from the center of the earth and compute its orbital time.

8. What initial speed in the direction tangent to the earth will cause a satellite to travel in an elliptical orbit of eccentricity 0.1?

9. A satellite travels around the earth in a circular orbit at speed 17,000 miles/hr. Find the altitude of the satellite.

10. A satellite travels around the earth in a circular orbit of radius 4360 miles. Find the orbital speed and the orbital period.

11. A satellite situated 400 miles above the earth's surface is given an initial velocity parallel to the plane that is tangent to the earth at the point directly under the satellite. Find the orbital speed and the escape speed.

12. The Vanguard satellite was launched at altitude 400 miles parallel to the earth's surface at initial speed 18,000 mph. Compute (a) the eccentricity of the elliptical orbit, (b) the maximum altitude achieved, and (c) the orbital time.

13. Show that the escape speed for a rocket fired vertically upward at the surface of the earth has value $\sqrt{2gR} \approx 6.9$ miles/sec, or 25,019 mph.

6.6 Vibrations of Coupled Systems

The system shown in Figure 6.9 oscillates vertically in such a manner that Hooke's law holds when the springs, having spring constants k_1 and k_2, are in either tension or compression. The system is said to have two degrees of freedom, since two coordinates, x and y, are required to specify the positions of the masses m_1 and m_2 at time t. (Vibrating systems with a single degree of freedom were discussed in Section 5.1.)

Applying Newton's second law to the two diagrams in Figure 6.9(b), we obtain the system of DE

$$k_2(y - x) - k_1 x = m_1 \frac{d^2 x}{dt^2} \tag{6.41}$$

$$-k_2(y - x) = m_2 \frac{d^2 y}{dt^2} \tag{6.42}$$

In Figure 6.9(b), $x > 0$ and $y - x > 0$. At the time t shown, the upper spring is stretched x units and the lower spring is stretched $y - x$ units. It is easy to show that (6.41) and (6.42) hold for all t, that is, even if $x \leq 0$ or $y - x \leq 0$. If, for example, $x < 0$, the force having magnitude $|k_1 x|$ reverses

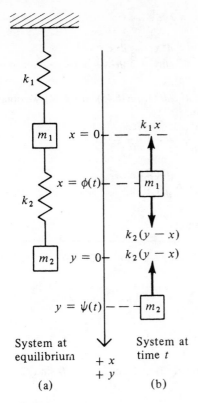

System at
equilibrium

+ x
+ y

(a)

System at
time t

(b)

Fig. 6.9

its direction. Similarly, if $y - x < 0$, the directions of the two forces having magnitude $|k_2(y - x)|$ are reversed.

A solution of system (6.41) and (6.42) consists of two functions given by $x = \phi(t)$ and $y = \psi(t)$, whose common domain is usually taken to be $t \geq 0$. By writing the system in operator notation, we can show that a solution contains four essential arbitrary constants. See Problem 5.

Except for algebraic difficulties, it is easy to solve the system by elimination. If (6.41) is solved for y, then d^2y/dt^2 can be found from y. Substituting y and d^2y/dt^2 into (6.42) yields a fourth-order DE in which ψ is the unknown function. For simplicity, we present an example involving fairly simple numerical values.

EXAMPLE 1. Solve the system given by (6.41) and (6.42) with $k_1 = 2$, $k_2 = 1$, $m_1 = 2$, and $m_2 = 1$.

Solution: Writing the system in the form

$$y = 2\frac{d^2x}{dt^2} + 3x \qquad (6.43)$$

$$\frac{d^2y}{dt^2} = -y + x \qquad (6.44)$$

we find from (6.43) that

$$\frac{d^2y}{dt^2} = 2\frac{d^4x}{dt^4} + 3\frac{d^2x}{dt^2} \tag{6.45}$$

Substituting y from (6.43) and d^2y/dt^2 from (6.45) into (6.44), we obtain

$$2\frac{d^4x}{dt^4} + 3\frac{d^2x}{dt^2} = -2\frac{d^2x}{dt^2} - 3x + x$$

or

$$2\frac{d^4x}{dt^4} + 5\frac{d^2x}{dt^2} + 2x = 0 \tag{6.46}$$

From the characteristic equation

$$2r^4 + 5r^2 + 2 = (2r^2 + 1)(r^2 + 2) = 0$$

we obtain

$$r^2 = -\frac{1}{2}, -2; \qquad r = \pm i/\sqrt{2}, \pm i\sqrt{2}$$

This yields a complete solution

$$x = A\cos(t/\sqrt{2}) + B\sin(t/\sqrt{2}) + C\cos\sqrt{2}t + D\sin\sqrt{2}t \tag{6.47}$$

of (6.46).

To find y, we note that

$$\frac{dx}{dt} = \frac{-A}{\sqrt{2}}\sin\frac{t}{\sqrt{2}} + \frac{B}{\sqrt{2}}\cos\frac{t}{\sqrt{2}} - \sqrt{2}C\sin\sqrt{2}t + \sqrt{2}D\cos\sqrt{2}t \tag{6.48}$$

and

$$\frac{d^2x}{dt^2} = \frac{-A}{2}\cos\frac{t}{\sqrt{2}} - \frac{B}{2}\sin\frac{t}{\sqrt{2}} - 2C\cos\sqrt{2}t - 2D\sin\sqrt{2}t$$

Hence, from (6.43)

$$y = 2\frac{d^2x}{dt^2} + 3x = -A\cos\frac{t}{\sqrt{2}} - B\sin\frac{t}{\sqrt{2}} - 4C\cos\sqrt{2}t - 4D\sin\sqrt{2}t$$

$$+ 3A\cos\frac{t}{\sqrt{2}} + 3B\sin\frac{t}{\sqrt{2}} + 3C\cos\sqrt{2}t + 3D\sin\sqrt{2}t$$

$$= 2A\cos\frac{t}{\sqrt{2}} + 2B\sin\frac{t}{\sqrt{2}} - C\cos\sqrt{2}t - D\sin\sqrt{2}t \tag{6.49}$$

EXAMPLE 2. Find the particular solution of (6.43) and (6.44) for which $x = 0$, $y = 2$, $dx/dt = 0$, and $dy/dt = \sqrt{2}$ when $t = 0$.

Solution: From

$$\frac{dy}{dt} = -\sqrt{2}A\sin\frac{t}{\sqrt{2}} + \sqrt{2}B\cos\frac{t}{\sqrt{2}} + \sqrt{2}C\sin\sqrt{2}t - \sqrt{2}D\cos\sqrt{2}t$$

we obtain

$$\sqrt{2} = \sqrt{2}B - \sqrt{2}D \qquad (6.50)$$

From (6.48) we obtain

$$0 = \frac{B}{\sqrt{2}} + D\sqrt{2} \qquad (6.51)$$

Equations (6.50) and (6.51) yield

$$B = \frac{2}{3}, \quad D = -\frac{1}{3}$$

From (6.47), $0 = A + C$, and from (6.49)

$$2 = 2A - C$$

These two equations in A and C yield

$$A = \frac{2}{3}, \quad C = -\frac{2}{3}$$

Thus,

$$x = \frac{2}{3}\left(\cos\frac{t}{\sqrt{2}} + \sin\frac{t}{\sqrt{2}}\right) - \frac{2}{3}\left(\cos\sqrt{2}t + \frac{1}{2}\sin\sqrt{2}t\right)$$

and

$$y = \frac{4}{3}\left(\cos\frac{t}{\sqrt{2}} + \sin\frac{t}{\sqrt{2}}\right) + \frac{2}{3}\left(\cos\sqrt{2}t + \frac{1}{2}\sin\sqrt{2}t\right)$$

In general, the system (6.41) and (6.42) will have solutions of the form

$$x = c_1\sin(\omega_1 t + \alpha_1) + c_2\sin(\omega_2 t + \alpha_2)$$

$$y = \lambda_1 c_1\sin(\omega_1 t + \alpha_1) + \lambda_2 c_2\sin(\omega_2 t + \alpha_2)$$

Thus, x (and also y) is the superposition of two simple harmonic motions, one with frequency $\omega_1/(2\pi)$ and the other with frequency $\omega_2/(2\pi)$. If the initial conditions are such that $c_2 = 0$, then x and y describe simple harmonic motions in the *principal, or normal, mode* corresponding to the normal frequency $\omega_1/(2\pi)$. Similarly, if $c_1 = 0$, then x and y describe simple harmonic motions in another principal, or normal, mode corresponding to the normal frequency $\omega_2/(2\pi)$. The positive constants λ_1 and λ_2 are called *amplitude ratios*.

6.6 Vibrations of Coupled Systems

EXAMPLE 3. Find the particular solution of (6.43) and (6.44) for which $x = 1$ and $y = 2$ when $t = 0$ and when $t = \pi/\sqrt{2}$.

Solution: Setting $t = 0$, we obtain from (6.47) and (6.49)

$$1 = A + C, \quad 2 = 2A - C$$

yielding $A = 1$ and $C = 0$.

Setting $t = \pi/\sqrt{2}$ in (6.47) and (6.49), we obtain

$$1 = B + D, \quad 2 = 2B - D$$

yielding $B = 1$ and $D = 0$.

The required solution is given by

$$x = \phi(t) = \cos \frac{t}{\sqrt{2}} + \sin \frac{t}{\sqrt{2}} = \sqrt{2} \sin \left(\frac{t}{\sqrt{2}} + \frac{\pi}{4} \right)$$

and

$$y = \psi(t) = 2 \cos \frac{t}{\sqrt{2}} + 2 \sin \frac{t}{\sqrt{2}} = 2\sqrt{2} \sin \left(\frac{t}{\sqrt{2}} + \frac{\pi}{4} \right)$$

The two masses undergo simple harmonic motions in the principal mode corresponding to the normal frequency $(1/\sqrt{2}) \cdot (1/2\pi) = \sqrt{2}/4\pi$, the amplitude ratio having value 2.

More general systems containing n degrees of freedom can be expected to have n principal modes of vibration. In designing structures or mechanisms such as a building or an airplane, care must be exercised that resonance effects are not produced by periodic external forces having frequencies equal to or close to the various normal frequencies of the systems. Another complication in the study of vibrating systems arises when frictional forces are taken into account.

Mechanical systems having more than one degree of freedom can be simulated by designing corresponding systems of coupled electric circuits. This approach is often fruitful when the complexity of the system renders a mathematical analysis difficult.

For more complete details see Reference 6.9.

Problem List 6.5

1. Given that $k_1 = 2$, $k_2 = 1$, $m_1 = 3$, and $m_2 = 1$, show that the system (6.41) and (6.42) has a complete solution given by

$$x = A \cos \omega_1 t + B \sin \omega_1 t + C \cos \omega_2 t + D \sin \omega_2 t$$

and

$$y = \sqrt{3}\,(A\cos\omega_1 t + B\sin\omega_1 t) - \sqrt{3}\,(C\cos\omega_2 t + D\sin\omega_2 t)$$

where $\omega_1 = (1 - 1/\sqrt{3})^{1/2} \approx 0.650$ and $\omega_2 = (1 + 1/\sqrt{3})^{1/2} \approx 1.256$.

2. In Problem 1 find the particular solution for which $x = 0$, $y = 2\sqrt{3}$, $dx/dt = 0$, and $dy/dt = 2\sqrt{3}\,\omega_2$ when $t = 0$.

3. In Problem 1 find the particular solution for which $x = 1$ and $y = \sqrt{3}$ when $t = 0$ and when $t = \pi/(2\omega_1)$.

4. In Problem 1 find the particular solution for which $x = 1$, $y = \sqrt{3}$, $dx/dt = 0$, and $dy/dt = 0$ when $t = 0$.

5. Write system (6.41) and (6.42) in operator notation and show that a solution contains four essential arbitrary constants.

6. (a) Eliminate y between (6.41) and (6.42) to obtain the fourth-order DE

$$\frac{d^4x}{dt^4} + \left(\frac{k_1 + k_2}{m_1} + \frac{k_2}{m_2}\right)\frac{d^2x}{dt^2} + \frac{k_1 k_2}{m_1 m_2}\,x = 0$$

(b) Show that the squares of the roots of the characteristic equation of the DE of part (a) are negative, and hence that the four roots have the forms $\pm i\omega_1$, $\pm i\omega_2$ where $\omega_1 > 0$ and $\omega_2 > 0$.

(c) Show that a general solution of (6.41) and (6.42) is given by

$$x = c_1 \cos\omega_1 t + c_2 \sin\omega_1 t + c_3 \cos\omega_2 t + c_4 \sin\omega_2 t$$

$$y = A(c_1 \cos\omega_1 t + c_2 \sin\omega_1 t) + B(c_3 \cos\omega_2 t + c_4 \sin\omega_2 t)$$

where

$$A = \frac{k_1 + k_2}{k_2} - \frac{m_1\omega_1^2}{k_2} \quad \text{and} \quad B = \frac{k_1 + k_2}{k_2} - \frac{m_1\omega_2^2}{k_2}$$

7. Find the amplitude ratios in Problem 2.

8. Derive the system of DE governing the motions of the two masses in Figure 6.10. Assume that x and y are measured from the equilibrium positions of m_1 and m_2 and that no frictional forces act. How is this system of DE related to system (6.41) and (6.42)?

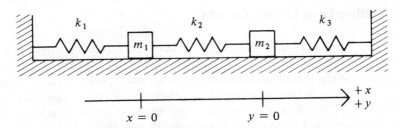

Fig. 6.10

9. Derive the system of DE governing the motions of the three masses in Figure 6.11. Assume that x, y, and z are measured from the equilibrium positions of m_1, m_2 and m_3.

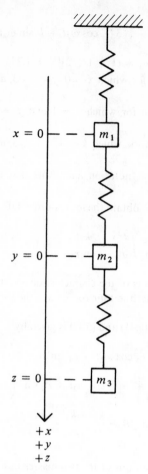

$x = 0$ — m_1

$y = 0$ — m_2

$z = 0$ — m_3

$+x$
$+y$
$+z$

Fig. 6.11

6.7 Multiple-Loop Electric Circuits

Kirchhoff's voltage law applies to the flow of electricity in a single-loop circuit (Sections 3.12 and 5.6). This law states that the algebraic sum of the voltage drops around any closed circuit is zero, with a voltage increase due to a battery or generator regarded as the negative of a voltage decrease.

The flow of current in a multiple-loop circuit or network such as that shown in Figure 6.12(a) is governed by a system of DE.

A network consists of *branches* that meet at *branch points*, or *nodes*. The network in Figure 6.12(a) has nodes B and M, and branches $MDAB$, BM, and $MFKB$. Let the currents in these branches be denoted by $i_1, i_2,$ and i_3 as shown, and let us assume that these currents flow in the directions indicated by arrows. This network consists of the three loops $ABMDA$, $BMFKB$, and $ADFKA$, shown in Figure 6.12(b).

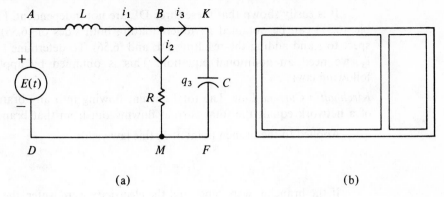

(a)   (b)

Fig. 6.12

Kirchhoff's voltage law applies to each loop, provided that a voltage drop in a branch is regarded as positive in the assumed direction of current flow. For example, loop $ABMDA$ yields the DE

$$L\frac{di_1}{dt} + Ri_2 - E = 0 \tag{6.52}$$

Similarly, loop $BMFKB$ yields

$$C^{-1}q_3 - Ri_2 = 0 \tag{6.53}$$

where q_3 satisfies $i_3 = dq_3/dt$, and loop $ADFKA$ yields

$$L\frac{di_1}{dt} + C^{-1}q_3 - E = 0 \tag{6.54}$$

Differentiating (6.53) and (6.54) with respect to t and replacing dq_3/dt by i_3, we obtain the following system of three DE in i_1, i_2, and i_3:

$$L\frac{di_1}{dt} + Ri_2 = E \tag{6.55}$$

$$C^{-1}i_3 - R\frac{di_2}{dt} = 0 \tag{6.56}$$

$$L\frac{d^2i_1}{dt^2} + C^{-1}i_3 = \frac{dE}{dt} \tag{6.57}$$

These DE are correct even if one (or more) of the assumed directions of current flow is incorrect. It is only necessary that our adopted sign convention for voltage drops be applied consistently.

It is easily shown that these three DE are not independent. For example, (6.57) can be obtained by differentiating both sides of (6.55) with respect to t and adding the resulting DE and (6.56). To determine i_1, i_2, and i_3, we need an additional equation. This is obtained by applying the following law:

Kirchhoff's Current Law. The total current flowing into any branch point of a network equals the total current flowing out from that branch point.

Applied to the branch point at B, this law yields

$$i_1 = i_2 + i_3 \tag{6.58}$$

If the branches were pipes and the electricity were water, the interpretation of (6.58) would be that the rate at which water flowed out from B would equal the rate at which water flowed into B.

To determine i_1, i_2, and i_3, we use the algebraic equation (6.58) and any two of the three DE (6.55), (6.56), and (6.57).

EXAMPLE 1. Find i_1, i_2, and i_3 in Figure 6.12(a) if $L = 2$ henrys, $R = 10$ ohms, $C = 0.05$ farad, and $E = 60$ volts. Assume that when $t = 0$, $i_1 = 0$ and $i_2 = 0$.

Solution: From (6.56), $2di_1/dt + 10i_2 = 60$, and from (6.56) and (6.58), $20(i_1 - i_2) - 10 \, di_2/dt = 0$.

In operator notation these DE become

$$Di_1 + 5i_2 = 30$$

$$2i_1 - (D + 2)i_2 = 0$$

Eliminating i_2,

$$[D(D + 2) + 10]i_1 = 60$$

$$(D^2 + 2D + 10)i_1 = 60$$

and

$$r^2 + 2r + 10 = 0, \quad r = -1 \pm 3i$$

$$i_1 = e^{-t}(c_1 \sin 3t + c_2 \cos 3t) + 6 \tag{6.59}$$

Setting $t = 0$ and $i_1 = 0$, we find that $c_2 = -6$. From (6.58), $i_3 = 0$ when $t = 0$. From (6.59), (6.55), (6.56), and (6.58), it is determined (see Problem 1) that

$$i_1 = 6 + e^{-t}[8 \sin 3t - 6 \cos 3t]$$

$$i_2 = 6 - e^{-t}[2 \sin 3t + 6 \cos 3t]$$

$$i_3 = i_1 - i_2 = 10e^{-t} \sin 3t$$

Multiple-loop circuits are often constructed as analogs of mechanical systems having two or more degrees of freedom. Designing appropriate networks for simulating such systems often requires considerable ingenuity.

Problem List 6.6

1. Find i_1, i_2, and i_3 in Example 1 if $i_1 = i_2 = 0$ when $t = 0$.
   In Problems 2–5 set up but do not solve a set of independent equations that are sufficient to find the unknown currents in terms of arbitrary constants.

2.

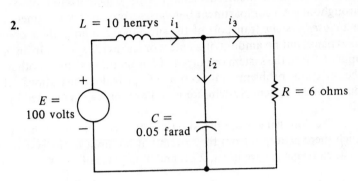

Fig. 6.13

3.

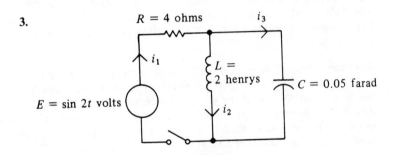

Fig. 6.14

4.

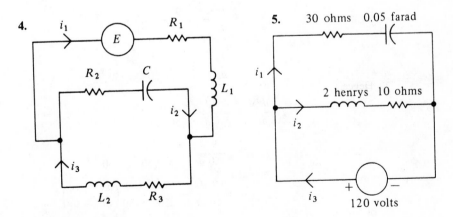

Fig. 6.15

Fig. 6.16

6.8 Compartmental Systems

Let $X_1, X_2, \ldots, X_n$ denote a system of n separate compartments and let x_i denote the amount of a substance S in compartment X_i at time t. If the substance is dissolved in a solvent, the solution in each compartment is assumed to be well-stirred. That is, the concentration of S is uniform throughout each compartment but varies from one compartment to another. In a *closed system*, transport of the substance S among the n compartments takes place, but no amount of S enters or leaves the system. In an *open system*, input of S into the system or output of S from the system or both is permitted. The mixture problems in Section 3.6 provide illustrations of a single compartment system. Systems for $n > 1$ are generally governed by systems of DE.

The following example shows a two-compartment open system in which there is output from the system but no input to the system. For notational convenience, we use x, X, y, and Y in place of x_1, X_1, x_2, and X_2.

EXAMPLE 1. Vat X contains 50 gal brine, in which 12 lb salt are dissolved, and vat Y contains 50 gal water containing no salt. Water containing no salt flows into X at 8 gal/min; the mixture (well-stirred) in X flows from X to Y at 9 gal/min; the mixture in Y (also well-stirred) flows back into X at 1 gal/min and out of the system at 8 gal/min. Let $x(t)$ and $y(t)$ denote the amounts of salt in pounds in X and Y at time t. Find $x(t)$ and $y(t)$.

Solution: No salt flows into the 2-compartment system. Since the concentration of salt in X at time t is $x/50$ lb/gal and 9 gal solution flows from X to Y per min, salt flows from X to Y at the rate $9(x/50)$ lb/min. Similarly, salt flows from Y to X at $1(y/50)$ lb/min, and from Y out of the system at $8(y/50)$ lb/min. The transport in gallons per minute of fluid is shown in Figure 6.17(a), the transport of salt in pounds per minute in Figure 6.17(b).

Applying the equation of continuity to the amount of salt in X and in Y, we obtain

$$\frac{dx}{dt} = -\frac{9}{50}x + \frac{1}{50}y$$

$$\frac{dy}{dt} = \frac{9}{50}x - \frac{1}{50}y - \frac{8}{50}y$$

(6.60)

Hence,

$$\frac{d^2x}{dt^2} = \frac{-9}{50}\frac{dx}{dt} + \frac{1}{50}\frac{dy}{dt}$$

$$= -\frac{9}{50}\frac{dx}{dt} + \frac{1}{50}\left(\frac{9}{50}x - \frac{9}{50}y\right)$$

$$= -\frac{9}{50}\frac{dx}{dt} + \frac{9x}{2500} - \frac{9}{50}\left(\frac{dx}{dt} + \frac{9x}{50}\right)$$

or

$$\frac{d^2x}{dt^2} + \frac{18}{50}\frac{dx}{dt} + \frac{72x}{2500} = 0$$

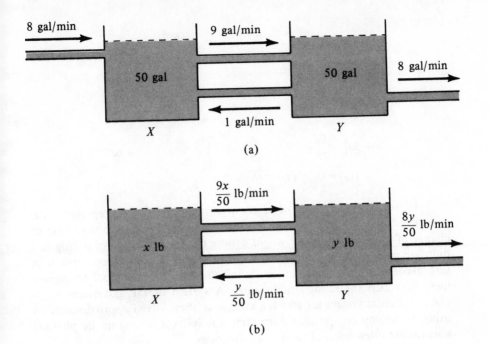

8 gal/min

9 gal/min

50 gal

50 gal

8 gal/min

1 gal/min

X

Y

(a)

$\dfrac{9x}{50}$ lb/min

x lb

y lb

$\dfrac{8y}{50}$ lb/min

$\dfrac{y}{50}$ lb/min

X

Y

(b)

Fig. 6.17

From the characteristic equation

$$r^2 + \frac{18}{50}r + \frac{72}{2500} = \left(r + \frac{3}{25}\right)\left(r + \frac{6}{25}\right) = 0$$

we obtain $r_1 = -\frac{3}{25}$ and $r_2 = -\frac{6}{25}$. Therefore,

$$x(t) = Ae^{-3t/25} + Be^{-6t/25}$$

and

$$x'(t) = \frac{dx}{dt} = -\frac{3}{25}Ae^{-3t/25} - \frac{6}{25}Be^{-6t/25}$$

Using $x(0) = 12$, $y(0) = 0$, and from (6.60),

$$x'(0) = -\frac{9}{50}x(0) + \frac{1}{50}y(0)$$

$$= -\frac{9}{25}(12) + 0 = -\frac{54}{25}$$

we obtain

$$12 = A + B \quad \text{and} \quad \frac{-54}{25} = -\frac{3A}{25} - \frac{6B}{25}$$

from which we find that $A = B = 6$. Thus,

$$x(t) = 6e^{-3t/25} + 6e^{-6t/25}$$

From

$$x'(t) = \frac{dx}{dt} = \frac{-18}{25} e^{-3t/25} - \frac{36}{25} e^{-6t/25}$$

and (6.60), we obtain

$$y(t) = 50 \frac{dx}{dt} + 9x$$

$$= 50 \left(-\frac{18}{25} e^{-3t/25} - \frac{36}{25} e^{-6t/25} \right) + 9(6e^{-3t/25} + 6e^{-6t/25})$$

$$= 18e^{-3t/25} - 18e^{-6t/25}$$

In some applications of n compartment systems, x_i represents the concentration of substance S in X_i and dx_i/dt represents the time rate of change of x_i. For example, x_i might denote the concentration of a drug in a diffusion process. In cancer chemotherapy, the drug concentration varies over several compartments that connect the injection compartment to the compartment in which the malignancy occurs. A system of DE governing an n-compartmental system for which n is large is often solved approximately by using an analog computer. Compartmental systems arising in the physical sciences are often referred to as *block diagrams*.

The system (6.60) is a linear autonomous system of DE inasmuch as the right members do not involve t and are linear in x and y. The following equations illustrate a nonlinear autonomous system of DE:

$$\frac{dx}{dt} = ax - bxy \qquad\qquad \textbf{(6.61)}$$

$$\frac{dy}{dt} = -cy + dxy \qquad\qquad \textbf{(6.62)}$$

Equations (6.61) and (6.62) are known as *Volterra's prey-predator equations*, after the Italian mathematician Vito Volterra (1860–1940), who formulated them during his numerous investigations in mathematical biology. The variable x denotes the size of the prey population in a given environment, the variable y the size of the predator population in the same environment. In a well-known application, the predator population consists of foxes and the prey population consists of rabbits. The constants a, b, c, and d are nonnegative; the constant a is the birthrate of the prey and the constant c is the deathrate of the predator. The constants b and d depend on the intensity of the interaction between the two populations.

See Reference 6.1 for an interesting application of Volterra's equations.

It is much more difficult to solve nonlinear systems than to solve linear systems. Although it is difficult to solve Equations (6.61) and (6.62) for x and y in terms of t, it is easy to find a relation holding between x and y. See Problem 3. Problem 4 indicates the way in which (6.61) and (6.62) can be linearized and approximated in a certain sense by a linear system of DE.

Problem List 6.7

1. In Example 1, find the maximum value assumed by $y(t)$ and the number of minutes before this maximum is assumed.

2. Solve Example 1 if water flows into X at 15 gal/min, the mixture in X flows to Y at 16 gal/min, and the mixture in Y flows back to X at 1 gal/min and out of the system at 15 gal/min.

3. Derive a relation of the form $g(x, y) = 0$ where $x = x(t)$ and $y = y(t)$ satisfy the system (6.61) and (6.62).

4. Show that $dx/dt = 0$ when $y = a/b$, and $dy/dt = 0$ when $x = c/d$, if $x(t)$ and $y(t)$ satisfy (6.61) and (6.62). These values of x and y at which the population rates of change are zero are called *equilibrium populations*. The points $(0, 0)$ and $(c/d, a/b)$ are critical points of the system of DE.

   Introduce the variables $u = x - c/d, v = y - a/b$, and find the system of DE satisfied by u and v. Linearize this system by neglecting terms involving uv. Show that for the linearized system, $ad^2u^2 + b^2cv^2 = k^2 = $ constant.

5. In Figure 6.18, x, y, and z denote the concentrations in grams per cubic centimeter of a substance S in compartments X, Y, and Z. Given that $A, B, C, D, E,$ and F denote rates of transport of the solvent containing S in cm^3/min, write a system of DE governing the amounts x, y, and z at time t.

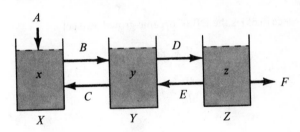

Fig. 6.18

6.9 Domar Burden-of-Debt Model

Let $D = D(t)$ denote the total national debt (in dollars) at time t, and $Y = Y(t)$ the total national annual income (in dollars per year) at time t. The *Domar burden-of-debt model* (see Reference 6.3) assumes that the rate of borrowing $D'(t) = dD/dt$ is proportional to the yearly annual income rate $Y(t)$, that is, that

$$\frac{dD}{dt} = kY \qquad (6.63)$$

where the constant k satisfies $0 < k < 1$.

This is equivalent to the assumption that a constant fraction of the national income becomes additional public debt.

The DE (6.63) involves two unknown functions of t, namely, D and Y. To determine $D(t)$ and $Y(t)$, we need an additional equation, or DE, of the form

$$f\left(t, Y, \frac{dY}{dt}, D, \frac{dD}{dt}\right) = 0 \tag{6.64}$$

together with initial conditions.

Let $T = T(t)$ denote the interest-induced tax (in dollars per year) and assume that the effective yearly interest rate i at which the interest on the debt is computed remains constant. Then $T = iD$.

The *debt burden* $B = B(t)$ is defined by the ratio $B = T/Y = iD/Y$.

It is clear that if B increases too rapidly with time, an untenable situation will ultimately prevail. To determine $\lim_{t \to +\infty} B(t)$, we must know or at least have some information about (6.64).

EXAMPLE 1. Find $\lim_{t \to +\infty} B(t)$ if (6.64) is given by

$$\frac{dY}{dt} = cY \tag{6.65}$$

where the constant c satisfies $0 < c < 1$. Assume that $D(0) = D_0$ and $Y(0) = Y_0$.

Solution: Since (6.65) is the DE of organic growth, we get

$$Y = Y(t) = Y_0 e^{ct}$$

Hence, from (6.63),

$$\frac{dD}{dt} = kY_0 e^{ct}$$

and

$$D = \frac{kY_0}{c} e^{ct} + C$$

Setting $t = 0$ and $D = D_0$, we find that

$$C = D_0 - \frac{kY_0}{c}, \quad \text{and hence,} \quad D(t) = \frac{kY_0}{c} e^{ct} + D_0 - \frac{kY_0}{c}$$

Therefore,

$$B(t) = \frac{iD(t)}{Y(t)}$$

$$= \frac{ikY_0 c^{-1} e^{ct} + iD_0 - ikY_0 c^{-1}}{Y_0 e^{ct}}$$

$$= \frac{ik}{c} + \left(\frac{iD_0 - ikY_0 c^{-1}}{Y_0}\right) e^{-ct}$$

and

$$\lim_{t \to +\infty} B(t) = \frac{ik}{c}$$

The conclusion is that if the national income per year grows exponentially, the debt burden will approach a finite limit as $t \to +\infty$ and will not get out of control.

Problem List 6.8

1. Find $B = B(t)$ and $\lim_{t \to +\infty} B(t)$ if:

(a) $\dfrac{dY}{dt} = 0$

(b) $\dfrac{dY}{dt} = a = $ constant

(c) $\dfrac{dY}{dt} = at$ (a constant)

(d) $\dfrac{dY}{dt} = at + b$ (a and b constant)

(e) $\dfrac{dY}{dt} = at^2 + bt + c$ (a, b, and c constant)

2. Show that $\dfrac{dB}{dt} = i\left(k - \dfrac{B}{y} \dfrac{dY}{dt} \right)$.

3. Find $D(t)$ in Example 1 by solving the second-order DE

$$\frac{d^2 D}{dt^2} = k \frac{dY}{dt} = kcY = c \frac{dD}{dt}$$

References

6.1 Braun. M. 1975. *Differential Equations and Their Applications.* New York: Springer-Verlag.

6.2 Christie, Dan E. 1964. *Vector Mechanics,* 2nd ed. New York: McGraw-Hill.

6.3 Domar, E. D. 1957. *Essays in the Theory of Economic Growth.* Fairlawn, N.J.: Oxford University Press.

6.4 Frauenthal, James C. 1980. *Introduction to Population Modeling.* Boston: Birkhauser.

6.5 Ince, E. L. 1956. *Ordinary Differential Equations.* New York: Dover.

6.6 Kaplan, Wilfred. 1962. *Ordinary Differential Equations.* Reading, Mass.: Addison-Wesley.

6.7 Shames, I. H. 1960. *Engineering Mechanics, Statics and Dynamics.* Englewood Cliffs, N.J.: Prentice-Hall.

6.8 Tierney, John A. 1979. *Calculus and Analytic Geometry,* 4th ed. Boston: Allyn and Bacon.

6.9 Yeh, H., and J. I. Abrams. 1960. *Principles of Mechanics of Solids and Fluids,* vol. 1. York, Pa.: Maple Press.

7

The Laplace
Transform

7.1 Introduction

The differential operator D has as domain the set of differentiable functions (Chapter 4). The Heaviside operational calculus (Section 4.9) uses the operator D to solve DE. The integral operator $\mathscr{L}$, known as the Laplace transform, provides another method of solving DE; it is particularly effective in solving initial-value problems involving linear DE with constant coefficients.

The Laplace transform is named for the renowned French mathematician Pierre Simon de Laplace (1749–1827), who used transforms in his classical treatise on the theory of probability. Whereas originally the Heaviside operational calculus did not receive wide acceptance in the mathematical community, the later justification for many of Heaviside's techniques was based on the theory of the Laplace transform.

Definition Let F be a real-valued function defined for $0 \leq t < +\infty$. The Laplace transform $\mathscr{L}$, if it exists, is defined by

$$\mathscr{L}[F(t)] = \int_0^{+\infty} e^{-st}F(t)\,dt = f(s) \tag{7.1}$$

The transform $\mathscr{L}$ is a function or operator that maps the function F into the function f, and we often write

$$\mathscr{L}F = f \tag{7.2}$$

An element of the domain of F is customarily denoted by t, since in many applications F is a function of the time t. We emphasize that F in (7.1) has domain $[0, +\infty)$; this will be understood henceforth if not mentioned explicitly. The variable of integration in (7.1), although a dummy variable, is usually denoted by t. The improper integral in (7.1) is defined by

$$\lim_{T \to +\infty} \int_0^T e^{-st}F(t)\, dt = f(s) \tag{7.3}$$

provided that the limit exists. Since t is a dummy variable, the limit, if it exists, will generally involve s. The domain of f consists of the totality of values of s for which the limit in (7.3) exists. This is generally a set of complex numbers. For simplicity, we consider the domain of f the set of *real* values of s for which the limit in (7.3) exists. Thus, the domain and range of $\mathscr{L}$ consist of real functions, with each element (function) in the domain of $\mathscr{L}$ having its own domain $[0, +\infty)$.

Our preference is to regard the Laplace transform $\mathscr{L}$ as the operator, or function, that maps or transforms F into f. It is customary, although technically inaccurate, to refer to $\mathscr{L}F = f$ or $\mathscr{L}[F(t)] = f(s)$ as the Laplace transform of F or of $F(t)$. We shall follow the useful convention of capital letters for elements of the domain of $\mathscr{L}$ and corresponding lowercase letters for elements of the range of $\mathscr{L}$, as in $\mathscr{L}F = f$ or $\mathscr{L}[G(t)] = g(s)$.

When we introduced the differential operator D, it was clear that the domain of D consisted of the set of differentiable functions. Hence it is natural that we ask ourselves the nature of the domain of $\mathscr{L}$; that is, we ask ourselves what real functions F defined on $[0, +\infty)$ have Laplace transforms. Before we consider this, we use Definition (7.1) to find the Laplace transforms of a few simple functions.

EXAMPLE 1. Find $\mathscr{L}[F(t)]$ where $F(t) \equiv 1$ on $[0, +\infty)$.

Solution:

$$\mathscr{L}[1] = \int_0^{+\infty} e^{-st}\, dt$$

$$= \left(\frac{-1}{s}\right) \lim_{T \to +\infty} \int_0^T e^{-st}(-s)\, dt$$

$$= \left(-\frac{1}{s}\right) \lim_{T \to +\infty} [e^{-st}]_0^T = \left(-\frac{1}{s}\right) \lim_{T \to +\infty} [e^{-sT} - 1]$$

$$= \left(-\frac{1}{s}\right)[0 - 1] = \frac{1}{s} = f(s)$$

provided $s > 0$. If $s = 0$, $\int_0^{+\infty} e^{-st}\, dt = \int_0^{+\infty} dt$ diverges, and if $s < 0$, $\lim_{T \to +\infty} e^{-sT} = +\infty$. Thus, the transform $f(s) = 1/s$ of $F(t) \equiv 1$ has domain $(0, +\infty)$. This result is easily extended to show that $\mathscr{L}[c] = c/s$ where c is constant.

EXAMPLE 2.
$$\mathscr{L}[e^{at}] = \int_0^{+\infty} e^{-st} e^{at}\, dt$$

$$= \lim_{T \to +\infty} \int_0^T e^{(a-s)t}\, dt = \lim_{T \to +\infty} \left[\frac{e^{(a-s)t}}{a-s} \right]_0^T$$

$$= \lim_{T \to +\infty} \left[\frac{e^{(a-s)T}}{a-s} - \frac{1}{a-s} \right] = \frac{-1}{a-s} = \frac{1}{s-a}$$

for all $s > a$. NOTE: This result holds for all a. If $a = 0$, $\mathscr{L}[e^{at}] = \mathscr{L}[1] = 1/s$, agreeing with the result in Example 1.

EXAMPLE 3.
$$\mathscr{L}[\sin at] = \int_0^{+\infty} e^{-st} \sin at\, dt$$

$$= \lim_{T \to +\infty} \left[\frac{e^{-st}}{s^2 + a^2} (-s \sin at - a \cos at) \right]_0^T$$

$$= \lim_{T \to +\infty} \left[\frac{-1}{s^2 + a^2} \frac{s \sin aT + a \cos aT}{e^{sT}} + \frac{a}{s^2 + a^2} \right]$$

$$= \frac{a}{s^2 + a^2}$$

for all a and all $s > 0$. NOTE:

$$\left| \frac{s \sin aT + a \cos aT}{e^{sT}} \right| \leq \frac{|s| + |a|}{e^{sT}}$$

EXAMPLE 4.
$$\mathscr{L}[t] = \int_0^{+\infty} e^{-st} t\, dt = \lim_{T \to +\infty} \int_0^T e^{-st} t\, dt$$

$$= \lim_{T \to +\infty} \left[\frac{e^{-st}}{s^2} (-st - 1) \right]_0^T = \lim_{T \to +\infty} \left[\frac{-1}{s^2} \frac{sT + 1}{e^{sT}} + \frac{1}{s^2} \right]$$

$$= \frac{1}{s^2}$$

for all $s > 0$. NOTE: It is easily shown by L'Hôpital's rule that

$$\lim_{T \to +\infty} (sT + 1)/e^{sT} = 0$$

EXAMPLE 5. If we are given that $F(t) = t^n$ for $t > 0$ and $F(0) = 0$, then

$$\mathscr{L}[t^n] = \int_0^{+\infty} e^{-st} t^n\, dt$$

Let $t = u/s$. Then $dt = du/s$ and u ranges from 0 to $+\infty$ provided that $s > 0$.

Therefore,

$$\mathscr{L}[t^n] = \int_0^{+\infty} e^{-u} \frac{u^n}{s^n} \frac{du}{s} = \frac{1}{s^{n+1}} \int_0^{+\infty} e^{-u} u^n \, du \qquad (7.4)$$

The integral in (7.4) can be expressed in terms of the well-known *gamma function*, defined by

$$\Gamma(x) = \int_0^{+\infty} e^{-t} t^{x-1} \, dt, \quad x > 0 \qquad (7.5)$$

The gamma function is important in both pure and applied mathematics and has been extensively tabulated. The improper integral in (7.5) converges, provided that $x > 0$, and hence by (7.4) and (7.5),

$$\mathscr{L}[t^n] = \frac{1}{s^{n+1}} \int_0^{+\infty} e^{-u} u^{(n+1)-1} \, du = \frac{\Gamma(n+1)}{s^{n+1}} \qquad (7.6)$$

for $n + 1 > 0$ or $n > -1$, and $s > 0$.

It is a simple exercise to verify that $\Gamma(1) = 1$ (see Problem 4) and to establish the recurrence relation (see Problem 5)

$$x\Gamma(x) = \Gamma(x + 1) \qquad (7.7)$$

for $x > 1$.

Then $\Gamma(2) = 1\Gamma(1) = 1 = 1!$, $\Gamma(3) = 2\Gamma(2) = 2!$, $\Gamma(4) = 3\Gamma(3) = 3!$, and by mathematical induction

$$\Gamma(n + 1) = n! \quad n = 1, 2, 3, \ldots \qquad (7.8)$$

It is owing to (7.8) that the gamma function is also referred to as the *generalized factorial function*. Thus, for n a positive integer,

$$\mathscr{L}[t^n] = \frac{n!}{s^{n+1}} \qquad (7.9)$$

for $s > 0$.

EXAMPLE 6. Find $\mathscr{L}[F(t)]$, given that

$$F(t) = \begin{cases} 1 & \text{for } 0 \le t \le 1 \\ 0 & \text{for } t > 1 \end{cases}$$

Solution:

$$\mathscr{L}[F(t)] = \int_0^{+\infty} e^{-st} F(t) \, dt$$

$$= \int_0^1 e^{-st} \, dt$$

$$= \begin{cases} \left[\dfrac{e^{-st}}{-s} \right]_0^1 = \dfrac{1 - e^{-s}}{s} & \text{for } s \ne 0 \\ \displaystyle\int_0^1 dt = 1 & \text{for } s = 0 \end{cases}$$

EXAMPLE 7. It is easy to extend the result in Example 6 to show that if

$$F(t) = \begin{cases} 0 & \text{for } 0 \le t < a \\ 1 & \text{for } a \le t < b \\ 0 & \text{for } t \ge b \end{cases}$$

then

$$\mathcal{L}[F(t)] = \begin{cases} \dfrac{e^{-as} - e^{-bs}}{s} & \text{for } s \ne 0 \\ b - a & \text{for } s = 0 \end{cases}$$

Problem List 7.1

1. Find $\mathcal{L}[F(t)]$ directly from Definition (7.1). Give the domain of $f = \mathcal{L}F$ in each instance.
   (a) $F(t) \equiv 0$　　　　　　　(b) $F(t) \equiv 2$
   (c) $F(t) = e^t$　　　　　　　(d) $F(t) = e^{-t}$
   (e) $F(t) = e^{3t}$　　　　　　(f) $F(t) = \cos at$
   (g) $F(t) \equiv a = $ constant

2. Find $\mathcal{L}[F(t)]$ by using the results of Examples 2–7.
   (a) $F(t) = e^{4t}$　　　(b) $F(t) = e^{-5t}$
   (c) $F(t) = \sin 2t$　　(d) $F(t) = -\sin 3t$
   　　　　　　　　　　　$= \sin(-3t)$
   (e) $F(t) = t^{1/4}$　　(f) $F(t) = t^2$
   (g) $F(t) = t^3$　　　(h) $F(t) = t^5$

3. Solve Example 4 by Formula (7.9).
4. Show that $\Gamma(1) = 1$.
5. Show that $x\Gamma(x) = \Gamma(x + 1)$ for $x > 1$.
6. Evaluate (a) $\Gamma(3)$; (b) $\Gamma(5)$; and (c) $\Gamma(7)$.
7. Show that the transform $\mathcal{L}F = f$ of Example 6 is a continuous function on $(-\infty, +\infty)$.
8. Find $\mathcal{L}[F(t)]$, given that

$$F(t) = \begin{cases} 1 & \text{for } 0 \le t \le 2 \\ 0 & \text{for } t > 2 \end{cases}$$

9. Find $\mathcal{L}[F(t)]$, given that

$$F(t) = \begin{cases} t & \text{for } 0 \le t \le 1 \\ 1 & \text{for } t > 1 \end{cases}$$

10. Find $\mathcal{L}[F(t)]$, given that

$$F(t) = \begin{cases} 2 & \text{for } 0 \le t \le 1 \\ 3 & \text{for } t > 1 \end{cases}$$

7.2 The Domain and Range of $\mathscr{L}$

We return to the question of what real functions defined on $[0, +\infty)$ have Laplace transforms. This difficult question is the fundamental problem of Laplace transform theory and is treated in textbooks such as Reference 7.11. We content ourselves with characterizing a large class of functions that do have Laplace transforms. A few definitions enable us to describe this class of functions.

Definition   If $\lim_{t \to a} F(t)$ exists and is not equal to $F(a)$, the function F has a *removable discontinuity* at $t = a$.

Definition   Let a be contained in the domain of F. If

$$\lim_{t \to a^+} F(t) = L_1 \quad \text{and} \quad \lim_{t \to -a} F(t) = L_2 \neq L_1$$

then F has a *jump discontinuity* at $t = a$. The difference $L_1 - L_2$ is called the *jump in the function*.

Definitions   Removable and jump discontinuities are called *ordinary discontinuities*. A function that is continuous on an interval I except for a finite number of ordinary discontinuities is said to be *piecewise continuous*, or *sectionally continuous*, on I.

At an endpoint of I, a one-sided limit is intended. For example, since we are interested in a function F defined on $[0, +\infty)$, we assume that $\lim_{t \to 0^+} F(t)$ exists. If this limit is not equal to $F(0)$, then F has a removable discontinuity at $t = 0$.

Definition   The function F is of *exponential order* as $t \to +\infty$ if there exist constants M, k, and t_0 such that $|F(t)| < Me^{kt}$ for all $t \geq t_0$.

EXAMPLE 1.   Show that the function F given by $F(t) = 4e^{3t}$ is of exponential order.

Solution: $|4e^{3t}| = 4e^{3t} < 4e^{5t}$ for $t \geq 1$. We have chosen $M = 4$, $k = 5$, and $t_0 = 1$.

If F is sectionally continuous on $[0, t_0]$ where t_0 is an arbitrary positive number, it can be shown that $\int_0^{t_0} e^{-st}F(t)\,dt$ exists. If F is sectionally continuous on $[t_0, +\infty)$ and is of exponential order as $t \to +\infty$ it can be shown that $\int_{t_0}^{+\infty} e^{-st}F(t)\,dt$ converges for at least one value of s. The convergence is due to the fact that $e^{-st}F(t)$ approaches zero rapidly enough as

$t \to +\infty$, which in turn happens since F is of exponential order as $t \to +\infty$. These results are combined in Theorem 7-I, proved in Reference 7.11, which states a set of conditions sufficient for the Laplace transform to exist.

Definition   If F is sectionally continuous on $[0, +\infty)$ and is of exponential order as $t \to +\infty$, then F is said to belong to class A.

Theorem 7-I   If F belongs to class A, the Laplace transform of F exists.

Laplace transforms of many of the functions arising in applications exist by Theorem 7-I. For example, the functions in Examples 1, 2, 3, 4, and 6 of Section 7.1 have Laplace transforms by Theorem 7-I. The theorem also applies to $F(t) = t^n$, $t > 0$, $F(0) = 0$ in Example 5 provided that $n \geq 0$. If $-1 < n < 0$, then F has a Laplace transform even though F is not sectionally continuous on $[0, +\infty)$. The reason is that the discontinuity at $t = 0$ is not ordinary. The existence of the Laplace transform in this case does not contradict Theorem 7-I, since the theorem states conditions sufficient (but not necessary) for $\mathscr{L}F$ to exist.

EXAMPLE 2.   Find $\mathscr{L}[F(t)]$ where $F(t) = t^{-1/2}$ for $t > 0$ and $F(0) = 0$.

Solution:

$$\mathscr{L}[F(t)] = \int_0^{+\infty} e^{-st} t^{-1/2}\, dt = \frac{\Gamma(1/2)}{s^{1/2}}$$

for $s > 0$ by (7.6). Since $\Gamma(\tfrac{1}{2}) = \sqrt{\pi}$ (see Problem 4), this result can also be written as $(\pi/s)^{1/2}$.

EXAMPLE 3.   Show that $\mathscr{L}[\exp(t^2)]$ does not exist.

Solution:

$$\int_0^{+\infty} \exp(-st) \exp(t^2)\, dt = \int_0^{+\infty} \exp\left(t^2 - st + \frac{s^2}{4} - \frac{s^2}{4}\right) dt$$

$$= \exp\left(-\frac{s^2}{4}\right) \int_0^{+\infty} \exp\left(\left[t - \frac{s}{2}\right]^2\right) dt$$

There is no value of s for which the latter integral converges, since $\lim_{t \to +\infty} \exp([t - s/2]^2) = +\infty$ for every s. In Problem 2, the student is asked to show that $\exp(t^2)$ is not of exponential order as $t \to +\infty$.

An element of the domain of $\mathscr{L}$ has domain $[0, +\infty)$ whereas elements of the range of $\mathscr{L}$ do not all have the same domain. By Example 2 of Section 7.1, $\mathscr{L}[e^{-t}] = 1/(s + 1)$ has domain $s > -1$ and $\mathscr{L}[e^{2t}] = 1/(s - 2)$ has domain $s > 2$. In Reference 7.11, it is shown (1) that $\mathscr{L}[F(t)] = f(s)$ exists for all s, or (2) that there is no s for which $\mathscr{L}[F(t)]$ exists, or (3) that there exists a real number s_c, called the *abscissa of convergence*, such that $\mathscr{L}[F(t)]$ exists for all $s > s_c$ and does not exist for $s \leq s_c$.

EXAMPLE 4. If we are given $F(t) \equiv 0$ on $[0, +\infty)$, then

$$\mathscr{L}[F(t)] = \int_0^{+\infty} e^{-st}0 \, dt = f(s) \equiv 0$$

for all s. That is, the domain of f is $(-\infty, +\infty)$.

Example 6 of Section 7.1 furnishes another example in which the domain of f is $(-\infty, +\infty)$. This example also exhibits a function that is discontinuous at $t = 1$ and that has a Laplace transform.

We conclude this section by stating two theorems that furnish information regarding the range of $\mathscr{L}$. Proofs are given in Reference 7.12.

Theorem 7-II If F belongs to class A, then $\mathscr{L}[F(t)] = f(s) \to 0$ as $s \to +\infty$.

This theorem is also true for arbitrary Laplace transforms f. It rules out as possible Laplace transforms simple expressions such as $s^2 + s + 1, \cos 2s$, $\sin 3s, e^{4s}, \ln s$, and so on.

Theorem 7-III If F belongs to class A, then $s\mathscr{L}[F(t)] = sf(s)$ is bounded as $s \to +\infty$.

We say that $sf(s)$ is bounded as $s \to +\infty$ if and only if there exists an M and an s_0 such that $|sf(s)| \leq M$ for all $s > s_0$. Theorem 7-III rules out $f(s) = s^{-1/2}$ as a Laplace transform of a function F of class A since $|ss^{-1/2}| = \sqrt{s}$ can be made greater than any fixed M by choosing s sufficiently large.

See Reference 7.9 for examples of various types of discontinuities and a treatment of improper integrals.

Problem List 7.2

1. Show that F is of exponential order as $t \to +\infty$ if:
   (a) $F(t) = t$ (b) $F(t) = t^2$
   (c) $F(t) = \sin t$ (d) $F(t) = e^{-2t}$
   (e) $F(t) = t^x$, x real   (f) $F(t) = \cosh at$
   (g) $F(t) = \sinh at$ (h) $F(t) = t^n$, $n \geq 0$
2. Show that the function F given by $F(t) = e^{t^2}$ is not of exponential order as $t \to +\infty$.
3. Show the $\mathscr{L}[e^{-t^2}]$ exists and has domain $-\infty < s < +\infty$.
4. Show that $\Gamma[\frac{1}{2}] = \sqrt{\pi}$. Hint: Show that $\Gamma[\frac{1}{2}] = 2\int_0^{+\infty} e^{-x^2} \, dx = 2\int_0^{+\infty} e^{-y^2} \, dy$, multiply these integrals, and use polar coordinates.
5. Show that $\mathscr{L}[t^{-1}]$ does not exist.

7.3 Properties of Laplace Transforms

Numerous interesting properties of Laplace transforms are expressed in the theorems of this section. Many transforms can be found more simply by using these properties than by applying the basic definition (7.1).

The first theorem states that $\mathscr{L}$ is a linear operator.

Theorem 7-IV Let $\mathscr{L}[F_1(t)] = f_1(s)$ for $s \in S_1$, let $\mathscr{L}[F_2(t)] = f_2(s)$ for $s \in S_2$, and let c_1 and c_2 be arbitrary constants. Then

$$\mathscr{L}[c_1 F_1(t) + c_2 F_2(t)] = c_1 \mathscr{L}[F_1(t)] + c_2 \mathscr{L}[F_2(t)]$$

for $s \in S_1 \cap S_2$.

The proof is left to Problem 12. Theorem 7-IV is readily extended to any finite number of functions.

EXAMPLE 1.
$$\mathscr{L}[3e^{2t} + 4\sin 5t] = 3\mathscr{L}[e^{2t}] + 4\mathscr{L}[\sin 5t]$$

$$= \frac{3}{s-2} + \frac{20}{s^2 + 25}$$

for $s > \max(2, 0) = 2$.

EXAMPLE 2.
$$\mathscr{L}[\cosh at] = \mathscr{L}\left[\frac{e^{at}}{2} + \frac{e^{-at}}{2}\right]$$

$$= \frac{1}{2}\mathscr{L}[e^{at}] + \frac{1}{2}\mathscr{L}[e^{-at}]$$

$$= \left(\frac{1}{2}\right)\frac{1}{s-a} + \left(\frac{1}{2}\right)\frac{1}{s+a} = \frac{s}{s^2 - a^2}$$

for $s > \max(a, -a) = |a|$.

EXAMPLE 3.
$$\mathscr{L}[\sin^2 bt] = \mathscr{L}\left[\frac{1}{2} - \frac{1}{2}\cos 2bt\right]$$

$$= \left(\frac{1}{2}\right)\left(\frac{1}{s}\right) - \left(\frac{1}{2}\right)\frac{s}{s^2 + 4b^2}$$

$$= \frac{2a^2}{s(s^2 + 4b^2)}$$

for $s > 0$.

NOTE: See Problem 1(f) of List 7.1 for a derivation of the formula

$$\mathscr{L}[\cos at] = \frac{s}{s^2 + a^2} \quad (s > 0)$$

The next theorem, known as the *first shifting theorem*, provides a simple method of obtaining $\mathscr{L}[e^{at}F(t)]$ from $\mathscr{L}[F(t)] = f(s)$.

Theorem 7-V   If $\mathscr{L}[F(t)] = f(s)$, then $\mathscr{L}[e^{at}F(t)] = f(s - a)$.

Proof:

$$\mathscr{L}[e^{at}F(t)] = \int_0^{+\infty} e^{-st}e^{at}F(t)\,dt$$

$$= \int_0^{+\infty} e^{-(s-a)t}F(t)\,dt$$

$$= f(s - a)$$

This first shifting theorem states that the Laplace transform of $e^{at}F(t)$ can be obtained by replacing s by $s - a$ in the Laplace transform $f(s)$ of $F(t)$. It will be particularly useful when we discuss inverse Laplace transforms.

EXAMPLE 4.
$$\mathscr{L}[e^{at} \cos bt] = \frac{s - a}{(s - a)^2 + b^2}$$

EXAMPLE 5.
$$\mathscr{L}[e^{-2t}t^3] = \frac{6}{(s + 2)^4}$$

The next property provides a method of obtaining $\mathscr{L}[tF(t)]$ from $\mathscr{L}[F(t)] = f(s)$; its generalization provides a method of obtaining $\mathscr{L}[t^nF(t)]$ from $\mathscr{L}[F(t)]$ when n is a positive integer.

Theorem 7-VI $\mathscr{L}[tF(t)] = -f'(s) = -\dfrac{df}{ds}$

Proof: Differentiation of both sides of

$$\mathscr{L}[F(t)] = \int_0^{+\infty} e^{-st}F(t)\,dt = f(s)$$

with respect to s yields

$$\int_0^{+\infty} e^{-st}(-t)F(t)\,dt = -\int_0^{+\infty} e^{-st}[tF(t)]\,dt$$

$$= -\mathscr{L}[tF(t)] = f'(s)$$

The required result follows when both sides are multiplied by -1. For the justification for using Leibniz's rule for differentiating under the integral sign, see Reference 7.5.

EXAMPLE 6.
$$\mathcal{L}[te^{2t}] = -\frac{d}{ds}\left(\frac{1}{s-2}\right)$$
$$= -\left[\frac{-1}{(s-2)^2}\right]$$
$$= (s-2)^{-2}$$

EXAMPLE 7.
$$\mathcal{L}[t\sin t] = -\frac{d}{ds}\left(\frac{1}{s^2+1}\right)$$
$$= -\left[\frac{-2s}{(s^2+1)^2}\right]$$
$$= \frac{2s}{(s^2+1)^2}$$

By Theorem 7-VI,

$$\mathcal{L}[t^2 F(t)] = \mathcal{L}[t(tF(t))]$$
$$= -\frac{d}{ds}\mathcal{L}[tF(t)]$$
$$= -\frac{d}{ds}\left(-\frac{df}{ds}\right)$$
$$= \frac{d^2 f}{ds^2} = (-1)^2 f''(s) = f''(s)$$

Similarly, $\mathcal{L}[t^3 F(t)] = (-1)^3 f'''(s)$, and so on. The following theorem can be proved by mathematical induction.

Theorem 7-VII $\mathcal{L}[t^n F(t)] = (-1)^n f^{(n)}(s) = (-1)^n \dfrac{d^n f}{ds^n}$
for $n = 1, 2, 3, \ldots$

EXAMPLE 8. $\mathcal{L}[t^2 e^{-t}] = (-1)^2 \dfrac{d^2}{ds^2}\left(\dfrac{1}{s+1}\right) = 2(s+1)^{-3}$

The following companion theorem to Theorem 7-VI is proved in Reference 7.11.

Theorem 7-VIII If $\mathcal{L}[F(t)] = f(s)$ for $s > s_0$, and G belonging to class A is given by

$$G(t) = \begin{cases} t^{-1} F(t) & \text{for } t > 0 \\ G_0 & \text{for } t = 0 \end{cases}$$

then

$$\mathcal{L}[G(t)] = \int_s^{+\infty} f(u)\, du, \quad \text{for } s > s_0.$$

EXAMPLE 9. Find $\mathcal{L}[G(t)]$ if

$$G(t) = \begin{cases} t^{-1} \sin t & \text{for } t > 0 \\ 1 & \text{for } t = 0 \end{cases}$$

Solution: From $\mathcal{L}[\sin t] = 1/(s^2 + 1)$ and Theorem 7-VIII, we obtain

$$\mathcal{L}[G(t)] = \int_s^{+\infty} \frac{du}{u^2 + 1} = \left. \text{Tan}^{-1} u \right]_s^{+\infty} = \frac{\pi}{2} - \text{Tan}^{-1} s = \text{Tan}^{-1} \frac{1}{s}$$

The next theorem expresses the Laplace transform of the derivative F' of a function F in terms of $\mathcal{L}F$ and the value of F at $t = 0$. We shall see in Sections 7.4 and 7.5 that Laplace transform methods of solving DE are based mainly on this theorem and its extension Theorem 7-X.

Theorem 7-IX Let F be continuous on $[0, +\infty)$ and let F and F' belong to class A. Then

$$\mathcal{L}[F'(t)] = s\mathcal{L}[F(t)] - F(0) = sf(s) - F(0)$$

Proof: $\mathcal{L}[F'(t)] = \lim\limits_{T \to +\infty} \int_0^T e^{-st} F'(t)\, dt$

Integrating by parts with $u = e^{-st}$, $du = -se^{-st}\, dt$, $dv = F'(t)\, dt$, and $v = F(t)$, we obtain

$$\mathcal{L}[F'(t)] = \lim\limits_{T \to +\infty} [e^{-st} F(t)]_0^T + s \int_0^{+\infty} e^{-st} F(t)\, dt$$

$$= s\mathcal{L}[F(t)] + \lim\limits_{T \to +\infty} [e^{-st} F(t)] - F(0)$$

$$= s\mathcal{L}[F(t)] - F(0)$$

We stated earlier that $\lim_{T \to +\infty} [e^{-st} F(T)] = 0$. This follows from the fact that F is of exponential order as $t \to +\infty$.

EXAMPLE 10. Find $\mathcal{L}[\cos t]$ by Theorem 7-IX.

Solution: Since $(d/dt)(\sin t) = \cos t$, $\mathscr{L}[\sin t] = 1/(s^2 + 1)$, and $\sin(0) = 0$, it follows from Theorem 7-IX that

$$\mathscr{L}[\cos t] = s\left(\frac{1}{s^2 + 1}\right) - 0 = \frac{s}{s^2 + 1}$$

With suitable restrictions on F, F', F'', and F''', Theorem 7-IX yields

$$\begin{aligned}
\mathscr{L}[F''(t)] &= s\mathscr{L}[F'(t)] - F'(0) \\
&= s[sf(s) - F(0)] - F'(0) \\
&= s^2 f(s) - sF(0) - F'(0) \quad\quad (7.10)
\end{aligned}$$

$$\begin{aligned}
\mathscr{L}[F'''(t)] &= s\mathscr{L}[F''(t)] - F''(0) \\
&= s^3 f(s) - s^2 F(0) - sF'(0) - F''(0) \quad\quad (7.11)
\end{aligned}$$

and so on.

The following generalization of Theorem 7-IX can be proved by mathematical induction.

Theorem 7-X Let F, F', ..., $F^{(n-1)}$ be continuous on $[0, +\infty)$ and let F, F', ..., $F^{(n)}$ belong to class A. Then

$$\mathscr{L}[F^{(n)}(t)] = s^n f(s) - \sum_{i=0}^{n-1} s^{n-1-i} F^{(i)}(0)$$

where $F^{(0)}(0) = F(0)$.

NOTE: Some authors, in defining the Laplace transform of F, assume that F is defined on $(0, +\infty)$ instead of $[0, +\infty)$. In Theorem 7-X, $F^{(i)}(0)$ then becomes $F^{(i)}(0+)$ where $F^{(i)}(0+) = \lim_{t \to 0+} F^{(i)}(t)$.

A short table of Laplace transforms is found at the end of this chapter on pages 310–311. A more elaborate table is contained in Reference 7.4.

Problem List 7.3

1. Find:
   (a) $\mathscr{L}[3t^2 + 2\cos 2t]$ (b) $\mathscr{L}[t - 4\sin 3t]$
   (c) $\mathscr{L}[t^3 - \sin t + 4]$ (d) $\mathscr{L}[\sinh at]$
   (e) $\mathscr{L}[\cos^2 at]$ (f) $\mathscr{L}[3t^4 - \sqrt{t}]$
   (g) $\mathscr{L}\left[\sin\left(t + \frac{\pi}{4}\right)\right]$ (h) $\mathscr{L}[(t + 2)^2]$
   (i) $\mathscr{L}[t^3 + 2t - 4]$ (j) $\mathscr{L}[3 + 5e^{-t} + 2\sin t - 6\cos 2t]$
2. Find:
   (a) $\mathscr{L}[e^{at}\sin bt]$ (b) $\mathscr{L}[e^{at}t^n]$

(c) $\mathcal{L}[e^{-t}t]$ (d) $\mathcal{L}[e^{-2t}t^4]$

(e) $\mathcal{L}[e^{-3t}\cos 4t]$ (f) $\mathcal{L}[t^2 + e^{-2t}\sin 3t]$

3. Find $\mathcal{L}[e^{at}]$ by applying the first shifting theorem.

4. Find:

(a) $\mathcal{L}[te^{-t}]$ (b) $\mathcal{L}[t \sin at]$

(c) $\mathcal{L}[t \cos at]$ (d) $\mathcal{L}[t \sin 3t]$

(e) $\mathcal{L}[t^2 e^{4t}]$

5. Find:

(a) $\mathcal{L}[t^2 \sin at]$ (b) $\mathcal{L}[t^2 \cos at]$

(c) $\mathcal{L}[e^{-t}\cos t]$ (d) $\mathcal{L}[t \sin^2 at]$

6. Find $\mathcal{L}[t^2 e^{3t}]$ by (a) the first shifting theorem, and (b) Theorem 7-VII.

7. Find $\mathcal{L}[G(t)]$ if $G(t) = t^{-1}(e^{3t} - e^{2t})$ for $t > 0$ and $G(0) = 1$.

8. Find $\mathcal{L}[G(t)]$ if $G(t) = (\cos at - 1)/t$ for $t > 0$ and $G(0) = 0$.

9. Find $\mathcal{L}[\sin t]$ by applying Theorem 7-IX and the formula $\mathcal{L}[\cos t] = s/(s^2 + 1)$.

10. Obtain $\mathcal{L}[t^n] = n!/s^{n+1}$ from Theorem 7-VII.

11. If $\mathcal{L}[F(t)] = f(s)$ for $s > s_0$, establish the change of scale property

$$\mathcal{L}[F(at)] = \frac{1}{a} f\left(\frac{s}{a}\right)$$

for $s > as_0, a > 0$.

12. Prove Theorem 7-IV.

7.4 Inverse Laplace Transforms

If $\mathcal{L}[F(t)] = f(s)$ exists, we know that f is unique since $\mathcal{L}$ is a function. If F were the only function for which $\mathcal{L}F = f$, then $\mathcal{L}$ would have an inverse $\mathcal{L}^{-1}$, and we should write

$$\mathcal{L}^{-1}[f(s)] = F(t) \tag{7.12}$$

or

$$\mathcal{L}^{-1}f = F \tag{7.13}$$

That this is not the case is shown by the following example.

EXAMPLE 1. If $F(t) = t$ on $[0, +\infty)$, then $\mathcal{L}[F(t)] = f(s) = s^{-2}$ for $s > 0$. Find a function $G \neq F$ such that $\mathcal{L}[G(t)] = f(s)$.

Solution: Let $G(t) = F(t)$ on $[0, +\infty]$ except for $t = 1$, and let $G(1) = 2$. Then $\mathcal{L}[G(t)] = f(s)$. This follows, since the Laplace transform is defined by an integral, and two definite integrals have the same value if the integrands are equal except at a finite number of points.

This example makes it clear that, given any Laplace transform f of a function F, one can construct an infinite number of distinct functions, each having Laplace transform f. We merely make use of the fact that $\mathcal{L}[G(t)]$ is independent of the value of G at a fixed point $t = t_0$.

The following theorem, proved in Reference 7.11, enables us to define $\mathscr{L}^{-1}$ in a meaningful and useful way.

Theorem 7-XI *Lerch's theorem:* Let F and G belong to class A, and let $s = s_0$ be a real number such that $\mathscr{L}[F(t)] = f(s) = \mathscr{L}[G(t)]$ for all $s > s_0$. Then, with the possible exception of points of discontinuity, $F(t) = G(t)$ for all $t > 0$.

Motivated by Lerch's theorem, we agree that (7.12) defines the "unique" function F of class A, if it exists, for which $\mathscr{L}[F(t)] = f(s)$. In other words, we regard a set of functions of class A as identical if they agree in value at all points except at points at which they are discontinuous. With this understanding, $\mathscr{L}^{-1}$ is a function. This convention will also suit our purposes since we shall use (7.12) to produce $F(t)$, where F satisfies a DE on $[0, +\infty)$. A solution of a DE, since it is differentiable, is automatically continuous.

There exist advanced methods for constructing $F(t)$ from $f(s)$. However, we shall content ourselves with using a table of Laplace transforms, and also a few techniques to be illustrated by examples. Whenever we discover a Laplace transform formula, we obtain an inverse Laplace transform formula. The method is analogous to using differentiation formulas to obtain corresponding integral formulas.

EXAMPLE 2. $\mathscr{L}^{-1}\left[\dfrac{1}{s}\right] = 1$ since $\mathscr{L}[1] = \dfrac{1}{s}$

EXAMPLE 3. $\mathscr{L}^{-1}\left[\dfrac{1}{s^2 + 1}\right] = \sin t$ since $\mathscr{L}[\sin t] = \dfrac{1}{s^2 + 1}$

EXAMPLE 4. $\mathscr{L}^{-1}\left[\dfrac{s}{s^2 + 4}\right] = \cos 2t$

EXAMPLE 5. $\mathscr{L}^{-1}\left[\dfrac{4}{s^3}\right] = 2\mathscr{L}^{-1}\left[\dfrac{2}{s^3}\right] = 2t^2$

EXAMPLE 6. $\mathscr{L}^{-1}\left[\dfrac{3 - 4s}{s^2 + 25}\right] = \mathscr{L}^{-1}\left[\left(\dfrac{3}{5}\right)\dfrac{5}{s^2 + 25} - 4\dfrac{s}{s^2 + 25}\right] = \dfrac{3}{5}\sin 5t - 4\cos 5t$

Examples 5 and 6 use the formula

$$\mathscr{L}^{-1}[c_1 f_1(s) + c_2 f_2(s)] = c_1 \mathscr{L}^{-1}[f_1(s)] + c_2 \mathscr{L}^{-1}[f_2(s)]$$

In other words, $\mathscr{L}^{-1}$ is a linear operator. The linearity of $\mathscr{L}^{-1}$ is readily deduced from the linearity of $\mathscr{L}$. See Problem 1.

Inverting $\mathscr{L}[e^{at}F(t)] = f(s - a)$, we obtain

$$\mathscr{L}^{-1}[f(s - a)] = e^{at}F(t) = e^{at}\mathscr{L}^{-1}[f(s)]$$

EXAMPLE 7. $\mathscr{L}^{-1}\left[\dfrac{5}{(s - 1)^2}\right] = 5e^t\mathscr{L}^{-1}\left[\dfrac{1}{s^2}\right] = 5te^t$

EXAMPLE 8. $\mathscr{L}^{-1}\left[\dfrac{8}{s^2 + 6s + 10}\right] = 8\mathscr{L}^{-1}\left[\dfrac{1}{(s + 3)^2 + 1}\right]$

$$= 8e^{-3t}\mathscr{L}^{-1}\left[\dfrac{1}{s^2 + 1}\right]$$

$$= 8e^{-3t}\sin t$$

Inversion of $\mathscr{L}[tF(t)] = -f'(s)$ yields

$$\mathscr{L}^{-1}[f'(s)] = -tF(t) = -t\mathscr{L}^{-1}[f(s)]$$

EXAMPLE 9. Find $\mathscr{L}^{-1}\left[\dfrac{4}{(s - 3)^2}\right]$.

Solution: Since $\dfrac{4}{(s - 3)^2} = \dfrac{d}{ds}\left(\dfrac{-4}{s - 3}\right)$,

$$\mathscr{L}^{-1}\left[\dfrac{4}{(s - 3)^2}\right] = -t\mathscr{L}^{-1}\left[\dfrac{-4}{s - 3}\right]$$

$$= 4te^{3t}$$

EXAMPLE 10. Find $\mathscr{L}^{-1}\left[\dfrac{6s}{(s^2 + 4)^2}\right]$.

Solution: Since $\dfrac{6s}{(s^2 + 4)^2} = \dfrac{d}{ds}\left(\dfrac{-3}{s^2 + 4}\right)$,

$$\mathscr{L}^{-1}\left[\dfrac{6s}{(s^2 + 4)^2}\right] = -t\mathscr{L}^{-1}\left[\dfrac{-3}{s^2 + 4}\right]$$

$$= \dfrac{3t}{2}\mathscr{L}^{-1}\left[\dfrac{2}{s^2 + 4}\right]$$

$$= \dfrac{3t}{2}\sin 2t$$

Inversion of $\mathscr{L}[t^nF(t)] = (-1)^nf^{(n)}(s)$ yields

$$\mathscr{L}^{-1}[f^{(n)}(s)] = (-1)^nt^nF(t)$$
$$= (-1)^nt^n\mathscr{L}^{-1}[f(s)]$$

EXAMPLE 11. Find $\mathscr{L}^{-1}\left[\dfrac{1}{(s + 2)^5}\right]$.

Solution: Since $\dfrac{1}{(s + 2)^5} = \dfrac{d^4}{ds^4}\left[\dfrac{1}{4!(s + 2)}\right]$,

$$\mathscr{L}^{-1}\left[\frac{1}{(s + 2)^5}\right] = (-1)^4 t^4 \mathscr{L}^{-1}\left[\frac{1}{4!(s + 2)}\right] = \frac{t^4 e^{-2t}}{24}$$

We shall see in the following section that when Laplace transform methods are used to solve DE, it is often necessary to find $\mathscr{L}^{-1}[N(s)/D(s)]$, where $N(s)$ and $D(s)$ are polynomials. Since $\lim_{s \to +\infty}[N(s)/D(s)] = 0$, the degree of $D(s)$ is greater than the degree of $N(s)$. The basic method of finding the inverse Laplace transform of a rational function of this type is to expand $N(s)/D(s)$ into its partial fractions and use the linearity of $\mathscr{L}^{-1}$. For more details about partial fraction expansions, see References 7.7 and 7.9.

EXAMPLE 12. Find $\mathscr{L}^{-1}\left[\dfrac{2}{s^2 - 1}\right]$.

Solution: Let $\dfrac{2}{s^2 - 1} = \dfrac{A}{s - 1} + \dfrac{B}{s + 1}$, $s \neq 1, -1$.

Then $2 = A(s + 1) + B(s - 1)$ for all s. Setting $s = 1, -1$, we obtain $A = 1, B = -1$. Thus

$$\mathscr{L}^{-1}\left[\frac{2}{s^2 - 1}\right] = \mathscr{L}^{-1}\left[\frac{1}{s - 1} - \frac{1}{s + 1}\right]$$

$$= \mathscr{L}^{-1}\left[\frac{1}{s - 1}\right] - \mathscr{L}^{-1}\left[\frac{1}{s + 1}\right]$$

$$= e^t - e^{-t}$$

$$= 2 \sinh t$$

NOTE: The result also follows from $\mathscr{L}[\sinh at] = a/(s^2 - a^2)$.

EXAMPLE 13. $\mathscr{L}^{-1}\left[\dfrac{3s - 1}{s(s^2 + 1)}\right] = \mathscr{L}^{-1}\left[-\dfrac{1}{s} + \dfrac{s + 3}{s^2 + 1}\right]$

$$= -\mathscr{L}^{-1}\left[\frac{1}{s}\right] + \mathscr{L}^{-1}\left[\frac{s}{s^2 + 1}\right] + 3\mathscr{L}^{-1}\left[\frac{1}{s^2 + 1}\right]$$

$$= -1 + \cos t + 3 \sin t$$

EXAMPLE 14. $\mathscr{L}^{-1}\left[\dfrac{2s^2 - 3s + 4}{s^3 - 3s^2 + 2s}\right] = \mathscr{L}^{-1}\left[\dfrac{2s^2 - 3s + 4}{s(s - 1)(s - 2)}\right]$

$$= \mathscr{L}^{-1}\left[\frac{2}{s} - \frac{3}{s - 1} + \frac{3}{s - 2}\right]$$

$$= 2 - 3e^t + 3e^{2t}$$

EXAMPLE 15. $\mathscr{L}^{-1}\left[\dfrac{12}{s^2(s^2+4)}\right] = \mathscr{L}^{-1}\left[\dfrac{3}{s^2} - \dfrac{3}{s^2+4}\right]$

$$= 3t - \dfrac{3}{2}\sin 2t$$

Problem List 7.4

1. Prove that $\mathscr{L}^{-1}$ is a linear operator.

2. Find:

(a) $\mathscr{L}^{-1}\left[\dfrac{-2}{s}\right]$

(b) $\mathscr{L}^{-1}\left[\dfrac{2s}{s^2+9}\right]$

(c) $\mathscr{L}^{-1}\left[\dfrac{6}{s^2}\right]$

(d) $\mathscr{L}^{-1}\left[\dfrac{s-4}{s^2+4}\right]$

(e) $\mathscr{L}^{-1}\left[\dfrac{4}{(s+2)^2}\right]$

(f) $\mathscr{L}^{-1}\left[\dfrac{5}{s^2+2s+5}\right]$

(g) $\mathscr{L}^{-1}\left[\dfrac{5s}{s^2+2s+5}\right]$

(h) $\mathscr{L}^{-1}\left[\dfrac{12s}{(s^2+1)^2}\right]$

(i) $\mathscr{L}^{-1}\left[\dfrac{3s+8}{s^2+4}\right]$

(j) $\mathscr{L}^{-1}\left[\dfrac{3s-1}{s^2+2s+5}\right]$

(k) $\mathscr{L}^{-1}\left[\dfrac{s+5}{s^2+10s+26}\right]$

(l) $\mathscr{L}^{-1}\left[\dfrac{15}{s^2+4s+13}\right]$

3. Find:

(a) $\mathscr{L}^{-1}\left[\dfrac{1}{s(s+1)}\right]$

(b) $\mathscr{L}^{-1}\left[\dfrac{1}{s^2(s+1)}\right]$

(c) $\mathscr{L}^{-1}\left[\dfrac{2}{s(s^2-1)}\right]$

(d) $\mathscr{L}^{-1}\left[\dfrac{4}{s^2(s^2-1)}\right]$

(e) $\mathscr{L}^{-1}\left[\dfrac{1}{s(s^2+1)}\right]$

(f) $\mathscr{L}^{-1}\left[\dfrac{1}{s^2(s^2+1)}\right]$

(g) $\mathscr{L}^{-1}\left[\dfrac{18}{(s-1)(s+2)}\right]$

(h) $\mathscr{L}^{-1}\left[\dfrac{12}{(s+1)(s+2)(s-1)}\right]$

(i) $\mathscr{L}^{-1}\left[\dfrac{2s^2+15s+7}{(s+1)^2(s-2)}\right]$

(j) $\mathscr{L}^{-1}\left[\dfrac{4s-8}{s(s^2+4)}\right]$

(k) $\mathscr{L}^{-1}\left[\dfrac{3s+1}{s^2(s-1)}\right]$

(l) $\mathscr{L}^{-1}\left[\dfrac{5s^2-9s+2}{(s-3)(s^2+1)}\right]$

(m) $\mathscr{L}^{-1}\left[\dfrac{18}{s(s^2+9)}\right]$

(n) $\mathscr{L}^{-1}\left[\dfrac{s+4}{(s-1)(s+2)}\right]$

4. Derive the following formulas.

(a) $\mathscr{L}^{-1}\left[\dfrac{1}{(s+a)^2+b^2}\right] = \dfrac{e^{-at}}{b}\sin bt$

(b) $\mathscr{L}^{-1}\left[\dfrac{s}{(s+a)^2+b^2}\right] = \dfrac{e^{-at}}{b}(b\cos bt - a\sin bt)$

(c) $\mathscr{L}^{-1}\left[\dfrac{1}{s^n}\right] = \dfrac{t^{n-1}}{(n-1)!}, n = 1, 2, 3, \ldots$

(d) $\mathcal{L}^{-1}\left[\dfrac{1}{(s-a)^n}\right] = \dfrac{t^{n-1}e^{at}}{(n-1)!}, \ n = 1, 2, 3, \ldots$

(e) $\mathcal{L}^{-1}[f(s)] = -t^{-1}\mathcal{L}^{-1}[f'(s)]$

(f) $\mathcal{L}^{-1}\left[\dfrac{s}{s^2 - a^2}\right] = \cosh at$

(g) $\mathcal{L}^{-1}\left[\dfrac{f(s)}{s-a}\right] = e^{at}\displaystyle\int_0^t F(u)\,du$

(h) $\mathcal{L}^{-1}\left[\dfrac{1}{(s-a)(s-b)}\right] = \dfrac{1}{a-b}(e^{at} - e^{bt})$

(i) $\mathcal{L}^{-1}\left[\dfrac{s}{(s-a)(s-b)}\right] = \dfrac{1}{a-b}(ae^{at} - be^{bt})$

(j) $\mathcal{L}^{-1}\left[\dfrac{3a^2}{s^3 + a^3}\right] = e^{-at} - e^{-at/2}\left(\cos\dfrac{at\sqrt{3}}{2} - \sqrt{3}\sin\dfrac{at\sqrt{3}}{2}\right)$

(k) $\mathcal{L}^{-1}[sf(s)] = F'(t) \ (\mathcal{L}^{-1}[sf(s)] \text{ of class } A)$

(l) $\mathcal{L}^{-1}\left[\dfrac{f(s)}{s}\right] = \displaystyle\int_0^t F(u)\,du \ \ \left(\mathcal{L}^{-1}\left[\dfrac{f(s)}{s}\right]\right) \text{ of class } A$

5. Find $\mathcal{L}^{-1}\left[\dfrac{1}{s(s^2 + a^2)}\right]$ by the formula of Problem 4(1).

6. Find $\mathcal{L}^{-1}\left[\dfrac{1}{s^2(s^2 + a^2)}\right]$ by the formula of Problem 4(1).

7. Find $\mathcal{L}^{-1}\left[\dfrac{s}{s^2 + a^2}\right]$ by the formula of Problem 4(k).

8. Show that if F and F' belong to class A, then

$$\lim_{s \to +\infty} sf(s) = F(0+) = \lim_{t \to 0^+} F(t)$$

9. Show that $\mathcal{L}^{-1}\left[\dfrac{s^3 + 7s}{(s^2 + 9)(s^2 + 1)}\right] = \cos^3 t$.

10. (a) Show that $\mathcal{L}[t\cos at] = \dfrac{s^2 - a^2}{(s^2 + a^2)^2}$.

   (b) Find $\mathcal{L}^{-1}\left[\dfrac{1}{(s^2 + a^2)^2}\right]$ from $\mathcal{L}[t\cos at] = \dfrac{s^2 - a^2}{(s^2 + a^2)^2}$.

7.5 Laplace Transform Solutions of Initial-Value Problems

To outline the general approach to solving differential equations by Laplace transform methods, let us consider the initial-value problem

$$aY''(t) + bY'(t) + cY(t) = F(t); \quad Y(0) = Y_0, \quad Y'(0) = Y_0'$$

Taking the Laplace transform of both sides of the DE, we obtain

$$a\mathcal{L}[Y''(t)] + b\mathcal{L}[Y'(t)] + c\mathcal{L}[Y(t)] = \mathcal{L}[F(t)]$$

Theorem 7-IX and Equation (7.10) then yield

$$a[s^2 y(s) - sY_0 - Y_0'] + b[sy(s) - Y_0] + cy(s) = f(s)$$

Solving for $y(s)$, we get

$$y(s) = \frac{f(s) + (as + b)Y_0 + aY_0'}{as^2 + bs + c}$$

The required solution is then given by

$$Y(t) = \mathcal{L}^{-1}[y(s)]$$

Two major questions arise: Does the procedure work? Is it advantageous? It is first of all necessary that Y, Y', Y'', and F have Laplace transforms. In most applications, F belongs to class A and it can be shown that this implies that Y, Y', and Y'' are also in class A. In many cases, existence Theorem 4-I asserts that the given DE has a unique solution, at least on an interval $[0, T]$, where T is a positive number. Since the guaranteed solution must be continuous, it is the one given by $\mathcal{L}^{-1}[y(s)]$. Although these theoretical questions are interesting, we are free to overlook them since we can always verify that $\mathcal{L}^{-1}[y(s)]$ is indeed a solution of the given initial-value problem. Of more practical concern is obtaining the inverse Laplace transform of $y(s)$. This can present difficulties not easily overcome even by experience, ingenuity, and a comprehensive table of inverse transforms. Despite the shortcomings of the Laplace transform method, it is often simpler to apply than the methods of Chapters 4 and 6. It is not necessary to solve the homogeneous DE, to produce a general solution of the nonhomogeneous DE, and then to determine the arbitrary constants so that the initial conditions are satisfied. In the Laplace transform method, the required particular solution is given immediately by $Y(t) = \mathcal{L}^{-1}[y(s)]$. Although the initial conditions, given at the particular value $t = 0$, seem very special, these are the conditions that are known at time $t = 0$ in most applications. The method is also easily modified to fit conditions on Y and its derivatives at $t = t_1 \neq 0$, or to apply to boundary-value problems.

EXAMPLE 1. Solve the initial-value problem $Y'(t) - 2Y(t) = 4$, $Y(0) = 3$.

Solution:

$$\mathcal{L}[Y'(t)] - 2\mathcal{L}[Y(t)] = \mathcal{L}[4]$$

$$[sy(s) - 3] - 2y(s) = \frac{4}{s}$$

$$y(s) = \frac{(4/s) + 3}{s - 2} = \frac{4 + 3s}{s(s - 2)} = \frac{-2}{s} + \frac{5}{s - 2}$$

$$Y(t) = \mathcal{L}^{-1}\left[\frac{-2}{s} + \frac{5}{s - 2}\right] = -2 + 5e^{2t}$$

It is easy to verify that $Y(t) = -2 + 5e^{2t}$ defines a solution of the initial-value problem. A unique solution on $[0, +\infty)$ is guaranteed by Peano's existence theorem (Theorem 2-IV) and Picard's uniqueness theorem (Theorem 2-V).

A comment on the general nature of the Laplace transform method is in order. To solve a DE in the t space, we transform both sides of the DE to obtain an algebraic equation involving $y(s)$ in the s space. After solving this algebraic equation for $y(s)$, we return to the t space by finding the inverse Laplace transform of $y(s)$, thereby obtaining the required solution given by $Y(t)$. By solving an algebraic equation in a new setting (the s space), we indirectly solve a DE in the t space. We have in a sense circumvented the limiting process inherent in the DE.

This type of procedure is very common in mathematics. To multiply a and b, we sometimes move from the multiplication space of a and b to the addition space containing the logarithms of a and b. After adding the logarithms of a and b, we return to the multiplication space by finding the antilogarithm of $[\log a + \log b] = \log (ab)$. This yields the required product ab.

EXAMPLE 2. Solve the initial-value problem

$$Y''(t) + 4Y'(t) + 4Y(t) = 0; \quad Y(0) = 0, \quad Y'(0) = 5$$

Solution:

$$[s^2 y(s) - 0s - 5] + 4[sy(s)] + 4[y(s)] = 0$$

$$y(s) = \frac{5}{s^2 + 4s + 4} = \frac{5}{(s + 2)^2}$$

$$Y(t) = \mathcal{L}^{-1}\left[\frac{5}{(s + 2)^2}\right] = 5te^{-2t}$$

EXAMPLE 3. Solve the initial-value problem

$$Y''(t) + 6Y'(t) + 25Y = 0, \quad Y(0) = 2, \quad Y'(0) = 3$$

Solution:

$$[s^2 y(s) - 2s - 3] + 6[sy(s) - 2] + 25y(s) = 0$$

$$(s^2 + 6s + 25)y(s) = 2s + 15$$

$$y(s) = \frac{2s + 15}{(s + 3)^2 + (4)^2} = \frac{2(s + 3)}{(s + 3)^2 + (4)^2} + \frac{9}{4}\left[\frac{4}{(s + 3)^2 + (4)^2}\right]$$

$$Y(t) = 2e^{-3t} \cos 4t + \frac{9}{4}e^{-3t} \sin 4t$$

EXAMPLE 4. Solve the initial-value problem

$$Y''(t) + 9Y(t) = \sin 3t; \quad Y(0) = 2, \quad Y'(0) = 1$$

Solution:
$$[s^2 y(s) - 2s - 1] + 9y(s) = \frac{3}{s^2 + 9}$$

$$(s^2 + 9)y(s) = 2s + 1 + \frac{3}{s^2 + 9}$$

$$y(s) = \frac{2s + 1}{s^2 + 9} + \frac{3}{(s^2 + 9)^2}$$

$$Y(t) = 2 \cos 3t + \frac{1}{3} \sin 3t + \frac{3}{2(27)} (\sin 3t - 3t \cos 3t)$$

$$= 2 \cos 3t + \frac{7}{18} \sin 3t - \frac{1}{6} t \cos 3t$$

NOTE: A transform table was used to obtain $\mathcal{L}^{-1}[3/(s^2 + 9)^2]$. An interesting method of finding this inverse Laplace transform will be presented in Section 7.8. Or see Problem 10 of List 7.4.

EXAMPLE 5. Solve the initial-value problem

$$Y'''(t) + 3Y'(t) = 9t^2 - 12t + 6; \quad Y(0) = 3, \quad Y'(0) = 0, \quad Y''(0) = -4$$

Solution: Applying Theorem 7-IX and Theorem 7-X with $n = 3$, we obtain

$$[s^3 y(s) - 3s^2 - 0s + 4] + 3[sy(s) - 3] = \frac{18}{s^3} - \frac{12}{s^2} + \frac{6}{s}$$

$$(s^3 + 3s)y(s) = 3s^2 + 5 + \frac{18}{s^3} - \frac{12}{s^2} + \frac{6}{s}$$

$$= \frac{3s^5 + 5s^3 + 6s^2 - 12s + 18}{s^3}$$

$$y(s) = \frac{(s^2 + 3)(3s^3 - 4s + 6)}{s^4(s^2 + 3)} = \frac{3}{s} - \frac{4}{s^3} + \frac{6}{s^4}$$

and
$$Y(t) = 3 - 2t^2 + t^3$$

The Laplace transform method is less effective when the coefficients in the DE are nonconstant. If the coefficients are polynomials in t, the transform of the left member can be found by applying Theorem 7-VII. This leads to a DE in y and s rather than an algebraic equation. In the following example, it is about as easy to solve the new DE as it is the original DE, but this is not generally the case.

EXAMPLE 6. Solve the initial-value problem

$$tY''(t) + Y'(t) = 4t^2, \quad Y(0) = 1$$

Solution: Substituting $t = 0$ in the DE, we find that $Y'(0) = 0$. Then

$$\mathcal{L}[tY''(t)] + \mathcal{L}[Y'(t)] = 4\mathcal{L}[t^2]$$

$$-\frac{d}{ds}\mathcal{L}[Y''(t)] + [sy(s) - 1] = \frac{8}{s^3}$$

$$-\frac{d}{ds}[s^2 y(s) - s - 0] + sy(s) - 1 = \frac{8}{s^3}$$

$$-s^2 \frac{dy}{ds} - 2sy + 1 + sy - 1 = \frac{8}{s^3}$$

and hence s and $y = y(s)$ satisfy the DE

$$\frac{dy}{ds} + \frac{1}{s}y = -\frac{8}{s^5}$$

Using the integrating factor $\exp[\int ds/s] = \exp(\ln s) = s$, we obtain

$$sy = \int -8s^{-4}\,ds = \frac{8}{3s^3} + c$$

or

$$y = y(s) = \frac{8}{3s^4} + \frac{c}{s}$$

Hence,

$$Y(t) = \mathcal{L}^{-1}[y(s)] = \frac{8t^3}{3(6)} + c$$

Since $Y(0) = 1$, it follows that $c = 1$, and finally $Y(t) = (4t^3/9) + 1$.

Problem List 7.5

1. Solve the initial-value problems:
   (a) $Y'(t) - 2Y(t) = e^{3t}$, $Y(0) = 3$
   (b) $Y''(t) - 2Y'(t) + Y(t) = 0$; $Y(0) = 0$, $Y'(0) = 4$
   (c) $Y''(t) - 2Y'(t) + Y(t) = 0$; $Y(0) = 1$, $Y'(0) = 1$
   (d) $Y''(t) - 2Y'(t) - 3Y(t) = 0$; $Y(0) = 3$, $Y'(0) = 1$
   (e) $Y''(t) - 4Y'(t) + 3Y(t) = 0$; $Y(0) = 3$, $Y'(0) = 5$
   (f) $Y''(t) + 10Y'(t) + 25Y(t) = 0$; $Y(0) = 0$, $Y'(0) = 10$
   (g) $Y''(t) + 8Y'(t) + 25Y(t) = 100$; $Y(0) = 2$, $Y'(0) = 20$
   (h) $Y''(t) + 4Y(t) = 4$; $Y(0) = 1$, $Y'(0) = 0$
   (i) $Y''(t) - 3Y'(t) + 2Y(t) = 4$; $Y(0) = 2$, $Y'(0) = 3$
   (j) $Y''(t) + 4Y'(t) = 16t$; $Y(0) = 3$, $Y'(0) = -6$
   (k) $Y''(t) + 4Y(t) = 8\sin 2t$; $Y(0) = 0$, $Y'(0) = 2$
   (l) $Y''(t) + Y(t) = 4e^t$; $Y(0) = 0$, $Y'(0) = 0$
   (m) $Y'''(t) - Y''(t) - Y'(t) + Y(t) = 6e^t$; $Y(0) = Y'(0) = Y''(0) = 0$

2. Find a complete solution of $Y''(t) - 2Y'(t) + Y(t) = 0$ by the Laplace transform method.

3. Solve $Y''(t) + \omega^2 Y(t) = 0$; $Y(0) = Y_0$, $Y'(0) = Y'_0$ by the Laplace transform method.

4. Solve $tY''(t) - Y'(t) = t^2$, $Y(0) = 2$ by the Laplace transform method.

5. Solve $Y'(t) + 2Y(t) = 4t$, $Y(1) = 2$ by the Laplace transform method.

6. Solve $Y''(t) - 2Y'(t) + Y(t) = 2e^t$; $Y(0) = 0$, $Y(1) = 2e$ by the Laplace transform method.

7. Solve $Y''(t) + tY'(t) - 2Y(t) = 4$, $Y(0) = Y'(0) = 0$ by the Laplace transform method. Assume that $\lim_{s \to +\infty} y(s) = 0$.

8. Given that J_0, Bessel's function of the first kind of order zero, satisfies $tJ_0''(t) + J_0'(t) + tJ_0(t) = 0$, and that $J_0(0) = 1$:

(a) Assuming that J_0 is of class A and that $\lim_{s \to +\infty} sj_0(s) = J_0(0) = 1$, show that

$$j_0(s) = \mathscr{L}[J_0(t)] = (s^2 + 1)^{-1/2} = s^{-1}\left(1 + \frac{1}{s^2}\right)^{-1/2}$$

(b) Expand $[1 + (1/s^2)]^{-1/2}$ by the binomial theorem to find a series expansion for $j_0(s)$. Find $J_0(t)$, assuming that the inverse Laplace transform of the infinite series can be obtained by replacing each term in the series by its inverse Laplace transform. (The justification for the suggested formalisms is given in Reference 7.10.)

7.6 Laplace Transform Solutions of Differential Equations with Discontinuous Right Members

Many important physical systems are governed by constant coefficient differential equations of the form

$$a_0 Y^{(n)}(t) + a_1 Y^{(n-1)}(t) + \cdots + a_{n-1} Y'(t) + a_n Y(t) = F(t), \quad a_0 \neq 0 \tag{7.14}$$

The right member $F(t)$ is the input to the system and the solution $Y(t)$ is the output, or response, of the system to the input. The input $F(t)$ is often discontinuous. This situation can be handled by previously described methods but solutions are frequently quite complicated. Laplace transform solutions often present a considerable advantage.

In finding $\mathscr{L}[F(t)]$ when F is discontinuous, the Heaviside unit step function, or the H function, defined by

$$H(t - a) = \begin{cases} 0 & \text{for } t < a \\ 1 & \text{for } t \geq a \end{cases} \tag{7.15}$$

is very useful. The graph of H for $t \geq 0$ is shown in Figure 7.1. Since we are interested in the interval $t \geq 0$, we assume that $a \geq 0$.

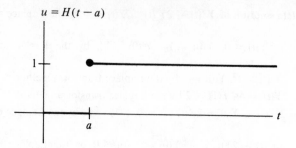

$$u = H(t - a)$$

Fig. 7.1

From Definition 7.1,

$$\mathscr{L}[H(t - a)] = \int_0^{+\infty} e^{-st} H(t - a)\, dt = \int_a^{+\infty} e^{-st}\, dt$$

$$= \lim_{T \to +\infty} \left[\frac{e^{-st}}{-s} \right]_a^T = \lim_{T \to +\infty} \left[\frac{e^{-sT}}{-s} + \frac{e^{-as}}{s} \right]$$

Therefore,

$$\mathscr{L}[H(t - a)] = \frac{e^{-as}}{s}, \quad s > 0 \tag{7.16}$$

and

$$\mathscr{L}^{-1}\left[\frac{e^{-as}}{s} \right] = H(t - a) \tag{7.17}$$

We next assume that F is of class A and observe that the graph of $F(t - a)$ can be constructed by translating the graph of $F(t)$ a units to the right. See Figure 7.2. Thus, the graph of $F(t - a)H(t - a)$ has height zero when

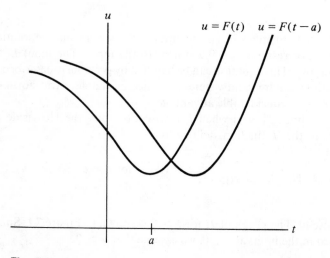

$$u = F(t) \quad u = F(t - a)$$

Fig. 7.2

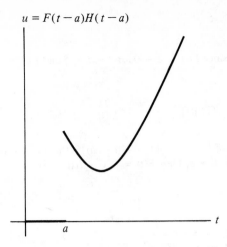

$u = F(t - a)H(t - a)$

a

t

Fig. 7.3

$t < a$ and coincides with the graph of $F(t - a)$ when $t \geq a$. See Figure 7.3. The method of finding $\mathcal{L}[F(t)]$ when F is discontinuous is based on the following important theorem, known as the *second shifting theorem*.

Theorem 7-XII $\quad \mathcal{L}[F(t - a)H(t - a)] = e^{-as}\mathcal{L}[F(t)], \quad a \geq 0 \qquad (7.18)$

Proof: $\quad \mathcal{L}[F(t - a)H(t - a)] = \displaystyle\int_0^{+\infty} e^{-st}F(t - a)H(t - a)\,dt$

$$= \int_a^{+\infty} e^{-st}F(t - a)\,dt$$

Let $v = t - a$; then $dt = dv$, and the last integral becomes

$$\int_0^{+\infty} e^{-s(v + a)}F(v)\,dv = e^{-as}\int_0^{+\infty} e^{-sv}F(v)\,dv = e^{-as}\mathcal{L}[F(t)]$$

To use (7.18) to find $\mathcal{L}[F(t)H(t - a)]$, we let $F(t) = F([t - a] + a) = G(t - a)$. Then $\mathcal{L}[G(t - a)H(t - a)] = e^{-as}\mathcal{L}[G(t)]$ by (7.18). But $G(t) = F(t + a)$ and hence (7.18) can be written in the alternative form,

$$\mathcal{L}[F(t)H(t - a)] = e^{-as}\mathcal{L}[F(t + a)], \quad a \geq 0 \qquad (7.19)$$

We now present a few examples in which we find $\mathcal{L}[F(t)]$ after first expressing $F(t)$ analytically by the H function.

EXAMPLE 1. Let $F(t) = \begin{cases} 0 & \text{for } 0 \leq t < 5 \\ 2 & \text{for } t \geq 5 \end{cases}$

Then $F(t) = 2H(t - 5)$ and $\mathcal{L}[F(t)] = e^{-5s}\mathcal{L}[2] = e^{-5s}\left(\dfrac{2}{s}\right)$

EXAMPLE 2. Let $F(t) = \begin{cases} 2 & \text{for } 0 \le t < 5 \\ 0 & \text{for } t \ge 5 \end{cases}$

Then $F(t) = 2 - 2H(t - 5)$ since $F(t) = 2 - 0$ for $0 \le t < 5$ and $F(t) = 2 - 2 = 0$ for $t \ge 5$. We then obtain

$$\mathscr{L}[F(t)] = \frac{2}{s} - e^{-5s}\left(\frac{2}{s}\right)$$

NOTE: On $[0, +\infty)$ the graph of $h(t) = g(t)[1 - H(t - a)]$ is the same as the graph of $g(t)$ for $t < a$ with $h(t) = 0$ for $t \ge a$. Thus $F(t) = 2[1 - H(t - 5)]$.

EXAMPLE 3. Let $F(t) = \begin{cases} t^2 & \text{for } 0 \le t < 2 \\ 5 & \text{for } t \ge 2 \end{cases}$

Then $F(t) = t^2[1 - H(t - 2)] + 5H(t - 2) = t^2 + (5 - t^2)H(t - 2)$
Replacing t by $t + 2$ in $5 - t^2$, we obtain $5 - (t + 2)^2 = 1 - t^2 - 4t$.
Applying (7.19) we obtain

$$\mathscr{L}[F(t)] = \mathscr{L}[t^2] + \mathscr{L}[(5 - t^2)H(t - 2)]$$

$$= \frac{2}{s^3} + e^{-2s}\mathscr{L}[1 - t^2 - 4t]$$

$$= \frac{2}{s^3} + e^{-2s}\left(\frac{1}{s} - \frac{2}{s^3} - \frac{4}{s^2}\right)$$

EXAMPLE 4. Let $F(t) = \begin{cases} t & \text{for } 2 \le t < 3 \\ 0 & \text{elsewhere} \end{cases}$

On $[0, +\infty)$ the graph of $h(t) = g(t)[H(t - a) - H(t - b)]$, where $0 < a < b$, is the same as the graph of $g(t)$ for $a \le t < b$, with $h(t) = 0$ for $0 \le t < a$ and for $t \ge b$.
Thus,

$$F(t) = t[H(t - 2) - H(t - 3)]$$
$$= tH(t - 2) - tH(t - 3)$$

and

$$\mathscr{L}[F(t)] = e^{-2s}\mathscr{L}[t + 2] - e^{-3s}\mathscr{L}[t + 3]$$

$$= e^{-2s}\left(\frac{1}{s^2} + \frac{2}{s}\right) - e^{-3s}\left(\frac{1}{s^2} + \frac{3}{s}\right)$$

Examples 1–4 suggest that when F is discontinuous in (7.14) and we take the Laplace transform of both sides of the DE, then $y(s)$ will contain terms involving e^{-ks} where k is a positive constant. The following theorem is very useful in finding $Y(t) = \mathscr{L}^{-1}[y(s)]$. It is obtained by taking the inverse Laplace transform of both sides of (7.18), the second shifting theorem.

Theorem 7-XIII $\quad \mathscr{L}^{-1}[e^{-as}f(s)] = F(t - a)H(t - a)$ $\qquad$ (7.20)

EXAMPLE 5. Find $\mathscr{L}^{-1}\left[e^{-2s}\dfrac{4}{s+2}\right]$.

Solution: Since $F(t) = \mathscr{L}^{-1}[4/(s+2)] = 4e^{-2t}$, it follows that $F(t-2) = 4e^{-2(t-2)}$, and (7.20) yields

$$\mathscr{L}^{-1}\left[e^{-2s}\frac{4}{s+2}\right] = 4e^{-2(t-2)}H(t-2)$$

EXAMPLE 6. Solve the initial-value problem

$$Y''(t) + 9Y(t) = \begin{cases} 1 & \text{for } 0 \leq t < 1 \\ 0 & \text{for } t \geq 1 \end{cases}, \quad Y(0) = Y'(0) = 0$$

Solution: Taking the Laplace transform of each side of

$$Y''(t) + 9Y(t) = 1 - H(t-1)$$

we obtain

$$s^2 y(s) - 0s - 0 + 9y(s) = \frac{1}{s} - e^{-s}\frac{1}{s}$$

Then

$$y(s) = \frac{1}{s(s^2+9)} - e^{-s}\frac{1}{s(s^2+9)}$$

and

$$Y(t) = \mathscr{L}^{-1}\left[\frac{1}{s(s^2+9)}\right] - \mathscr{L}^{-1}\left[e^{-s}\frac{1}{s(s^2+9)}\right]$$

$$= \frac{1}{9}[1 - \cos 3t] - \frac{1}{9}[1 - \cos 3(t-1)]H(t-1)$$

$$= \begin{cases} \dfrac{1}{9}[1 - \cos 3t] \text{ for } 0 \leq t < 1 \\ \dfrac{1}{9}[\cos 3(t-1) - \cos 3t] \text{ for } t \geq 1 \end{cases}$$

From

$$Y'(t) = \frac{1}{3}\sin 3t - \left[\frac{1}{3}\sin 3(t-1)\right]H(t-1)$$

and

$$Y''(t) = \cos 3t - [\cos 3(t-1)]H(t-1)$$

we verify that $Y''(t) + 9Y(t) = 1 - H(t-1)$.

We also verify that $Y(0) = Y'(0) = 0$. Since $\lim_{t \to 1^-} Y'(t) = (\sin 3)/3 = \lim_{t \to 1^+} Y'(t)$, it follows that $Y'(1) = (\sin 3)/3$. Thus, Y is differentiable for $t \geq 0$.

When $n = 2$, DE (7.14) has the form

$$a_0 Y''(t) + a_1 Y'(t) + a_2 Y(t) = F(t) \tag{7.21}$$

If F is of class A, it can be shown (see Reference 7.6) that Y and Y' are continuous on $[0, +\infty)$. Hence, if F is discontinuous at $t = t_0$, Y'' must be discontinuous at $t = t_0$ since the second and third terms on the left of (7.21) are continuous. See Problem 5 for a verification of this result in Example 6.

The situation is different when $n = 1$ in (7.14). If F is of class A in

$$a_0 Y'(t) + a_1 Y(t) = F(t) \tag{7.22}$$

and F is discontinuous at $t = t_0$, then Y and Y' cannot both be continuous at $t = t_0$, since the left member of (7.22) would then be continuous at $t = t_0$. Thus, if Y is continuous at $t = t_0$, Y' must be discontinuous at $t = t_0$. Hence Y cannot be differentiable at $t = t_0$ since differentiability at a point implies continuity at that point. We get around this difficulty by agreeing to call Y a solution of (7.22) on $[0, +\infty)$ if Y is continuous on $[0, +\infty)$ and satisfies (7.22) at every point of $[0, +\infty)$ at which F is continuous.

EXAMPLE 7. Solve the initial-value problem

$$Y'(t) + Y(t) = \begin{cases} 1 & \text{for } 0 \le t < 2 \\ 0 & \text{for } t \ge 2 \end{cases}, \quad Y(0) = 0$$

Solution: Taking the Laplace transform of each side of

$$Y'(t) + Y(t) = 1 - H(t - 2) \tag{7.23}$$

we obtain

$$sy(s) + y(s) = \frac{1}{s} - e^{-2s}\frac{1}{s}$$

or

$$y(s) = \frac{1}{s(s + 1)} - e^{-2s}\frac{1}{s(s + 1)}$$

Hence

$$Y(t) = \mathcal{L}^{-1}\left[\frac{1}{s} - \frac{1}{s + 1}\right] - \mathcal{L}^{-1}\left[e^{-2s}\left(\frac{1}{s} - \frac{1}{s + 1}\right)\right]$$

$$= 1 - e^{-t} - [1 - e^{-(t - 2)}]H(t - 2)$$

$$= \begin{cases} 1 - e^{-t} & \text{for } 0 \le t < 2 \\ e^{-(t - 2)} - e^{-t} = e^{-t}(e^2 - 1) & \text{for } t \ge 2 \end{cases}$$

It is easy to see that Y is continuous at $t = 2$ with $Y(2) = 1 - e^{-2}$. However, $\lim_{t \to 2^-} Y'(t) = \lim_{t \to 2^-} (e^{-t}) = e^{-2}$, whereas

$$\lim_{t \to 2^+} Y'(t) = \lim_{t \to 2^+} [-e^{-t}(e^2 - 1)] = -1 + e^{-2}$$

Thus, Y' is not defined at $t = 2$ and yet we agree to call Y a solution of (7.23) on $[0, +\infty)$.

In summary, a discontinuous input F of class A in (7.14) has a differentiable output for $n = 2$ and a continuous output for $n = 1$.

Problem List 7.6

1. Find $L[1]$ from (7.16).

2. Express $F(t)$ in terms of the H function and find $\mathscr{L}[F(t)]$ for the following.

(a) $F(t) = \begin{cases} 0 & \text{for } 0 \le t < 8 \\ 3 & \text{for } t \ge 8 \end{cases}$

(b) $F(t) = \begin{cases} 4 & \text{for } 0 \le t < 1 \\ 0 & \text{for } t \ge 1 \end{cases}$

(c) $F(t) = \begin{cases} t & \text{for } 0 \le t < 4 \\ 3 & \text{for } t \ge 4 \end{cases}$

(d) $F(t) = \begin{cases} t^2 & \text{for } 1 \le t < 2 \\ 0 & \text{elsewhere} \end{cases}$

(e) $F(t) = \begin{cases} 1 & \text{for } 0 \le t < 3 \\ t & \text{for } t \ge 3 \end{cases}$

(f) $F(t) = \begin{cases} \sin 2t & \text{for } 0 \le t < \pi \\ 0 & \text{for } t \ge \pi \end{cases}$

(g) $F(t) = \begin{cases} \sin t & \text{for } 0 \le t < \pi/2 \\ 0 & \text{for } t \ge \pi/2 \end{cases}$

(h) $F(t) = \begin{cases} e^{-2t} & \text{for } 0 \le t < 1 \\ 0 & \text{for } t \ge 1 \end{cases}$

3. Find:

(a) $\mathscr{L}^{-1}\left[e^{-3s}\dfrac{1}{s} \right]$

(b) $\mathscr{L}^{-1}\left[e^{-2s}\dfrac{3}{s^2} \right]$

(c) $\mathscr{L}^{-1}\left[e^{-4s}\dfrac{3}{s-2} \right]$

(d) $\mathscr{L}^{-1}\left[\dfrac{e^{-s}}{s} + \dfrac{e^{-3s}}{s^2} \right]$

4. Solve the following initial-value problems.

(a) $Y''(t) + 4Y(t) = 2 - 2H(t - 3);\ Y(0) = Y'(0) = 0$

(b) $Y''(t) + 2Y'(t) + Y(t) = \begin{cases} e^{-t} & \text{for } 0 \le t < 2 \\ 0 & \text{for } t \ge 2 \end{cases}, \quad Y(0) = Y'(0) = 0$

(c) $Y''(t) + Y(t) = \begin{cases} 2\sin t & \text{for } 0 \le t < \pi \\ 0 & \text{for } t \ge \pi \end{cases}, \quad Y(0) = 0, \quad Y'(0) = 0$

5. Show that Y'' is discontinuous at $t = 1$ in Example 6.

6. Verify that the solution Y in Example 7 satisfies the DE and the initial conditions.

7.7 Laplace Transforms of Periodic Functions

Definition   A function F is *periodic* with *period* p, $p > 0$, if and only if $F(t + p) = F(t)$ for every t in the domain D of F. The smallest p for which $F(t + p) = F(t)$ is called the *fundamental period* of F.

This definition implies that $t + p \in D$ whenever $t \in D$. Since we are interested in $\mathscr{L}[F(t)]$, the domain D will be the interval $[0, +\infty)$ in this section. Since the right member $F(t)$ of DE (7.14) defines a periodic function in many important applications, it is convenient to have a formula for the Laplace transform of a periodic function. Such a formula is provided by the following theorem.

Theorem 7-XIV   Let the function F of period p be defined on $[0, +\infty)$ and let $\mathscr{L}[F(t)]$ exist. Then

$$\mathscr{L}[F(t)] = f(s) = \frac{\int_0^p e^{-st} F(t)\, dt}{1 - e^{-ps}} \tag{7.24}$$

Proof: Let $G(t) = F(t) - F(t)H(t - p)$. Then

$$G(t) = \begin{cases} F(t) & \text{for } 0 \leq t < p \\ 0 & \text{for } t \geq p \end{cases}$$

Employing form (7.19) of the second shifting theorem, we obtain

$$\begin{aligned}
\mathcal{L}[G(t)] &= \mathcal{L}[F(t)] - e^{-ps}\mathcal{L}[F(t + p)] \\
&= \mathcal{L}[F(t)] - e^{-ps}\mathcal{L}[F(t)] \\
&= \int_0^p e^{-st} F(t)\, dt
\end{aligned}$$

Formula (7.24) is obtained by solving for $\mathcal{L}[F(t)] = f(s)$.

EXAMPLE 1. Find $\mathcal{L}[\sin t]$.

Solution: By (7.24), with $p = 2\pi$,

$$\begin{aligned}
\mathcal{L}[\sin t] &= \frac{\int_0^{2\pi} e^{-st} \sin t\, dt}{1 - e^{-2\pi s}} \\
&= \frac{1}{1 - e^{-2\pi s}} \frac{[e^{-st}(-s \sin t - \cos t)]_0^{2\pi}}{s^2 + 1} \\
&= \frac{1}{s^2 + 1}
\end{aligned}$$

In the next example, F is periodic and has an infinite number of discontinuities on $[0, +\infty)$.

EXAMPLE 2. Let F be defined on $[0, +\infty)$ by

$$F(t) = \begin{cases} 1 & \text{for } 0 \leq t < L \\ -1 & \text{for } L \leq t < 2L \end{cases}$$

with $F(t + 2L) = F(t)$. Then, by (7.24)

$$\begin{aligned}
\mathcal{L}[F(t)] &= \frac{\int_0^L e^{-st}\, dt + \int_L^{2L} - e^{-st}\, dt}{1 - e^{-2Ls}} \\
&= \frac{e^{-st}/-s]_0^L - e^{-st}/-s]_L^{2L}}{1 - e^{-2Ls}} \\
&= \frac{-e^{-Ls} + 1 + e^{-2Ls} - e^{-Ls}}{s(1 - e^{-2Ls})} \\
&= \frac{1 - 2e^{-Ls} + e^{-2Ls}}{s(1 - e^{-2Ls})} = \frac{(1 - e^{-Ls})^2}{s(1 - e^{-2Ls})} \\
&= \frac{1 - e^{-Ls}}{s(1 + e^{-Ls})} = f(s)
\end{aligned}$$

The numerator and denominator are sometimes multiplied by $e^{Ls/2}$ to yield $f(s) = (1/s) \tanh (Ls/2)$.

The graph of F, known as the square-wave function, is shown in Figure 7.4.

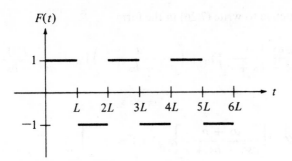

Fig. 7.4 Square-wave Function.

EXAMPLE 3. Solve the initial-value problem $Y'(t) + Y(t) = F(t)$, $Y(0) = 0$, where F is the square-wave function of Example 2, with $L = 1$.

Solution: Taking the Laplace transform of both sides,

$$sy(s) + y(s) = \frac{1 - e^{-s}}{s(1 + e^{-s})} = \frac{1}{s}[1 - 2e^{-s} + 2e^{-2s} - 2e^{-3s} + 2e^{-4s} - \cdots]$$

$$y(s) = \frac{1}{s(s + 1)}[1 - 2e^{-s} + 2e^{-2s} - 2e^{-3s} + 2e^{-4s} - \cdots]$$

$$= \left[\frac{1}{s} - \frac{1}{s + 1}\right][1 - 2e^{-s} + 2e^{-2s} - 2e^{-3s} + 2e^{-4s} - \cdots]$$

By (7.20),

$$Y(t) = [1 - e^{-t}] - 2[1 - e^{-(t-1)}]H(t - 1) + 2[1 - e^{-(t-2)}]H(t - 2)$$
$$-2[1 - e^{-(t-3)}]H(t - 3) + 2[1 - e^{-(t-4)}]H(t - 4) - \cdots$$

$$= 1 - e^{-t} + 2\sum_{k=1}^{+\infty} (-1)^k[1 - e^{-(t-k)}]H(t - k)$$

For a justification of the use of the series expansion of $(1 - e^{-s})/(1 + e^{-s})$ see Reference 7.11.

7.8 The Convolution Theorem for the Laplace Transform

In Section 7.5 we considered the initial-value problem

$$aY''(t) + bY'(t) + cY(t) = F(t); \quad Y(0) = Y_0, \quad Y'(0) = Y_0' \quad (7.25)$$

and found that $\mathcal{L}[Y(t)] = y(s)$ is given by

$$y(s) = \frac{f(s) + (as + b)Y_0 + aY_0'}{as^2 + bs + c} \tag{7.26}$$

It is instructive to write (7.26) in the form

$$y(s) = \left(\frac{as + b}{as^2 + bs + c}\right)Y_0 + \left(\frac{a}{as^2 + bs + c}\right)Y_0' + \frac{f(s)}{as^2 + bs + c}$$

and to let

$$Y_1(t) = \mathcal{L}^{-1}\left[\frac{as + b}{as^2 + bs + c}\right], \quad Y_2(t) = \mathcal{L}^{-1}\left[\frac{a}{as^2 + bs + c}\right],$$

and

$$Y_3(t) = \mathcal{L}^{-1}\left[\frac{f(s)}{as^2 + bs + c}\right]$$

Then

$$Y(t) = Y_0 Y_1(t) + Y_0' Y_2(t) + Y_3(t) \tag{7.27}$$

The first two terms on the right side of (7.27) depend on the initial values Y_0 and Y_0', and the third term on the input $F(t)$, since $f(s) = \mathcal{L}[F(t)]$. Also, $Y_p(t) = Y_3(t)$ defines the particular solution of $aY''(t) + bY'(t) + cY(t) = F(t)$ for which $Y(0) = Y'(0) = 0$.

We next observe that

$$\frac{f(s)}{as^2 + bs + c} = [f(s)]\left[\frac{a}{a(as^2 + bs + c)}\right] = \mathcal{L}[F(t)]\mathcal{L}\left[\frac{Y_2(t)}{a}\right]$$

If we had a formula for the product of two Laplace transforms, the inverse Laplace transform of the product would immediately yield $Y_3(t)$. A first conjecture might be that the product of the Laplace transforms of two functions is the Laplace transform of the product of the two functions, that is, that $(\mathcal{L}[F])(\mathcal{L}[G]) = \mathcal{L}[FG]$. It is easy to show by a counterexample that this conjecture is false. See Problem 6.

The following important theorem, known as the *convolution theorem for the Laplace transform*, provides a method of finding the product of two Laplace transforms.

Theorem 7-XV If F and G belong to class A on $[0, +\infty)$, then

$$\mathcal{L}[F(t)]\mathcal{L}[G(t)] = f(s)g(s) = \mathcal{L}\left[\int_0^t F(t - u)G(u)\, du\right] \tag{7.28}$$

Proof:

$$\mathscr{L}\left[\int_0^t F(t-u)G(u)\,du\right] = \int_0^{+\infty} e^{-st} \int_0^t F(t-u)G(u)\,du\,dt$$

$$= \int_0^{+\infty} \int_0^t e^{-s(t-u)}F(t-u)e^{-su}G(u)\,du\,dt$$

The region of the tu plane described by $0 \le u \le t, 0 \le t < +\infty$ can also be described by $u \le t < +\infty, 0 \le u < +\infty$.

Changing the order of integration, this iterated integral becomes

$$\int_0^{+\infty} e^{-su}G(u) \int_u^{+\infty} e^{-s(t-u)} F(t-u)\,dt\,du$$

We now let $v = t - u$, to obtain

$$\int_0^{+\infty} e^{-su}G(u) \int_0^{+\infty} e^{-sv} F(v)\,dv\,du$$

$$= \left[\int_0^{+\infty} e^{-su}G(u)\,du\right]\left[\int_0^{+\infty} e^{-sv}F(v)\,dv\right]$$

$$= \mathscr{L}[F(t)]\mathscr{L}[G(t)]$$

For a more detailed discussion of the formal manipulations including the change in order of integration, see References 7.10 and 7.11.

The integral in the right member of the *convolution formula* (7.28) is called the *convolution* of F and G and is denoted by $F * G$. The operation of forming the convolution of F and G is in a sense a generalization of multiplication.

Returning to DE (7.25), we are now able to write the solution (7.27) in the form

$$Y(t) = Y_0 Y_1(t) + Y'_0 Y_2(t) + [f(s)] * \left[\frac{1}{as^2 + bs + c}\right]$$

or

$$Y(t) = Y_0 Y_1(t) + Y'_0 Y_2(t) + \frac{1}{a}\int_0^t F(t-u)Y_2(u)\,du \qquad (7.29)$$

There is an advantage in expressing the output $Y(t)$ of a system to an input $F(t)$ by means of (7.29). Changing the input $F(t)$ affects only the third term of $Y(t)$, whereas $Y_1(t)$ and $Y_2(t)$ are determined by the constants a, b, and c in the system. Thus, Equation (7.29) enables us to study various outputs corresponding to changing inputs. Hence the Laplace transform method of solving initial-value problems is advantageous when we are interested primarily in steady-state solutions. Or one can study the effect on the output of changes in the initial values Y_0 and Y_0'. Finally, one can study the effect of varying the parameters a, b, and c. It is interesting to note that if we set $Y_0 = Y_0' = 0$, the solution of (7.29) is the particular solution obtained by the method of variation of parameters. In many instances, (7.29) is simpler to apply than the variation of parameters method of Section 4.8.

EXAMPLE 1. Solve the initial-value problem

$$Y''(t) + 16Y(t) = \sin 4t, \quad Y(0) = Y'(0) = 0$$

Solution: In (7.29), $a = 1$ and

$$Y_2(t) = \mathscr{L}^{-1}\left[\frac{1}{s^2 + 16}\right] = \frac{1}{4}\sin 4t$$

Hence,

$$Y(t) = \int_0^t \sin(4t - 4u)\frac{1}{4}\sin 4u \, du$$

$$= \frac{1}{4}\int_0^t (\sin 4t \cos 4u - \cos 4t \sin 4u)\sin 4u \, du$$

$$= \frac{\sin 4t}{4}\int_0^t \sin 4u \cos 4u \, du - \frac{\cos 4t}{4}\int_0^t \sin^2 4u \, du$$

$$= \frac{\sin 4t}{4}\left[\frac{\sin^2 4u}{8}\right]_0^t - \frac{\cos 4t}{16}\left[2u - \frac{\sin 8u}{4}\right]_0^t$$

$$= \frac{\sin^3 4t}{32} - \frac{t \cos 4t}{8} + \frac{\sin 8t \cos 4t}{64}$$

$$= \sin^2 4t\left(\frac{\sin 4t}{32}\right) + \cos^2 4t\left(\frac{\sin 4t}{32}\right) - \frac{t \cos 4t}{8}$$

$$= \frac{1}{32}[\sin 4t - 4t \cos 4t]$$

A useful result is obtained by equating the inverse Laplace transforms of both sides of (7.28). This yields

$$\mathscr{L}^{-1}[f(s)g(s)] = \int_0^t F(t - u)G(u) \, du = [F(t)] * [G(t)] = (F * G)t$$

$$(7.30)$$

This formula asserts that the inverse Laplace transform of the product of two functions equals the convolution of the inverse Laplace transforms of the two functions. It could be used to find $\mathscr{L}^{-1}[f(s)/(as^2 + bs + c)]$ when another factoring of the quantity in brackets besides $[f(s)][1/(as^2 + bs + c)]$ is found to be convenient.

NOTE: The convolution $(F * G)(t)$ is sometimes written $F * G(t)$.

EXAMPLE 2.

$$\mathcal{L}^{-1}\left[\frac{1}{(s-1)(s+2)}\right] = \mathcal{L}^{-1}\left[\left(\frac{1}{s-1}\right)\left(\frac{1}{s+2}\right)\right]$$

$$= \int_0^t e^{t-u}e^{-2u}\,du$$

$$= e^t \int_0^t e^{-3u}\,du$$

$$= \frac{e^t e^{-3u}}{-3}\Bigg]_0^t$$

$$= e^t\left(\frac{e^{-3t}-1}{-3}\right)$$

$$= \frac{1}{3}(e^t - e^{-2t})$$

EXAMPLE 3.

$$\mathcal{L}^{-1}\left[\frac{1}{(s^2+a^2)^2}\right] = \mathcal{L}^{-1}\left[\frac{1}{s^2+a^2}\cdot\frac{1}{s^2+a^2}\right]$$

$$= \frac{1}{a^2}\int_0^t [\sin(at-au)\sin au]\,du$$

$$= \frac{1}{a^2}\int_0^t \sin au(\sin at \cos au - \cos at \sin au)\,du$$

$$= \frac{\sin at}{a^2}\int_0^t \sin au \cos au\,du - \frac{\cos at}{a^2}\int_0^t \sin^2 au\,du$$

$$= \frac{\sin at}{a^2}\left[\frac{\sin^2 au}{2a}\right]_0^t - \frac{\cos at}{a^2}\left[\frac{2au - \sin 2au}{4a}\right]_0^t$$

$$= \frac{1}{2a^3}\left[\sin^3 at - at\cos at + \cos at\left(\frac{1}{2}\sin 2at\right)\right]$$

$$= \frac{1}{2a^3}[\sin^3 at - at\cos at + \sin at\cos^2 at]$$

$$= \frac{1}{2a^3}[\sin at - at\cos at]$$

Another useful result is obtained by setting $g(s) = 1/s$ in

$$\mathcal{L}^{-1}[g(s)f(s)] = \int_0^t G(t-u)F(u)\,du$$

Then $G(t) \equiv 1$, and we obtain

$$\mathcal{L}^{-1}\left[\frac{f(s)}{s}\right] = \int_0^t F(u)\,du \qquad (7.31)$$

EXAMPLE 4. (a) $\mathcal{L}^{-1}\left[\dfrac{1}{s(s^2+1)}\right] = \mathcal{L}^{-1}\left[\dfrac{(s^2+1)^{-1}}{s}\right] = \displaystyle\int_0^t \sin u\, du = -\cos u\Big]_0^t = 1 - \cos t$

(b) $\mathcal{L}^{-1}\left[\dfrac{1}{s^2(s^2+1)}\right] = \mathcal{L}^{-1}\left[\dfrac{1}{s}\,\dfrac{1}{s(s^2+1)}\right]$

$$= \int_0^t (1 - \cos u)\, du = u - \sin u\Big]_0^t$$

$$= t - \sin t$$

(c) $\mathcal{L}^{-1}\left[\dfrac{1}{s^3(s^2+1)}\right] = \mathcal{L}^{-1}\left[\dfrac{1}{s}\,\dfrac{1}{s^2(s+1)}\right]$

$$= \int_0^t (u - \sin u)\, du = \frac{u^2}{2} + \cos u\Big]_0^t$$

$$= \frac{t^2}{2} + \cos t - 1$$

EXAMPLE 5.   Solve the initial-value problem

$$Y''(t) + 100Y(t) = 100; \quad Y(0) = 2, \quad Y'(0) = 0$$

Solution:

$$s^2 y(s) - 2s + 100y(s) = \frac{100}{s}$$

$$y(s) = \frac{2s}{s^2+100} + 10\left[\frac{1}{s}\cdot\frac{10}{s^2+100}\right]$$

$$Y(t) = 2\mathcal{L}^{-1}\left[\frac{s}{s^2+100}\right] + 10\int_0^t 1\sin 10u\, du$$

$$= 2\cos 10t - \cos 10u]_0^t$$

$$= 2\cos 10t - (\cos 10t - 1)$$

$$= 1 + \cos 10t$$

The operation of forming the convolution of two functions is commutative, that is, $(F * G)(t) = (G * F)(t)$, or

$$\int_0^t F(t-u)G(u)\, du = \int_0^t G(t-u)F(u)\, du \tag{7.32}$$

This result is readily obtained by making the substitution $v = t - u$ (see Problem 7). It asserts that the integrand in the convolution integral is the product of either function evaluated at u and the other function evaluated at $t - u$.

EXAMPLE 6.

$$\mathcal{L}^{-1}\left[\left(\frac{1}{s^2+1}\right)\left(\frac{1}{s+1}\right)\right] = \int_0^t e^{-t+u} \sin u \, du$$

$$= e^{-t} \int_0^t e^u \sin u \, du$$

$$= e^{-t}\left[\frac{e^u}{2}(\sin u - \cos u)\right]_0^t$$

$$= e^{-t}\left[\frac{e^t}{2}(\sin t - \cos t) + \frac{1}{2}\right]$$

$$= \frac{1}{2}(e^{-t} + \sin t - \cos t)$$

It is simpler to evaluate $\int_0^t e^{-t+u} \sin u \, du$ than to evaluate $\int_0^t \sin(t-u)e^{-u} \, du$. See Problem 8.

The convolution theorem is centrally important in the theoretical development of the Laplace transform. It is also useful in many advanced applications of DE, particularly those involving responses of systems to external stimuli.

Problem List 7.7

1. Find $\mathcal{L}[F(t)]$, given that F has period p on $[0, +\infty)$, for the following.
  (a) $F(t) = \cos t, p = 2\pi$
  (b) $F(t) = \begin{cases} 1 & \text{for } 0 \le t < p/2 \\ 0 & \text{for } p/2 \le t < p \end{cases}$ (meander function)
  See Figure 7.5.

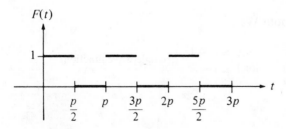

Fig. 7.5 Meander Function.

  (c) $F(t) = \begin{cases} \sin t & \text{for } 0 \le t < \pi \\ 0 & \text{for } \pi \le t < 2\pi \end{cases}$ (half-wave rectification of the sine function)

  See Figure 7.6 ($p = 2\pi$).

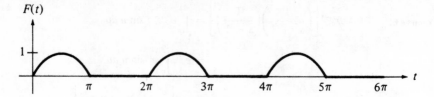

Fig. 7.6 Half-wave Rectification of the Sine Function.

(d) $F(t) = |\sin t|$ (full-wave rectification of the sine function)
    See Figure 7.7 ($p = \pi$).

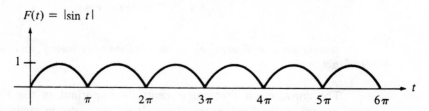

Fig. 7.7 Full-wave Rectification of the Sine Function.

(e) $F(t) = \dfrac{k}{p} t$ for $0 \leq t < p$ (saw-tooth wave)
    See Figure 7.8.

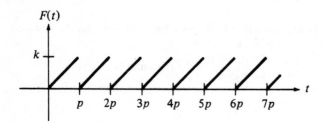

Fig. 7.8 Saw-tooth Wave.

(f) $F(t) = \begin{cases} t & \text{for } 0 \leq t < 1 \\ 2 - t & \text{for } 1 \leq t < 2 \end{cases}$ (triangle-wave function)
    See Figure 7.9.

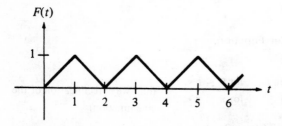

Fig. 7.9 Triangle-wave Function.

2. Find $\mathscr{L}[F(t)]$ where $F(t) = [t]$, the largest integer less than or equal to t (greatest integer, or staircase, function). See Figure 7.10. *Hint:* Use the result in Problem 1(e).

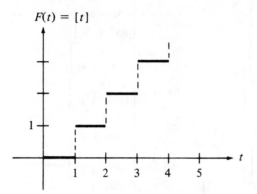

$F(t) = [t]$

Fig. 7.10 Greatest Integer, or Staircase, Function.

3. Find $\mathscr{L}[|\sin t|]$ by using the result of Problem 1(c).

4. Solve Problem 1(b) (meander function) by using the result of Example 2 of Section 7.7.

5. Solve the initial-value problem

$$Y''(t) + Y(t) = F(t), \quad Y(0) = Y'(0) = 0$$

where F is the meander function of Problem 1(b) with $p = 2$. See Figure 7.5.

6. Show that $\mathscr{L}[F]\mathscr{L}[G] \neq \mathscr{L}[FG]$ by letting $F(t) = t$ and $G(t) = t^2$.

7. Prove that $(F * G)(t) = (G * F)(t)$, that is, that

$$\int_0^t F(t - u)G(u)\,du = \int_0^t G(t - u)F(u)\,du$$

8. Find $\mathscr{L}^{-1}\left[\dfrac{1}{(s^2 + 1)(s + 1)}\right]$ in Example 6 by evaluating $\int_0^t \sin(t - u)e^{-u}\,du$.

9. Solve each initial-value problem using the convolution theorem.
  (a) $Y''(t) + Y(t) = 2e^{-t}$, $Y(0) = Y'(0) = 0$
  (b) $Y''(t) + 4Y(t) = 8$; $Y(0) = 1$, $Y'(0) = 1$
  (c) $Y''(t) + 2Y'(t) + 2Y(t) = 4$, $Y(0) = Y'(0) = 0$

10. Express $Y(t)$ in terms of $F(t)$ using the convolution theorem:

$$Y''(t) + k^2 Y(t) = F(t); \quad Y(0) = Y'(0) = 0$$

11. Find:

(a) $\mathscr{L}^{-1}\left[\dfrac{1}{s(s + 1)}\right]$ (b) $\mathscr{L}^{-1}\left[\dfrac{1}{s^2(s + 1)}\right]$

(c) $\mathcal{L}^{-1}\left[\dfrac{1}{s^2(s^2-1)}\right]$ (d) $\mathcal{L}^{-1}\left[\dfrac{1}{s(s+1)^2}\right]$

(e) $\mathcal{L}^{-1}\left[\dfrac{2}{s(s^2+4)}\right]$ (f) $\mathcal{L}^{-1}\left[\dfrac{1}{(s-2)(s-3)}\right]$

(g) $\mathcal{L}^{-1}\left[\dfrac{1}{s(s-3)}\right]$ (h) $\mathcal{L}^{-1}\left[\dfrac{4s-8}{s(s^2+4)}\right]$

(i) $\mathcal{L}^{-1}\left[\dfrac{1}{s(s+5)}\right]$ (j) $\mathcal{L}^{-1}\left[\dfrac{1}{s^2(s^2+4)}\right]$

(k) $\mathcal{L}^{-1}\left[\dfrac{s}{(s^2+1)^2}\right]$ (l) $\mathcal{L}^{-1}\left[\dfrac{s}{(s^2+a^2)^2}\right]$

(m) $\mathcal{L}^{-1}\left[\dfrac{1}{(s^2+1)(s-1)}\right]$ (n) $\mathcal{L}^{-1}\left[\dfrac{1}{s^3(s^2+1)}\right]$

(o) $\mathcal{L}^{-1}\left[\dfrac{s}{s^2-a^2}\right]$ (p) $\mathcal{L}^{-1}\left[\dfrac{s^2}{(s^2+a^2)^2}\right]$

(q) $\mathcal{L}^{-1}\left[\dfrac{1}{\{(s+a)^2+b^2\}^2}\right]$ (r) $\mathcal{L}^{-1}\left[\dfrac{f(s)}{s-a}\right]$

12. Find $\mathcal{L}^{-1}[18/\{s^2(s^2+9)\}]$, (a) by the method of partial fractions; (b) by the convolution method.

7.9 Laplace Transform Solutions of Systems of Differential Equations

The Laplace transform method of solving a DE extends readily to the solution of a system of linear DE with constant coefficients. The system of DE in the t space is transformed into a system of algebraic equations in the s space. This system is solved algebraically in the s space; inverse Laplace transforms then yield the required solution functions in the t space. The Laplace transform method is generally applied to systems of DE when initial conditions are known. One of the main advantages of the Laplace method is that arbitrary constants do not appear in the solution. This avoids the complication, seen in Chapter 6, of having more arbitrary constants than are essential appear in a general solution.

EXAMPLE 1. Solve the initial-value problem

$$X'(t) - Y(t) = \frac{t^2}{2}, \quad X(t) + Y'(t) = 0; \quad X(0) = 0, \quad Y(0) = 1$$

Solution:

$$sx(s) - 0 - y(s) = \frac{1}{s^3}$$

$$x(s) + sy(s) - 1 = 0$$

Solving for $x(s)$ and $y(s)$ gives

$$(s^2 + 1)x(s) - 1 = \frac{1}{s^2}$$

$$x(s) = \frac{(1/s^2) + 1}{s^2 + 1} = \frac{1}{s^2}$$

and

$$y(s) = \frac{1 - x(s)}{s} = \frac{1 - (1/s^2)}{s} = \frac{s^2 - 1}{s^3} = \frac{1}{s} - \frac{1}{s^3}$$

Thus,

$$X(t) = \mathcal{L}^{-1}\left[\frac{1}{s^2}\right] = t \quad \text{and} \quad Y(t) = \mathcal{L}^{-1}\left[\frac{1}{s} - \frac{1}{s^3}\right] = 1 - \frac{t^2}{2}$$

NOTE: After finding $X(t) = t$, we can find $Y(t)$ from

$$Y(t) = X'(t) - \frac{t^2}{2} = 1 - \frac{t^2}{2}$$

EXAMPLE 2. Solve the initial-value problem

$$X'(t) + X(t) + Y'(t) - Y(t) = 2, \quad X''(t) + X'(t) - Y'(t) = \cos t;$$
$$X(0) = 0, \quad X'(0) = 2, \quad Y(0) = 1$$

Solution:

$$sx(s) - 0 + x(s) + sy(s) - 1 - y(s) = \frac{2}{s}$$

$$s^2x(s) - 0s - 2 + sx(s) - sy(s) + 1 = \frac{s}{s^2 + 1}$$

$$(s + 1)x(s) + (s - 1)y(s) = \frac{2}{s} + 1 \tag{7.33}$$

$$s(s + 1)x(s) - sy(s) = \frac{s}{s^2 + 1} + 1 \tag{7.34}$$

Multiplying (7.33) by s and subtracting (7.34) from the result, we obtain

$$[s(s - 1) + s]y(s) = 2 + s - \frac{s}{s^2 + 1} - 1$$

$$s^2y(s) = 1 + s - \frac{s}{s^2 + 1}$$

$$y(s) = \frac{1}{s^2} + \frac{1}{s} - \frac{1}{s(s^2 + 1)} \tag{7.35}$$

and

$$Y(t) = t + 1 - (1 - \cos t) = t + \cos t$$

Substituting $y(s)$ from (7.35) into (7.34) yields

$$s(s + 1)x(s) = \frac{1}{s} + 1 - \frac{1}{s^2 + 1} + \frac{s}{s^2 + 1} + 1$$

$$= \frac{2s^3 + 2s^2 + s + 1}{s(s^2 + 1)}$$

$$= \frac{2s^2(s + 1) + (s + 1)}{s(s^2 + 1)}$$

$$x(s) = \frac{2}{s^2 + 1} + \frac{1}{s^2(s^2 + 1)}$$

$$= \frac{2}{s^2 + 1} + \left(\frac{1}{s^2} - \frac{1}{s^2 + 1} \right)$$

$$= \frac{1}{s^2} + \frac{1}{s^2 + 1}$$

and

$$X(t) = t + \sin t$$

EXAMPLE 3. Solve the initial-value problem

$$X''(t) - 4X(t) + Y'(t) = 0, \quad -4X'(t) + Y''(t) + 2Y(t) = 0;$$

$$X(0) = 1, \quad X'(0) = 2; \quad Y(0) = -4, \quad Y'(0) = 8$$

Solution:

$$s^2x(s) - s - 2 - 4x(s) + sy(s) + 4 = 0$$

$$-4[sx(s) - 1] + s^2y(s) + 4s - 8 + 2y(s) = 0$$

$$(s^2 - 4)x(s) + sy(s) = -2 + s$$

$$-4sx(s) + (s^2 + 2)y(s) = 4 - 4s$$

$$[(s^2 - 4)(s^2 + 2) - (-4s)s]x(s) = (s - 2)(s^2 + 2) - s(4 - 4s)$$

$$(s^4 - 2s^2 - 8 + 4s^2)x(s) = s^3 + 2s^2 - 2s - 4$$

$$(s^2 + 4)(s^2 - 2)x(s) = s^2(s + 2) - 2(s + 2) = (s^2 - 2)(s + 2)$$

$$x(s) = \frac{s}{s^2 + 4} + \frac{2}{s^2 + 4}$$

$$X(t) = \cos 2t + \sin 2t$$

From $Y'(t) = 4X(t) - X''(t)$, we obtain

$$Y'(t) = 4(\sin 2t + \cos 2t) - (-4 \sin 2t - 4 \cos 2t)$$

$$= 8 \sin 2t + 8 \cos 2t$$

$$Y(t) = -4 \cos 2t + 4 \sin 2t + c$$

$$-4 = -4 + c, \quad c = 0$$

and hence

$$Y(t) = -4 \cos 2t + 4 \sin 2t$$

7.10 Laplace Transform Solutions of Applied Problems

In this section we apply Laplace transform methods to a variety of physical situations.

EXAMPLE 1. A 64-lb body falls from rest under the influence of gravity. It is also acted on by air resistance, given by $R = 8\,V$ lb, where V is the velocity in ft/sec. Find V in terms of the time t. Use $g = 32$ ft/sec^2.

Solution: Applying Newton's second law, we obtain

$$\frac{64}{32} V'(t) = 64 - 8V(t), \quad V(0) = 0$$

or

$$V'(t) + 4V(t) = 32$$

Hence,

$$sv(s) + 4v(s) = \frac{32}{s}$$

$$v(s) = \frac{32}{s(s + 4)} = \frac{8}{s} - \frac{8}{s + 4}$$

and

$$V(t) = 8 - 8e^{-4t}$$

EXAMPLE 2. In Section 3.5, Newton's law of cooling led to the DE $dT/dt = k(T - \tau)$, where T is the temperature of the cooling body at time t, where $\tau > 0$ is the constant temperature of the surrounding medium, and $k < 0$ a constant of proportionality. Find T in terms of t.

Solution:

$$T'(t) - kT(t) = -k\tau, \quad T(0) = T_0$$

Letting $\mathscr{L}T = u$,

$$su(s) - T_0 - ku(s) = -\frac{k\tau}{s}$$

$$u(s) = \frac{T_0}{s - k} - \frac{k\tau}{s(s - k)} = \frac{T_0}{s - k} + \frac{\tau}{s} - \frac{\tau}{s - k}$$

$$T(t) = T_0 e^{kt} + \tau - \tau e^{kt}$$
$$= \tau + (T_0 - \tau)e^{kt}$$

EXAMPLE 3. An 8-lb weight is hung on the end of a vertically suspended spring, thereby stretching the spring 6 in. The weight is raised 3 in. above its equilibrium position and released from rest at time $t = 0$. Find the displacement X of the weight from its equilibrium position at time t. Use $g = 32$ ft/sec^2.

Solution: By Hooke's law, $8 = k(\frac{1}{2})$, and $k = 16$ lb/ft (spring constant). By Newton's second law,

$$\frac{8}{32}\frac{d^2X}{dt^2} = -16X; \quad X(0) = -\frac{1}{4}, \quad X'(0) = 0$$

where X is measured positively downward. From $X''(t) + 64X(t) = 0$, we obtain

$$s^2x(s) + \frac{s}{4} + 64x(s) = 0$$

$$(s^2 + 64)x(s) = -\frac{s}{4}$$

$$x(s) = -\frac{1}{4}\frac{s}{s^2 + 64}$$

and

$$X(t) = -\frac{1}{4}\cos 8t$$

EXAMPLE 4. A spring suspended vertically from a fixed support has spring constant $k = 50$ lb/ft. A 64-lb weight is placed on the spring and an external force given by $F(t) = 4 \sin 4t$ is applied to the weight. Find the displacement X in terms of t if $X(0) = 0$ and $X'(0) = 0$. Measure X positively downward and use $g = 32$ ft/sec^2.

Solution: By Newton's second law,

$$\frac{64}{32}X''(t) = -50X(t) + 4 \sin 4t, \quad X(0) = X'(0) = 0$$

or

$$X''(t) + 25X(t) = 2 \sin 4t$$

Hence,

$$s^2x(s) + 25x(s) = \frac{8}{s^2 + 16}$$

$$x(s) = \frac{8}{(s^2 + 16)(s^2 + 25)} = \frac{8}{9}\left[\frac{1}{s^2 + 16} - \frac{1}{s^2 + 25}\right]$$

and

$$X(t) = \frac{8}{9}\left(\frac{1}{4}\sin 4t - \frac{1}{5}\sin 5t\right)$$

$$= \frac{2}{45}(5 \sin 4t - 4 \sin 5t)$$

The following problem was solved in Section 5.1.

EXAMPLE 5. The motion of a block undergoing a rectilinear motion is governed by the DE

$$X''(t) + 8X'(t) + 36X(t) = 72 \cos 6t$$

Find X in terms of t if $X(0) = \frac{1}{2}$ and $X'(0) = 0$.

Solution:

$$s^2 x(s) - \frac{s}{2} + 8sx(s) - 4 + 36x(s) = \frac{72s}{s^2 + 36}$$

$$x(s) = \frac{s + 8}{2(s^2 + 8s + 36)} + \frac{72s}{(s^2 + 8s + 36)(s^2 + 36)}$$

$$= \frac{1}{2} \frac{s + 4 + 4}{(s + 4)^2 + 20} + \frac{9}{s^2 + 36} - \frac{9}{(s + 4)^2 + 20}$$

$$= \frac{1}{2} \frac{s + 4}{(s + 4)^2 + 20} + \frac{2}{2\sqrt{5}} \frac{2\sqrt{5}}{(s + 4)^2 + 20} + \frac{3}{2} \left[\frac{6}{s^2 + 36} \right]$$

$$- \frac{9}{2\sqrt{5}} \frac{2\sqrt{5}}{(s + 4)^2 + 20}$$

$$X(t) = \frac{e^{-4t}}{2} \cos 2\sqrt{5} t + \frac{2e^{-4t}}{2\sqrt{5}} \sin 2\sqrt{5} t$$

$$+ \frac{3}{2} \sin 6t - \frac{9e^{-4t}}{2\sqrt{5}} \sin 2\sqrt{5} t$$

$$= e^{-4t} \left(\frac{-7}{2\sqrt{5}} \sin 2\sqrt{5} t + \frac{1}{2} \cos 2\sqrt{5} t \right) + \frac{3}{2} \sin 6t$$

The following problem was presented in Section 5.9 of Chapter 5.

EXAMPLE 6. To find an equation of the elastic curve of a cantilever beam l feet long built in horizontally at the left end and unsupported at the right end, we solve the initial-value problem

$$EIY''(x) = M(x) = \frac{-w}{2} (l - x)^2 = \frac{-w}{2}(l^2 - 2lx + x^2), \quad Y(0) = Y'(0) = 0$$

Hence

$$EIs^2 y(s) = \frac{-w}{2} \left(\frac{l^2}{s} - \frac{2l}{s^2} + \frac{2}{s^3} \right)$$

$$y(s) = \frac{-w}{2EI} \left(\frac{l^2}{s^3} - \frac{2l}{s^4} + \frac{2}{s^5} \right)$$

and

$$Y(x) = \frac{-w}{2EI} \left(\frac{l^2 x^2}{2} - \frac{lx^3}{3} + \frac{x^4}{12} \right)$$

$$= \frac{-w}{24EI} (x^4 - 4lx^3 + 6l^2 x^2)$$

EXAMPLE 7. In Figure 7.11 the initial charge $Q(0)$ on the capacitor is one coulomb. Find the charge Q on the capacitor and the current I at time t, if t sec after the switch is closed, (a) $E(t) = 120$; (b) $E(t) = 120 \cos t$; (c) $E(t) = 120[1 - H(t - 10)]$.

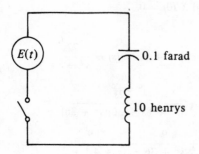

Fig. 7.11

Solution: Since no resistor is present,

$$L\frac{d^2Q}{dt^2} + R\frac{dQ}{dt} + \frac{Q}{C} = E(t) \qquad (5.44)$$

becomes $10\, d^2Q/dt^2 + Q/0.1 = E(t)$.
(a) $Q''(t) + Q(t) = 12$; $Q(0) = 1$, $Q'(0) = I(0) = 0$

$$s^2q(s) - s + q(s) = \frac{12}{s}$$

$$q(s) = \frac{s}{s^2 + 1} + \frac{12}{s(s^2 + 1)}$$

$$Q(t) = \cos t + \int_0^t 12 \sin u \, du$$

$$= \cos t + 12[-\cos u]_0^t$$

$$= 12 - 11 \cos t$$

The current $I(t) = Q'(t) = 11 \sin t$.
(b) $Q''(t) + Q(t) = 12 \cos t$; $Q(0) = 1$, $Q'(0) = I(0) = 0$

$$s^2q(s) - s + q(s) = \frac{12s}{s^2 + 1}$$

$$q(s) = \frac{s}{s^2 + 1} + \frac{12s}{(s^2 + 1)^2}$$

$$Q(t) = \cos t + 6t \sin t$$

$$I(t) = Q'(t) = -\sin t + 6 \sin t + 6t \cos t$$

$$= 5 \sin t + 6t \cos t$$

(c) $Q''(t) + Q(t) = 12 - 12H(t - 10)$

$$s^2 q(s) - s + q(s) = \frac{12}{s} - \frac{12e^{-10s}}{s}$$

$$q(s) = \frac{s}{s^2 + 1} + \frac{12}{s(s^2 + 1)} - \frac{12e^{-10s}}{s(s^2 + 1)}$$

$$Q(t) = \cos t + [12 - 12 \cos t] - [12 - 12 \cos (t - 10)]H(t - 10)$$
$$= 12 - 11 \cos t - 12[1 - \cos (t - 10)]H(t - 10)$$
$$= \begin{cases} 12 - 11 \cos t & \text{for } 0 \le t < 10 \\ 12 \cos (t - 10) - 11 \cos t & \text{for } t \ge 10 \end{cases}$$

$$I(t) = Q'(t) = 11 \sin t - 12[\sin (t - 10)]H(t - 10)$$
$$= \begin{cases} 11 \sin t & \text{for } 0 \le t < 10 \\ 11 \sin t - 12 \sin (t - 10) & \text{for } t \ge 10 \end{cases}$$

We note that $Q(t)$ and $I(t)$ are the same in parts (a) and (c) for $0 \le t < 10$. In this interval the system reacts as if no abrupt change in $E(t)$ is about to take place.

EXAMPLE 8. For the network shown in Figure 7.12, $I_1(0) = I_2(0) = I_3(0) = 0$. Find I_1, I_2, and I_3 at time t sec after the switch is closed.

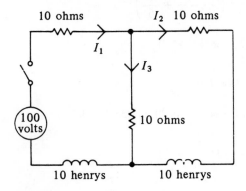

Fig. 7.12

Solution: Applying Kirchhoff's laws to the left circuit, we obtain

$$10I_1(t) + 10I_3(t) + 10I'_1(t) = 100$$

From the right circuit,

$$10I_2(t) + 10I'_2(t) - 10I_3(t) = 0$$

Using $I_3(t) = I_1(t) - I_2(t)$, we obtain

$$2I_1(t) - I_2(t) + I'_1(t) = 10$$

$$2I_2(t) + I'_2(t) - I_1(t) = 0$$

Thus,

$$2i_1(s) - i_2(s) + si_1(s) = \frac{10}{s}$$

$$2i_2(s) + si_2(s) - i_1(s) = 0$$

$$(2 + s)i_1(s) - i_2(s) = \frac{10}{s}$$

$$-i_1(s) + (2 + s)i_2(s) = 0$$

$$[-1 + (2 + s)^2]i_2(s) = \frac{10}{s}$$

$$i_2(s) = \frac{10}{s[(s + 2)^2 - 1]}$$

and

$$I_2(t) = 10e^{-2t}\mathscr{L}^{-1}\left[\frac{1}{(s - 2)(s^2 - 1)}\right]$$

$$= 10e^{-2t}\mathscr{L}^{-1}\left[\frac{1}{3}\frac{1}{s - 2} - \frac{1}{2}\frac{1}{s - 1} + \frac{1}{6}\frac{1}{s + 1}\right]$$

$$= 10e^{-2t}\left(\frac{e^{2t}}{3} - \frac{e^t}{2} + \frac{e^{-t}}{6}\right)$$

$$= \frac{5}{3}(2 - 3e^{-t} + e^{-3t})$$

Then,

$$I_1(t) = 2I_2(t) + I_2'(t)$$

$$= \frac{10}{3}(2 - 3e^{-t} + e^{-3t}) + \frac{5}{3}(3e^{-t} - 3e^{-3t})$$

$$= \frac{5}{3}(4 - 3e^{-t} - e^{-3t})$$

and

$$I_3(t) = I_1(t) - I_2(t)$$

$$= \frac{5}{3}(4 - 3e^{-t} - e^{-3t}) - \frac{5}{3}(2 - 3e^{-t} + e^{-3t})$$

$$= \frac{10}{3}(1 - e^{-3t})$$

EXAMPLE 9. In Section 6.5 the motion of two masses moving in a coupled system was governed by DE (6.41) and (6.42). See Figure 6.9. Set $k_1 = 2$, $k_2 = 1$, $m_1 = 3$, and $m_2 = 1$ to obtain the system

$$3X''(t) + 3X(t) - Y(t) = 0, \quad Y''(t) + Y(t) - X(t) = 0$$

Find $X(t)$ and $Y(t)$ if $X(0) = 1$, $X'(0) = 0$, $Y(0) = \sqrt{3}$, and $Y'(0) = 0$.

Solution:

$$3s^2x(s) - 3s + 3x(s) - y(s) = 0$$

$$s^2y(s) - \sqrt{3}\,s + y(s) - x(s) = 0$$

$$3(s^2 + 1)x(s) - y(s) = 3s$$

$$-x(s) + (s^2 + 1)y(s) = \sqrt{3}\,s$$

$$[3(s^2 + 1)(s^2 + 1) - 1]x(s) = s[3(s^2 + 1) + \sqrt{3}]$$

$$x(s) = \frac{s(3s^2 + 3 + \sqrt{3})}{3s^4 + 6s^2 + 2} = \frac{s}{s^2 + 1 - 1/\sqrt{3}}$$

$$X(t) = \cos\sqrt{1 - 1/\sqrt{3}}\,t$$

$$y(s) = 3(s^2 + 1)x(s) - 3s = \frac{3s(s^2 + 1)}{s^2 + 1 - 1/\sqrt{3}} - 3s$$

$$= \frac{3s/\sqrt{3}}{s^2 + 1 - 1/\sqrt{3}}$$

$$Y(t) = \sqrt{3}\cos\sqrt{1 - 1/\sqrt{3}}\,t$$

The Laplace transform method is particularly advantageous when applied to nonhomogeneous linear DE with constant coefficients, with initial (at $t = 0$) values given. As observed in Example 9, the method does not avoid the need for solving the characteristic equation. Other integral transforms, not restricted to application on $[0, +\infty)$, are discussed in Reference 7.4.

Problem List 7.8

Solve the following initial-value problems by the Laplace transform method.
1. $X'(t) - 2X(t) + 6Y(t) = 0$, $Y'(t) - Y(t) + X(t) = 0$; $X(0) = 2$, $Y(0) = 1$
2. $X(t) - Y'(t) = 2t - 1$, $X'(t) + Y(t) = t - t^2$; $X(0) = 0$, $Y(0) = -1$
3. $2X(t) - Y'(t) = 1$, $X'(t) - 2Y(t) = 1$; $X(0) = \frac{3}{2}$, $Y(0) = \frac{1}{2}$
4. $X'(t) - Y(t) = 0$, $2X'(t) - X(t) - Y'(t) = 0$; $X(0) = 0$, $Y(0) = 1$
5. $X(t) - Y'(t) = 0$, $X'(t) - 2Y'(t) + Y(t) = 0$; $X(0) = 2$, $Y(0) = 1$
6. An 8-lb body is dropped from rest, subject to air resistance of 0.05 V lb, where V is the velocity of the body in ft/sec at time t. Find V in terms of t. Use $g = 32$ ft/sec^2.
7. A 32-lb body is thrown vertically downward from the top of a high building with an initial speed of 40 ft/sec. The force R of air resistance acting on the body is given by $R = |2V|$, where R is in pounds and $V = dX/dt$ is the instantaneous velocity of the body in ft/sec. Find V in terms of t. Use $g = 32$ ft/sec^2.

8. The motion of a body moving in a straight line is governed by the DE

$$\frac{d^2X}{dt^2} = \frac{dV}{dt} = \frac{-1}{2}\frac{dX}{dt}$$

If $V(0) = 50$ and $X(0) = 0$, what are V and X in terms of t?

9. A weight undergoes a simple harmonic motion governed by the DE $d^2X/dt^2 + 64X = 0$. If $X(0) = \frac{1}{3}$ and $X'(0) = 2$, what is X in terms of t (X in ft, t in sec)? Also find the amplitude, period, and frequency of the motion.

10. A spring, suspended vertically from a fixed support, has spring constant 72 lb/ft. A particle weighing 16 lb is attached to the lower end. After it comes to rest, the particle is pulled down 2 in. and released from rest. Find the displacement X, measured positively downward, in terms of the time t. Use $g = 32$ ft/sec^2.

11. A 32-lb weight is suspended by a vertical spring with spring constant 4 lb/ft. A vertical force given by $F(t) = 15 \sin t$ is applied. If at $t = 0$ the weight is 6 in. below the equilibrium position and has upward velocity 3 ft/sec, what is the displacement X in terms of t? Use $g = 32$ ft/sec^2.

12. A 16-lb weight is attached to a vertical spring whose spring constant is 32 lb/ft. A force given by $F(t) = 4 \sin 2t$ is applied. Find the displacement X in terms of t if at $t = 0$ the weight is at rest at the equilibrium position. Explain why resonance does not occur.

13. The growth of A dollars compounded continuously at a nominal rate r percent per annum is governed by the DE $dA/dt = 0.01rA$. Find A in terms of t if $A(0) = A_0$.

14. The DE of the elastic curve of an l-ft-long cantilever beam, built in horizontally at the left end, and carrying a load of W lb suspended from its right end, is $EIY''(x) = M(x) = -W(l - x)$. Find an equation of the elastic curve. (The weight of the beam is neglected.)

15. An l-ft-long beam, built in horizontally at the left end and simply supported at the right end, carries a uniformly distributed load of w lb/ft. Find an equation of the elastic curve, given that this curve has DE

$$EIY''(x) = M(x) = -\frac{w(l - x)^2}{2} + R(l - x)$$

where R is the number of pounds acting upward at $(l, 0)$.

16. An inductor of 0.5 H is in series with a resistor of 5 ohms, a capacitor of 0.08 farad, and an emf of 20 volts. If $Q(0) = 0$ and $I(0) = Q'(0) = 0$, what are Q and I in terms of t?

17. An inductor of 3 H and a 6-Ω resistor are connected in series with a generator having emf $150e^{-2t} \cos 25t$ V. Find the current I in terms of t if $I(0) = 20$ A.

18. An inductor of 0.5 H is in series with a resistor of 5 Ω and an emf in volts given by $E(t) = 100[1 - H(t - 8)]$. Find the current I in terms of t if $I(0) = 0$.

19. For the network shown in Figure 7.13, $I_1(0) = I_2(0) = I_3(0) = 0$. Find I_1, I_2, and I_3 at time t sec after the switch is closed.

20. The motion of masses m_1 and m_2 moving in the coupled system of Figure 6.7 was

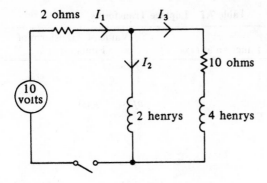

Fig. 7.13

governed by the system of DE

$$m_1 X''(t) = k_2[Y(t) - X(t)] - k_1 X(t)$$

$$m_2 Y''(t) = -k_2[Y(t) - X(t)]$$

Find $X(t)$ and $Y(t)$ if $m_1 = m_2 = 1$, $k_1 = 16$, $k_2 = 8$, $X(0) = X'(0) = Y(0) = 0$, and $Y'(0) = 4$.

References

7.1 Churchill, R. V. 1963. *Operational Mathematics*. New York: McGraw-Hill.

7.2 Doetsch, G. 1950, 1955, 1956. *Handbuch der Laplace Transformation*, vols. 1, 2, 3. Basel: Birkhäuser-Verlag.

7.3 Doetsch, G. 1961. *Guide to the Application of Laplace Transforms*. New York: Van Nostrand Reinhold.

7.4 Erdelyi, A., W. Magnus, F. Oberhettinger, and F. Tricomi. 1954. *Tables of Integral Transforms*, vol. 1. New York: McGraw-Hill.

7.5 Kaplan, Wilfred. 1959. *Advanced Calculus*. Reading, Mass.: Addison-Wesley.

7.6 Kaplan, Wilfred. 1962. *Ordinary Differential Equations*. Reading, Mass.: Addison-Wesley.

7.7 Osgood, W. F. 1928. *Advanced Calculus*. New York: Macmillan.

7.8 Standard Mathematical Tables, 20th ed. 1972. Cleveland: Chemical Rubber Co.

7.9 Tierney, John A. 1979. *Calculus and Analytic Geometry*, 4th ed. Boston: Allyn and Bacon.

7.10 Widder, D. V. 1961. *Advanced Calculus*, 2nd ed. Englewood Cliffs, N.J.: Prentice-Hall.

7.11 Widder, D. V. 1941. *The Laplace Transformation*. Princeton: Princeton University Press.

7.12 Wylie, C. R., Jr. 1966. *Advanced Engineering Mathematics*, 3rd ed. New York: McGraw-Hill.

Table 7.1   Laplace Transforms

Value of Function at t	Value of Transform of Function at s
$F(t)$	$\int_0^{+\infty} e^{-st}F(t)\, dt = f(s)$
$aF(t) + bG(t)$	$af(s) + bg(s)$
1	$\dfrac{1}{s}$
t	$\dfrac{1}{s^2}$
$t^n,\ n = 1, 2, 3, \ldots$	$n!/s^{n+1}$
$t^a,\ a > -1$	$\dfrac{\Gamma(a+1)}{s^{a+1}}$
e^{at}	$\dfrac{1}{s-a}$
$\cos at$	$\dfrac{s}{s^2 + a^2}$
$\sin at$	$\dfrac{a}{s^2 + a^2}$
$\cosh at$	$\dfrac{s}{s^2 - a^2}$
$\sinh at$	$\dfrac{a}{s^2 - a^2}$
$t^n e^{at},\ n = 1, 2, 3, \ldots$	$\dfrac{n!}{(s-a)^{n+1}}$
$t \sin at$	$\dfrac{2as}{(s^2 + a^2)^2}$
$t \cos at$	$\dfrac{s^2 - a^2}{(s^2 + a^2)^2}$
$e^{at}F(t)$	$f(s-a)$
$tF(t)$	$-f'(s) = -\dfrac{df}{ds}$
$t^n F(t),\ n = 1, 2, 3, \ldots$	$(-1)^n f^{(n)}(s) = (-1)^n \dfrac{d^n f}{ds^n}$
$F'(t)$	$sf(s) - F(0)$
$F''(t)$	$s^2 f(s) - sF(0) - F'(0)$
$F^{(n)}(t),\ n = 1, 2, 3, \ldots$	$s^n f(s) - \displaystyle\sum_{i=0}^{n-1} s^{n-1-i} F^{(i)}(0)$
$H(t-a)$	$\dfrac{e^{-as}}{s}$
$F(t)H(t-a)(a \geq 0)$	$e^{-as}\,\mathscr{L}[F(t+a)]$
$\displaystyle\int_0^t F(t-u)G(u)\, du$	$\mathscr{L}[F(t)]\mathscr{L}[G(t)] = f(s)g(s)$
$F(t)\ \{F \text{ of period } p \text{ on } [0, +\infty)\}$	$\dfrac{\int_0^p e^{-st}F(t)\, dt}{1 - e^{-ps}}$

Table 7.1 (*Contd.*)

Value of Function at t	Value of Transform of Function at s
$\displaystyle\int_0^t F(u)\,du$	$\dfrac{f(s)}{s}$
$\dfrac{F(t)}{t}$	$\displaystyle\int_s^{+\infty} f(u)\,du$
$F(t-a)$	$e^{-as}f(s)$
$\dfrac{t^{n-1}}{(n-1)!},\; n=1,2,3,\ldots$	$\dfrac{1}{s^n}$
te^{at}	$\dfrac{1}{(s-a)^2}$
$\dfrac{1}{a-b}(e^{at}-e^{bt})$	$\dfrac{1}{(s-a)(s-b)}$
$\dfrac{1}{a-b}(ae^{at}-be^{bt})$	$\dfrac{s}{(s-a)(s-b)}$
$\dfrac{1}{a^2}(1-\cos at)$	$\dfrac{1}{s(s^2+a^2)}$
$\dfrac{1}{a^3}(at-\sin at)$	$\dfrac{1}{s^2(s^2+a^2)}$
$\dfrac{1}{2a^3}(\sin at - at\cos at)$	$\dfrac{1}{(s^2+a^2)^2}$
$\dfrac{t}{2a}\sin at$	$\dfrac{s}{(s^2+a^2)^2}$
$\dfrac{1}{2a}(\sin at + at\cos at)$	$\dfrac{s^2}{(s^2+a^2)^2}$
$\dfrac{a\sin bt - b\sin at}{ab(a^2-b^2)}$	$\dfrac{1}{(s^2+a^2)(s^2+b^2)},\; a^2\neq b^2$
$\dfrac{\cos at - \cos bt}{b^2-a^2}$	$\dfrac{s}{(s^2+a^2)(s^2+b^2)},\; a^2\neq b^2$
$\dfrac{1}{2a^3}(\sinh at - \sin at)$	$\dfrac{1}{s^4-a^4}$
$\dfrac{1}{2a^2}(\cosh at - \cos at)$	$\dfrac{s}{s^4-a^4}$
$\dfrac{e^{bt}-e^{at}}{t}$	$\ln\dfrac{s-a}{s-b}$
$\dfrac{\sin at}{t}$	$\tan^{-1}\dfrac{a}{s}$

8

Series Solutions
of Differential
Equations

8.1 Power Series Solutions of Differential Equations; Method of Undetermined Coefficients

The initial-value problem

$$y'' + y = 0; \quad y(0) = 0, \quad y'(0) = 1 \tag{8.1}$$

has a unique solution on $(-\infty, +\infty)$ by the basic existence and uniqueness Theorem 4-I. By previous methods we know that this solution is given by

$$y = \sin x \tag{8.2}$$

If we replace $\sin x$ in (8.2) by the power series

$$\sum_{n=0}^{+\infty} \frac{(-1)^n x^{2n+1}}{(2n+1)!} = x - \frac{x^3}{3!} + \frac{x^5}{5!} - \cdots \tag{8.3}$$

known as the Maclaurin series for $\sin x$, the solution of (8.1) is given by

$$y = x - \frac{x^3}{3!} + \frac{x^5}{5!} - \cdots \tag{8.4}$$

The solution of (8.1) given by (8.4) is said to be in "series form," whereas the solution given by (8.2) is said to be in "closed form." The right member of

(8.4) is just another way of writing sin x, and (8.1) has a closed form solution because the power series in (8.3) defines a familiar function that has been given the name "sin." Many DE have series solutions that do not have familiar designations. If we know that a given DE has a solution, the DE and the associated initial or boundary conditions may be regarded as defining a function, namely, the function known to be a solution of the DE. If the DE has considerable applied or theoretical importance, the solution is often given a name such as sin, tan, exp, or ln, and is tabulated for convenience. Newton introduced power series solutions in the seventeenth century.

The properties of a function are often best studied by investigating the DE defining the function. In Reference 8.1, the properties of the sine function are developed from (8.1).

The basic facts about power series are found in References 8.2, 8.6, and 8.8.

One of the following possibilities holds for every power series $\sum_{n=0}^{+\infty} c_n x^n$:

 (i) It converges for all values of x.

 (ii) It converges for $x = 0$ and diverges for $x \neq 0$.

 (iii) There exists a positive number R such that the series converges for $|x| < R$ and diverges for $|x| > R$.

The number R in (iii) is called the *radius of convergence* of the series. The series may converge for $x = R$ and $x = -R$, or may converge for only one of these values of x, or may diverge for both $x = R$ and $x = -R$. The set of values for which the series converges is called the *interval of convergence* of the series.

If the power series $\sum_{n=0}^{+\infty} c_n x^n$ defines a function f given by $f(x) = \sum_{n=0}^{+\infty} c_n x^n$ in an interval of convergence of radius $R > 0$, then the power series $\sum_{n=1}^{+\infty} nc_n x^{n-1}$ also has radius of convergence R and converges to $f'(x)$ for $-R < x < R$. That is, term-by-term differentiation is valid inside the interval of convergence.

Before considering conditions under which a DE has a power series solution, we investigate the technique for producing such a solution. Let us assume that we are unfamiliar with the sine function, and see whether we can obtain (8.4) from (8.1).

We assume that a solution of (8.1), valid on some interval $-R < x < R$, is given by

$$y(x) = \sum_{n=0}^{+\infty} c_n x^n \tag{8.5}$$

Since term-by-term differentiation is valid,

$$y'(x) = \sum_{n=1}^{+\infty} nc_n x^{n-1} \tag{8.6}$$

and

$$y''(x) = \sum_{n=2}^{+\infty} n(n-1)c_n x^{n-2} \tag{8.7}$$

Substituting (8.7) and (8.5) into (8.1), we obtain

$$\sum_{n=2}^{+\infty} n(n-1)c_n x^{n-2} + \sum_{n=0}^{+\infty} c_n x^n = 0 \qquad (8.8)$$

For the left member to be identically zero on an interval, the coefficient of every power of x must be zero. This result depends on the uniqueness theorem for power series. See Reference 8.8. To obtain the coefficient of x^n, we replace n by $n + 2$ in the first summation and sum from 0 to $+\infty$ instead of from 2 to $+\infty$. This is equivalent to letting $m = n - 2$ to obtain

$$\sum_{m=0}^{+\infty} (m+2)(m+2-1)c_{m+2} x^m$$

and then changing the dummy summation index from m to n. Equation (8.8) then becomes

$$\sum_{n=0}^{+\infty} (n+2)(n+1)c_{n+2} x^n + \sum_{n=0}^{+\infty} c_n x^n = 0$$

or

$$\sum_{n=0}^{+\infty} [(n+2)(n+1)c_{n+2} + c_n] x^n = 0$$

Setting the coefficient of x^n equal to 0, we obtain

$$(n+2)(n+1)c_{n+2} + c_n = 0$$

or

$$c_{n+2} = \frac{-c_n}{(n+2)(n+1)}$$

This formula, known as a *recursion formula*, enables us to express c_n in terms of c_0 for n even and in terms of c_1 for n odd. Thus, setting $n = 0, 2, 4, \ldots$ in the recursion formula, we obtain

$$c_2 = \frac{-c_0}{2 \cdot 1}$$

$$c_4 = \frac{-c_2}{4 \cdot 3} = \frac{c_0}{4!}$$

$$c_6 = \frac{-c_4}{6 \cdot 5} = \frac{-c_0}{6!}$$

$$c_{2n} = \frac{(-1)^n c_0}{(2n)!}$$

and setting $n = 1, 3, 5, \ldots$ in the recursion formula, we obtain

$$c_3 = \frac{-c_1}{3 \cdot 2}$$

$$c_5 = \frac{-c_3}{5 \cdot 4} = \frac{c_1}{5!}$$

$$c_7 = \frac{-c_5}{7 \cdot 6} = \frac{-c_1}{7!}$$

$$c_{2n+1} = \frac{(-1)^n c_1}{(2n+1)!}$$

We now rewrite (8.5) in the form

$$y = c_0 \sum_{n=0}^{+\infty} \frac{(-1)^n x^{2n}}{(2n)!} + c_1 \sum_{n=0}^{+\infty} \frac{(-1)^n x^{2n+1}}{(2n+1)!} \tag{8.9}$$

or

$$y = c_0 \left(1 - \frac{x^2}{2!} + \frac{x^4}{4!} - \cdots \right) + c_1 \left(x - \frac{x^3}{3!} - \frac{x^5}{5!} - \cdots \right) \tag{8.10}$$

The rearrangement of terms is valid since a power series converges absolutely inside its radius of convergence. We have shown that if $y'' + y = 0$ has a power series solution, then the solution must be given by (8.9), with c_0 and c_1 arbitrary. It is possible to show by the ratio test that each series in (8.9) converges for all x. See Problem 4. If one computes y' and y'' from (8.9) by term-by-term differentiation, it is not difficult to show by direct substitution into $y'' + y = 0$ that (8.9) defines a solution of the DE for all c_0 and c_1. See Problem 5.

Setting $x = 0$ and $y = 0$ in (8.9), we find that $c_0 = 0$. Setting $x = 0$ and $y' = 1$ in $y' = c_1(1 - x^2/2! + x^4/4! - \cdots)$, we find that $c_1 = 1$. We conclude that

$$y = \sum_{n=0}^{+\infty} \frac{(-1)^n x^{2n+1}}{(2n+1)!} = x - \frac{x^3}{3!} + \frac{x^5}{5!} - \cdots \tag{8.11}$$

is a solution of initial-value problem (8.1). That is, the power series solution is the unique solution guaranteed by Theorem 4-I.

Nowhere in this approach did we make use of the fact that (8.9) can be written as $y = c_0 \cos x + c_1 \sin x$, and (8.11) as $y = \sin x$.

Although series solutions of linear DE with nonconstant coefficients can sometimes be expressed in terms of the elementary functions used in the calculus, this is the exception rather than the rule.

If $y, y', y'', \ldots$ are specified at $x = x_0$ instead of at $x = 0$, one can seek solutions of a DE that are power series in $x - x_0$; that is, solutions of the form $y = \sum_{n=0}^{+\infty} c_n(x - x_0)^n$.

This problem can be reduced to the one we have considered by making the substitution $t = x - x_0$. Then

$$\frac{dy}{dx} = \frac{dy}{dt}, \quad \frac{d^2y}{dx^2} = \frac{d^2y}{dt^2}, \ldots$$

and values of $y, dy/dt, d^2y/dt^2, \ldots$ are then known at $t = 0$.

8.2 Normal Linear Differential Equations with Analytic Coefficients

Definition   A linear DE of the form

$$\frac{d^ny}{dx^n} + a_{n-1}(x)\frac{d^{n-1}y}{dx^{n-1}} + \cdots + a_0(x)y = F(x) \tag{8.12}$$

is called a *normal linear DE*.

It should be noted that the coefficient of d^ny/dx^n in (8.12) is $+1$.

Definitions   A function f that can be represented in an open interval I containing x_0 by a convergent power series of the form

$$f(x) = \sum_{n=0}^{+\infty} c_n(x - x_0)^n$$

is said to be *analytic* at $x = x_0$. A function is analytic on an open interval if and only if it is analytic at every point of the interval. A function is said to be an analytic function if and only if it is analytic at every point in its domain.

Except for complicated functions, all the elementary functions that are defined on an open interval I are analytic on I. For example, the functions given by $\sin x$, $\cos x$, $\tan x$, $\exp x$, $\ln x$, and $P(x)$, where P is a polynomial function, are analytic functions. The function given by $f(x) = \sqrt{x}$ is analytic on $(0, +\infty)$.

Definitions   A point x_0 at which $a_{n-1}, a_{n-2}, \ldots, a_0$ and F of (8.12) are analytic is called an *ordinary point* of the DE. Points that are not ordinary points are called *singular points* of the DE.

The following basic theorem is proved in Reference 8.3.

Theorem 8-I Let x_0 be an ordinary point of the nth-order normal linear DE (8.12), and let R be a positive number such that the power series expansions of $a_{n-1}, a_{n-2}, \ldots, a_0$ and F converge in the interval $(x_0 - R, x_0 + R)$. Then *every* solution f of (8.12) that is defined at x_0 is analytic at x_0, and the power series for f at x_0 converges at least in the interval $(x_0 - R, x_0 + R)$.

For the DE $y'' + y = 0$, we find that $a_0(x) \equiv 1$ and $F(x) \equiv 0$. Since a_0 and F are analytic at all x, according to Theorem 8-I the unique solution satisfying $y(0) = 0$ and $y'(0) = 1$, known to exist by Theorem 4-1, must be a power series solution. Furthermore, this power series, found to be given by $y = x - x^3/3! + x^5/5! - \cdots$ converges for all x.

The point x_0 in Theorem 8-I can be taken to be $x_0 = 0$ without loss of generality. This normalization is accomplished by making the substitution $t = x - x_0$.

The special case of (8.12), in which $n = 2$ and $F(x) \equiv 0$, has particular importance. This second-order DE is usually written in the form

$$y'' + p_1(x)y' + p_2(x)y = 0 \tag{8.13}$$

Many DE of the form (8.13) arise in both pure and applied mathematics. A large part of the theory of DE was developed by mathematicians making detailed investigations of special functions defined by DE of the form (8.13).

EXAMPLE 1. Solve the initial-value problem $y'' + xy = 2$; $y(0) = 1$, $y'(0) = 1$.

Solution: By Theorems 4-I and 8-I, a unique power series solution exists and it is valid for all x. From (8.5), (8.6), and (8.7),

$$y = \sum_{n=0}^{+\infty} c_n x^n, \quad y' = \sum_{n=1}^{+\infty} n c_n x^{n-1},$$

and

$$y'' = \sum_{n=2}^{+\infty} n(n-1) c_n x^{n-2}$$

Substituting for y and y'' into the DE yields

$$\sum_{n=2}^{+\infty} n(n-1) c_n x^{n-2} + x \sum_{n=0}^{+\infty} c_n x^n = 2$$

or

$$\sum_{n=2}^{+\infty} n(n-1) c_n x^{n-2} + \sum_{n=0}^{+\infty} c_n x^{n+1} = 2$$

Replacing n by $n + 2$ in the first summation and n by $n - 1$ in the second summation, we obtain

$$\sum_{n=0}^{+\infty} (n+2)(n+1) c_{n+2} x^n + \sum_{n=1}^{+\infty} c_{n-1} x^n = 2$$

Writing the first summation as

$$(2)(1)c_2x^0 + \sum_{n=1}^{+\infty} (n + 2)(n + 1)c_{n+2}x^n$$

yields

$$\sum_{n=1}^{+\infty} [(n + 2)(n + 1)c_{n+2} + c_{n-1}]x^n + 2c_2 = 2$$

Hence,

$$c_0 = y(0) = 1$$

$$c_1 = y'(0) = 1$$

$$2c_2 = 2$$

$$c_2 = 1$$

and

$$c_{n+2} = \frac{-c_{n-1}}{(n + 2)(n + 1)}$$

for $n \geq 1$. From this recursion formula,

$$c_3 = \frac{-c_0}{3 \cdot 2} = \frac{-1}{6}$$

$$c_4 = \frac{-c_1}{4 \cdot 3} = \frac{-1}{12}$$

$$c_5 = \frac{-c_2}{5 \cdot 4} = \frac{-1}{20}$$

$$c_6 = \frac{-c_3}{6 \cdot 5} = \frac{1}{180}$$

$$\vdots$$

The solution is given by

$$y = 1 + x + x^2 - \frac{x^3}{6} - \frac{x^4}{12} - \frac{x^5}{20} + \frac{x^6}{180} + \cdots$$

The general term is somewhat difficult to express in terms of n. Another situation in which the general term is often difficult to find is that in which a three-term recursion formula occurs. For example, c_{n+2} might be given in terms of both c_{n+1} and c_n. (See Problem 15 in Problem List 8.1.) In many applications, a few terms of a power series are sufficient to carry out required numerical calculations.

EXAMPLE 2. Solve Legendre's DE,

$$y'' - \frac{2x}{1 - x^2} y' + \frac{p(p + 1)}{1 - x^2} y = 0 \tag{8.14}$$

This DE, named after the French mathematician Adrien Marie Legendre (1752–1833), has many significant applications, particularly in the physical sciences. The parameter p is usually a nonnegative constant. The only solutions defined at $x = 0$ are analytic since all points except $x = \pm 1$ are ordinary points.

Solution: Substituting from (8.5), (8.6), and (8.7) into (8.14), and multiplying both sides of (8.14) by $1 - x^2$, we obtain

$$(1 - x^2) \sum_{n=2}^{+\infty} n(n-1)c_n x^{n-2} - 2x \sum_{n=1}^{+\infty} nc_n x^{n-1} + p(p+1) \sum_{n=0}^{+\infty} c_n x^n = 0$$

or

$$\sum_{n=2}^{+\infty} n(n-1)c_n x^{n-2} - \sum_{n=2}^{+\infty} n(n-1)c_n x^n - \sum_{n=1}^{+\infty} 2nc_n x^n + \sum_{n=0}^{+\infty} p(p+1)c_n x^n = 0$$

$$(8.15)$$

If we shift the summation index in the first summation by replacing n by $n + 2$, write separately the terms corresponding to $n = 0$ and $n = 1$, and simplify, the last equation becomes (see Problem 6)

$$[2c_2 + p(p+1)c_0] + [(p+2)(p-1)c_1 + (3 \cdot 2)c_3]x$$

$$+ \sum_{n=2}^{+\infty} [(n+2)(n+1)c_{n+2} + (p+n+1)(p-n)c_n]x^n = 0 \quad (8.16)$$

Setting the coefficient of x^n equal to zero for $n = 0, 1, 2, \ldots$, we obtain

$$2c_2 + p(p+1)c_0 = 0$$

$$(p+2)(p-1)c_1 + (3 \cdot 2)c_3 = 0$$

and

$$(n+2)(n+1)c_{n+2} + (p+n+1)(p-n)c_n = 0, \quad n \geq 2$$

or

$$c_{n+2} = \frac{-(p+n+1)(p-n)}{(n+2)(n+1)} c_n, \quad n \geq 0$$

The recursion formula makes it possible to express c_n in terms of c_0 when n is even and in terms of c_1 when n is odd. Specifically,

$$c_2 = \frac{-(p+1)p}{2} c_0$$

$$c_4 = \frac{-(p+3)(p-2)}{4 \cdot 3} c_2 = \frac{(p+3)(p+1)p(p-2)}{4!} c_0$$

$$c_{2n} = \frac{(-1)^n(p+2n-1)(p+2n-3)\cdots(p+1)p(p-2)\cdots(p-2n+2)}{(2n)!} c_0$$

for $n > 0$, and

$$c_3 = \frac{-(p + 2)(p - 1)}{3!} c_1$$

$$c_5 = \frac{-(p + 4)(p - 3)}{5 \cdot 4} c_3 = \frac{(p + 4)(p + 2)(p - 1)(p - 3)}{5!} c_1$$

$$\vdots$$

$$c_{2n+1} = \frac{(-1)^n(p + 2n)(p + 2n - 2) \cdots (p + 2)(p - 1)(p - 3) \cdots (p - 2n + 1)}{(2n + 1)!} c_1$$

$$\vdots$$

for $n > 0$.

The solution of (8.14) can be written in the form

$$y = y(x) = c_0 y_0(x) + c_1 y_1(x) \tag{8.17}$$

where $c_0 = y(0)$, $c_1 = y'(0)$,

$$y_0(x) = 1 - \frac{(p + 1)p}{2!} x^2 + \frac{(p + 3)(p + 1)p(p - 2)}{4!} x^4 - \cdots$$

and

$$y_1(x) = x - \frac{(p + 2)(p - 1)}{3!} x^3 + \frac{(p + 4)(p + 2)(p - 1)(p - 3)}{5!} x^5 - \cdots$$

Equation (8.17) is called a complete solution of Legendre's equation of order p. It is easy to show that the coefficients in Legendre's equation (8.14) have power series expansions valid in the interval $(-1, +1)$ (see Problem 7). Hence, by Theorem 8-I, every solution of the form (8.17) converges at least in $(-1, +1)$. The functions defined by $y_0(x)$ and $y_1(x)$ are called *Legendre functions*. When p is a nonnegative integer n, then $c_0 y_0(x)$ and $c_1 y_1(x)$ have only a finite number of nonzero terms and hence are polynomials $P_n(x)$. For particular choices of c_0 and c_1, these are called *Legendre polynomials*, and like all polynomials, they converge for all x. Thus, we have an example in which a power series solution of a DE converges in a larger interval than the power series expansions of the coefficients in the DE do. This result does not contradict Theorem 8-I, which asserts that a power series solution converges in an interval *at least* as large as one in which the power series expansions of the coefficients converge.

It is sometimes necessary to replace $p_1(x)$ or $p_2(x)$ in (8.13) by a power series.

EXAMPLE 3. Find the first six terms of the power series solution of the initial-value problem

$$y'' + (\cos x)y = 0; \quad y(0) = 1, \quad y'(0) = 2$$

Solution: From

$$y = \sum_{n=0}^{+\infty} c_n x^n = c_0 + c_1 x + c_2 x^2 + c_3 x^3 + c_4 x^4 + c_5 x^5 + \cdots$$

$$y' = \sum_{n=1}^{+\infty} n c_n x^{n-1} = c_1 + 2c_2 x + 3c_3 x^2 + 4c_4 x^3 + 5c_5 x^4 + \cdots$$

$$y'' = \sum_{n=2}^{+\infty} n(n-1) c_n x^{n-2} = 2c_2 + 6c_3 x + 12c_4 x^2 + 20c_5 x^3 + \cdots$$

we find that $c_0 = y(0) = 1$ and $c_1 = y'(0) = 2$.

Replacing y'', $\cos x$, and y in the DE by their power series, we obtain

$$\sum_{n=2}^{+\infty} n(n-1) c_n x^{n-2} + \left[\sum_{n=0}^{+\infty} \frac{(-1)^n x^{2n}}{n!} \right]\left[\sum_{n=0}^{+\infty} c_n x^n \right] = 0$$

or

$$2c_2 + 6c_3 x + 12c_4 x^2 + 20c_5 x^3 + 30c_6 x^4 + \cdots$$

$$+ \left[1 - \frac{x^2}{2} + \frac{x^4}{24} - \frac{x^6}{720} + \cdots \right][1 + 2x + c_2 x^2 + c_3 x^3 + c_4 x^4 + \cdots] = 0$$

Multiplying the two power series for $\cos x$ and y as if they were polynomials (see Reference 8.8) yields

$$2c_2 + 6c_3 x + 12c_4 x^2 + 20c_5 x^3 + 30c_6 x^4 + \cdots$$

$$+ \left[1 + 2x + \left(c_2 - \frac{1}{2} \right) x^2 + (c_3 - 1)x^3 + \cdots \right] = 0$$

Equating to zero the coefficients of x^0, x, x^2, and x^3, we obtain

$$2c_2 + 1 = 0 \qquad c_2 = -\frac{1}{2}$$

$$6c_3 + 2 = 0 \qquad c_3 = -\frac{1}{3}$$

$$12c_4 + c_2 - \frac{1}{2} = 0 \qquad c_4 = \frac{1}{12}$$

$$20c_5 + c_3 - 1 = 0 \qquad c_5 = \frac{1}{15}$$

and

$$y = 1 + 2x - \frac{1}{2} x^2 - \frac{x^3}{3} + \frac{x^4}{12} + \frac{x^5}{15} + \cdots$$

By Theorems 4-I and 8-I the solution is valid for all x.

Problem List 8.1

1. Find power series solutions of:
   (a) $y' - y = 0$ (b) $y' + y = 0$
   (c) $y' - 2y = 0$ (d) $y'' - y = 0$
   (e) $y' - xy = 0$ (f) $(1 + x^2)y'' + 2xy' = 0$
   (g) $y'' + xy' + y = 0$ (h) $y'' - xy' + y = 0$
   (i) $y' + 3x^2 y = 0$

2. Solve each of the following initial-value problems by the method of power series.
   (a) $y' + 2y = 0,\ y(0) = 4$
   (b) $y' - y + x = 0,\ y(0) = 3$
   (c) $y'' + xy' + y = 0;\ y(0) = 1,\ y'(0) = 0$

3. Find the first five nonzero terms of the power series expansion of y in terms of x if $y' = x^2 y + 2$ and $y(0) = 2$.

4. Use the ratio test to show that the power series for $\sin x$ and $\cos x$ converge for all x.

5. Show by direct substitution into $y'' + y = 0$ that (8.9) defines a solution of the DE for all c_0 and c_1.

6. Carry out the calculations required to obtain (8.16) from (8.15).

7. Show that the power series for the coefficients $-2x/(1 - x^2)$ and $p(p + 1)/(1 - x^2)$, $p \neq 0$ or -1, in Legendre's DE both have radius of convergence $R = 1$.

8. Show that every nonpolynomial solution of Legendre's DE has radius of convergence $R = 1$.

9. Obtain the Legendre polynomials $P_n(x)$ for $n = 0, 1, 2, 3, 4$, and 5. Assign to c_0 (or c_1) the reciprocal of the coefficient of c_0 (or c_1) when $x = 1$. For example, when $p = 1, c_1 = 1$, and when $p = 2, c_0 = [1 - 3(2)/2]^{-1} = -1/2$.

10. Show that Legendre's DE can be written in the form

$$\frac{d}{dx}\left[(1 - x^2)\frac{dy}{dx}\right] + \lambda y = 0$$

11. In Reference 8.4 it is shown that the Legendre polynomials are given by

$$P_n(x) = \frac{1}{2^n n!}\frac{d^n}{dx^n}(x^2 - 1)^n$$

This formula is known as *Rodrigues' formula* after the French mathematician Olinde Rodrigues (1794–1851). Use Rodrigues' formula to find $P_n(x)$ for $n = 1, 2$, and 3.

12. The solution of the initial-value problem $(x^2 + 5)y'' - 6xy' + 12y = 0;\ y(0) = 25$, $y'(0) = 0$, is a polynomial of degree 4. Find this solution by the method of power series.

13. The DE $y'' + xy = 0$, known as *Airy's equation*, after the British astronomer George Biddell Airy (1801–1892), arises in the theory of refraction. Find the power series solution for which $y(0) = 1$ and $y'(0) = 0$.

14. The DE $y'' - 2xy' + \lambda y = 0$, known as the *Hermite equation*, after the French mathematician Charles Hermite (1822–1901), has important applications in mathematical physics. Show that the method of power series leads to the recursion

formula

$$c_{n+2} = \frac{(2n - \lambda)c_n}{(n+2)(n+1)}, \quad n = 0, 1, 2, \ldots$$

Show that when λ is a nonnegative integer, the solutions of Hermite's DE are polynomials, These solutions, after multiplication by suitable constants, are called *Hermite polynomials*.

15. Apply the method of power series to the initial-value problem $y'' + y' + xy = 0$; $y(0) = y'(0) = 1$ to obtain the three-term recursion formula

$$c_{n+2} = \frac{-(n+1)c_{n+1} - c_{n-1}}{(n+2)(n+1)}$$

Find the first six nonzero terms of the solution.

16. Find the first five nonzero terms in the power series solution of $y'' = e^x y$ if $y(0) = y'(0) = 1$.

17. Find the first four nonzero terms of the power series solution of the initial-value problem

$$y'' + (\sin x)y = 0; \quad y(0) = 1, \quad y'(0) = 0$$

8.3 Solutions Near a Regular Singular Point; Method of Frobenius

The differential equation

$$y'' + p_1(x)y' + p_2(x)y = 0 \tag{8.13}$$

has power series solutions about an ordinary point x_0. If x_0 is a singular point, that is, if one of p_1 or p_2 (or both p_1 and p_2) is not analytic at x_0, solutions similar to power series solutions may exist near x_0 if p_1 and p_2 resemble analytic functions at x_0. The following definition divides the singular points of (8.13) into two classes.

Definition A singular point x_0 of (8.13) is called *regular* if and only if the functions defined by $(x - x_0)p_1(x)$ and $(x - x_0)^2 p_2(x)$ are analytic at x_0. A point that is not a regular singular point of (8.13) is termed an *irregular singular point*.

EXAMPLE 1. The differential equation

$$y'' + x(x - 3)y' + \frac{1}{x^3(x - 3)^2} y = 0$$

has singular points at $x = 0$ and at $x = 3$.

The point $x = 3$ is a regular singular point since

$$(x - 3)[x(x - 3)] = x(x - 3)^2 \quad \text{and} \quad (x - 3)^2 \left[\frac{1}{x^3(x - 3)^2} \right] = \frac{1}{x^3}$$

are both analytic at $x = 3$.

The point $x = 0$ is an irregular singular point since

$$x^2 \left[\frac{1}{x^3(x - 3)^2} \right] = \frac{1}{x(x - 3)^2}$$

is not analytic at $x = 0$.

EXAMPLE 2. Legendre's DE

$$y'' - \frac{2x}{1 - x^2} y' + \frac{p(p + 1)}{1 - x^2} y = 0$$

has singular points at $x = 1$ and at $x = -1$. The point $x = 1$ is regular since

$$(x - 1) \left[\frac{-2}{1 - x^2} \right] = \frac{2x}{x + 1} \quad \text{and} \quad (x - 1)^2 \left[\frac{p(p + 1)}{1 - x^2} \right] = \frac{(1 - x)p(p + 1)}{1 + x}$$

are both analytic at $x = 1$. In similar fashion, the point $x = -1$ is easily shown to be a regular singular point.

Without loss of generality, the study of regular singular points can be restricted to a regular singular point at $x = 0$. If $x = x_0 \neq 0$ is a regular singular point of a DE, the substitution $t = x - x_0$ transforms the DE into a second DE having a regular singular point at $t = 0$.

Although most DE encountered in practice have few singular points, the behavior of solutions near a singular point is nevertheless extremely important. In a physical problem a singular point often corresponds to a location where something unusual happens, and primary interest is often centered on what happens in the neighborhood of such a location.

If $xp_1(x)$ and $x^2p_2(x)$ define analytic functions at $x = 0$, it is reasonable to suspect that (8.13) may possess solutions near $x = 0$ of the form

$$y = x^r \sum_{n=0}^{+\infty} c_n x^n \tag{8.18}$$

where r is a real number. The method of producing solutions of the form (8.18) is similar to the power series method of undetermined coefficients. It was introduced by the German mathematician Lazarus Fuchs (1833–1902) and later refined by Ferdinand Georg Frobenius (1849–1917), also a German. The method is called the *method of Frobenius* and solutions of the form (8.18) are referred to as *Frobenius series*, or *Frobenius-type*, *solutions*.

We make the customary assumption that $c_0 \neq 0$ in (8.18). This involves no loss of generality and merely assumes that in the standard form of a

Frobenius series, the lowest power of x appearing has been factored from the series. For example,

$$x^{1/2} + x^{3/2} + \frac{x^{5/2}}{2!} + \cdots = x^{1/2}e^x$$

would, in standard Frobenius form, be written

$$x^{1/2}\left(1 + x + \frac{x^2}{2!} + \cdots\right)$$

with $c_0 = 1 \neq 0$.

We shall seek Frobenius-type solutions valid on $0 < x < h$ where $h > 0$. Solutions on $-h < x < 0$ can be investigated by transforming the given DE by the substitution $t = -x$ and studying solutions of the transformed DE on $t > 0$.

Before stating an existence theorem for Frobenius-type solutions, we present an example illustrating the method of Frobenius.

EXAMPLE 3. By the method of Frobenius, solve

$$y'' - \frac{1}{2x}y' + \frac{1 - x^2}{2x^2}y = 0 \tag{8.19}$$

Solution: Since

$$x\left(\frac{-1}{2x}\right) = -\frac{1}{2} \quad \text{and} \quad x^2\left(\frac{1 - x^2}{2x^2}\right) = \frac{1 - x^2}{2}$$

are analytic at $x = 0$, $x = 0$ is a regular singular point of (8.19). Assume that (8.19) has a solution of the form

$$y = x^r \sum_{n=0}^{+\infty} c_n x^n = \sum_{n=0}^{+\infty} c_n x^{r+n}, \quad c_0 \neq 0 \tag{8.20}$$

Substituting y from (8.20),

$$y' = \sum_{n=0}^{+\infty} c_n(r + n)x^{r+n-1} \quad \text{and} \quad y'' = \sum_{n=0}^{+\infty} c_n(r + n)(r + n - 1)x^{r+n-2}$$

into (8.19), we obtain

$$2x^2y'' - xy' + y - x^2y = \sum_{n=0}^{+\infty} 2c_n(r + n)(r + n - 1)x^{r+n}$$

$$- \sum_{n=0}^{+\infty} c_n(r + n)x^{r+n} + \sum_{n=0}^{+\infty} c_n x^{r+n}$$

$$- \sum_{n=0}^{+\infty} c_n x^{r+n+2} = 0 \tag{8.21}$$

The coefficient of x^r, the lowest-appearing power of x, is obtained by setting $n = 0$ in the first three summations. Setting this coefficient equal to zero, we obtain the equation

$$2c_0r(r - 1) - c_0r + c_0 = 0$$

known as the *indicial equation* of the DE (8.19). Writing the indicial equation in the form $c_0(2r - 1)(r - 1) = 0$, we conclude that since $c_0 \neq 0$, the only possible values of r for which a Frobenius solution can exist are $r_1 = 1$ and $r_2 = \frac{1}{2}$. The roots r_1 and r_2 of the indicial equation are called the *indicial roots*, or the *exponents*, of the DE.

Setting $n = 1$ in the first three summations of (8.21), and equating the coefficient of x^{r+1} to zero, we obtain

$$2c_1(r + 1)r - c_1(r + 1) + c_1 = 0$$

or

$$c_1r(2r + 1) = 0 \tag{8.22}$$

Since r and $2r + 1$ are nonzero for both $r = 1$ and $r = \frac{1}{2}$, we must conclude that $c_1 = 0$.

To obtain a recursion formula for the Frobenius coefficients, we rewrite (8.21) in the form

$$[2c_0r(r - 1) - c_0r + r_0]x^r$$
$$+ [2c_1(r + 1)r - c_1(r + 1) + c_1]x^{r+1}$$

$$+ \sum_{n=2}^{+\infty} [2c_n(r + n)(r + n - 1) - c_n(r + n) + c_n]x^{r+n} - \sum_{n=0}^{+\infty} c_nx^{r+n+2} = 0$$

Replacing n by $n + 2$ in the summation involving x^{r+n}, and setting the coefficient of x^{r+n+2} equal to zero, we obtain

$$2c_{n+2}(r + n + 2)(r + n + 1) - c_{n+2}(r + n + 2) + c_{n+2} - c_n = 0$$

or

$$c_{n+2} = \frac{c_n}{2(r + n + 2)(r + n + 1) - (r + n + 2) + 1} \tag{8.23}$$

The constant c_0 can be assigned any value different from zero. The coefficients $c_2, c_4, c_6, \ldots$ are determined from recursion formula (8.23). Since $c_1 = 0$, it follows from (8.23) that $c_3, c_5, c_7, \ldots$ are all zero.

Solutions y_1 and y_2, corresponding to $r_1 = 1$ and $r_2 = \frac{1}{2}$, are now found separately. With $r = 1$, (8.23) becomes

$$c_{n+2} = \frac{c_n}{2(n + 3)(n + 2) - (n + 3) + 1}$$

from which we obtain

$$c_2 = \frac{c_0}{10}$$

$$c_4 = \frac{c_2}{36} = \frac{c_0}{360}$$

$$c_6 = \frac{c_4}{78} = \frac{c_0}{28,080}$$

yielding, after we set $c_0 = A$,

$$y_1(x) = Ax\left(1 + \frac{x^2}{10} + \frac{x^4}{360} + \frac{x^6}{28,080} + \cdots\right)$$

With $r = \frac{1}{2}$, (8.23) becomes

$$c_{n+2} = \frac{c_n}{2\left(n + \frac{5}{2}\right)\left(n + \frac{3}{2}\right) - \left(n + \frac{5}{2}\right) + 1}$$

from which we obtain

$$c_2 = \frac{c_0}{6}$$

$$c_4 = \frac{c_2}{28} = \frac{c_0}{168}$$

$$c_6 = \frac{c_4}{66} = \frac{c_0}{11,088}$$

$$\vdots$$

yielding, after we set $c_0 = B$,

$$y_2(x) = Bx^{1/2}\left(1 + \frac{x^2}{6} + \frac{x^4}{168} + \frac{x^6}{11,088} + \cdots\right)$$

The solution y_1 is a power series and could have been obtained by the method of power series. This situation prevails whenever an exponent of the DE is a nonnegative integer.

To verify that a Frobenius series is a solution of a DE on an interval $0 < x < h$, we can find the interval of convergence of the series and substitute the series for y, y', and y'' into the DE as in the verification of a power series solution. Term-by-term differentiation of a Frobenius series inside its interval of convergence can be shown to be valid.

Under certain conditions, it can be shown that a Frobenius-type solution exists even when the exponents of DE (8.13) are complex numbers. We omit consideration of this possibility and state an existence theorem that holds when r_1 and r_2 are real.

Theorem 8-II   Let $x = 0$ be a regular singular point of DE (8.13), assume that the power series expansions of $xp_1(x)$ and $x^2p_2(x)$ converge in an interval $[0, R)$ where $R > 0$, and let the indicial equation of (8.13) have real roots r_1 and r_2, with $r_1 \geq r_2$. Then,
(i) If $r_1 - r_2$ is not zero or a positive integer, the DE has two linearly independent Frobenius-type solutions

$$y_1(x) = x^{r_1} \sum_{n=0}^{+\infty} a_n x^n \quad \text{and} \quad y_2(x) = x^{r_2} \sum_{n=0}^{+\infty} b_n x^n$$

both valid on $(0, R)$.

(ii) If $r_1 = r_2$, the DE has a Frobenius-type solution

$$y_1(x) = x^{r_1} \sum_{n=0}^{+\infty} c_n x^n$$

valid on $(0, R)$.

(iii) If $r_1 - r_2$ is a positive integer, the DE has a Frobenius-type solution

$$y_1(x) = x^{r_1} \sum_{n=0}^{+\infty} c_n x^n$$

valid on $(0, R)$.

In case i, if we set $a_0 = 1$ and $b_0 = 1$, a linear combination $y(x) = Ay_1(x) + By_2(x)$ of y_1 and y_2 defines a complete solution of the DE. In Example 3, $xp_1(x) = -\frac{1}{2}$ and $x^2 p_2(x) = \frac{1}{2}(1 - x^2)$ are analytic for all x. Hence, with $c_0 = 1$ in $y_1(x)$ and in $y_2(x)$, $y = Ay_1(x) + By_2(x)$ defines a complete solution of DE (8.19) and is valid on $(0, +\infty)$.

In case ii, it is clear that only one Frobenius-type solution is available since the indicial equation is satisfied by a single number r_1. It can be shown that a second solution y_2, such that y_1 and y_2 are linearly independent, can be found by the formula of Section 4.5:

$$y_2(x) = y_1(x) \int \frac{\exp\left(\int - p_1(x)\, dx\right)}{[y_1(x)]^2}\, dx \qquad (8.24)$$

More specifically, it can be shown that $y_2(x)$ is of the form

$$y_2(x) = y_1(x) \ln x + x^{r_1 + 1} \sum_{n=0}^{+\infty} d_n x^n$$

valid at least on $(0, R)$. A complete solution of the DE is then given by $y(x) = Ay_1(x) + By_2(x)$, also valid at least on $(0, R)$.

Simple illustrations of Theorem 8-II are furnished by the Euler DE

$$x^2 y'' + \alpha x y' + \beta y = 0$$

studied in Section 4.11. Writing the DE in normal form

$$y'' + \alpha x^{-1} y' + \beta x^{-2} y = 0$$

we see that $x = 0$ is a regular singular point. Let us seek solutions of the Euler DE of the form $y = x^r$. A solution of this type is a single-term Frobenius-type solution $y = x^r \sum_{n=0}^{+\infty} c_n x^n$ with $c_0 = 1$ and $c_n = 0$ for $n > 1$. Substitution of

$y = x^r$, $y' = rx^{r-1}$, and $y'' = r(r-1)x^{r-2}$ into the Euler DE yields

$$x^2[r(r-1)x^{r-2}] + \alpha x(rx^{r-1}) + \beta(x^r) = 0$$

Setting the coefficient of x^r equal to zero, we obtain

$$r(r-1) + \alpha r + \beta = 0$$

It is easy to see that this is the same equation as the indicial equation we would have obtained if we had sought solutions of the form $y = x^r \sum_{n=0}^{+\infty} c_n x^n$ instead of solutions of the form $y = x^r$. The lowest power of x appearing would be x^r and it is the coefficient of that power that we have equated to zero.

To illustrate case ii of Theorem 8-II, we write the equation for r in the form

$$r^2 + (\alpha - 1)r + \beta = 0$$

Setting $\alpha = 5$ and $\beta = 4$, we obtain the equation

$$r^2 + 4r + 4 = (r + 2)^2 = 0$$

having the repeated root $r_1 = r_2 = -2$.

It is a simple exercise to verify that $y_1 = x^{-2}$ is a solution of the DE

$$x^2 y'' + 5xy' + 4y = 0$$

on $(0, +\infty)$.

Applying the method of reduction of order to find a second solution y_2, we set $y_1(x) = x^{-2}$ and $p_1(x) = 5x^{-1}$ in (8.24) to obtain

$$y_2(x) = x^{-2} \int \frac{\exp \int -5x^{-1}\, dx}{(x^{-2})^2}\, dx$$

$$= x^{-2} \int \frac{\exp(-5\ln x)}{x^{-4}}\, dx = x^{-2} \int \frac{x^{-5}\, dx}{x^{-4}} = x^{-2} \ln x$$

A complete solution of $x^2 y'' + 5xy' + 4y = 0$ is given by $y = Ax^{-2} + Bx^{-2} \ln x$, valid on $(0, +\infty)$.

To illustrate case iii, we choose $\alpha = 4$ and $\beta = -4$ to obtain $r^2 + 3r - 4 = (r + 4)(r - 1) = 0$.

Then $r_1 = 1$, $r_2 = -4$, $y_1 = x$, $y_2 = x^{-4}$, and a complete solution of $x^2 y'' + 4xy' - 4y = 0$ is given by $y = Ax + Bx^{-4}$, valid on $(0, +\infty)$. It is interesting to note that y_1 and y_2 behave quite differently near $x = 0$. As $x \to 0^+$, $y_1 \to 0$ whereas $y_2 \to +\infty$. It is sometimes possible in an application to rule out one part of a complete solution because of physical considerations.

In case iii, the larger indicial root r_1 always yields a Frobenius-type solution y_1. The smaller root r_2 may yield a complete solution of the DE, that is, a linear combination of two linearly independent solutions of the DE. On the other hand, using r_2 in the method of Frobenius may fail to produce a solution or may merely reproduce the solution y_1, corresponding to r_1. In this case, a solution y_2 linearly independent of y_1 can be found from (8.24), as in case ii. Then $y_2(x)$ has the form

$$y_2(x) = x^{r_2} \sum_{n=0}^{+\infty} k_n x^n + Cy_1(x) \ln x, \quad k_1 = 1$$

where the constant C may be zero.

The solution y_2 and the complete solution given by $y(x) = Ay_1(x) + By_2(x)$ are valid on $(0, R)$.

For a proof of Theorem 8-II and a more detailed discussion of cases i, ii, and iii, see Reference 8.3.

EXAMPLE 4. Apply the method of Frobenius to the DE $xy'' + y' + 4xy = 0$, which has a regular singular point at $x = 0$.

Solution:

$$xy'' + y' + 4xy = x \sum_{n=0}^{+\infty} c_n(r + n)(r + n - 1)x^{r+n-2}$$

$$+ \sum_{n=0}^{+\infty} c_n(r + n)x^{r+n-1} + 4x \sum_{n=0}^{+\infty} c_n x^{r+n}$$

$$= \sum_{n=0}^{+\infty} c_n(r + n)(r + n - 1)x^{r+n-1}$$

$$+ \sum_{n=0}^{+\infty} c_n(r + n)x^{r+n-1} + \sum_{n=0}^{+\infty} 4c_n x^{r+n+1} = 0$$

The indicial equation is obtained by setting $n = 0$ in the first two summations and equating to zero the coefficient of x^{r-1}, the lowest power of x appearing. This yields

$$c_0 r(r - 1) + c_0 r = 0$$

or

$$c_0 r^2 = 0$$

from which we conclude that $r_1 = r_2 = 0$.

The recursion formula is obtained by replacing n by $n + 2$ in the first two summations, and then equating to zero the coefficient of x^{r+n+1}. This yields

$$c_{n+2}(r + n + 2)(r + n + 1) + c_{n+2}(r + n + 2) + 4c_n = 0$$

or

$$c_{n+2} = \frac{-4c_n}{(r + n + 2)(r + n + 1) + (r + n + 2)}$$

$$= \frac{-4c_n}{(r + n + 2)^2}$$

If we set $r = 0$, the recursion formula becomes

$$c_{n+2} = \frac{-4c_n}{(n + 2)^2}$$

from which we obtain

$$c_2 = \frac{-4c_0}{2^2} = -c_0, \quad c_4 = \frac{-4c_2}{4^2} = \frac{4^2 c_0}{4^2 \cdot 2^2}$$

$$c_6 = \frac{-4c_4}{6^2} = \frac{-4^3 c_0}{6^2 \cdot 4^2 \cdot 2^2}, \ldots, c_{2n} = \frac{(-1)^n 4^n c_0}{2^{2n}(n!)^2} = \frac{(-1)^n c_0}{(n!)^2}, \ldots$$

It is easy to show that, as in Example 3, $c_n = 0$ when n is odd. See Problem 12. Hence,

$$y_1(x) = c_0 \sum_{n=0}^{+\infty} \frac{(-1)^n x^{2n}}{(n!)^2} = c_0 \left(1 - x^2 + \frac{x^4}{4} - \frac{x^6}{36} + \cdots\right)$$

This example illustrates case ii of Theorem 8-II. A solution y_2, independent of y_1, could be obtained from Formula (8.24).

The next example illustrates case iii of Theorem 8-II.

EXAMPLE 5.   The origin is a regular singular point of the DE

$$x^2 y'' + xy' + (x^2 - 4)y = 0 \tag{8.25}$$

Show that the exponents of the DE differ by a positive integer and find the first three nonzero terms of the Frobenius-type solution corresponding to the larger of the two exponents of the DE.

Solution:

$$x^2 y'' + xy' + x^2 y - 4y = \sum_{n=0}^{+\infty} c_n(r + n)(r + n - 1)x^{r+n-2+2}$$

$$+ \sum_{n=0}^{+\infty} c_n(r + n)x^{r+n-1+1} + \sum_{n=0}^{+\infty} c_n x^{r+n+2}$$

$$+ \sum_{n=0}^{+\infty} -4c_n x^{r+n} = 0$$

Setting $n = 0$ in the first, second, and fourth summations, and equating to zero the coefficient of x^r, we obtain the indicial equation

$$c_0 r(r - 1) + c_0 r - 4c_0 = 0$$

or

$$c_0(r^2 - 4) = 0$$

Thus, $r_1 = 2, r_2 = -2$, and the exponents differ by the positive integer $r_1 - r_2 = 4$.

Setting $n = 1$ in the same three summations, and equating to zero the coefficient of x^{r+1}, we obtain

$$c_1(r + 1)r + c_1(r + 1) - 4c_1 = c_1(r^2 + 2r - 3) = 0$$

which implies, since r must be 2 or -2, that $c_1 = 0$.

Replacing n by $n + 2$ in the first, second, and fourth summations, we have

$$\sum_{n=-2}^{+\infty} c_{n+2}(r + n + 2)(r + n + 1)x^{r+n+2} + \sum_{n=-2}^{+\infty} c_{n+2}(r + n + 2)x^{r+n+2}$$

$$+ \sum_{n=0}^{+\infty} c_n x^{r+n+2} + \sum_{n=-2}^{+\infty} -4c_{n+2}x^{r+n+2}$$

Equating to zero the coefficient of x^{r+n+2} in all four summations, we obtain

$$c_{n+2}(r + n + 2)(r + n + 1) + c_{n+2}(r + n + 2) + c_n - 4c_{n+2} = 0$$

leading to the recursion formula

$$c_{n+2} = \frac{-c_n}{(r + n + 2)(r + n + 1) + (r + n + 2) - 4}$$

$$= \frac{-c_n}{(r + n + 2)^2 - 4}$$

For $r = 2$, the recursion formula becomes

$$c_{n+2} = \frac{-c_n}{(n + 4)^2 - 4}$$

Thus, with c_0 nonzero but otherwise arbitrary,

$$c_2 = \frac{-c_0}{12}, \quad c_4 = \frac{-c_2}{32} = \frac{c_0}{384}, \dots$$

Since $c_1 = 0$, it follows that $c_3, c_5, c_7, \dots$ are zero.

Hence, the Frobenius series corresponding to $r = 2$ is given by

$$y = x^2\left(c_0 - \frac{c_0 x^2}{12} + \frac{c_0 x^4}{384} - \cdots\right)$$

$$= c_0\left(x^2 - \frac{x^4}{12} + \frac{x^6}{384} - \cdots\right)$$

The Frobenius-type solution corresponding to $r = -2$ is the same as the solution we have found. See Problem 13. The DE (8.25) is a special case of Bessel's DE, to be discussed briefly in the next section.

In general, the DE $y'' + p_1(x)y' + p_2(x)y = 0$ does not have a power series solution about an irregular singular point $x = x_0$.

8.4 Bessel's Differential Equation

In Section 1.2 we mentioned briefly the DE

$$x^2 y'' + xy' + (x^2 - p^2)y = 0 \tag{8.26}$$

This equation was studied by Daniel Bernoulli (1700–1782) while he was investigating the oscillations of a hanging chain, and also by Leonard Euler (1707–1783) in his analysis of a vibrating membrane. It is called Bessel's equation of order p, however, after Friedrich Wilhelm Bessel (1784–1846), the German astronomer and mathematician, who encountered it while he was studying planetary motion. The parameter p is a nonnegative constant.

Writing (8.26) in the normal form

$$y'' + \frac{1}{x} y' + \frac{x^2 - p^2}{x^2} y = 0$$

we note that

$$x\left(\frac{1}{x}\right) = 1 \quad \text{and} \quad x^2\left(\frac{x^2 - p^2}{x^2}\right) = x^2 - p^2$$

are analytic for all x, and hence $x = 0$ is a regular singular point.

Applying the method of Frobenius, we obtain

$$x^2 y'' + xy' + x^2 y - p^2 y = \sum_{n=0}^{+\infty} c_n(r+n)(r+n-1)x^{r+n}$$

$$+ \sum_{n=0}^{+\infty} c_n(r+n)x^{r+n} + \sum_{n=0}^{+\infty} c_n x^{r+n+2} + \sum_{n=0}^{+\infty} - p^2 c_n x^{r+n} = 0 \quad \textbf{(8.27)}$$

As in Example 5 of Section 8.3, this leads to the indicial equation $c_0(r^2 - p^2) = 0$, and hence the exponents of the DE are $r_1 = p$ and $r_2 = -p$. Again as in Example 5 of Section 8.3, the recursion formula is

$$c_{n+2} = \frac{-c_n}{(r+n+2)^2 - p^2} = \frac{-c_n}{(r+n+2-p)(r+n+2+p)} \quad \textbf{(8.28)}$$

which for $r = p$ becomes

$$c_{n+2} = \frac{-c_n}{(n+2)(2p+n+2)}$$

When we set $n = 1$ in (8.27) and equate to zero the coefficient of x^{r+1}, we obtain

$$c_1(r+1)r + c_1(r+1) - p^2 c_1 = 0 \quad \textbf{(8.29)}$$

When $r = p$, this becomes

$$c_1(2p+1) = 0$$

and hence $c_1 = 0$ for all $p \neq -\frac{1}{2}$. Even in the case $p = -\frac{1}{2}$, it is permissible to take $c_1 = 0$. The solution we are about to obtain will also be valid when $p = -\frac{1}{2}$, as we shall be able to verify directly in Example 1. The constant c_0 is nonzero but otherwise arbitrary. Since $c_1 = 0$, the recursion formula implies that $c_3, c_5, c_7, \ldots$ are all zero. We now obtain

$$c_2 = \frac{-c_0}{2(2p + 2)}$$

$$c_4 = \frac{-c_2}{4(2p + 4)} = \frac{c_0}{2 \cdot 4(2p + 2)(2p + 4)}$$

$$c_6 = \frac{-c_4}{6(2p + 6)} = \frac{-c_0}{2 \cdot 4 \cdot 6(2p + 2)(2p + 4)(2p + 6)}$$

leading to the following Frobenius solution guaranteed by Theorem 8-II:

$$\begin{aligned}
y &= c_0 x^p \left[1 - \frac{x^2}{2^2(p + 1)} + \frac{x^4}{2^4 2!(p + 1)(p + 2)} \right. \\
&\quad \left. - \frac{x^6}{2^6 3!(p + 1)(p + 2)(p + 3)} + \cdots \right] \\
&= c_0 x^p + c_0 x^p \sum_{n = 1}^{+\infty} \frac{(-1)^n x^{2n}}{2^{2n} n!(p + 1)(p + 2) \ldots (p + n)}
\end{aligned} \tag{8.30}$$

To give (8.30) a more convenient form, we usually choose $c_0 = 1/[2^p \Gamma(p + 1)]$, where Γ denotes the generalized factorial, or gamma, function. The resulting function is denoted by

$$J_p(x) = \sum_{n = 0}^{+\infty} \frac{(-1)^n(x/2)^{2n + p}}{n!\Gamma(p + n + 1)} \tag{8.31}$$

and is termed the *Bessel function of the first kind of order p*. The function J_p is a solution of Bessel's DE on $(-\infty, +\infty)$ for all $p \geq 0$.

If $p = 0$, a second solution independent of J_0 can be obtained by (8.24) using

$$y_2(x) = J_0(x) \int \frac{dx}{x[J_0(x)]^2}$$

If $p > 0$ and p is not an integer, the Frobenius series corresponding to $r = -p$ can be obtained by replacing p by $-p$ in (8.31). The resulting function is denoted by

$$J_{-p}(x) = \sum_{n = 0}^{+\infty} \frac{(-1)^n(x/2)^{2n - p}}{n!\Gamma(-p + n + 1)} \tag{8.32}$$

It is possible to define $\Gamma(-p + n + 1)$ appropriately in (8.32) when p is not an integer. The functions J_p and J_{-p} are linearly independent, even in the case where p is half a positive integer and the exponents differ by an integer.

This can be shown by using the fact that $J_{-p}(x)$, unlike $J_p(x)$, contains negative powers of x. Hence, J_{-p} is unbounded near $x = 0$, whereas J_p is bounded. Thus, the ratio $J_{-p}(x)/J_p(x)$ cannot be constant on any interval, $0 < x < h$.

It follows from Theorem 8-II that when p is not an integer, a complete solution of Bessel's DE is given by

$$y(x) = AJ_p(x) + BJ_{-p}(x) \tag{8.33}$$

When p is an integer, it is customary to denote $J_p(x)$ and $J_{-p}(x)$ by $J_n(x)$ and $J_{-n}(x)$. From (8.31) and (8.32) it can be shown (see Problem 14) that

$$J_{-n}(x) = (-1)^n J_n(x) \tag{8.34}$$

Thus, the ratio $J_{-n}(x)/J_n(x)$ is either $+1$ or -1, and hence J_{-n} and J_n are not linearly independent.

It can be shown (see Reference 8.9) that the unusual choices $A = \cot p\pi$, $B = -\csc p\pi$, in (8.33) make it possible to express a complete solution of Bessel's DE in a simple way. The function denoted by

$$Y_p(x) = \frac{\cos p\pi J_p(x) - J_{-p}(x)}{\sin p\pi} \tag{8.35}$$

is independent of J_p and is called the *Bessel function of the second kind of order p.*

A complete solution of Bessel's DE valid for all $x \neq 0$ and *all p* is given by

$$y(x) = c_1 J_p(x) + c_2 Y_p(x) \tag{8.36}$$

where $Y_p(x)$ is defined by (8.35) when p is not an integer and by $Y_n(x) = \lim_{p \to n} Y_p(x)$ when p is an integer n.

The literature on Bessel's DE and Bessel's functions is extensive. The classic treatise on these subjects is Reference 8.9. For less detailed treatments see References 8.4, 8.7, and 8.10.

The Bessel functions of orders 0 and 1 have many important applications. These functions are given by

$$J_0(x) = \sum_{n=0}^{+\infty} \frac{(-1)^n (x/2)^{2n}}{(n!)^2} = 1 - \frac{x^2}{2^2} + \frac{x^4}{2^2 \cdot 4^2} - \frac{x^6}{2^2 \cdot 4^2 \cdot 6^2} + \cdots \tag{8.37}$$

and

$$J_1(x) = \sum_{n=0}^{+\infty} \frac{(-1)^n (x/2)^{2n+1}}{n!(n+1)!} = \frac{x}{2} - \frac{1}{1!2!} \left(\frac{x}{2}\right)^3 + \frac{1}{2!3!} \left(\frac{x}{2}\right)^5 - \cdots \tag{8.38}$$

The graphs of J_0 and J_1, shown in Figure 8.1, resemble damped cosine and sine curves. The zeros of these functions, given graphically by the x intercepts in Figure 8.1, are significant in applications.

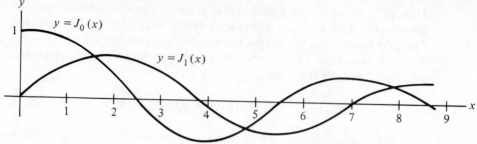

Fig. 8.1

The Bessel functions are related by numerous remarkable formulas and identities, such as

$$J'_0(x) = -J_1(x) \quad \text{and} \quad \frac{d}{dx}[xJ_1(x)] = xJ_0(x)$$

a few of which are exploited in the Problem List.

EXAMPLE.   Find a complete solution of Bessel's DE of order $\frac{1}{2}$

$$x^2y'' + xy' + \left(x^2 - \frac{1}{4}\right)y = 0 \qquad\qquad \textbf{(8.39)}$$

Solution: Setting $p^2 = \frac{1}{4}$ and $r = -\frac{1}{2}$, the smaller of the exponents $\frac{1}{2}$ and $-\frac{1}{2}$, in (8.29) we obtain $c_1(0) = 0$, and hence we may allow c_1 as well as c_0 to be an arbitrary constant.

  With $r = -\frac{1}{2}$ and $p = \frac{1}{2}$, the recursion formula (8.28) becomes

$$c_{n+2} = \frac{-c_n}{(n+2)(n+1)}$$

from which we obtain

$$c_2 = \frac{-c_0}{2!}$$

$$c_4 = \frac{-c_2}{4 \cdot 3} = \frac{c_0}{4!}$$

$$c_6 = \frac{-c_4}{6 \cdot 5} = \frac{-c_0}{6!}$$

$$c_{2n} = \frac{(-1)^n c_0}{(2n)!}$$

and

$$c_3 = \frac{-c_1}{3!}$$

$$c_5 = \frac{-c_3}{5 \cdot 3} = \frac{c_1}{5!}$$

$$c_7 = \frac{-c_5}{7 \cdot 6} = \frac{-c_1}{7!}$$

$$c_{2n+1} = \frac{(-1)^n c_1}{(2n+1)!}$$

The function given by

$$y(x) = c_0 x^{-1/2} \sum_{n=0}^{+\infty} \frac{(-1)^n x^{2n}}{(2n)!} + c_1 x^{-1/2} \sum_{n=0}^{+\infty} \frac{(-1)^n x^{2n+1}}{(2n+1)!}$$

$$= c_0 x^{-1/2} \cos x + c_1 x^{-1/2} \sin x \tag{8.40}$$

is a complete solution of (8.39), valid for all $x \neq 0$. Equation (8.40) is a linear combination of two linearly independent functions since the ratio $(x^{-1/2} \sin x)/(x^{-1/2} \cos x) = \tan x$ when $x^{-1/2} \cos x \neq 0$ is not constant on any interval. This example furnishes an illustration of case iii of Theorem 8-II, in which the smaller exponent yields a complete solution of the DE. It is interesting to note that (8.40) defines an elementary function.

We shall see in Chapter 10 how functions such as Bessel functions arise in applications. A physical situation or process is often governed by a partial DE. Solving the partial DE frequently requires solving an ordinary DE like the Bessel DE or the Legendre DE.

Problem List 8.2

1. List (i) the singular points and (ii) the regular singular points of the following DE.
   (a) $x^3(x - 4)y'' + 2y' + 5y = 0$
   (b) $x(x + 1)^3 y'' - (x + 1)^3 y' + 6y = 0$
   (c) $(1 + x^2)y'' - 2y' - (1 + x^2)y = 0$
   (d) $x^2 y'' - 3y' + 2y = 0$
   (e) $x^2(x^2 - 1)(x + 1)y'' + x(x - 1)^2 y' - (x + 1)y = 0$
   (f) $xy'' + (1 - x)y' + \lambda y = 0$ (Laguerre's DE)
2. Given the DE $x^2(x - 1)^2 y'' + (x - 1)y' + y = 0$, classify each of the points $x = 0$, $x = 1$, and $x = 3$ as ordinary, regular singular, or irregular singular.
3. Find the exponents for a Frobenius-type solution of the DE $x(1 - x)y'' + 2y' + 2y = 0$. Do not solve the DE.

4. Use the method of Frobenius and the positive exponent $r = \frac{1}{2}$ of the DE $x^2y'' + xy' + (x^2 - \frac{1}{4})y = 0$ to find a solution of the DE. Compare the result with that of Example 1 of Section 8.4.

5. Obtain a Frobenius-type series solution of the DE $xy'' + y' + xy = 0$. Write three nonzero terms of the series.

6. Show that the DE $xy'' + 2y' + 2xy = 0$ has exponents 0 and -1. Use the exponent -1 to obtain a Frobenius-type solution. Write the first four nonzero terms of a complete solution of the DE.

7. Find the exponents of the DE $2xy'' - y' - xy = 0$. Do not solve the DE.

8. Find the exponents of the DE $9x^2y'' + (x + 2)y = 0$. Do not solve the DE.

9. Use the larger of the exponents of $xy'' - 2y = 0$ to find a Frobenius-type solution of the DE. Write four nonzero terms of the solution.

10. Show that except for the trivial solution given by $y \equiv 0$, the DE $x^3y'' + y = 0$ has no Frobenius-type solution.

11. Show that $r(r - 1) + p_1(0)r + p_2(0) = 0$ is the indicial equation associated with the DE

$$y'' + \frac{p_1(x)}{x} y' + \frac{p_2(x)}{x^2} y = 0$$

12. Show in Example 4 of Section 8.3 that $c_n = 0$ when n is odd.

13. Verify in Example 5 of Section 8.3 that the Frobenius solution corresponding to the exponent $r = -2$ is the same as the solution corresponding to the exponent $r = 2$.

14. Show that $J_{-n}(x) = (-1)^n J_n(x)$.

15. Show that $J_0'(x) = -J_1(x)$.

16. Show that $(d/dx)[x^p J_p(x)] = x^p J_{p-1}(x)$. Assume that term-by-term differentiation is valid.

17. Show that the DE $y'' + x^{-1}y' + \lambda^2 y = 0$ has a solution given by $y(x) = cJ_0(\lambda x)$.

18. Show that the exponents of the DE $y'' + x^{-1}y' + x^{-2}(x + 1)y = 0$ are pure imaginary numbers.

19. Solve $y'' + 2x^{-1}y' = 0$ by the method of Frobenius.

20. Solve $2xy' - y = 0$ by the method of Frobenius.

21. Solve $xy' + y = 0$ by the method of Frobenius.

22. Find a complete solution of $4xy'' + 2y' + y = 0$.

23. Find $J_0(1)$ and $J_1(1)$ correct to four decimal places.

24. Find the smallest positive root of $J_0(x) = 0$ correct to one decimal place.

8.5 Series Solutions by the Taylor or Maclaurin Series Method

The power series method of Section 8.1 of solving a DE consisted of assuming that a solution of the form $y(x) = \sum_{n=0}^{+\infty} c_n(x - x_0)^n$ existed, substituting this series and those for $y'(x)$, $y''(x)$, ... into the DE, and finding the coefficients c_n from recursion relations. An alternative procedure, known as the *Taylor series method*, computes the coefficients in the Taylor series

(Maclaurin series if $x_0 = 0$)

$$c_n = \frac{y^{(n)}(x_0)}{n!}$$

directly from the DE. The Taylor series method is often simpler than the power series method, especially when initial-value problems are being solved and nonlinear DE are involved. The method will be illustrated by examples.

NOTE: The expression $y^{(0)}(x_0)/0!$ denotes $y(x_0)/1 = y(x_0)$.

EXAMPLE 1. Solve $y' = y$ by the Taylor series method, with $x_0 = 0$.

Solution: Assume $y(0) = y_0$. From

$$y'(x) = y(x), \qquad\qquad y'(0) = y(0) = y_0$$

$$y''(x) = y'(x) = y(x), \quad y''(0) = y(0) = y_0$$

$$y^{(n)}(x) = y(x), \qquad\qquad y^{(n)}(0) = y(0) = y_0$$

we obtain

$$y(x) = \sum_{n=0}^{+\infty} \frac{y^{(n)}(0)}{n!} x^n$$

$$= \sum_{n=0}^{+\infty} \frac{y_0 x^n}{n!}$$

$$= y_0\left(1 + x + \frac{x^2}{2!} + \frac{x^3}{3!} + \cdots\right)$$

$$= y_0 e^x$$

EXAMPLE 2. Solve the initial-value problem

$$y'' = 2x - y; \quad y(0) = 2, \quad y'(0) = 0$$

Solution:

$$y''(x) = 2x - y(x) \quad \text{implies that} \quad y''(0) = -y(0) = -2$$

$$y'''(x) = 2 - y'(x) \quad \text{implies that} \quad y'''(0) = 2 - y'(0) = 2$$

$$y^{(iv)}(x) = -y''(x) \quad \text{implies that} \quad y^{(iv)}(0) = -y''(0) = 2$$

$$y^{(v)}(x) = -y'''(x) \quad \text{implies that} \quad y^{(v)}(0) = -y'''(0) = -2$$

$$y^{(n)}(x) = -y^{(n-2)}(x) \quad \text{implies that} \quad y^{(n)}(0) = -y^{(n-2)}(0) \quad \text{for } n \geq 2$$

Hence,

$$y(x) = \sum_{n=0}^{+\infty} \frac{y^{(n)}(0)}{n!} x^n = 2\left(1 - \frac{x^2}{2!} + \frac{x^3}{3!} + \frac{x^4}{4!} - \frac{x^5}{5!} - \frac{x^6}{6!} + \frac{x^7}{7!} + \frac{x^8}{8!} - \cdots\right)$$

It is fairly easy to show (see Problem 7) that $y(x) = 2(x + \cos x - \sin x)$.

EXAMPLE 3. Obtain the first five terms of the Taylor series solution about $x = 1$ of the DE $y'' = y + 2x$, given that $y(1) = 2$ and $y'(1) = 4$.

Solution:

$$y''(x) = y(x) + 2x \quad \text{implies that} \quad y''(1) = y(1) + 2 = 4$$
$$y'''(x) = y'(x) + 2 \quad \text{implies that} \quad y'''(1) = y'(1) + 2 = 6$$
$$y^{(iv)}(x) = y''(x) \quad \text{implies that} \quad y^{(iv)}(1) = y''(1) = 4$$

Hence,

$$y(x) = \sum_{n=0}^{+\infty} \frac{y^{(n)}(1)}{n!}(x-1)^n$$

$$= 2 + 4(x-1) + \frac{4(x-1)^2}{2!} + \frac{6(x-1)^3}{3!} + \frac{4(x-1)^4}{4!} + \cdots$$

NOTE: If values of $y(x)$, $y'(x)$, $y''(x)$, ... are given at $x = x_0 \neq 0$, a Taylor series solution about $x = x_0$ is sought. If the substitution $t = x - x_0$ is made, values of $y(t)$, $y'(t)$, $y''(t)$, ... are then specified at $t = 0$, and a Maclaurin solution of the transformed DE is sought. Hence, x_0 can be taken as zero without loss of generality.

Examples 1, 2, and 3 can be solved by the methods of Chapter 4. The next example applies the Taylor series method to a linear DE with nonconstant coefficients.

EXAMPLE 4. Find seven terms of the Maclaurin solution of the initial-value problem (Example 1 of Section 8.2) $y'' + xy = 2$; $y(0) = 1$, $y'(0) = 1$.

Solution:

$$y''(x) = -xy(x) + 2 \text{ implies that } y''(0) = 2$$
$$y'''(x) = -xy'(x) - y(x) \text{ implies that } y'''(0) = -y(0) = -1$$
$$y^{(iv)}(x) = -xy''(x) - 2y'(x) \text{ implies that } y^{(iv)}(0) = -2y'(0) = -2$$
$$y^{(v)}(x) = -xy'''(x) - 3y''(x) \text{ implies that } y^{(v)}(0) = -3y''(0) = -6$$
$$y^{(vi)}(x) = -xy^{(iv)}(x) - 4y'''(x) \text{ implies that } y^{(vi)}(0) = -4y'''(0) = 4$$
$$y^{(vii)}(x) = -xy^{(v)}(x) - 5y^{(iv)}(x) \text{ implies that } y^{(vii)}(0) = -5y^{(iv)}(0) = 10$$

$$y^{(n)}(x) = -xy^{(n-2)}(x) - (n-2)y^{(n-3)}(x) \text{ implies that } y^{(n)}(0)$$
$$= -(n-2)y^{(n-3)}(0) \quad [n \geq 3]$$

Hence,

$$y(x) = \sum_{n=0}^{+\infty} \frac{y^{(n)}(0)}{n!} x^n$$

$$= 1 + x + \frac{2x^2}{2!} - \frac{x^3}{3!} - \frac{2x^4}{4!} - \frac{6x^5}{5!} + \frac{4x^6}{6!} + \cdots$$

which agrees with the result obtained in Section 8.2.

The Taylor series method does have certain drawbacks. In many cases it is difficult if not impossible to obtain a simple expression for $c_n = y^{(n)}(x_0)/n!$.

Another difficulty is that even if a formal Taylor series solution is obtained, the question of the size of the interval of convergence of the solution is a difficult one. In certain cases, Theorem 8-I provides an answer. The situation is much more complicated when one is dealing with nonlinear DE. To indicate the complexity of convergence, we present an existence theorem and an example involving a nonlinear DE. For a proof of the theorem and a more detailed discussion, see Reference 8.5.

Theorem 8-III Let f be a polynomial in x and y. Then the initial-value problem $y' = f(x, y)$, $y(0) = y_0$, has a unique analytic solution given by a Maclaurin series of the form

$$y(x) = y_0 + \sum_{n=1}^{+\infty} c_n x^n$$

valid in some interval $|x| < R$ where $R > 0$.

The coefficients c_n in Theorem 8-III can be found by the Taylor series method. The real difficulty is that it is not easy to determine or even to estimate the number R. The following example, in which R can be made arbitrarily small, illustrates this situation.

EXAMPLE 5. Solve the DE $y' = 1 + y^2$, given that $y(0) = y_0$.

Solution: If $y_0 = 0$,

$$y' = 1 + y^2$$

$$y'' = 2yy' = 2y(1 + y^2)$$

$$y''' = 2y(2yy') + 2y'(1 + y^2) = y'(4y^2 + 2 + 2y^2) = 2(1 + y^2)(1 + 3y^2)$$

$$y^{(iv)} = 2(1 + y^2)(6yy') + 2(1 + 3y^2)(2yy')$$

$$= 4yy'[3(1 + y^2) + (1 + 3y^2)]$$

$$= 8y(1 + y^2)(2 + 3y^2) = 16y + 40y^3 + 24y^5$$

and

$$y^{(v)} = (16 + 120y^2 + 120y^4)y' = (16 + 120y^2 + 120y^4)(1 + y^2)$$

yield

$$y(0) = 0, \qquad y'(0) = 1, \qquad y''(0) = 0$$

$$y'''(0) = 2, \qquad y^{(iv)}(0) = 0, \qquad y^{(v)}(0) = 16$$

Values of $y^{(n)}(0)$ for $n > 5$ can be computed similarly, but the required computations become increasingly complex. It is not easy to find a general expression for $y^{(n)}(x)$. The series solution guaranteed by Theorem 8-III is given, up to and including the term involving x^5, by

$$y(x) = y_0(0) + \frac{y'(0)}{1!}x + \frac{y''(0)}{2!}x^2 + \frac{y'''(0)}{3!}x^3 + \frac{y^{(iv)}(0)}{4!}x^4 + \frac{y^{(v)}(0)}{5!}x^5 + \cdots$$

$$= 0 + x + 0 + \frac{1}{3}x^3 + 0 + \frac{2}{15}x^5 + \cdots \tag{8.41}$$

The DE $y' = 1 + y^2$, with $y(0) = y_0$, is easily solved by separating the variables. From

$$\int_{y_0}^{y} \frac{du}{1 + u^2} = \int_0^x dv \quad \text{we obtain} \quad \text{Tan}^{-1}\, y - \text{Tan}^{-1}\, y_0 = x$$

Taking the tangent of both sides of $\text{Tan}^{-1}\, y = x + \text{Tan}^{-1}\, y_0$ yields

$$y = y(x) = \frac{\text{Tan}\, x + y_0}{1 - y_0 \,\text{Tan}\, x} \tag{8.42}$$

If $y_0 = 0$, then $y(x) = \tan x$, and hence the series in (8.41) is the Maclaurin series for tan x. This series is known to have radius of convergence $R = \pi/2$, the minimum distance from the origin to a point at which tan x is undefined. If we choose $y_0 > 0$, the solution to the initial-value problem is undefined when the denominator in (8.42) is zero. This occurs when $\tan x = 1/y_0$ or $x = \text{Tan}^{-1}(1/y_0)$. The series solution of the initial-value problem then has radius of convergence $R = \text{Tan}^{-1}(1/y_0)$, which can be made arbitrarily small by choosing y_0 sufficiently large. These results for $y_0 = 0$ and for $y_0 > 0$ are not apparent from examination of the well-behaved polynomial $f(x, y) = 1 + y^2$.

In this example, it is possible to base an analysis of the size of the interval of convergence of the series solution of the given DE on the closed form solution given by (8.42). In general, no closed form solution is available.

Problem List 8.3

1. Solve each DE by the Taylor series method. Give the answer in closed form.
   (a) $y' + y = 0$ (b) $y'' + y = 0$ (c) $y'' - y = 0$
2. Apply the Taylor series method to each initial-value problem. Write the number of terms requested for each series solution.
   (a) $y' = x - y$, $y(0) = 1$ (5 terms)
   (b) $y'' = x^2 + 2y$; $y(0) = 1$, $y'(0) = 1$ (5 terms)
   (c) $y'' = 2x - y'$; $y(0) = 2$, $y'(0) = 0$ (4 nonzero terms)
   (d) $y' = e^x - y$, $y(0) = 4$ (4 terms)
   (e) $y' = e^y$, $y(0) = 0$ (5 nonzero terms)
   (f) $y' = 1 + y^2$, $y(0) = 1$ (4 terms)
   (g) $y'' = x^2 - y^2$; $y(0) = 1$, $y'(0) = 0$ (4 nonzero terms)
   (h) $y'' = xy$; $y(0) = 1$, $y'(0) = 0$ (3 nonzero terms)
   (i) $y'' = 3y^2 - 2xy'$; $y(0) = 1$, $y'(0) = -1$ (5 terms)
   (j) $y'' - (1 - x^2)y' = -x$; $y(0) = -1$, $y'(0) = 1$ (5 terms)

3. Apply the Taylor series method to Legendre's DE

$$(1 - x^2)y'' - 2xy' + p(p + 1)y = 0$$

for (a) $p = 0$ and $y'(0) = 0$; and (b) $p = 3$ and $y(0) = 0$.

4. Apply the Taylor series method to the initial-value problem

$$(x^2 + 5)y'' - 6xy' + 12y = 0; \quad y(0) = 0, \quad y'(0) = -5$$

5. Solve Hermite's DE $y'' - 2xy' + \lambda y = 0$ by the Taylor series method for $\lambda = 4$, given that $y(0) = 1$ and $y'(0) = 0$.

6. (a) Solve the initial-value problem $y' - y = 2x$, $y(1) = -1$, by the Taylor series method. Write four terms of the solution.
   (b) Find the Maclaurin series solution by the same method after making the substitution $t = x - 1$.

7. Show that the solution of illustrative Example 2 can be written in the closed form $y = 2(x + \cos x - \sin x)$.

8. Solve the initial-value problem

$$(1 - x^2)y'' - xy' + 16y = 0; \quad y(0) = 1, \quad y'(0) = 0$$

Hint: Differentiate both sides of $(1 - x^2)y'' = xy' - 16y$ $(n - 2)$ times using Leibniz's formula for the nth derivative of a product

$$D^n[u(x)v(x)] = \sum_{k=0}^{n} \frac{n!}{k!(n - k)!} D^k u(x) D^{n-k} v(x)$$

References

8.1 Birkhoff, G., and G.-C. Rota., 1969. *Ordinary Differential Equations*, 2nd ed. New York: Ginn.

8.2 Bromwich, T. J. I'A. 1942. *An Introduction to the Theory of Infinite Series.* London: Macmillan.

8.3 Coddington, E. A. 1961. *An Introduction to Ordinary Differential Equations.* Englewood Cliffs, N.J.: Prentice-Hall.

8.4 Derrick, W. R., and S. I. Grossman, 1976. *Elementary Differential Equations with Applications.* Reading, Mass.: Addison-Wesley.

8.5 Kaplan, Wilfred. 1962. *Ordinary Differential Equations.* Reading, Mass.: Addison-Wesley.

8.6 Knopp, K. 1951. *Theory and Application of Infinite Series*, trans. R. C. Young. New York: Hafner.

8.7 Simmons, G. F. 1972. *Differential Equations with Applications and Historical Notes.* New York: McGraw-Hill.

8.8 Tierney, John A. 1979. *Calculus and Analytic Geometry*, 4th ed. Boston: Allyn and Bacon.

8.9 Watson, G. N. 1948. *A Treatise on the Theory of Bessel Functions.* London: Cambridge University Press.

8.10 Wylie, C. R., Jr. 1966. *Advanced Engineering Mathematics*, 3rd ed. New York: McGraw-Hill.

9

Approximate
Solutions
of Differential
Equations;
Numerical Methods

9.1 Approximate Solution of an Initial-Value Problem

Let us assume that a unique solution of the initial-value problem

$$y' = \frac{dy}{dx} = F(x, y), \quad y(x_0) = y_0 \tag{9.1}$$

has been guaranteed by basic existence-uniqueness theorems such as Peano's Theorem II-4 and Picard's Theorem II-5. Only in exceptional cases is it possible to find the guaranteed solution, denoted by $y = y(x)$, in closed form. When a closed form solution is not available, various methods can be used for approximating $y(x)$ in the neighborhood of $x = x_0$. One method is to analyze the DE and to deduce various properties of the solution without solving the DE. Another method, discussed in Section 2.7, is to draw the direction field of the DE and examine the integral curves that pass through or near the point (x_0, y_0). This approach is very valuable and highly informative for first-order DE and usually provides a good approximation to the integral curve through (x_0, y_0). A disadvantage is that it often takes much time and effort to draw the direction field. Also, if we wish to know $y(a)$ correct to some number of decimal places for a given number $x = a \neq x_0$, the direction field will not help us.

The Taylor series method (Chapter 8) furnishes still another way of approximating the solution of a DE. The value of $y(x)$ at $x = x_0 + h$ is given by

$$y(x_0 + h) = y_0 + \frac{y'(x_0)}{1!} h + \frac{y''(x_0)}{2!} h^2 + \frac{y'''(x_0)}{3!} h^3 + \cdots$$

We know that $y'(x_0) = F(x_0, y_0)$, and we can compute the higher derivatives y'', y''', ... at x_0 by implicit differentiation and the chain rule. For example,

$$y'' = \frac{d}{dx} F(x, y)$$

$$= F_x \frac{dx}{dx} + F_y \frac{dy}{dx} = F_x + F_y F$$

$$y''' = \frac{d}{dx} (F_x + F_y F)$$

$$= \frac{\partial}{\partial x} (F_x + F_y F) \frac{dx}{dx} + \frac{\partial}{\partial y} (F_x + F_y F) \frac{dy}{dx}$$

$$= (F_{xx} + F_y F_x + F F_{yx})1 + (F_{xy} + F_y F_y + F F_{yy})F$$

which, assuming that $F_{yx} = F_{xy}$, yields

$$y''' = F_{xx} + 2F F_{xy} + F^2 F_{yy} + F_x F_y + F F_y^2$$

after which the right members are evaluated at (x_0, y_0). A series solution can be approximated at $x = x_0 + h$ for small h if the convergence is sufficiently rapid and the remainder after n terms is bounded. It is even possible to program a computer to carry out the necessary differentiations, but the series approach has serious disadvantages. The differentiations may be extremely complicated or even impossible if some of the required higher partial derivatives of F do not exist. Also, the radius of convergence may be unknown, it may be difficult to find a bound on the remainder after n terms, convergence may be slow, and so on. Furthermore, as we have seen previously, not all DE possess Taylor series solutions.

The series method is sometimes used to advantage to approximate $y(x)$ for a few values of x near x_0. These values are then used in a recursion formula as part of a numerical method to find additional approximate values of $y(x)$. The series method is then referred to as a *starting method* for the numerical method.

In 1931, the distinguished American scientist and electrical engineer Vannevar Bush (1890–1974) invented a calculating machine known as a *differential analyzer*. The modern form of this device, known as an *analog computer*, solves DE approximately and displays integral curves of DE on a screen. By simple changes in the computer components, it is easy to change the initial conditions and to examine many integral curves in a short period.

The accuracy obtainable by this method falls short of that obtainable using a numerical method on a digital computer. For additional information on analog computers, see References 9.8 and 9.14.

An inertial guidance system measures the components of the acceleration of a ship, aircraft, rocket, missile, and the like. The element of the system known as an integrator mechanically integrates the acceleration components to find the instantaneous velocity and position, thus solving approximately the Newtonian DE of motion. Both analog and digital computers are used in inertial guidance systems.

In the following section, we present a special method for solving DE approximately. Although it is rarely used for computation, it has great theoretical importance and is frequently used in the portion of the theory of DE that treats existence and uniqueness questions.

9.2 Picard's Method of Successive Approximations

Integrating the DE of the differential system (9.1) between corresponding limits for x and y, we obtain the equivalent *integral equation*

$$y(x) = y_0 + \int_{x_0}^{x} F(u, y(u)) \, du \tag{9.2}$$

Equation (9.2) is called an integral equation because the unknown function y appears under the integral sign. We say that (9.1) and (9.2) are equivalent because any solution of (9.1), automatically continuous, is a solution of (9.2), and any continuous solution of (9.2) is a solution of (9.1).

We cannot proceed with the integration in (9.2) since we do not know the function y in the integrand. Picard's method begins by assuming an approximation $y_0(x)$ to $y(x)$. Although other approximations are permissible, the usual one employed is $y_0(x) \equiv y_0$. This is due to the simplicity of the curve (straight line) $y_0(x) \equiv y_0$ and the fact that the curve passes through the point (x_0, y_0).

After $y(x)$ is replaced in the right member of (9.2) by $y_0(x) \equiv y_0$, the integration is carried out to obtain

$$y_1(x) = y_0 + \int_{x_0}^{x} F(u, y_0) \, du$$

We hope that $y_1(x)$ will be a better approximation to $y(x)$ than $y_0(x)$. Replacing $y(x)$ in the right member of (9.2) by $y_1(x)$ and integrating a second time, we obtain a third approximation to $y(x)$, given by

$$y_2(x) = y_0 + \int_{x_0}^{x} F(u, y_1(u)) \, du$$

After $(n - 1)$ integrations have been performed, the nth approximation is given by

$$y_n(x) = y_0 + \int_{x_0}^x F\big(u, y_{n-1}(u)\big)\, du$$

Thus, the method produces a sequence of functions $\{y_n(x)\}$ in which the elements of the sequence are known as the *Picard successive approximations* to $y(x)$. The eminent French mathematician Emile Picard (1856–1941) used these approximations to prove that under suitable hypotheses on the function F, the differential system (9.1) has a unique solution on some interval $I = [x_0 - \delta, x_0 + \delta]$ where $\delta > 0$. The sequence $\{y_n(x)\}$ converges to $y(x)$ in I if and only if $\lim_{n \to +\infty} y_n(x) = y(x)$ for every x in I. For a proof of Picard's result, see Reference 9.13. In Theorem 2-V, we presented a uniqueness theorem due to Picard.

EXAMPLE 1. Apply Picard's method to the initial-value problem

$$y' = y, \quad y(0) = 1 \tag{9.3}$$

Solution: From $y_0(x) \equiv 1$, we obtain

$$y_1(x) = 1 + \int_0^x 1\, du = 1 + x$$

$$y_2(x) = 1 + \int_0^x (1 + u)\, du = 1 + x + \frac{x^2}{2!}$$

$$y_3(x) = 1 + \int_0^x \left(1 + u + \frac{u^2}{2!}\right) du = 1 + x + \frac{x^2}{2!} + \frac{x^3}{3!}$$

$$\vdots$$

$$y_n(x) = 1 + \int_0^x \left(1 + u + \cdots + \frac{u^{n-1}}{(n-1)!}\right) du = 1 + x + \frac{x^2}{2!} + \cdots + \frac{x^n}{n!}$$

$$\vdots$$

Since the solution of (9.3) is given by $y(x) = e^x$, and

$$\lim_{n \to +\infty} y_n(x) = 1 + x + \frac{x^2}{2!} + \cdots + \frac{x^n}{n!} + \cdots = e^x$$

we see directly in this example that the Picard successive approximations converge to the solution given by $y = y(x) = e^x$. The convergence is not particularly rapid. By hand calculator, $y_7(1) \approx 2.718253968$, $y_8(1) \approx 2.71827877$, and $y(1) = e \approx 2.718281828$. Thus, nine Picard approximations were required to yield $y(1)$ correct to 5 decimal places.

EXAMPLE 2. Apply Picard's method to the initial-value problem

$$y' = x + y, \quad y(0) = 1 \tag{9.4}$$

Solution: From $y_0(x) \equiv 1$, we obtain

$$y_1(x) = 1 + \int_0^x (u + 1) \, du = \frac{x^2}{2} + x + 1$$

$$y_2(x) = 1 + \int_0^x \left(u + \frac{u^2}{2} + u + 1 \right) du = \frac{x^3}{6} + x^2 + x + 1$$

$$y_3(x) = 1 + \int_0^x \left(u + \frac{u^3}{6} + u^2 + u + 1 \right) du$$

$$= \frac{x^4}{24} + \frac{x^3}{3} + x^2 + x + 1$$

and

$$y_4(x) = 1 + \int_0^x \left(u + \frac{u^4}{24} + \frac{u^3}{3} + u^2 + u + 1 \right) du$$

$$= \frac{x^5}{120} + \frac{x^4}{12} + \frac{x^3}{3} + x^2 + x + 1$$

By using the integrating factor e^{-x}, we can easily show that the solution of $y' - y = x$ with $y(0) = 1$ is given by $y(x) = 2e^x - x - 1$.

The accompanying table shows how closely the first five Picard iterates approximate $y(x) = 2e^x - x - 1$. Entries are correct to five decimal places or are exact.

x	$y_0(x)$	$y_1(x)$	$y_2(x)$	$y_3(x)$	$y_4(x)$	$y(x)$
0.0	1.0	1.0	1.0	1.0	1.0	1.0
0.2	1.0	1.22	1.24133	1.24273	1.24280	1.24281
0.4	1.0	1.48	1.57067	1.5824	1.58356	1.58365
0.6	1.0	1.78	1.996	2.0374	2.04345	2.04424
0.8	1.0	2.12	2.52533	2.62773	2.64753	2.65108
1.0	1.0	2.5	3.16667	3.375	3.425	3.43656

From

$$\frac{|y(0.2) - y_4(0.2)|}{y(0.2)} = \frac{0.00001}{1.24281} \approx 0.000008$$

we see that $y_4(0.2)$ approximates $y(0.2)$ with relative error 0.0008%. From

$$\frac{|y(1) - y_4(1)|}{y(1)} = \frac{0.01156}{3.43656} \approx 0.0034$$

we see that $y_4(1)$ approximates $y(1)$ with relative error 0.34%.

Although the first few Picard approximations are often quite accurate, it frequently becomes extremely difficult if not impossible to perform the required integrations. It would be quite difficult, for example, to apply Picard's method to the initial-value problem

$$y' = \frac{2y - x}{y + x}, \quad y(0) = 1$$

An interesting feature of the method is that it is self-correcting. That is, if an error is made in finding one of the Picard approximations, the sequence $\{y_n(x)\}$ generally converges to $y(x)$ in spite of the error.

Problem List 9.1

1. Apply Picard's method to find $y_1(x)$, $y_2(x)$, and $y_3(x)$. Find the exact solution given by $y = y(x)$ and compare $y_3(0.2)$, $y_3(0.5)$, and $y_3(1)$ with $y(0.2)$, $y(0.5)$, and $y(1)$. Give numerical values to five decimal places.
   (a) $y' = -y$, $\quad y(0) = 1$
   (b) $y' = xy$, $\quad y(0) = 1$ } Use $y_0(x) \equiv 1$.
   (c) $y' = -y^2$, $\quad y(0) = 1$
   (d) $y' = x - y$, $\quad y(0) = 0$ Use $y_0(x) \equiv 0$.
2. Apply Picard's method to the initial-value problem $y' = y^2$, $y(0) = 1$. Find the exact solution given by $y = y(x)$ and compare $y_2(0.1)$ with $y(0.1)$. Give values to five decimal places. Note that $y(1)$ is undefined.
3. Apply Picard's method to the initial-value problem $y' = x^2 + y^2$, $y(0) = 1$. Find $y_2(0.1)$ and $y_2(0.2)$. Compare with approximations obtained by using four terms of the Maclaurin series for the solution. Give values to five decimal places.
4. Apply Picard's method to the initial-value problem $y' = 1 + y^2$, $y(0) = 0$. Find $y_3(0.1)$ and $y_3(0.2)$ to five decimal places and compare these values with the values of the exact solution.
5. With $y_0(x) \equiv 0$, apply Picard's method to the initial-value problem $y' = x^2 - y^2$, $y(0) = 0$, to find $y_1(x)$, $y_2(x)$, and $y_3(x)$.
6. On the same set of axes draw the graphs on $0 \le x \le 1$ of $y_0(x)$, $y_1(x)$, $y_2(x)$, $y_3(x)$, $y_4(x)$, and $y(x)$ in illustrative Example 2.
7. Find $y_n(x)$ in illustrative Example 2 and show that $\lim_{n \to +\infty} y_n(x) = y(x)$.
8. In illustrative Example 1, find $y_3(x)$, using $y_0(x) = 1 + x$ instead of $y_0(x) \equiv 1$. Show that $y_3(1)$ is closer to $y(1) = e$ than the value of $y_3(1)$ obtained using $y_0(x) \equiv 1$.

9.3 Numerical Methods

Let us assume that the initial-value problem

$$y' = \frac{dy}{dx} = F(x, y), \quad y(x_0) = y_0 \tag{9.1}$$

is known to possess a unique solution on some interval $[x_0, a]$ where

$a > x_0$. We denote this solution by $y = y(x)$. Let us assume also that no closed form solution of (9.1) is available. We are interested in finding $y(x_m)$ correct to a specified number of decimal places where $x_0 < x_m \le a$. Such a numerical approximation to $y(x_m)$ will be denoted by y_m. Another basic problem is that of finding numerical approximations to $y(x)$ at a discrete set of *mesh points* $x_1, x_2, \ldots, x_m$, where

$$x_0 < x_1 < x_2 < \cdots < x_{m-1} < x_m$$

These approximations will be denoted by $y_1, y_2, \ldots, y_m$. Thus, $y_n \approx y(x_n)$ for $n = 1, 2, 3, \ldots, m$. After $y_1, y_2, \ldots, y_m$ have been determined by a numerical method, approximations to other values of $y(x)$ on $[x_0, x_m]$ are sometimes computed by an appropriate method of interpolation.

If the mesh points are equally spaced, that is, if $x_1 - x_0 = x_2 - x_1 = \cdots = x_m - x_{m-1} = h$, the numerical method is said to use the *constant step size h*. In some applications to fields like ballistics, astronomy, and orbit theory, it is sometimes advantageous to use a *variable step size*. For simplicity, we restrict ourselves to a constant step size h.

It is easy to modify a numerical method to find approximate values of $y(x)$ for $x < x_0$. Also, there is no loss of generality in assuming that $x_0 = 0$, since substituting $t = x - x_0$ transforms a DE with initial value specified at $x = x_0$ into a second DE with initial value specified at $t = 0$.

An important simple type of numerical method uses a recursion formula to compute y_{n+1} from y_n for $n = 0, 1, 2, \ldots, m - 1$. This kind of method is called a *one-step method*, since the computation of y_{n+1} uses the value y_n but no preceding values such as y_{n-1}. Another kind of numerical method, known as a *multistep*, or *k-step*, *method*, $k > 1$, uses a recursion formula to compute y_{n+1} from the k preceding values $y_n, y_{n-1}, y_{n-2}, \ldots$. The well-known method of Milne (see Reference 9.11) is a four-step method. The advantage of a multistep method over a one-step method is that in finding y_{n+1}, one uses the history of the solution prior to y_n, whereas one ignores it in a one-step method. On the other hand, a k-step method cannot be used until k values of y_n are available. The $(k - 1)$ values besides y_0 must be found by some other method, referred to as a *starting method* for the k-step method. The starting method "starts" the solution for the multistep method, which is then termed a *continuing method*. A one-step method needs no starting method. This permits a change in step size at any step, an advantage in some applications.

The question of which method to use in a given problem is difficult. The answer depends on several factors, including the type of DE involved. In some applications, prior knowledge of properties possessed by the solution may affect the choice of numerical method. The computing device (tables, hand calculator, desk calculator, digital computer, and so forth) to be used is important, and so is the desired degree of accuracy in the results. When a digital computer is available, the cost of computer time is sometimes an important consideration.

When a numerical method is first studied or introduced, it is a good idea to try it out on a simple problem whose solution is known. For example, the simple initial-value problem

$$y' = y, \quad y(0) = 1 \qquad (9.5)$$

has exact solution given by $y(x) = e^x$, and values of e^x are readily available to many decimal places. The results of the numerical method can be compared with values known in advance to be correct. A problem such as (9.5) is referred to as a *test problem*. It is standard practice to check a computer program by a test problem.

The approximate solution of a DE by a numerical method is a vast subject and constitutes a large and important part of the branch of mathematics known as *numerical analysis*. Our objective in this chapter is to give the reader an idea of some of the significant considerations in numerical solutions, and to present a few simple numerical methods.

Note, as a matter of interest, that numerical methods are sometimes preferred even when closed form solutions of DE are available. For example, the initial-value problem

$$y' = \frac{y - x}{y + x}, \quad y(1) = 1 \qquad (9.6)$$

can be solved by the usual substitution $y = vx$. This yields the exact analytical solution given implicitly by

$$\ln (x^2 + y^2) + 2 \, \text{Tan}^{-1} \frac{y}{x} = \frac{\pi}{2} + \ln 2 \qquad (9.7)$$

To find values of y for values of x near $x = 1$ from (9.7), it is necessary to solve a complicated transcendental equation. This presents a formidable numerical problem. It is simpler to apply a numerical method directly to (9.6).

9.4 Euler's Method

The oldest and simplest numerical method of solving a DE approximately is due to the famous Swiss mathematician Leonard Euler (1707–1783). About a century after Euler introduced the method, it was used by the great French mathematician Augustine-Louis Cauchy (1789–1857) in developing constructive existence proofs for first-order differential equations. The method is also referred to as the *Euler-Cauchy method*, the *tangent method*, and the *constant-slope method*. Although not accurate enough for most practical applications, the method serves as an excellent introduction to the study of numerical methods for solving DE.

To find an approximation y_1 to $y(x_1)$ for the initial-value problem

$$y' = \frac{dy}{dx} = F(x, y), \quad y(x_0) = y_0 \qquad (9.1)$$

we replace

$$y'(x_0) = \lim_{h \to 0} \frac{y(x_0 + h) - y(x_0)}{h} = \lim_{h \to 0} \frac{y(x_1) - y_0}{x_1 - x_0}$$

by the difference quotient

$$\frac{y(x_1) - y_0}{x_1 - x_0} = \frac{y(x_1) - y_0}{h} \approx y'(x_0) = F(x_0, y_0)$$

to obtain $y(x_1) \approx y_0 + hF(x_0, y_0)$.

We take the linear approximation $y_0 + hF(x_0, y_0)$ as our approximation y_1 to $y(x_1)$, and write

$$y_1 = y_0 + hF(x_0, y_0) \tag{9.8}$$

As shown in Figure 9.1, we are approximating the height of the curve $y = y(x)$ at $x = x_1$ by the height of the line tangent to $y = y(x)$ at the point (x_0, y_0). To find $y_2 \approx y(x_2)$, we replace x_0 by x_1 and y_0 by y_1 in the right member of (9.8) to obtain $y_2 = y_1 + hF(x_1, y_1)$.

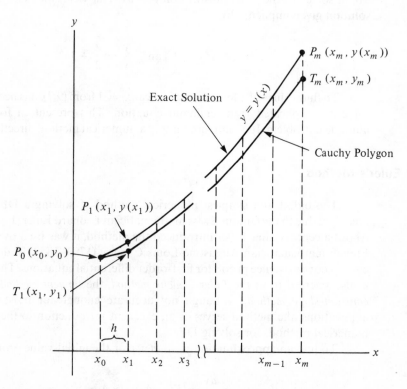

Fig. 9.1

The process is repeated to find y_i from y_{i-1} for $i = 3, 4, \ldots, m$ by the recursion formula

$$y_{n+1} = y_n + hF(x_n, y_n) \tag{9.9}$$

The error due to the Euler linear approximation at the nth step of the Euler process is given by

$$e_n = y_n - y(x_n) \tag{9.10}$$

This error e_n is known as the *discretization error*, since (9.9) uses the function F at a discrete number of points, whereas in (9.1) F varies continuously over the interval $[x_0, x_m]$. The discretization error is also known as the *truncation error*, since the right member of (9.8) can be written

$$y(x_0) + \frac{y'(x_0)}{1!} h$$

the two-term truncated Taylor series for $y(x)$ evaluated at $x = x_0$. In Figure 9.1, the discretization error at each step is negative because of the upward concavity of the integral curve $y = y(x)$, the graph of the exact solution. The absolute value $|e_n|$ of e_n, called the *absolute error*, increases from $|T_1 P_1|$ at step 1 to $|T_m P_m|$ at step m. The polygonal path from P_0 to T_m is called the *Cauchy polygon* approximation to the solution $y = y(x)$. In Figure 9.1, a constant step size h is used. This is not essential, as we pointed out previously, since Euler's method is a one-step method.

An upper bound that $|e_n|$ may assume is very important in any numerical method. It is shown in Reference 9.2 that if the unique solution of (9.1) given by $y = y(x)$ is contained in a rectangle $R: x_0 \leq x \leq x_m, y_0 - b \leq y \leq y_0 + b$ where $b > 0$; and if F, F_x, and F_y are continuous in R, there exists a positive constant K such that e_n, the discretization error in Euler's method, satisfies $|e_n| = |y_n - y(x_n)| \leq Kh$, where K is independent of h.

If the step size h is halved, then $|e_n| \leq K(h/2) = \frac{1}{2}(Kh)$; that is, the maximum possible $|e_n|$ would then be half what it was with step size h.

If in a numerical method there exists a constant $K > 0$ such that

$$|e_n| = |y_n - y(x_n)| \leq Kh^r$$

the method is said to have *order of accuracy* h^r and to be a method *of order* h^r. Thus, Euler's method has order of accuracy $h^1 = h$. The discretization error e_n depends on the concavity of the integral curve, the step size h, and the direction field of $y' = F(x, y)$. If the integral curves of the direction field for which $y(x_0) \approx y_0$ diverge (separate) as x increases from x_0 to x_m, $|e_n|$ tends to be large; if they converge (come closer together) as x increases from x_0 to x_m, then $|e_n|$ tends to be small.

In the Euler recursion formula (9.9) it should be noted that $F(x_n, y_n)$ denotes the slope of the solution curve $y = y(x)$ at $P_0(x_0, y_0)$ when $n = 0$. For $n > 0$, $F(x_n, y_n)$ denotes the slope at the point (x_n, y_n) of the direction field defined by $y' = F(x, y)$. In general, $F(x_n, y_n) \neq F(x_n, y(x_n))$; in fact, we do not have the value of $y(x_n)$ available. We are assuming that $F(x_n, y_n)$ approximates $F(x_n, y(x_n))$ since we are also assuming that the points (x_n, y_n) and $(x_n, y(x_n))$ are close together.

EXAMPLE 1. Apply Euler's method to the test problem $y' = y$, $y(0) = 1$, to find $y_1 \approx y(0.2)$, $y_2 \approx y(0.4)$, and $y_3 \approx y(0.6)$.

Solution: From (9.9), we obtain

$$y_1 \approx y(0.2) = y_0 + hy_0 = 1 + (0.2)(1) = 1.2$$

$$y_2 \approx y(0.4) = y_1 + hy_1 = 1.2 + (0.2)(1.2) = 1.44$$

$$y_3 \approx y(0.6) = y_2 + hy_2 = 1.44 + (0.2)(1.44) = 1.728$$

When a numerical method is used, it is standard practice to present the results in tabular form. The accompanying table includes, for purposes of comparison, the exact values of $y(x) = e^x$ to four decimal places.

n	x_n	y_n	$F(x_n, y_n) = y_n$	$y_{n+1} = y_n + 0.2y_n$	$y(x_n) = e^{x_n}$
0	0.0	1.0	1.0	1.2	1.0000
1	0.2	1.2	1.2	1.44	1.2214
2	0.4	1.44	1.44	1.728	1.4918
3	0.6	1.728			1.8221

The relative error in $y_3 = 1.728 \approx y(0.6)$, given by

$$\frac{|1.728 - 1.8221|}{1.8221} = \frac{0.0941}{1.8221} \approx 0.0516$$

is approximately 5.2%.

EXAMPLE 2. Solve Example 1 with $h = 0.1$.

Solution: The results are displayed in the accompanying table. Values of $y(x)$ are correct to six decimal places; all other entries are exact.

n	x_n	$y_n = F(x_n, y_n)$	$y_{n+1} = y_n + 0.1y_n$	$y(x_n) = e^{x_n}$
0	0.0	1.0	1.1	1.000000
1	0.1	1.1	1.21	1.105171
2	0.2	1.21	1.331	1.221403
3	0.3	1.331	1.4641	1.349859
4	0.4	1.4641	1.61051	1.491825
5	0.5	1.61051	1.771561	1.648721
6	0.6	1.771561		1.822119

The relative error in $y_6 = 1.771561 \approx y(0.6)$, given by

$$\frac{|1.771561 - 1.822119|}{1.822119} \approx 0.0277$$

is approximately 2.8%. This result, although an improvement over the approximation in Example 1, is not accurate enough for most applications.

In the tables of Examples 1 and 2, $e_n = y_n - y(x_n)$ is always negative, and $|e_n|$ increases as x increases. This is partly because after step 1, we are using an initial point (x_n, y_n) that is on an integral curve different from the integral curve having equation $y = y(x) = e^x$. It is also partly because the curve $y = e^x$ is concave upward on $[0, 0.6]$. This concavity, incidentally, could be determined even if we were unable to solve $y' = y$ in closed form. From $y'' = y' = y$, it follows that $y''(x) > 0$ for $y > 0$.

We conclude this section by looking at Euler's method from a different point of view. As indicated earlier, solving the initial-value problem (9.1) is equivalent to solving the integral equation

$$y(x) = y_0 + \int_{x_0}^{x} F(u, y(u))\, du \qquad (9.2)$$

For the first step in the Euler method, the integral in (9.2) is equal to the area under the curve $y' = F(x, y(x))$ from x_0 to x_1. See Figure 9.2. Since we are interested in small h, we approximate this area by the shaded area in Figure 9.2. That is, we approximate $F(x, y(x))$ on $[x_0, x_1]$ by its value $F(x_0, y(x_0)) = F(x_0, y_0)$ at the left end of the interval $[x_0, x_1]$. Since the shaded area has value $hF(x_0, y_0)$, our approximation y_1 to $y(x_1)$ in (9.2) becomes

$$y_1 = y_0 + hF(x_0, y_0) \approx y(x_1)$$

in agreement with (9.8). It should be noted that the curve in Figure 9.2 is the graph of $y' = F(x, y) = F(x, y(x))$ whereas the curve in Figure 9.1 is the graph of $y = y(x)$.

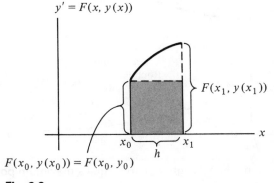

Fig. 9.2

We mention this interpretation because numerous other numerical methods for solving DE are obtained by using various approximations to the integral in (9.2).

9.5 Roundoff Error

If the number $\frac{2}{3}$ is given correct to four decimal places, it is written as 0.6667. Computations involving this approximation to $\frac{2}{3}$ contain an error known as *roundoff error*. In the illustrative examples of Section 9.4, the computations in Euler's method were performed exactly and hence no roundoff error was introduced. The error was entirely discretization error. This is the exception, since in most computations rounding off of intermediate or final results occurs.

Suppose, for example, that we use Euler's method with $y(0) = \pi$. No matter how powerful a digital computer we have at our disposal, an approximation to π must be used. If we use the approximation 3.141592654 (from a hand calculator), the point (0, 3.141592654) is not on the integral curve through (0, π) and roundoff error has been introduced at the outset.

It is theoretically true that Euler's method will yield accuracy to any required number of decimal places provided that a small enough h is chosen. The difficulty is that the smaller we make h, the more calculations we have to perform; and the larger the number of calculations the greater the possible roundoff error. To obtain four-place accuracy in solving a DE by Euler's method, it would not be unusual if 10,000–20,000 calculations were required. The usual situation in using a numerical method is depicted in Figure 9.3, in which we assume for simplicity that the discretization error and the roundoff

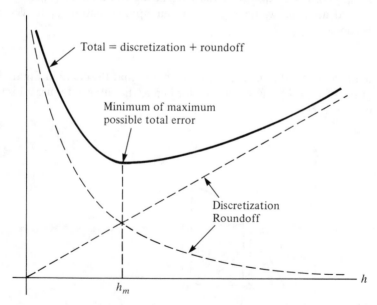

Fig. 9.3 Numerical method. Height of curve equals maximum possible total error

error are both nonnegative. For each problem there exists an $h = h_m$ for which the maximum possible total error is least. For $h < h_m$, the advantage of small step size is overcome by the possibility of a large roundoff error.

Unfortunately, determining h_m is a formidable problem. Henrici has made a statistical analysis of cumulative roundoff error. See Reference 9.6.

It is standard procedure to reduce the step size until the computed results no longer differ in the number of decimal places required. Although the accuracy of the results is not then absolutely guaranteed, many numerical analysts find this practical rule very useful. It is a simple procedure to halve the step size h in a digital computer program, but eventually too small a value of h will result in greatly increased roundoff error. In some applications, a very small h may require so many calculations that the cost of computer time may become a significant factor.

For an error analysis of various numerical methods, see References 9.1, 9.3, and 9.4. When the error associated with a particular numerical method is said to be of order h^r, this error is discretization error and does not involve roundoff error.

EXAMPLE. Apply Euler's method to the initial-value problem $y' = x + y$, $y(0) = 1.24$.

Find $y_5 \approx y(1)$ using $h = 0.2$. Round off tabular entries to four decimal places. Compare y_5 with $y(1)$ where $y(x) = 2.24e^x - x - 1$.

Solution:

n	x_n	y_n	$F(x_n, y_n) = x_n + y_n$	$y_{n+1} = y_n + (0.2)(x_n + y_n)$	$y(x_n)$
0	0.0	1.24	1.24	1.488	1.24
1	0.2	1.488	1.688	1.8256	1.5359
2	0.4	1.8256	2.2256	2.2707	1.9417
3	0.6	2.2707	2.8707	2.8448	2.4815
4	0.8	2.8448	3.6448	3.5738	3.1852
5	1.0	3.5738			4.0890

The relative error in $y_5 = 3.5738 \approx y(1)$, given by

$$\frac{|3.5738 - 4.0890|}{4.0890} = \frac{0.5152}{4.0890} \approx 0.1260$$

is approximately 12.6%. In Sections 9.6 and 9.7, we consider methods that are far more accurate than Euler's method.

Problem List 9.2

1. Apply Euler's method to the initial-value problem $y' = 2xy$, $y(0) = 1$. Approximate $y(1)$, using $h = 0.2$. Round off tabular entries to three decimal places. Find the exact solution and determine the relative error in the approximation to $y(1)$.

2. Apply Euler's method to the initial-value problem $y' = -y$, $y(0) = 1$. Approximate $y(1)$, using (a) $h = 0.5$; (b) $h = 0.25$; and (c) $h = 0.2$. Compare results with $y(1) = e^{-1}$. Use four decimal places.

3. Apply Euler's method to the initial-value problem $y' = x^2 + y^2$, $y(0) = 1$. Approximate $y(0.2)$, using $h = 0.1$. Reduce h until the approximations to $y(0.2)$ agree to two decimal places. Use four decimal places in tabular entries.

4. Write a computer program for Euler's method. Use the program to approximate $y(1)$ to three decimal places for the initial-value problem $y' = (x^2 + 1)^{-1}$, $y(0) = 0$, (a) with $h = 0.2$, and (b) with $h = 0.1$.

Find the exact value of $y(1)$, the value of $y(1)$ to three decimal places, and the relative error in the approximations to $y(1)$ in (a) and (b). Note that the error is approximately halved by halving the step size.

9.6 The Improved Euler Method

The Euler method could be referred to as the two-term Taylor series method, since (9.8), the Euler formula for approximating $y(x_0 + h)$, uses two terms of the Taylor series for $y(x)$ at $x = x_0$. That is,

$$y_1 = y_0 + hF(x_0, y_0) = y(x_0) + \frac{y'(x_0)}{1!} h$$

This suggests that greater accuracy might be obtained by using three terms of the Taylor series with

$$y_1 \approx y(x_0) + \frac{y'(x_0)}{1!} h + \frac{y''(x_0)}{2!} h^2$$

This method is sometimes used and is referred to as the *three-term Taylor series method*. The chief disadvantage is that $y''(x_0) = F_x + F_y F]_{(x_0, y_0)}$ is often complicated, particularly when one is solving a nonlinear DE. The method has order of accuracy h^2. Instead of illustrating this method, we present a method that is much simpler in application and yet has the same order of accuracy. The method is known as the *improved Euler method*.

To approximate the integral in

$$y(x) = y_0 + \int_{x_0}^{x} F(u, y(u)) \, du \tag{9.2}$$

for $x = x_1$, we approximate the area under the curve $y' = F(x, y(x))$ from x_0 to x_1 by the shaded trapezoidal area in Figure 9.4.
    This yields

$$y_1 = y_0 + \frac{h}{2} [F(x_0, y_0) + F(x_1, y(x_1))] \approx y(x_1) \tag{9.11}$$

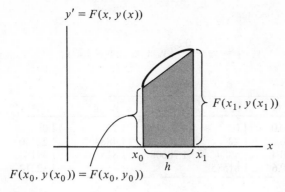

$y' = F(x, y(x))$

$F(x_1, y(x_1))$

x_0

h

x_1

x

$F(x_0, y(x_0)) = F(x_0, y_0))$

Fig. 9.4

from which we obtain the recursion formula

$$y_{n+1} = y_n + \frac{h}{2}[F(x_n, y_n) + F(x_{n+1}, y(x_{n+1}))] \qquad (9.12)$$

We cannot use (9.12) directly to find y_{n+1} since the quantity $y(x_{n+1})$ is unavailable. We circumvent this difficulty by using the Euler approximation

$$z_{n+1} = y_n + hF(x_n, y_n) \qquad (9.13)$$

for $y(x_{n+1})$ in the right member of (9.12). The improved Euler method consists in using (9.13) and

$$y_{n+1} = y_n + \frac{h}{2}[F(x_n, y_n) + F(x_{n+1}, z_{n+1})] \qquad (9.14)$$

alternately with $n = 0, 1, 2, \ldots, m - 1$.

Formula (9.14) is known as the *improved Euler formula* or the *Heun formula*. It is shown in Reference 9.9 that the improved Euler method has order of accuracy h^2, the same as that of the three-term Taylor series method. That is, Formula (9.14) can be written in the form

$$y_{n+1} = y_n + \frac{y'(x_n)}{1!}h + \frac{y''(x_n)}{2!}h^2 + \begin{array}{l}\text{(terms involving } h^3 \text{ and}\\ \text{higher powers of } h)\end{array}$$

EXAMPLE 1. Apply the improved Euler method to the test problem $y' = y, y(0) = 1$, to find $y_1 \approx y(0.2)$, $y_2 \approx y(0.4)$, and $y_3 \approx y(0.6)$. Round off tabular entries to five decimal places.

Solution: The accompanying table is constructed from (9.13) and (9.14) with $h = 0.2$.
Formula (9.13) becomes

$$z_{n+1} = y_n + 0.2y_n = 1.2y_n$$

and Formula (9.14) becomes

$$y_{n+1} = y_n + \frac{0.2}{2}[y_n + z_{n+1}] = 1.22y_n$$

n	x_n	y_n	z_{n+1}	y_{n+1}	$y(x_n) = e^{x_n}$
0	0.0	1.0	1.2	1.22	1.0
1	0.2	1.22	1.464	1.4884	1.22140
2	0.4	1.4884	1.78608	1.81585	1.49182
3	0.6	1.81585			1.82212

The relative error in $y_3 = 1.81585 \approx y(0.6)$, given by

$$\frac{|1.81585 - 1.82212|}{1.82212} = \frac{0.00627}{1.82212} \approx 0.0034$$

is approximately 0.3%, a marked improvement in accuracy over the 5.2% obtained by the Euler method in Example 1, Section 9.4.

EXAMPLE 2. Apply the improved Euler method to the initial-value problem $y' = x + y$, $y(0) = 1.24$.

Find $y_5 \approx y(1)$, using $h = 0.2$. Round off tabular entries to four decimal places. Compare results with those of the illustrative Example of Section 9.5.

Solution: Formula (9.13) becomes

$$z_{n+1} = y_n + 0.2(x_n + y_n) = 0.2x_n + 1.2y_n$$

and Formula (9.14) becomes

$$y_{n+1} = y_n + \frac{0.2}{2}(x_n + y_n + x_{n+1} + 0.2x_n + 1.2y_n)$$

$$= 0.12x_n + 0.1x_{n+1} + 1.22y_n$$

The tabulated values are seen in the accompanying table.

x_n	y_n	y_{n+1}	$y(x_n) = 2.24e^{x_n} - x_n - 1$
0.0	1.24	1.5328	1.24
0.2	1.5328	1.9340	1.5359
0.4	1.9340	2.4675	1.9417
0.6	2.4675	3.1623	2.4815
0.8	3.1623	4.0540	3.1852
1.0	4.0540		4.0890

The relative error in $y_5 = 4.0540 \approx y(1)$, given by

$$\frac{|4.0540 - 4.0890|}{4.0890} = \frac{0.0350}{4.0890} \approx 0.0086$$

is approximately 0.9%. This compares with 12.6% by Euler's method.

The following digital computer program written in BASIC computer language can be used to solve a DE approximately by the improved Euler method.

```
10  DEF    FNF(X, Y) = X * X + Y * Y
20  INPUT  X,Y,H,N
30  FOR    K = 0 TO N
40  PRINT  X,Y
50  LET    Z = Y + H * FNF(X, Y)
60  LET    Y = Y + H * (FNF(X, Y) + FNF(X + H, Z))/2
70  LET    X = X + H
80  NEXT   K
90  END
```

Instruction 10 defines the function F in $y' = F(x, y)$. In the program shown, $F(x, y) = x^2 + y^2$. Instruction 20 furnishes the data x_0, y_0, h, and $n = (x_m - x_0)/h$. Instruction 30 determines the number of iterations to be performed, and instruction 40 tells the computer to print out (x_0, y_0), $(x_1, y_1), \ldots, (x_m, y_m)$. Instructions 50 and 60 define the quantities z_{n+1} and y_{n+1} from Formulas (9.13) and (9.14) that must be computed. Instructions 70 and 80 replace x_n by x_{n+1} and repeat the computations the number of times specified in instruction 30.

To change the step size, we merely rerun the program, changing H and N. To apply the program to a new problem, we simply change the function in instruction 10.

In Figure 9.4 the height of the curve $y' = F(x, y(x))$ at $x = x_1$ is $F(x_1, y(x_1))$, where $y(x_1)$ is the exact value of the solution of the DE when $x = x_1$. Let us replace $y(x_1)$ by Y_1, where Y_1 is the approximation of $y(x_1)$ that satisfies the equation

$$Y_1 = y_0 + \frac{h}{2}[F(x_0, y_0) + F(x_1, Y_1)] \tag{9.15}$$

Thus, we replace (9.11) by (9.15) and (9.12) by

$$Y_{n+1} = Y_n + \frac{h}{2}[F(x_n, Y_n) + F(x_{n+1}, Y_{n+1})] \tag{9.16}$$

where $Y_0 = y_0$.

We then regard (9.16) as defining Y_{n+1} implicitly. That is, (9.16) may be solved for Y_{n+1} explicitly in terms of Y_n, x_n, and x_{n+1}, and the resulting recursion formula can be used to find approximate values of $y(x)$. This method is known as *trapezoidal integration* of DE. If $F(x, y)$ does not depend on y, that is, if $F(x, y) = F(x)$, then Y_{n+1} does not appear in the right member of (9.16), and there is no difference between the trapezoidal method and the improved Euler method. Methods requiring the solution of an algebraic equation to obtain an explicit recursion formula are known as *implicit methods*. These methods are often very accurate and frequently have other desirable characteristics. Their chief disadvantage is that solving the algebraic equation often introduces computational complexities. If, for example, (9.16) is obtained from a nonlinear DE, solving for Y_{n+1} might be extremely difficult.

The improved Euler method is classified as an *explicit method*, since (9.14) defines y_{n+1} explicitly in terms of y_n, x_n, and x_{n+1}. Although z_{n+1} appears in the right member of (9.14), z_{n+1} is defined explicitly in terms of x_n and y_n by (9.13). On the other hand, the method is sometimes regarded as a combination of implicit-explicit methods. The value z_{n+1} in (9.13) that is used to approximate $y(x_{n+1})$ is called the *predictor*.

Equation (9.14), in which the predictor is substituted, is called the *corrector formula*. The Euler approximation in (9.13) "predicts" the value of $y(x_{n+1})$, after which the corrector formula (9.14) "corrects" the value of $y(x_{n+1})$ and denotes the result by y_{n+1}.

The improved Euler method has order of accuracy h^2. That is, under the same assumptions on F as were made in the basic Euler method, it can be shown that there exists a constant $K > 0$ such that

$$|e_n| = |y_n - y(x_n)| \leq Kh^2 \tag{9.17}$$

where K is independent of h. If h is replaced by $h/2$, (9.17) becomes

$$|e_n| \leq K\left(\frac{h}{2}\right)^2 = \frac{1}{4}(Kh^2)$$

Thus, when the step size h is halved, the maximum possible $|e_n|$ is divided by 4.

Although the predictor uses a method having order of accuracy h, the improved Euler method using this predictor has order of accuracy h^2. See Reference 9.3 for more details. Predictor-corrector formulas are significant in solving differential equations numerically.

A slight variation exists for applying the improved Euler method. Let us assume that z_1 has been found from (9.13) and substituted into (9.14) to find y_1. Instead of proceeding to find y_2, one can substitute the value of y_1 into the right member of (9.14) as a second predictor, and a corrected value of y_1 is then determined. This process is continued until successive values of y_1 agree to the number of decimal places desired in the answer. A new predictor is then found by using the final y_1 in (9.13), after which a sequence of values for y_2 is determined similarly. This method of iterating the corrector is generally inferior to the practice of reducing the step size to improve accuracy.

The improved Euler method is very effective in many applications when one- or two-place accuracy is sufficient. It is also useful as a starting method to find a few approximations to $y(x)$ for x near x_0. In the following section, we consider a method that has a higher order of accuracy and is extremely effective when a digital computer is available.

The names by which various numerical methods for solving DE are known vary considerably in mathematical literature. The improved Euler method is referred to by many authors as the *modified trapezoidal method*, or the *Heun method*.

Problem List 9.3

Apply the improved Euler method in the following initial-value problems.

1. $y' = y$, $y(0) = 1$.

   Find $y_6 \approx y(0.6)$, using $h = 0.1$. Round tabular values off to four decimal places. Compare the relative error in y_6 with the 2.8% obtained in illustrative Example 2, Section 9.4.

2. $y' = -y$, $y(0) = 1$.

   Find $y_5 \approx y(1)$, using $h = 0.2$. Round tabular values off to four decimal places. Find the relative error in y_5.

3. $y' = y - x$, $y(0) = 2$.

   Find an approximation to $y(1)$, using $h = 0.2$. Find the relative error in this approximation by using $y(1)$ from $y(x) = e^x + x + 1$. Round off tabular values to four decimal places.

4. $y' = \dfrac{-y}{1 + x}$, $y(0) = 1$.

   Find an approximation to $y(1.2)$, using $h = 0.4$. Compare this approximation with $y(1.2)$, using $y(x) = 1/(1 + x)$. Use four decimal places.

5. $y' = (1 + x^2)^{-1}$, $y(0) = 0$.

   Find an approximation to $y(1)$, using $h = 0.2$. Find the relative error, using $y(x) = \text{Tan}^{-1} x$. Round off tabular values to four decimal places.

6. $y' = e^{-y}$, $y(0) = 0$.

   Find an approximation to $y(1)$, using $h = 0.2$. Compare this approximation with $y(1) = \ln 2$ from $y(x) = \ln(1 + x)$. Use four decimal places.

7. $y' = x^2 + y^2$, $y(0) = 1$.

   Find an approximation y_2 to $y(0.2)$, using $h = 0.1$. Show that $u(0.2) \le y_2 \le v(0.2)$, where u and v denote solutions of $u' = u^2$, $u(0) = 1$, and $v' = 1 + v^2$, $v(0) = 1$. Use four decimal places.

9.7 Runge-Kutta Methods

Let us review the Euler approximation y_1 of $y(x_1)$ given by

$$y_1 = y_0 + hF(x_0, y_0) \tag{9.8}$$

In Figure 9.5 the exact value of $y(x_1)$ is given by

$$y(x_1) = y_0 + hM = y_0 + h\left[\frac{y(x_1) - y_0}{x_1 - x_0}\right] \tag{9.18}$$

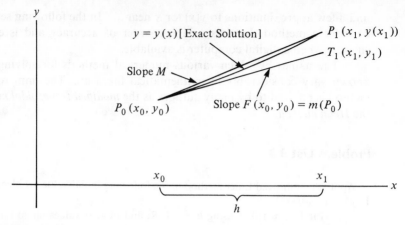

Fig. 9.5

where M is the slope of the line segment $P_0 P_1$. We do not know $y(x_1)$ and hence cannot find M. Hence, we approximate M by $F(x_0, y_0) = y'(x_0, y_0)$, the slope of the direction field of $y' = F(x, y)$ at $P_0(x_0, y_0)$. We denote this slope by $m(P_0)$ and henceforth denote the slope of the direction field at point Q by $m(Q)$. As noted previously, when we use the Euler recursion formula

$$y_{n+1} = y_n + hF(x_n, y_n) \tag{9.9}$$

we employ the slope $m(P_n)$ of the direction field $y' = F(x, y)$ at $P_n(x_n, y_n)$ where P_n is not on the solution curve having equation $y = y(x)$ but is a point near this curve.

For the solution curve depicted in Figure 9.5 it appears that we might obtain a better approximation for $y(x_1)$ by using the slope of the direction field at a point other than P_0. The point $A\left(x_0 + \frac{h}{2}, y\left(x_0 + \frac{h}{2}\right)\right)$ shown in Figure 9.6

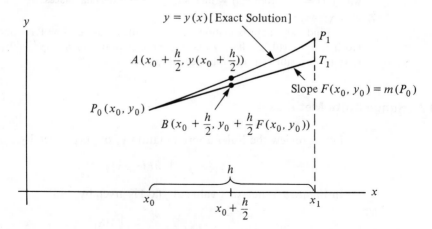

Fig. 9.6

would be a reasonable choice but we do not know $y(x_0 + \frac{h}{2})$. Instead of A we could use the point B on $P_0 T_1$, where B also has abscissa $x_0 + \frac{h}{2}$. It is easy to see that B has ordinate $y_0 + \frac{h}{2} F(x_0, y_0)$.

We then approximate $y(x_1)$ by

$$y_1 = y_0 + hm(B)$$

where $m(B)$ is the slope of the direction field of $y' = F(x, y)$ at point B. Iteration of this process yields the *midpoint method*, which has order of accuracy h^2.

In the improved Euler method we approximate $y(x_1)$ by

$$y_1 = y_0 + \frac{h}{2} [F(x_0, y_0) + F(x_1, z_1)]$$

$$= y_0 + \frac{h}{2} [F(x_0, y_0) + F(x_1, y_0 + hF(x_0, y_0))]$$

$$y_1 = y_0 + h \left[\frac{m(P_0) + m(T_1)}{2} \right] \tag{9.19}$$

where $m(P_0)$ and $m(T_1)$ denote the slopes of the direction field of $y' = F(x, y)$ at points P_0 and T_1 in Figure 9.5.

Thus, the method consists of replacing M in

$$y(x_1) = y_0 + hM \tag{9.18}$$

by

$$\frac{m(P_0) + m(T_1)}{2} = \frac{1}{2} m(P_0) + \frac{1}{2} m(T_1)$$

which may be regarded as a weighted average of the slopes $m(P_0)$ and $m(T_1)$. Since each weight has value $\frac{1}{2}$, the average slope is the arithmetic mean of the slopes $m(P_0)$ and $m(T_1)$.

This view of the improved Euler method suggests the possibility of obtaining greater accuracy by replacing M in (9.18) by a weighted average of the slopes of the direction field of $y' = F(x, y)$ at several points located near the portion between P_0 and P_1 of the solution curve $y = y(x)$. About 1894, the German mathematician C. D. T. Runge (1865–1927) employed this approach to devise a numerical method for solving DE approximately. A few years later, M. W. Kutta, another German mathematician, modified and extended Runge's results. Their combined efforts are known as *Runge-Kutta methods*. They are highly accurate and widely used, particularly when the computations are performed on a digital computer.

A Runge-Kutta method is called an *nth order Runge-Kutta* method if the expansion in powers of h of the right member $y_0 + hM$ of (9.18) agrees with the first $(n + 1)$ terms of the Taylor series for $y(x_0 + h) = y(x_1)$.

For example, in the improved Euler formula

$$y_1 = y_0 + h\left[\frac{1}{2}m(P_0) + \frac{1}{2}m(T_1)\right]$$ (9.19)

the right member can be written in the form

$$y_0 + \frac{y'(x_0)}{1!}h + \frac{y''(x_0)}{2!}h^2 + \text{(terms involving } h^3 \text{ and higher powers of } h)$$

We mentioned this previously when discussing the order of accuracy of the improved Euler method. A proof that a given Runge-Kutta method has a specified order is usually quite complicated. Thus the improved Euler method is a Runge-Kutta method of order two. It requires the evaluation of $F(x, y)$ at two points, P_0 and T_1. A Runge-Kutta method of order n requires the evaluation of $F(x, y)$ at n or more points.

We now give a brief description of a Runge-Kutta method of order three. In Figure 9.7 point A is the intersection of the line $x = x_1$ and the line through $P_0(x_0, y_0)$ having slope $m(P_0) = F(x_0, y_0)$. Point B is the midpoint of line segment $P_0 A$. The abscissa of B is $x_0 + \frac{h}{2}$; the ordinate of B is $y_0 + \frac{h}{2}m(P_0)$; and the slope of the direction field of $y' = F(x, y)$ at B is $m(B) = F\left(x_0 + \frac{h}{2}, y_0 + \frac{h}{2}m(P_0)\right)$. Point C is the intersection of the line $x = x_1$ and the line through $P_0(x_0, y_0)$ having slope $m(A) = F(x_1, y_0 + hm(P_0))$. The ordinate of C is $y_0 + hm(A)$ and the slope of the direction field of $y' = F(x, y)$ at C is $m(C) = F(x_1, y_0 + hm(A))$. We next replace M, the slope of the line through (x_0, y_0) and $(x_1, y(x_1))$ in Figure 9.7, in

$$y(x_1) = y_0 + hM$$ (9.18)

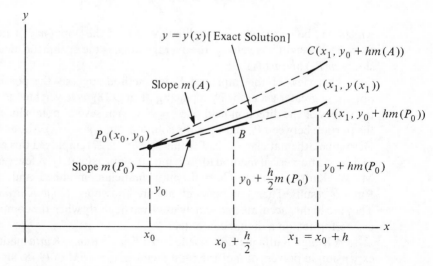

Fig. 9.7

by

$$\frac{1}{6} m(P_0) + \frac{2}{3} m(B) + \frac{1}{6} m(C)$$

to obtain the approximation of $y(x_1)$ given by

$$y_1 = y_0 + h\left[\frac{1}{6} m(P_0) + \frac{2}{3} m(B) + \frac{1}{6} m(C)\right] \qquad (9.20)$$

This approximation of M is a weighted average of the slopes of the direction field of $y' = F(x, y)$ at points $P_0, B,$ and C. The slopes at P_0 and C are given equal weights, and the slope at B is given four times the weight of the slope at P_0. The slope $m(A)$ of the direction field at point A is not involved in (9.20); the value $m(A)$ is employed only to locate point C. The slopes $m(P_0)$, $m(A)$, $m(B)$, and $m(C)$ of the direction field at points $P_0, A, B,$ and C are illustrated graphically in Figure 9.7 by line elements drawn at points $P_0, A, B,$ and C.

We assert without proof that the Runge-Kutta method presented has order three. That is, the right member of (9.20) can be written in the form

$$y_0 + \frac{y'(x_0)}{1!} h + \frac{y''(x_0)}{2!} h^2 + \frac{y'''(x_0)}{3!} h^3$$

$$+ \text{(terms involving } h^4 \text{ and higher powers of } h)$$

In (9.20) a weighted average of the slopes of the direction field of $y' = F(x, y)$ at three points is employed. This is not the only third-order Runge-Kutta method available. The wide latitude in the number of points chosen, the locations of these points, and the weights assigned to the slopes at the chosen points results in infinitely many Runge-Kutta methods of any given order.

One of the main advantages of Runge-Kutta methods over the method of Taylor series is that they avoid the evaluation of the partial derivatives of F and use instead the values of F at several points.

9.8 The Classical Runge-Kutta Method

In this section we present a fourth-order Runge-Kutta method. It is highly accurate and is one of the most widely used numerical methods, particularly when the computations are performed on a digital computer. It is known as *the classical Runge-Kutta method* and is also referred to as simply *the Runge-Kutta method* or *the fourth-order Runge-Kutta method*.

As depicted in Figure 9.8, the method uses values of $y' = F(x, y)$ at four points, the first of which is $P_0(x_0, y_0)$ at which $m(P_0) = F(x_0, y_0)$. The second point $B(x_0 + \frac{h}{2}, y_0 + \frac{h}{2} m(P_0))$ is located on the line through P_0 that is tangent to the solution curve $y = y(x)$ at P_0. The third point $C(x_0 + \frac{h}{2}, y_0 + \frac{h}{2} m(B))$ is located on the line through P_0 that has slope $m(B) = F(x_0 + \frac{h}{2}, y_0 + \frac{h}{2} m(P_0))$. That is, the slope of the direction field at the second point B is

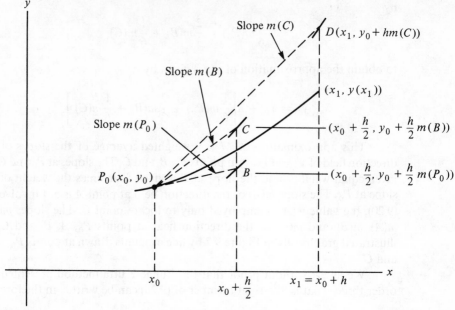

Fig. 9.8

employed to locate the third point C. Similarly, the slope $m(C)$ of the direction field at the third point C is employed to locate the fourth point $D(x_1, y_0 + hm(C))$. That is, D is on the line through P_0 that has slope

$$m(C) = F\left(x_0 + \frac{h}{2},\ y_0 + \frac{h}{2}m(B)\right)$$

We next replace M, the slope of the line through (x_0, y_0) and $(x_1, y(x_1))$, in

$$y(x_1) = y_0 + hM \tag{9.18}$$

by

$$\frac{1}{6}m(P_0) + \frac{1}{3}m(B) + \frac{1}{3}m(C) + \frac{1}{6}m(D)$$

to obtain

$$y_1 = y_0 + h\left[\frac{1}{6}m(P_0) + \frac{1}{3}m(B) + \frac{1}{3}m(C) + \frac{1}{6}m(D)\right] \tag{9.21}$$

The slopes of the direction field of $y' = F(x, y)$ at P_0, B, C, and D are illustrated graphically in Figure 9.8 by line elements drawn at these four points.

To find y_{n+1} from y_n we iterate the process of (9.21) for finding y_1 from y_0. We then compute the following four numbers: q_1, q_2, q_3, and q_4, after which y_{n+1} is found from (9.22).

$$q_1 = hF(x_n, y_n)$$

$$q_2 = hF\left(x_n + \frac{h}{2}, y_n + \frac{q_1}{2}\right)$$

$$q_3 = hF\left(x_n + \frac{h}{2}, y_n + \frac{q_2}{2}\right)$$

$$q_4 = hF(x_n + h, y_n + q_3)$$

$$y_{n+1} = y_n + \frac{1}{6}[q_1 + 2q_2 + 2q_3 + q_4] \qquad \text{(9.22)}$$

The ordinates of B, C, and D and the weights $\frac{1}{6}, \frac{1}{3}, \frac{1}{3}$, and $\frac{1}{6}$ are so chosen that a fourth-order Runge-Kutta method is obtained. That is, the choices are made in such a manner that the right member of (9.21) can be written in the form

$$y_0 + \frac{y'(x_0)}{1!}h + \frac{y''(x_0)}{2!}h^2 + \frac{y'''(x_0)}{3!}h^3 + \frac{y^{(iv)}(x_0)}{4!}h^4$$

$$+ \text{(terms involving } h^5 \text{ and higher powers of } h)$$

The proof that the classical Runge-Kutta method is a fourth-order method is not difficult. However, it is long and involves considerable algebraic manipulation. See Reference 9.4 for details. Problem 17 of Problem List 9.4, which asks for $y^{(iv)}(x_0)$ in terms of $F(x_0, y_0)$ and the partial derivatives of F at (x_0, y_0), gives an indication of the algebraic complications required in the proof in question.

The method has order of accuracy h^4; that is, with the same assumptions on F as were made in the improved Euler method, it can be shown that there exists a constant $K > 0$ such that the error $e_n = y_n - y(x_n)$ satisfies $|e_n| \leq Kh^4$, where K is independent of h. If h is replaced by $h/2$, the absolute error satisfies

$$|e_n| \leq K\left(\frac{h}{2}\right)^4 = \frac{1}{16}(Kh^4)$$

and has maximum value $\frac{1}{16}$ of what it was for step size h.

The method has many advantages. The high order of accuracy does not require a particularly small step size, and hence roundoff error and computer cost tend to be small. It is explicit and hence does not require the solution of an algebraic equation to find y_{n+1} from y_n. Like other one-step methods, it allows for a variable step size when desirable. It is readily applied to numerical solutions of systems of DE and hence to numerical solutions of higher-order DE.

The main disadvantage is that $F(x, y)$ must be evaluated at four points at each step in the process. This involves considerable effort when several steps are used in a hand computation.

It is interesting to note that when $F(x, y) = F(x)$, then $q_2 = q_3$, and (9.22) can be written in the form

$$y_{n+1} - y_n = \frac{h/2}{3}\left[F(x_n) + 4F\left(x_n + \frac{h}{2}\right) + F(x_n + h)\right]$$

Thus, in this special case, the Runge-Kutta method reduces to Simpson's rule.

EXAMPLE 1. Apply the Runge-Kutta method to the test problem $y' = y$, $y(0) = 1$, to approximate $y(0.6)$, using (a) $h = 0.6$, and (b) $h = 0.2$.

Solution: (a) $q_1 = 0.6(1) = 0.6$

$$q_2 = 0.6\left(1 + \frac{0.6}{2}\right) = 0.78$$

$$q_3 = 0.6\left(1 + \frac{0.78}{2}\right) = 0.834$$

$$q_4 = 0.6(1.834) = 1.1004$$

$$y_1 = 1 + \frac{1}{6}[0.6 + 2(0.78) + 2(0.834) + 1.1004] = 1.8214$$

Since $y(x) = e^x$ and $y(0.6) = e^{0.6} \approx 1.8221$, in a single step our approximation is correct to two decimal places. The relative error, given by

$$\frac{|1.8214 - 1.8221|}{1.8221} = \frac{0.0007}{1.8221} \approx 0.0004$$

is approximately 0.04%, compared with 0.03% using the improved Euler method with $h = 0.2$. (See illustrative Example 1, Section 9.6.)

(b) Since three applications of (9.22) are required, we use a table with entries rounded off to five decimal places.

x_n	y_n	q_1	q_2	q_3	q_4	y_{n+1}
0.0	1.0	0.2	0.22	0.222	0.2444	1.2214
0.2	1.2214	0.24428	0.26871	0.27115	0.29851	1.49182
0.4	1.49182	0.29836	0.32820	0.33118	0.36460	1.82211
0.6	1.82211					

Since $y(0.6) = e^{0.6} = 1.82212$ correct to five decimal places, the relative error, given by

$$\frac{0.00001}{1.82212} \approx 0.000005$$

is approximately 0.0005%.

This example illustrates the remarkable accuracy of the Runge-Kutta method, even when a fairly large step size is used. We also note that the calculations, particularly in part (b), are somewhat cumbersome. The method is not self-correcting, and a single error is perpetuated throughout the solution. When a numerical method is required in an applied problem, one does not have available values of the exact solution with which intermediate calculations can be compared. The table in part (b) was constructed with a hand calculator.

EXAMPLE 2. Apply the Runge-Kutta method to the initial-value problem $y' = x + y$, $y(0) = 1.24$, with $h = 1$ to approximate $y(1) = 2.24e - 2 \approx 4.0890$. $[y(x) = 2.24e^x - x - 1.]$

Solution:

$$q_1 = 1(0 + 1.24) = 1.24$$

$$q_2 = 1(0 + 0.5 + 1.24 + 0.62) = 2.36$$

$$q_3 = 1(0 + 0.5 + 1.24 + 1.18) = 2.92$$

$$q_4 = 1(0 + 1 + 1.24 + 2.92) = 5.16$$

$$y_1 = 1.24 + \left[\frac{1.24 + 2(2.36) + 2(2.92) + 5.16}{6} \right] \approx 4.0667$$

The relative error, given by

$$\frac{|4.0667 - 4.0890|}{4.0890} = \frac{0.0223}{4.0890} \approx 0.0055$$

is approximately 0.6%. This compares favorably with the relative error of approximately 0.9% with the improved Euler method (illustrative Example 2, Section 9.6), even using a Runge-Kutta step size five times that used in the improved Euler method.

The following BASIC program applies the Runge-Kutta method to the initial-value problem $y' = 2x + y$, $y(0) = 1$.

```
100   READ    X0, Y0, X1, N
110   DATA    0, 1, 0.5, 5
120   LET     H = (X1 - X0)/N
130   DEF     FNF(X, Y) = 2 * X + Y
140   LET     Y = Y0
150   FOR     J = 1 TO N
160   LET     X = X0 + (J - 1)H
170   LET     Q1 = H * FNF(X, Y)
```

```
180  LET     Q2 = H * FNF(X + H/2, Y + Q1/2)
190  LET     Q3 = H * FNF(X + H/2, Y + Q2/2)
200  LET     Q4 = H * FNF(X + H, Y + Q3)
210  LET     Y = Y + (Q1 + 2 * Q2 + 2 * Q3 + Q4)/6
220  NEXT    J
230  LET     X = X + H
240  PRINT   X, Y
250  END
```

Instruction 110 assigns the values $X0 = 0$, $Y0 = 1$, $X1 = 0.5$, and $N = 5$ to the variables defined in instruction 100. The program approximates $y(0.5)$ using five steps, each of step size $h = 0.1$. Instruction 240 tells the computer to print $x = 0.5$ and the Runge-Kutta approximation to $y(0.5)$. The value printed by the computer is 1.94616. From the exact solution given by $y(x) = 3e^x - 2x - 2$, we obtain $y(0.5) = 3\sqrt{e} - 3 \approx 1.94616$. Thus, the Runge-Kutta approximation is correct to five decimal places.

To change $Y0$ or H, it is only necessary to retype instruction 110 and assign new values to the variables defined in instruction 100. To apply the program to a new initial-value problem, instructions 110 and 130 must be changed. The change in 130 merely inserts the new $F(x, y)$. It is not difficult to modify the program slightly so that the step size is reduced automatically. If the computer is properly programmed, it will reduce the step size until two (or even three) consecutive approximations agree to the number of decimal places desired in the answer.

Numerous variations of the method of this section have been developed. See Reference 9.4.

The Runge-Kutta method is often used as a starting method when a multi-step method is employed. To illustrate, assume that we are employing the multi-step recursion formula

$$y_{n+4} = y_n + \frac{4h}{3}[2F(x_{n+1}, y_{n+1}) - F(x_{n+2}, y_{n+2}) + 2F(x_{n+3}, y_{n+3})]$$

$$(9.23)$$

due to W. E. Milne (1890–1971). See Section 9.3 and Reference 9.11. Setting $n = 0$ in the Milne formula, we obtain

$$y_4 = y_0 + \frac{4h}{3}[2F(x_1, y_1) - F(x_2, y_2) + 2F(x_3, y_3)] \qquad (9.24)$$

This recursion formula (9.24) cannot be used directly to find y_4 since, of the four numbers y_0, y_1, y_2, and y_3, only y_0 is known. We therefore "start" the Milne process by obtaining values of y_1, y_2, and y_3 by the Runge-Kutta method. Formula (9.24) is then used to find y_4, after which y_5 is obtained from

$$y_5 = y_1 + \frac{4h}{3}[2F(x_2, y_2) - F(x_3, y_3) + 2F(x_4, y_4)]$$

and each succeeding approximation of $y(x_n)$ is found from the preceding four approximations.

A multi-step method of the Milne type uses values of $F(x, y)$ that have been computed in previous steps whereas Runge-Kutta methods require the computation of intermediate values of $F(x, y)$. For example, the Runge-Kutta formulas (9.22) require four computations of $F(x, y)$ for each step, and these values of $F(x, y)$ are not used in succeeding steps. Thus multi-step methods often require less computing time and hence are generally less expensive to use. On the other hand, one-step methods have the advantage of simplicity and are suitable in many applications.

9.9 Numerical Solutions of Systems of Differential Equations

Methods like the improved Euler and the Runge-Kutta are readily extended to apply to systems of DE. Consider, for example, a system of the type

$$\frac{dy}{dt} = F(x, y, t), \quad y(t_0) = y_0 \tag{9.25}$$

$$\frac{dx}{dt} = G(x, y, t), \quad x(t_0) = x_0 \tag{9.26}$$

Each member of the system is treated separately. When the approximation to $y(t_0 + h)$ is obtained from (9.25), the approximation to $x(t_0 + h)$ must be obtained from (9.26) since it is needed to find the approximation to $y(t_0 + 2h)$.

Equations (9.13) and (9.14) of the improved Euler method become

$$\left. \begin{aligned} z_{n+1} &= y_n + hF(x_n, y_n, t_n) \\[2mm] w_{n+1} &= x_n + hG(x_n, y_n, t_n) \\[2mm] y_{n+1} &= y_n + \frac{h}{2}[F(x_n, y_n, t_n) + F(w_{n+1}, z_{n+1}, t_{n+1})] \\[2mm] x_{x+1} &= x_n + \frac{h}{2}[G(x_n, y_n, t_n) + G(w_{n+1}, z_{n+1}, t_{n+1})] \end{aligned} \right\} \tag{9.27}$$

and

EXAMPLE 1. Solve the system

$$\frac{dy}{dt} = 2x - 1, \quad y(0) = 0.5$$

$$\frac{dx}{dt} = 2y + 1, \quad x(0) = 3.5$$

to approximate $y(0.2)$ and $x(0.2)$ using $h = 0.1$. Use four decimal places.

Solution: Equations (9.27) become

$$z_{n+1} = y_n + 0.1(2x_n - 1)$$

$$w_{n+1} = x_n + 0.1(2y_n + 1)$$

$$y_{n+1} = y_n + 0.05[(2x_n - 1) + (2w_{n+1} - 1)]$$

and

$$x_{n+1} = x_n + 0.05[(2y_n + 1) + (2z_{n+1} + 1)]$$

from which we construct the accompanying table.

t_n	x_n	y_n	w_{n+1}	z_{n+1}	x_{n+1}	y_{n+1}
0.0	3.5	0.5	3.7	1.1	3.78	1.12
0.1	3.78	1.12	4.104	1.776	4.1695	1.8084
0.2	4.1695	1.8084				

From the exact solution of the system, given by

$$x(t) = 2e^{2t} + e^{-2t} + 0.5, \quad y(t) = 2e^{2t} - e^{-2t} - 0.5$$

we compute, for purposes of comparison,

$$x(0.2) \approx 4.1540, \quad y(0.2) \approx 1.8133$$

The formulas for solving a system of type (9.25) and (9.26) numerically by extending the Runge-Kutta method are given in Reference 9.12.

9.10 Numerical Solutions of Higher-Order Differential Equations

The standard procedure for solving an initial-value problem for a differential equation of order greater than or equal to two is to replace the problem by an initial-value problem for a system of DE. This kind of replacement was discussed in Section 6.1. For example, to solve the initial-value problem

$$y'' = -y; \quad y(0) = 0, \quad y'(0) = 1$$

where $y = y(t)$, we let $x = y'$ and solve the initial-value problem

$$y' = x, \quad y(0) = 0$$

$$x' = -y, \quad x(0) = 1$$

by the method of Section 9.9. With $F(x, y, t) = x$ and $G(x, y, t) = -y$ in (9.25) and (9.26), we obtain

$$z_{n+1} = y_n + hx_n$$

$$w_{n+1} = x_n + h(-y_n)$$

$$y_{n+1} = y_n + \frac{h}{2}[x_n + w_{n+1}] \tag{9.28}$$

and

$$x_{n+1} = x_n + \frac{h}{2}[-y_n - z_{n+1}]$$

EXAMPLE 1. Use Equations (9.28) to approximate $y(0.2)$, using $h = 0.1$.

Solution: The required approximation $y_2 = 0.199$ is obtained in the accompanying table.

t_n	x_n	y_n	w_{n+1}	z_{n+1}	x_{n+1}	y_{n+1}
0.0	1.0	0.0	1.0	0.1	0.995	0.1
0.1	0.995	0.1	0.985	0.1995	1.009975	0.199
0.2	1.009975	0.199				

The exact solution is given by $y(t) = \sin t$. The value of $\sin 0.2$ correct to six decimal places is 0.198669.

For more details on numerical solutions of higher-order DE, see Reference 9.12. It is sometimes advantageous, instead of replacing a higher-order DE by a system of DE, to apply a numerical method directly to the higher-order DE. For such direct methods, see Reference 9.4.

9.11 Numerical Solutions of Boundary-Value Problems

If a boundary-value problem has the form

$$\frac{d^2y}{dt^2} = y'' = G(y', y, t); \quad y(a) = A, \, y(b) = B$$

where $y = y(t)$, one method of obtaining a numerical solution assumes that the value of $y'(a) = m_0$. The number m_0 is an estimate of what the slope of the integral curve through (a, A) must be so that the same integral curve will pass through (b, B). Then, letting $x = y' = dy/dt$, as in Section 9.10, we solve the initial-value problem.

$$y' = x, \qquad y(a) = A$$

$$x' = G(x, y, t), \quad x(a) = m_0$$

Let this problem have solution given by $y = \phi(t)$ and let $\phi(b) = B_0$. If $B_0 = B$, scarcely to be expected, the boundary-value problem is solved with solution given by $y(t) = \phi(t)$. If $B_0 \neq B$, we repeat the process with a second guess m_1 replacing m_0 for $y'(a)$. Let the new solution be given by $y = \psi(t)$ and let $\psi(b) = B_1$. The next guess m_2 is found by linear interpolation from

$$m_2 = m_1 + \frac{B - B_1}{B_0 - B_1}(m_0 - m_1)$$

The formula is based on the assumption that the change in the slope of the integral curve through (a, A) is proportional to the change in the value of the height of the integral curve at $t = b$. The process is continued until an integral curve is obtained whose height at $t = b$ agrees with B to some desired number of decimal places. The numerical procedure is referred to as the *method of shooting*. This is suggested by the problem of determining the slope of a trajectory through (a, A) that will hit the point (b, B). See Reference 9.3.

The computations required to solve systems, higher-order equations, and boundary-value problems are often involved. Computer programs are available for applying various methods to these problems.

Problem List 9.4

Apply the Runge-Kutta method in Problems 1–12.
1. $y' = y$, $y(0) = 1$.

Approximate $y(0.5)$, using $h = 0.5$. Compare the result with $y(0.5) = e^{0.5}$. Use four decimal places.
2. $y' = x + y$, $y(0) = 1$.

Approximate $y(1)$ using (a) $h = 1$, and (b) $h = 0.5$. Compare results with the exact answer, using $y(x) = 2e^x - x - 1$. Use four decimal places.
3. $y' = -y$, $y(0) = 1$.

Approximate $y(0.25)$, using $h = 0.25$. Compare the result with $y(0.25) = e^{-0.25}$. Use five decimal places.
4. $y' = xy^2 - y$, $y(0) = 1$.

Approximate $y(0.3)$, using $h = 0.3$. Compare the result with $y(0.3) = \frac{10}{13}$ obtained from $y(x) = (1 + x)^{-1}$. Use four decimal places.
5. $y' = 2x + y$, $y(0) = 1$.

Approximate $y(0.5)$, using $h = 0.1$ (computer problem). Solve the DE exactly and compare the computer answer with $y(0.5)$.
6. $y' = -6y$, $y(0) = 100$.

Approximate $y(0.6)$, using (a) $h = 0.2$, (b) $h = 0.1$, and (c) $h = 0.02$ (computer problem). Solve the DE exactly and compare the computer answers with $y(0.6)$.
7. $y' = \dfrac{2x + y}{x}$, $y(1) = 1$.

Approximate $y(2)$, using (a) $h = 0.5$, and (b) $h = 0.1$ (computer problem). Solve the DE exactly and compare the computer answers with $y(2)$.

8. $y' = x^2 + y^2$, $y(0) = 1$.

Approximate $y(0.5)$ by decreasing h until two successive Runge-Kutta approximations agree to three decimal places (computer problem).

9. $y' = x^2 - y^2$, $y(1) = 1$.

Approximate $y(2)$ by decreasing h until two successive Runge-Kutta approximations agree to three decimal places (computer problem).

10. $y' = 1 - x^2 - y^2$, $y(0) = 0$.

Approximate $y(1)$ by decreasing h until three successive Runge-Kutta approximations agree to two decimal places (computer problem).

11. $y' = y^2 - x^2$, $y(0) = 0.5$.

Approximate $y(1)$ by decreasing h until two successive Runge-Kutta approximations agree to four decimal places (computer problem).

12. An emf $E(t) = 200\sqrt{t}e^{-4t}$ is connected in series with a 10-ohm (Ω) resistor and a 0.01-farad (F) capacitor. The charge Q in coulombs on the capacitor is zero at time $t = 0$ sec. Approximate the charge Q at $t = 0.5$ sec by decreasing h until two successive Runge-Kutta approximations agree to three decimal places (computer problem).

13. Apply the Runge-Kutta method to the initial-value problem $y' = (1 + x^2)^{-1}$, $y(0) = 0$, to approximate π to three decimal places.

14. Apply the improved Euler method to the system

$$\frac{dy}{dt} = -2x + 3y, \quad y(0) = 3$$

$$\frac{dx}{dt} = y, \qquad\qquad x(0) = 2$$

to approximate $y(0.2)$ and $x(0.2)$, using $h = 0.1$. Use four decimal places. Compare these approximations with the exact values of $y(0.2)$ and $x(0.2)$, rounded off to four decimal places.

15. Replace the initial-value problem $y'' = -y$; $y(0) = 1$, $y'(0) = 0$, by an initial-value problem for a system of DE. Apply the improved Euler method to the system to approximate $y(0.2)$ to four decimal places, using $h = 0.1$. Compare the approximation with the exact value of $y(0.2)$ rounded off to four decimal places.

16. The motion of a pendulum is governed by the DE $d^2\theta/dt^2 + 10 \sin \theta = 0$. If $\theta = 0.3$ rad and $d\theta/dt = 0$ rad/sec when $t = 0$ sec, what is the approximate value of θ as found by the improved Euler method for $t = 0.2$ sec? Use $h = 0.1$ and give the approximation to four decimal places.

17. Given $y' = F(x, y)$, show that

$$y^{(iv)} = F_{xxx} + 3F_x F_{xy} + F_{xx} F_y + F_x F_y^2 + 3FF_{xxy} + 5FF_{xy}F_y + 3FF_x F_{yy}$$
$$+ FF_y^3 + 3F^2 F_{xyy} + 4F^2 F_y F_{yy} + F^3 F_{yyy}$$

References

9.1 Birkhoff, G., and G.-C. Rota. 1969. *Ordinary Differential Equations*, 2nd ed. New York: Ginn.

9.2 Braun, M. 1978. *Differential Equations and Their Applications*, 2nd ed. New York: Springer-Verlag.

9.3 Daniel, J. W., and R. E. Moore. 1970. *Computation and Theory in Ordinary Differential Equations*. San Francisco: W. H. Freeman.

9.4 Gear, C. W. 1971. *Numerical Initial Value Problems in Ordinary Differential Equations*. Englewood Cliffs, N.J.: Prentice-Hall.

9.5 Henrici, P. 1967. *Elements of Numerical Analysis*. New York: John Wiley & Sons.

9.6 Henrici, P. 1961. *Numerical Integration of Ordinary Differential Equations*. New York: John Wiley & Sons.

9.7 Hildebrand, F. B. 1956. *Introduction to Numerical Analysis*. New York: McGraw-Hill.

9.8 Johnson, C. L. 1956. *Analogue Computer Techniques*. New York: McGraw-Hill.

9.9 Kaplan, Wilfred. 1962. *Ordinary Differential Equations*. Reading, Mass.: Addison-Wesley.

9.10 Levy, H., and E. Baggott. 1950. *Numerical Solutions of Differential Equations*. New York: Dover.

9.11 Milne, W. E. 1970. *Numerical Solution of Differential Equations*. New York: Dover.

9.12 Scarborough, J. B. 1966. *Numerical Mathematical Analysis*, 6th ed. Baltimore: Johns Hopkins University Press.

9.13 Simmons, G. F. 1972. *Differential Equations with Applications and Historical Notes*. New York: McGraw-Hill.

9.14 Soroka, W. W. 1954. *Analog Methods in Computation and Simulation*. New York: McGraw-Hill.

10

Partial Differential Equations; Boundary-Value Problems; Fourier Series

10.1 Basic Concepts and Definitions

A partial differential equation is defined as a DE involving one or more unknown functions and their partial derivatives. In our study, we shall restrict ourselves to partial DE involving a single unknown real function f of two or more real variables. For example, if $u = f(x, y)$, a second-order partial DE satisfied by u will have the form

$$F\left(x, y, u, \frac{\partial u}{\partial x}, \frac{\partial u}{\partial y}, \frac{\partial^2 u}{\partial x^2}, \frac{\partial^2 u}{\partial y^2}, \frac{\partial^2 u}{\partial y\, \partial x}\right) = 0 \qquad (10.1)$$

We assume that (10.1) involves at least one of the partial derivatives of u. We also assume that mixed partial derivatives of u are independent of the order of differentiation; for example $\partial^2 u/\partial x\, \partial y = \partial^2 u/\partial y\, \partial x$.

By a solution of (10.1), we mean a real function f, given by $u = f(x, y)$, defined on a set S of nonoverlapping regions in the xy plane such that (10.1) is satisfied at every point of S. For simplicity, we restrict ourselves to domains of f consisting of a single region R.

By a region R in the xy plane, we mean an open connected set. We say that R is open if for every point P in R there exists some circle in the xy plane, with center at P, which lies entirely in R. We say that R is connected if any two points in R can be joined by a polygonal path consisting entirely of points of R.

To illustrate, R might be the entire xy plane, the points in the first quadrant, the interior of a circle, an ellipse, or a rectangle, the points inside the circle $x^2 + y^2 = (r + 1)^2$ and outside the circle $x^2 + y^2 = r^2$, the points in the infinite strip $0 < y < 1$, and so on.

For a partial DE involving a function u of more than two variables, the domain of a solution is defined similarly as a region R in the space of three or more variables.

EXAMPLE 1. Show that $u = f(x, y) = x^2 - y^2$ defines a solution of the partial DE

$$\frac{\partial^2 u}{\partial x^2} + \frac{\partial^2 u}{\partial y^2} = 0$$

Solution: From $\dfrac{\partial u}{\partial x} = 2x, \dfrac{\partial^2 u}{\partial x^2} = 2, \dfrac{\partial u}{\partial y} = -2y$, and $\dfrac{\partial^2 u}{\partial y^2} = -2$, we obtain

$$\frac{\partial^2 u}{\partial x^2} + \frac{\partial^2 u}{\partial y^2} = 2 - 2 \equiv 0$$

The domain of f is the entire xy plane.

EXAMPLE 2. Show that $u = f(x, y) = \ln(x^2 + y^2)$ defines a solution of the partial DE

$$\frac{\partial^2 u}{\partial x^2} + \frac{\partial^2 u}{\partial y^2} = 0$$

Solution: From $\dfrac{\partial u}{\partial x} = \dfrac{2x}{x^2 + y^2}, \dfrac{\partial^2 u}{\partial x^2} = \dfrac{2y^2 - 2x^2}{(x^2 + y^2)^2},$

$$\frac{\partial u}{\partial y} = \frac{2y}{x^2 + y^2} \quad \text{and} \quad \frac{\partial^2 u}{\partial y^2} = \frac{2x^2 - 2y^2}{(x^2 + y^2)^2}$$

we obtain

$$\frac{\partial^2 u}{\partial x^2} + \frac{\partial^2 u}{\partial y^2} = \frac{2y^2 - 2x^2}{(x^2 + y^2)^2} + \frac{2x^2 - 2y^2}{(x^2 + y^2)^2} \equiv 0$$

provided that $x^2 + y^2 \neq 0$. The domain of f is the xy plane with the origin deleted. NOTE: By symmetry, u_{yy} can be obtained by interchanging x and y in u_{xx}.

Solutions of partial DE tend to be very general. The quite dissimilar functions in Examples 1 and 2 satisfy the same partial DE. Another aspect of this great generality is illustrated in Examples 3 and 4.

EXAMPLE 3. Solve $z_x = 2xy$, where $z = f(x, y)$.

Solution: Holding y fixed and integrating with respect to x, we obtain $z = f(x, y) = x^2 y + \phi(y)$ where ϕ is an arbitrary function of y. The arbitrary function ϕ of y is added instead of a constant of integration, since $\partial \phi / \partial x = 0$ for arbitrary ϕ.

Thus, whereas solutions of ordinary DE contain arbitrary constants, solutions of partial DE contain arbitrary functions.

EXAMPLE 4. Solve $z_{yx} = 0$, where $z = f(x, y)$.

Solution: Holding y constant and integrating with respect to x, we obtain

$$z_y = \theta(y) \tag{10.2}$$

Holding x constant in (10.2) and integrating with respect to y, we obtain $z = f(x, y) = \int \theta(y)\, dy + \psi(x)$, which we write as

$$z = f(x, y) = \phi(y) + \psi(x) \tag{10.3}$$

where ϕ and ψ are arbitrary. No constant of integration is added in the right member of (10.3), since any constant may be included in either $\phi(y)$ or $\psi(x)$. Although we refer to functions such as ϕ and ψ as arbitrary, we assume tacitly that such functions possess whatever derivatives are required for them to occur in a solution of a given partial DE.

Definition The *order* of a partial DE is the order of the highest-order partial derivative appearing in the DE.

Examples 3 and 4 suggest that solving an nth-order partial DE will involve n arbitrary functions. However, the concept of "general" or "complete" solution of a partial DE is seldom useful in an applied problem. When initial or boundary conditions are known in a physical situation governed by an ordinary DE, the particular solution required is usually obtained by finding a general or complete solution, and then, by satisfying the given conditions, obtaining the required particular solution as a special case. When a physical situation is governed by a partial DE, it is more fruitful to construct the required particular solution by applying the initial and boundary conditions one at a time until the particular solution presents itself. Virtually no use is made of a "general" solution. The procedure will be developed in the ensuing sections.

Definitions A partial DE is termed *linear* if and only if it is linear in the unknown function u and the partial derivatives of u. All other partial DE are termed *nonlinear*.

If each term of a linear partial DE contains the unknown function u or one of the partial derivatives of u, the partial DE is *homogeneous*; otherwise it is *nonhomogeneous*.

EXAMPLE 5. A second-order linear partial DE for a function u of two variables can be written in the form

$$Au_{xx} + Bu_{xy} + Cu_{yy} + Du_x + Eu_y + Fu = G$$

where A, B, C, D, E, F, and G are functions of x and y with A, B, and C not all zero.

EXAMPLE 6. The second-order partial DE $\partial^2 u/\partial x^2 + \partial^2 u/\partial y^2 = 0$ is homogeneous.

EXAMPLE 7. The second-order DE $\partial^2 u/\partial x^2 + \partial^2 u/\partial y^2 = x$ is nonhomogeneous.

We next state a theorem that is very useful in building up or constructing a particular solution of a partial DE satisfying given initial and boundary conditions.

Theorem 10-I If $u_1 = u_1(x, y)$ and $u_2 = u_2(x, y)$ denote any two solutions of a linear homogeneous partial DE in a region R of the xy plane, then every linear combination

$$u(x, y) = c_1 u_1(x, y) + c_2 u_2(x, y)$$

defines a solution of the same DE in R.

The proof of Theorem 10-I is similar to the proof of Theorem 4-II. The principle of building solutions by superposition is easily extended to a linear combination

$$c_1 u_1(x, y) + c_2 u_2(x, y) + \cdots + c_n u_n(x, y)$$

of any *finite* number n of solutions $u_1, u_2, \ldots, u_n$.

As in the case of ordinary differential equations, partial DE arise when arbitrary constants are eliminated from a given relation by differentiation.

EXAMPLE 8. Find a partial DE of the relation

$$z = ax^2 + by^2, \quad ab > 0 \tag{10.4}$$

Solution: From $z_x = 2ax$ and $z_y = 2by$, we obtain

$$z = \frac{z_x}{2x} x^2 + \frac{z_y}{2y} y^2$$

or

$$xz_x + yz_y = 2z \tag{10.5}$$

Interpreting the result geometrically, we say that (10.5) is a partial DE of the two-parameter family (10.4) of elliptic paraboloids.

Partial DE may also be obtained by eliminating arbitrary functions from a relation.

EXAMPLE 9. Find a partial DE of the relation $z = f(x - y)$.

Solution: Let $z = f(u)$ where $u = x - y$. Then,

$$\frac{\partial z}{\partial x} = \frac{dz}{du} \frac{\partial u}{\partial x} = \frac{dz}{du} (1) \quad \text{and} \quad \frac{\partial z}{\partial y} = \frac{dz}{du} \frac{\partial u}{\partial y} = \frac{dz}{du} (-1)$$

Hence,

$$\frac{\partial z}{\partial x} + \frac{\partial z}{\partial y} = 0$$

EXAMPLE 10. A curve C in the zx plane, having equation $z = f(x)$, is revolved around the z axis to form a surface of revolution. The height of the surface at a point in the xy plane r units from the origin is given by

$$z = f(r) = f(\sqrt{x^2 + y^2}) = g(x^2 + y^2)$$

where g is an arbitrary function. Find a partial DE of this family of surfaces of revolution.

Solution: Let $z = g(u)$ where $u = x^2 + y^2$.
Then,

$$\frac{\partial z}{\partial x} = \frac{dz}{du}\frac{\partial u}{\partial x} = \frac{dz}{du}(2x) \quad \text{and} \quad \frac{\partial z}{\partial y} = \frac{dz}{du}\frac{\partial u}{\partial y} = \frac{dz}{du}(2y)$$

Hence,

$$y\frac{\partial z}{\partial x} - x\frac{\partial z}{\partial y} = 0 \tag{10.6}$$

Many important physical laws are expressed as partial DE. A large part of the theory of partial DE was developed by intensive study of the partial DE of mathematical physics. Some of these equations are special cases of the following partial DE in which u denotes a function of x, y, z, and t.

$$\frac{\partial^2 u}{\partial x^2} + \frac{\partial^2 u}{\partial y^2} + \frac{\partial^2 u}{\partial z^2} = au + b\frac{\partial u}{\partial t} + c\frac{\partial^2 u}{\partial t^2} \tag{10.7}$$

If $a = b = 0$, then (10.7) reduces to the famous *wave equation* that governs wave propagation in a variety of vibratory phenomena. If $a = c = 0$, then (10.7) reduces to the *heat equation,* or *diffusion equation,* that governs many physical situations involving heat transfer in a solid or a liquid and various kinds of diffusion. If $a = b = c = 0$, then (10.7) reduces to *Laplace's equation,* or the *potential equation.* Named after Laplace, this partial DE governs so many diverse physical situations that it is regarded by many authorities as the most important of all partial DE.

As an illustration of the use of Laplace's equation, let u denote the temperature at the point $(x, y, z,)$ in a body at time t. If $\partial u/\partial t \equiv 0$, the heat equation reduces to Laplace's equation and the steady-state temperature satisfies Laplace's equation.

In applications, x, y, and z in (10.7) denote the space coordinates of a point and t denotes the time. The unknown function u may depend on any two, three, or four of these variables. For example, in the problem of the vibrating string to be studied in Section 10.6, u denotes the height of a vibrating string x

units from the fixed left end of the string at time t. In this case, (10.7) takes the form

$$\frac{\partial^2 u}{\partial t^2} = a^2 \frac{\partial^2 u}{\partial x^2}$$

referred to as the *one-dimensional wave equation*.

Second-order partial DE are by far the most important in applications. For an example of a fourth-order partial DE governing the transverse vibrations of a beam, see Reference 10.7.

Problem List 10.1

1. Show that the function defined by the given equation is a solution of the partial DE.
   (a) $u = e^{x+2t}$; $4u_{xx} - u_{tt} = 0$
   (b) $u = e^x \cos y$; $u_{xx} + u_{yy} = 0$
   (c) $u = \text{Tan}^{-1}(y/x)$; $u_{xx} + u_{yy} = 0$
   (d) $u = k \cos 3t \sin 3x$; $u_{tt} - u_{xx} = 0$
   (e) $u = y^4 - 2x^2y^2 + x^4$; $yu_x + xu_y = 0$
   (f) $u = e^{x/y}$; $xu_x + yu_y = 0$
   (g) $u = e^{-t} \sin x$; $u_t = u_{xx}$
   (h) $u = r^n \cos n\theta$; $u_{rr} + r^{-2}u_{\theta\theta} + r^{-1}u_r = 0$

2. Solve:
   (a) $z_y = 4x^3y$ \qquad (b) $z_x = 0$
   (c) $z_{xy} = \dfrac{\partial^2 z}{\partial y\, \partial z} = 0$

3. Prove Theorem 10-I for a second-order linear homogeneous partial DE.

4. Find a partial DE of the given relation. Assume $z = z(x, y)$.
   (a) $z = ax^2 + b \sin y$
   (b) $ay + b \ln z = xy$
   (c) $2z = ae^x + b \ln y$
   (d) $ax^2 + by^2 + z^2 = 1$; $a > 0, b > 0$ (two-parameter family of ellipsoids)

5. Solve $z_x - z_y = 0$ by letting $u = x + y$ and $v = x - y$.

6. Show that the partial DE $\partial^2 y/\partial u\, \partial v = 0$ becomes

$$\frac{\partial^2 y}{\partial t^2} = 4 \frac{\partial^2 y}{\partial x^2} \quad \text{if} \quad x = \frac{1}{2}(v + u) \quad \text{and} \quad t = \frac{1}{4}(v - u)$$

7. Show that the partial DE $\partial^2 y/\partial u\, \partial v = 0$ becomes

$$\frac{\partial^2 y}{\partial t^2} = a^2 \frac{\partial^2 y}{\partial x^2} \quad \text{if} \quad x = \frac{v + u}{2}, \quad t = \frac{v - u}{2a}$$

8. Show that $u(x, t) = f(x - at) + g(x + at)$, where f and g are arbitrary twice differentiable functions, defines a solution of the partial DE $u_{tt} = a^2 u_{xx}$.

9. Show that the following partial DE has no solution:

$$\left(\frac{\partial u}{\partial x}\right)^2 + 2\left(\frac{\partial u}{\partial y}\right)^2 + u^2 x^2 + 1 = 0$$

10. Show that if $x = r \cos \theta$, $y = r \sin \theta$, and $z = z$, Laplace's equation

$$\frac{\partial^2 u}{\partial x^2} + \frac{\partial^2 u}{\partial y^2} + \frac{\partial^2 u}{\partial z^2} = 0$$

transforms into

$$\frac{\partial^2 u}{\partial r^2} + \frac{1}{r}\frac{\partial u}{\partial r} + \frac{1}{r^2}\frac{\partial^2 u}{\partial \theta^2} + \frac{\partial^2 u}{\partial z^2} = 0$$

11. Given a vector field

$$\mathbf{V} = P(x, y, z)\mathbf{i} + Q(x, y, z)\mathbf{j} + R(x, y, z)\mathbf{k}$$

such that at any point A of the surface $z = f(x, y)$, the vector $\mathbf{V}_A$ is tangent to the surface, find a partial DE satisfied by f.

10.2 Boundary-Value Problems

The basic problem in applied partial DE is finding a solution u of a partial DE that is valid in a region R and satisfies certain conditions, known as boundary conditions, on the boundary B of R. The quantity u, a variable quantity such as temperature, is a real function of n real variables and R is an open connected set in the space of n variables.

To make these ideas more precise, let $n = 2$ and let $u = u(x, y)$ satisfy a partial DE at each point of the region R shown in Figure 10.1.

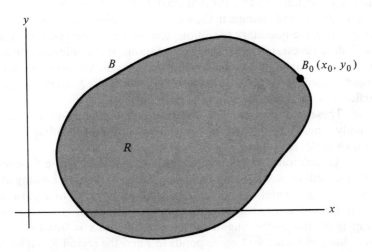

Fig. 10.1

The *boundary* B of R is the set of all points B_0 with the property that every circle centered at B_0 contains points in R and points not in R. A boundary-value problem is the determination of a function u satisfying certain boundary conditions on B. The solution, given by $u = f(x, y)$, need not satisfy the partial DE on B.

Additional assumptions must be made so that the boundary-value problem will be *well-posed*; that is, so that the boundary-value problem will provide a suitable model for studying a concrete physical situation. Suppose, for example, that a boundary condition asserts that $u = g(x, y)$ on B, where g is a prescribed function. If we could produce a solution in R given by $u = f(x, y)$ we could arbitrarily define u on B by $u = g(x, y)$. In a concrete situation, we invariably want $f(x, y)$ to be close to the value $g(x_0, y_0)$ of $g(x, y)$ at $B_0(x_0, y_0)$ on B when (x, y) in R is close to B_0. To ensure this, we make the following assumption.

Let $B_0(x_0, y_0)$ be an arbitrary point of B. Then, for every $\varepsilon > 0$, there exists a $\delta > 0$ such that $|f(x, y) - g(x_0, y_0)| < \varepsilon$ whenever (x, y) is in R and $\sqrt{(x - x_0)^2 + (y - y_0)^2} < \delta$.

We also assume that a solution of the boundary-value problem exists and that this solution is unique. One way of showing that a solution exists is to construct a solution satisfying all given boundary conditions and then to verify directly that this solution satisfies the given partial DE in R. To prove that a solution given by $u_1 = u_1(x, y)$ is unique, one can use the standard procedure of assuming that a second solution is defined by $u_2 = u_2(x, y)$ and then showing that $u_1(x, y) \equiv u_2(x, y)$ in and on the boundary of R. In our simple introduction to boundary-value problems, we shall not go into detail on these important and often difficult questions of existence and uniqueness. See References 10.3, 10.4, and 10.5. We assert without proof that each boundary-value problem that we consider possesses a unique solution.

Finally, for a boundary-value problem to be well-posed, it is assumed that the unique solution depends continuously on the boundary conditions. Loosely speaking, this means that small changes in the boundary conditions will result in small changes in the solution. This assumption is necessary since, in practice, the boundary conditions contain data based on observation, and such data necessarily involve approximations. If the solution corresponding to these approximate data did not approximate the solution corresponding to exact boundary values, then clearly the model used would not accurately reflect the physical situation.

These requirements for a well-posed boundary-value problem are readily generalized to situations in which a solution depending on more than two variables is sought.

As mentioned previously, in most applications u is a function of the space coordinates x, y, z and the time t. A condition on u holding at $t = 0$ is known as an *initial condition*. It is often convenient to regard an initial condition as a boundary condition. To illustrate, if u represents the displacement of a vibrating string that is fastened at $x = 0$ and at $x = 2$ ft, the variable y in Figure 10.1 corresponds to t and the region R is a semi-infinite strip defined by

$$R = \{(x, t): 0 < x < 2 \quad \text{and} \quad t > 0\}$$

The displacement u is a function of the time t and the distance x along the x axis measured from $x = 0$.

Another kind of boundary condition prescribes the values of one of the partial derivatives of u, say, $u_x(x, t)$, on all or part of B.

It is seldom that a partial DE or a boundary-value problem can be solved by direct integration using the method of illustrative Examples 3 and 4 of Section 10.1. In the following section, we present a method of solving boundary-value problems known as *separation of variables*. Although at first glance this method appears too special to be generally applicable, it turns out that the method is effective in solving many important boundary-value problems arising in applications.

10.3 Separation of Variables

Let $u = u(x_1, x_2, \ldots, x_n)$ denote the unique solution of a boundary-value problem. The method of separation of variables assumes that u can be written in the form

$$u = [X_1(x_1)][X_2(x_2)] \cdots [X_n(x_n)] \tag{10.8}$$

where each X_i is a real function of the single real variable x_i. Although this assumption may appear optimistic, it is valid in numerous important applied boundary-value problems. Substituting (10.8) into the partial DE reveals that each of the functions $X_1, X_2, \ldots, X_n$ satisfies an ordinary DE. These ordinary DE are solved and at the same time the boundary conditions are satisfied. The method is best illustrated by example.

EXAMPLE 1. Using the method of separation of variables, solve the boundary-value problem

$$u_x - 3u_y = 0, \quad x > 0$$

$$\tag{10.9}$$

$$u(0, y) = \frac{1}{2} e^{-2y}$$

Solution: We assume that a solution of the form

$$u = u(x, y) = X(x)\, Y(y) \tag{10.10}$$

exists, where X is a function of x alone and Y a function of y alone. Then $u_x = X'Y$ and $u_y = XY'$ where $X' = dX/dx$ and $Y' = dY/dy$.

Substituting for u_x and u_y in the partial DE (10.9), we obtain

$$X'Y - 3XY' = 0 \tag{10.11}$$

Writing (10.11) in the form

$$\frac{X'}{3X} = \frac{Y'}{Y} \tag{10.12}$$

we say that the variables have been "separated," since the left member involves x alone and the right member involves y alone. Equation (10.12) can hold only if each member is equal to the same constant k.

To see this, we note that

$$\frac{\partial}{\partial x}\left(\frac{X'}{3X}\right) = \frac{d}{dx}\left(\frac{X'}{3X}\right) = \frac{\partial}{\partial x}\left(\frac{Y'}{Y}\right) = 0$$

and hence $X'/(3X) = k_1 = $ constant. Similarly,

$$\frac{\partial}{\partial y}\left(\frac{X'}{3X}\right) = 0 = \frac{\partial}{\partial y}\left(\frac{Y'}{Y}\right) = \frac{d}{dy}\left(\frac{Y'}{Y}\right)$$

and hence $Y'/Y = k_2 = $ constant. By (10.12), $k_1 = k_2 = k$. From $X'/(3X) = k$ and $Y'/Y = k$, or $X' - 3kX = 0$ and $Y' - kY = 0$, we obtain

$$X = X(x) = Ae^{3kx} \quad \text{and} \quad Y = Y(y) = Be^{ky}$$

By (10.10),

$$u = u(x, y) = Ae^{3kx}Be^{ky} = Ce^{3kx+ky}$$

where $C = AB$.

We next note that $u(0, y) = Ce^{ky}$, and observe that the boundary condition is satisfied if we assign the constants C and k the values $C = \frac{1}{2}$ and $k = -2$.

We have shown that if the boundary-value problem has a solution of the form (10.10), then the solution is given by

$$u = u(x, y) = \left(\frac{1}{2}e^{-6x}\right)(e^{-2y}) = \frac{1}{2}e^{-6x-2y} \tag{10.13}$$

It is a simple matter to verify directly that (10.13) defines a solution of the boundary-value problem. We state without proof that this solution is unique.

EXAMPLE 2. Solve the boundary-value problem

$$u_t = u_{xx}; \quad u(0, t) = 0, \, u(\pi, t) = 0, \, u(x, 0) = 4\sin 3x$$

Solution: Assume $u = u(x, t) = X(x)T(t)$. Then

$$XT' = X''T \quad \text{or} \quad \frac{T'}{T} = \frac{X''}{X} = c = \text{constant}$$

from which we obtain

$$T' - cT = 0 \quad \text{and} \quad X'' - cX = 0$$

If $c = 0$, then $T' = 0$, $X'' = 0$, and hence $T = A = $ constant, and $X = Bx + C$.

In this case, u is given by $u = (Bx + C)A = Dx + E$, and it is clearly impossible for the boundary condition

$$u(x, 0) = 4 \sin 3x \qquad (10.14)$$

to hold.

Next, suppose $c > 0$. It is convenient to emphasize that c is positive by letting $c = \lambda^2$. Then $T = Ae^{\lambda^2 t}$, $X = Be^{\lambda x} + Ce^{-\lambda x}$, and u has the form $u = e^{\lambda^2 t}(De^{\lambda x} + Ee^{-\lambda x})$.

Again, it is clearly impossible to satisfy boundary condition (10.14).

Since we are assuming that X and T are real, c must be real, and the only remaining possibility is that $c < 0$. Let us make this assumption and let $c = -\lambda^2$. Then

$$T = Ae^{-\lambda^2 t}, \quad X = B \sin \lambda x + C \cos \lambda x$$

and u has the form

$$u = u(x, t) = e^{-\lambda^2 t}(D \sin \lambda x + E \cos \lambda x)$$

From $u(0, t) = 0$, we obtain $e^{-\lambda^2 t}(E) = 0$, from which we conclude, since $e^{-\lambda^2 t} > 0$, that $E = 0$.

From $u(\pi, t) = 0$ we obtain $e^{-\lambda^2 t}(D \sin \lambda \pi) = 0$, which can hold only if $D = 0$ or $\sin \lambda \pi = 0$.

If $D = 0$, then $u(x, t) \equiv 0$, and again it will be impossible to satisfy (10.14).

Setting $\sin \lambda \pi = 0$, we obtain $\lambda \pi = n\pi, n = 0, \pm 1, \pm 2, \ldots$, which holds when $\lambda = 0, \pm 1, \pm 2, \ldots$.

Thus, $u = u(x, t)$ has the form

$$u = u(x, t) = De^{-\lambda^2 t} \sin \lambda x \qquad (10.15)$$

where λ is an integer and $D \neq 0$.

From (10.15) we have $u(x, 0) = D \sin \lambda x$, and hence boundary condition (10.14) is satisfied by choosing $D = 4$ and $\lambda = 3$.

It can be shown directly that

$$u = u(x, t) = 4e^{-9t} \sin 3x$$

defines a solution of the given boundary-value problem. We state without proof that the solution is unique.

EXAMPLE 3. Solve Example 2 with $u(x, 0) = 4 \sin 3x$ replaced by

$$u(x, 0) = 5 \sin 2x + 7 \sin 6x \qquad (10.16)$$

Solution: Let $u_1 = u_1(x, t) = D_1 e^{-\lambda_1^2 t} \sin \lambda_1 x$, and

$$u_2 = u_2(x, t) = D_2 e^{-\lambda_2^2 t} \sin \lambda_2 x$$

where λ_1 and λ_2 are integers. Then u_1 and u_2 both satisfy the partial DE $u_t = u_{xx}$ and the boundary conditions

$$u(0, t) = 0, \quad u(\pi, t) = 0 \qquad (10.17)$$

By Theorem 10-I, the principle of superposition, $u(x, t) = u_1(x, t) + u_2(x, t)$ also satisfies $u_t = u_{xx}$. It is easy to see that in addition $u = u_1 + u_2$ also satisfies the boundary conditions (10.17). We now set $t = 0$ in

$$u(x, t) = D_1 e^{-\lambda_1^2 t} \sin \lambda_1 x + D_2 e^{-\lambda_2^2 t} \sin \lambda_2 x$$

to obtain

$$u(x, 0) = D_1 \sin \lambda_1 x + D_2 \sin \lambda_2 x$$

Boundary condition (10.16) is now satisfied by choosing $D_1 = 5$, $\lambda_1 = 2$, $D_2 = 7$, and $\lambda_2 = 6$.

The required solution is given by

$$u(x, t) = 5e^{-4t} \sin 2x + 7e^{-36t} \sin 6x$$

Since we have in Equation (10.15) an infinite number of choices (real numbers) available for D, and an infinite number of choices (integers) for λ, we can construct an infinite number of functions $u_1, u_2, \ldots$ that satisfy $u_t = u_{xx}$ and boundary conditions (10.17). By superposition, we can extend the method of Example 3 to construct a function given by

$$u(x, t) = \sum_{n=1}^{N} D_n e^{-n^2 t} \sin nx$$

that will satisfy $u_t = u_{xx}$, boundary conditions (10.17), and also the additional boundary condition

$$u(x, 0) = \sum_{n=1}^{N} D_n \sin nx \qquad\qquad \textbf{(10.18)}$$

where N is an arbitrary positive integer. For example, we could satisfy the boundary condition

$$u(x, 0) = 5 \sin x + 3 \sin 2x + 7 \sin 3x + \sqrt{2} \sin 4x + \pi \sin 5x$$

This gives us slightly more flexibility, but our procedure is still very limited. The natural question is how to proceed when the boundary condition along the boundary $t = 0$ has the form $u(x, 0) = f(x)$, where f is more or less arbitrary. The French mathematician Jean Baptiste Fourier (1768–1830), in his analysis of heat transfer in solid bodies, exploited the brilliant idea casually suggested earlier by the Swiss mathematician Daniel Bernoulli (1700–1782), of representing f by an infinite series of the form

$$f(x) = \sum_{n=1}^{+\infty} D_n \sin nx \qquad\qquad \textbf{(10.19)}$$

In other words, he considered the possibility of satisfying the boundary condition $u(x, 0) = f(x)$ by choosing the coefficients D_n appropriately and obtaining a solution given by an infinite series of the form

$$u(x, t) = \sum_{n=1}^{+\infty} D_n e^{-n^2 t} \sin nx$$

Fourier presented many of his ideas in 1822 in a treatise entitled *Théorie Analytique de la Chaleur* (The Analytic Theory of Heat). This was the first detailed exposition of the mathematical theory of heat flow. Many contemporaries of Fourier, including Lagrange, Laplace, and Legendre, criticized his methods as lacking in rigor. The main objections centered around the kind of function that could be expanded in a Fourier series (a generalization of (10.19) to be considered in Section 10.4) and the convergence of a Fourier series for a function throughout the domain of the function. The precise definition of a function was reexamined and eventually the theory of Fourier series was placed on a firm logical foundation. Today the work of Fourier is regarded as a landmark in the history of mathematics.

The theory of Fourier series is far from simple. In the next section we shall present the formal methods by which Fourier series for given functions are obtained. The remainder of this chapter will be devoted to deriving the partial DE that govern a few important physical situations, formulating associated boundary-value problems, and solving these boundary-value problems by the methods of separation of variables and Fourier series. The chapter concludes with a brief look at the expansion of a function in a series of orthogonal functions, an important generalization of the discoveries of Fourier.

Problem List 10.2

1. Solve each boundary-value problem by the method of separation of variables.
   (a) $u_x + u = u_y$, $u(x, 0) = 4e^{-3x}$
   (b) $u_x + u_t = 0$, $u(0, t) = 2e^{-4t}$
   (c) $u_t = 4u_{xx}$; $u(0, t) = 0$, $u(4, t) = 0$, $u(x, 0) = 5 \sin \pi x$
   (d) $25u_t = u_{xx}$; $u(0, t) = 0$, $u(100, t) = 0$, $u(x, 0) = 2 \sin \pi x + 3 \sin 2\pi x$
   (e) $u_{xx} + u_{yy} = 0$; $u(1, y) = 0$, $u(0, y) = 0$, $u(x, 0) = 0$, $u(x, 1) = 4 \sin \pi x$
   (f) $u_{xx} - u_{tt} = 0$; $u(0, t) = 0$, $u(1, t) = 0$, $u(x, 0) = 2 \sin \pi x$, $u_t(x, 0) = 0$
2. Use the method of separation of variables to solve the partial DE $u_x = 4u_t$ subject to these conditions on $u = u(x, t)$: $u(0, 0) = 10$, $u(0, 4) = 200$.
3. Solve each boundary-value problem by direct partial integration.

   (a) $\dfrac{\partial^2 u}{\partial x \, \partial y} = 4xy$; $u_y(0, y) = y$, $u(x, 0) = 2$

   (b) $\dfrac{\partial^2 u}{\partial x^2} = y \sin x$; $u_x(0, y) = 0$, $u(0, y) = 3y^2$

   (c) $\dfrac{\partial^2 u}{\partial x \, \partial y} = 2$; $u_y(0, y) = 2y$, $u(x, 0) = x^3$

   (d) $\dfrac{\partial^2 u}{\partial x^2} = 6xy$; $u_x(1, y) = 0$, $u(1, y) = 0$

   (e) $\dfrac{\partial^2 u}{\partial y \, \partial x} = xy$; $u_x(x, 0) = 1$, $u(0, y) = e^y$

   (f) $\dfrac{\partial^2 u}{\partial y^2} = 6xy + 4$; $u_y(x, 0) = 0$, $u(x, 0) = \ln x$

10.4 Fourier Series

The student knows that some functions have Taylor series expansions about $x = a$ given by

$$f(x) = \sum_{n=0}^{+\infty} \frac{f^{(n)}(a)}{n!} (x - a)^n \qquad (10.20)$$

In the interval I of convergence of the series, the right member of (10.20) is said to represent the function f. If the right member converges to $f(x)$ for every x in I, it is another manner of writing $f(x)$ in I.

In this section we consider the possibility of representing a function f by a series of the form

$$f(x) = \frac{a_0}{2} + \sum_{n=1}^{+\infty} \left(a_n \cos \frac{n\pi x}{p} + b_n \sin \frac{n\pi x}{p} \right) \qquad (10.21)$$

where a_0 is a constant and p is a positive constant. The reason for the coefficient $\frac{1}{2}$ of a_0 will soon become apparent.

The period of $\cos n\pi x/p$ and of $\sin n\pi x/p$ is

$$\frac{2\pi}{(n\pi)/p} = \frac{2p}{n}$$

and hence the terms in the summation have periods $2p/1, 2p/2, 2p/3, \ldots$. Thus, $f(x + 2p) = f(x)$, and hence if f can be represented by a series of the form (10.21), then f must be periodic with period $2p$. It is possible for f to have a fundamental period smaller than $2p$; that is, $f(x + \alpha) = f(x)$ might hold for some $\alpha < 2p$. The restriction that f should have period $2p$ is not serious. If we are interested in a function f defined on $[-p, p]$ in an application, we use the given definition of f on $[-p, p)$ and define f elsewhere by $f(x + 2p) = f(x)$.

We next consider the determination of a_n and b_n in (10.21). This problem is somewhat analogous to determining the Taylor coefficients $f^{(n)}(a)/n!$ in (10.20).

To find a_0, we assume that f is integrable on any interval $[x_1, x_2]$ and integrate both sides of (10.21) from $x = -p$ to $x = p$. This interval is selected for convenience; any interval $[k, k + 2p]$ will lead to the same result. We also assume that the term-by-term integration of the right member of (10.21) is permissible. Thus,

$$\int_{-p}^{p} f(x)\, dx = \frac{a_0}{2} \int_{-p}^{p} dx + a_n \sum_{n=1}^{+\infty} \int_{-p}^{p} \cos \frac{n\pi x}{p}\, dx$$

$$+ b_n \sum_{n=1}^{+\infty} \int_{-p}^{p} \sin \frac{n\pi x}{p}\, dx$$

It is left to Problem 1(a) to show that

$$\int_{-p}^{p} \cos \frac{n\pi x}{p} \, dx = \int_{-p}^{p} \sin \frac{n\pi x}{p} \, dx = 0$$

for $n = 1, 2, 3, \ldots$. Hence,

$$\int_{-p}^{p} f(x) \, dx = \frac{a_0}{2} [x]_{-p}^{p} = a_0 p$$

and

$$a_0 = \frac{1}{p} \int_{-p}^{p} f(x) \, dx \tag{10.22}$$

To find a_n, we multiply both sides of (10.21) by $\cos (n\pi x/p)$ and integrate both sides over $[-p, p]$, again assuming the validity of the term-by-term integration. Thus,

$$\int_{-p}^{p} f(x) \cos \frac{n\pi x}{p} \, dx = \frac{a_0}{2} \int_{-p}^{p} \cos \frac{n\pi x}{p} \, dx + a_1 \int_{-p}^{p} \cos \frac{\pi x}{p} \cos \frac{n\pi x}{p} \, dx + \cdots$$

$$+ a_n \int_{-p}^{p} \cos^2 \frac{n\pi x}{p} \, dx + \cdots$$

$$+ b_1 \int_{-p}^{p} \sin \frac{\pi x}{p} \cos \frac{n\pi x}{p} \, dx + \cdots$$

$$+ b_n \int_{-p}^{p} \sin \frac{n\pi x}{p} \cos \frac{n\pi x}{p} \, dx + \cdots$$

It is left to Problems 1 (a), (b), (c), and (e) to show that $\int_{-p}^{p} \cos^2 (n\pi x/p) \, dx = p$ for $n = 1, 2, 3, \ldots$, and that all other integrals on the right have value zero. Hence,

$$\int_{-p}^{p} f(x) \cos \frac{n\pi x}{p} \, dx = a_n p$$

and

$$a_n = \frac{1}{p} \int_{-p}^{p} f(x) \cos \frac{n\pi x}{p} \, dx \tag{10.23}$$

for $n = 0, 1, 2, 3, \ldots$.

If we set $n = 0$ in (10.23), we obtain

$$a_0 = \frac{1}{p} \int_{-p}^{p} f(x) \, dx$$

Since this agrees with the right member of (10.22), we can dispense with (10.22) and use (10.23) for $n = 0$ as well as for $n = 1, 2, 3, \ldots$. This is the reason we inserted the coefficient $\frac{1}{2}$ of a_0 in (10.21).

To obtain b_n, we proceed as we did in obtaining a_n except that we multiply both sides of (10.21) by $\sin (n\pi x/p)$ before integrating. Thus,

$$\int_{-p}^{p} f(x) \sin \frac{n\pi x}{p} \, dx = \frac{a_0}{2} \int_{-p}^{p} \sin \frac{n\pi x}{p} \, dx + a_1 \int_{-p}^{p} \cos \frac{\pi x}{p} \sin \frac{n\pi x}{p} \, dx + \cdots$$

$$+ a_n \int_{-p}^{p} \cos \frac{n\pi x}{p} \sin \frac{n\pi x}{p} \, dx + \cdots$$

$$+ b_1 \int_{-p}^{p} \sin \frac{\pi x}{p} \sin \frac{n\pi x}{p} \, dx + \cdots$$

$$+ b_n \int_{-p}^{p} \sin^2 \frac{n\pi x}{p} \, dx + \cdots$$

Since $\int_{-p}^{p} \sin^2 (n\pi x/p) \, dx = p$ and all other integrals on the right have value zero [see Problems 1 (a), (b), (d), and (e)], we obtain

$$b_n = \frac{1}{p} \int_{-p}^{p} f(x) \sin \frac{n\pi x}{p} \, dx \qquad (10.24)$$

for $n = 1, 2, 3, \ldots$.

The series (10.21), with a_n and b_n given by (10.23) and (10.24), is called the *Fourier series* of f. The coefficients a_n and b_n, if it is assumed that the integrals in (10.23) and (10.24) exist, are called the Fourier coefficients of f.

If f has period $2p$, is defined for all x, and is integrable, it can be shown that the integrands $f(x) \cos (n\pi x/p)$ and $f(x) \sin (n\pi x/p)$ in (10.23) and (10.24) are integrable and hence that a_n and b_n exist. This means that f has a Fourier series. It is natural to ask, given $x = x_0$, whether the Fourier series of f converges at $x = x_0$, and if so, whether or not it converges to $f(x_0)$. That is, does

$$\lim_{k \to +\infty} \left[\frac{a_0}{2} + \sum_{n=1}^{k} \left(a_n \cos \frac{n\pi x_0}{p} + b_n \sin \frac{n\pi x_0}{p} \right) \right]$$

exist, and if so, is this limit equal to $f(x_0)$? This is a very difficult question. We observe that if f is discontinuous at x_0, it can scarcely be expected that the Fourier series of f will converge to $f(x_0)$ when $x = x_0$. If the value of f is changed at $x = x_0$, the integrals used in computing Fourier coefficients are unchanged. Hence, the Fourier coefficients will remain the same, and the Fourier series could not converge to two different values.

EXAMPLE 1. The functions having period 2, given by

$$f(x) = \begin{cases} 0 & \text{for } -1 \le x \le 0 \\ 1 & \text{for } 0 < x < 1 \end{cases}$$

and

$$g(x) = \begin{cases} 0 & \text{for } -1 \le x < 0 \\ 1 & \text{for } 0 \le x < 1 \end{cases}$$

have the same Fourier series. At $x = 0$, this series cannot converge to $f(0) = 0$ and also to $g(0) = 1$.

For a discussion of various conditions on f under which the Fourier series for f represents f, see References 10.2 and 10.3. We shall cite a theorem stating one set of sufficient conditions under which a function is representable by a Fourier series. This theorem uses the following definitions:

Definitions The $\lim_{x \to a} + f(x) = f(a+)$ is called the *right-hand limit* of f at a.

The $\lim_{x \to a} - f(x) = f(a-)$ is called the *left-hand limit* of f at a.

Definition If $\lim_{x \to a} f(x)$ exists and does not equal the number $f(a)$, the function f has a *removable discontinuity* at $x = a$.

Definitions Let a be contained in the domain of f. If

$$\lim_{x \to a^+} f(x) = L_1 \quad \text{and} \quad \lim_{x \to a^-} f(x) = L_2$$

then f has a *jump discontinuity* at $x = 0$. The difference $L_1 - L_2$ is called the *jump in the function* at $x = a$.

Definitions Removable and jump discontinuities are called *ordinary discontinuities*. A function that is continuous on an interval I except for a finite number of ordinary discontinuities is said to be *piecewise continuous*, or *sectionally continuous*, on I.

Theorem 10-II Let f having period $2p$ be defined on $(-\infty, +\infty)$. If f has a finite number of points of ordinary discontinuity and a finite number of maxima and minima in the interval $-p \le x \le p$, then for every $x = x_0$,

$$\frac{a_0}{2} + \sum_{n=1}^{+\infty} \left(a_n \cos \frac{n\pi x_0}{p} + b_n \sin \frac{n\pi x_0}{p} \right) = \frac{f(x_0+) + f(x_0-)}{2}$$

The coefficients a_n and b_n are the Fourier coefficients of f.

This theorem is due to the German mathematician Peter Gustave Lejeune Dirichlet (1805–1859). A proof is given in Reference 10.3. The restrictions on f in the hypothesis are known as the *Dirichlet conditions*. These conditions are satisfied by most functions arising in applied mathematics. Since $f(x_0+) = f(x_0-)$ if f is continuous at x_0, then if f satisfies the Dirichlet conditions, the theorem asserts, the Fourier series of f converges to $f(x)$ at all points at which f is continuous and to the average of the right-hand and left-hand limits of f at each point at which f is discontinuous. Figure 10.2 displays the graph of a function f of period $2p = 8$ that satisfies the Dirichlet conditions.

The Fourier series of f converges to the height of the curve at each point in Figure 10.2 at which f is continuous and to the height of the heavy dot at each point at which f is discontinuous.

At $x = -4$, the Fourier series converges to $(1 + 3)/2 = 2$.
At $x = -3$, the Fourier series converges to 1.
At $x = 0$, the Fourier series converges to $(1 + 2)/2 = \frac{3}{2}$.
At $x = 2$, the Fourier series converges to 5.
At $x = 4$, the Fourier series converges to $(1 + 3)/2 = 2$.

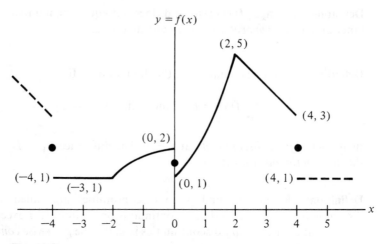

Fig. 10.2

EXAMPLE 2. Find the Fourier series of the periodic function defined in one period by

$$f(x) = \begin{cases} 0 & \text{for } -\pi < x < 0 \\ \pi & \text{for } \ \ 0 < x < \pi \end{cases}$$

Solution: By (10.23) and (10.24) with $p = \pi$,

$$a_0 = \frac{1}{\pi} \int_{-\pi}^{0} 0 \, dx + \frac{1}{\pi} \int_{0}^{\pi} \pi \, dx = \pi$$

$$a_n = \frac{1}{\pi} \int_{0}^{\pi} \pi \cos nx \, dx = 0, \quad n \geq 1$$

and

$$b_n = \frac{1}{\pi} \int_0^\pi \pi \sin nx \, dx$$

$$= -\frac{1}{n} \cos nx \big]_0^\pi = \frac{1}{n}(1 - \cos n\pi)$$

$$= \frac{1 - (-1)^n}{n}, \quad n \geq 1$$

Hence by (10.21),

$$f(x) = \frac{\pi}{2} + \sum_{n=1}^{+\infty} \frac{1 - (-1)^n}{n} \sin nx$$

$$= \frac{\pi}{2} + 2 \left(\frac{\sin x}{1} + \frac{\sin 3x}{3} + \frac{\sin 5x}{5} + \cdots \right) \qquad \textbf{(10.25)}$$

The graph of the function defined by

$$S_{n+1}(x) = \frac{\pi}{2} + \sum_{k=1}^n \frac{1 - (-1)^k}{k} \sin kx$$

approximates the graph of f very closely when n is a large positive integer. When $x = 0$, the Fourier series converges to $(\pi + 0)/2 = \pi/2$ by Theorem 10-II. This result also follows from (10.25) by inspection. Similarly, the Fourier series converges to $\pi/2$ when $x = -\pi$ and when $x = \pi$. We note also that $S_{n+1}(x)$ passes through $(-\pi, \pi/2)$, $(0, \pi/2)$, and $(\pi, \pi/2)$ for $n = 0, 1, 2, \ldots$, with

$$S_1(x) = \pi/2, \quad S_2(x) = S_3(x) = (\pi/2) + 2 \sin x,$$

$$S_4(x) = S_5(x) = \frac{\pi}{2} + 2 \sin x + \frac{2 \sin 3x}{3},$$

$$S_6(x) = S_7(x) = \frac{\pi}{2} + 2 \sin x + \frac{2 \sin 3x}{3} + \frac{2 \sin 5x}{5}, \cdots$$

Figure 10.3 displays the graphs of f and S_6 on the interval $[-\pi, \pi]$. We did not define $f(-\pi)$ or $f(0)$. The definitions of $f(-\pi)$ and $f(0)$ would not have changed the Fourier series (10.25) of f. By anticipating the conclusion of Theorem 10-II, we could have defined $f(-\pi) = f(0) = \pi/2$.

The graph of S_6 can be constructed by having a digital computer print a table of values of x and $S_6(x)$. It is interesting and instructive to note that the graph of S_6 can also be obtained by the method of composition of ordinates. See Reference 10.11. This procedure requires considerable time and effort.

Fourier series provide an interesting method of summing a variety of series of constant terms. For example, since f in Example 2 is continuous at $x = \pi/2$ and $f(\pi/2) = \pi$, the right member of (10.25) converges to π; we know this from Theorem 10-II. Hence, substituting $\pi/2$ into (10.25), we have

$$\pi = \frac{\pi}{2} + 2 \left(1 - \frac{1}{3} + \frac{1}{5} - \frac{1}{7} + \frac{1}{9} - \cdots \right)$$

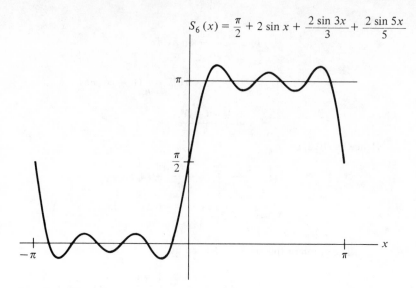

$$S_6(x) = \frac{\pi}{2} + 2 \sin x + \frac{2 \sin 3x}{3} + \frac{2 \sin 5x}{5}$$

Fig. 10.3

from which we obtain

$$1 - \frac{1}{3} + \frac{1}{5} - \frac{1}{7} + \frac{1}{9} - \cdots = \frac{\pi}{4}$$

Thus, we have summed Gregory's series, named for the Scottish mathematician James Gregory (1638–1675).

Summing the series $\sum_{n=1}^{+\infty} (1/n^2)$ was considered extremely difficult before Euler solved the problem about 1736. With an appropriate Fourier series, this becomes a simple exercise. See Problem 4. It often happens that a problem long regarded as challenging becomes almost trivial when a newly discovered method, theory, or technique is applied to it.

EXAMPLE 3. Find the Fourier series of the periodic function defined in one period by

$$f(x) = \begin{cases} 0 & \text{for } -1 < x < 0 \\ x & \text{for } 0 \leq x < 1 \end{cases}$$

Solution: Using $p = 1$, we get

$$a_0 = \frac{1}{1} \int_0^1 x \, dx = \frac{x^2}{2} \Big]_0^1 = \frac{1}{2}$$

$$a_n = \frac{1}{1} \int_0^1 x \cos n\pi x \, dx = \frac{1}{n^2 \pi^2} [\cos n\pi x + n\pi x \sin n\pi x]_0^1$$

$$= \frac{\cos n\pi - 1}{n^2 \pi^2} = \begin{cases} 0 & \text{for } n \text{ even} \\ \dfrac{-2}{n^2 \pi^2} & \text{for } n \text{ odd} \end{cases}$$

and

$$b_n = \frac{1}{1}\int_0^1 x \sin n\pi x \, dx = \frac{1}{n^2\pi^2}[\sin n\pi x - n\pi x \cos n\pi x]_0^1$$

$$= \frac{-n\pi \cos n\pi}{n^2\pi^2} = \frac{-\cos n\pi}{n\pi} = \begin{cases} -\dfrac{1}{n\pi} & \text{for } n \text{ even} \\[2mm] \dfrac{1}{n\pi} & \text{for } n \text{ odd} \end{cases}$$

Hence,

$$f(x) = \frac{1}{4} - \frac{2}{1^2\pi^2}\cos\pi x - \frac{2}{3^2\pi^2}\cos 3\pi x - \frac{2}{5^2\pi^2}\cos 5\pi x - \cdots$$

$$+ \frac{1}{1\pi}\sin\pi x - \frac{1}{2\pi}\sin 2\pi x + \frac{1}{3\pi}\sin 3\pi x - \cdots \qquad \textbf{(10.26)}$$

EXAMPLE 4. Evaluate

$$\sum_{n=1}^{+\infty} \frac{1}{(2n-1)^2}$$

Solution: Setting $x = 0$ in (10.26) gives

$$0 = \frac{1}{4} - \frac{2}{\pi^2}\left(\frac{1}{1^2} + \frac{1}{3^2} + \frac{1}{5^2} + \cdots\right)$$

and hence,

$$\frac{1}{1^2} + \frac{1}{3^2} + \frac{1}{5^2} + \cdots = \sum_{n=1}^{+\infty}\frac{1}{(2n-1)^2} = \frac{\pi^2}{8}$$

10.5 Fourier Sine and Cosine Series

Functions that have Fourier series and graphs symmetric with respect to the y axis or the origin have simple Fourier series.

Definitions A function f is called *even* if and only if its graph is symmetric with respect to the y axis. A function is called *odd* if and only if its graph is symmetric with respect to the origin.

Let f be given by $y = f(x)$ and let x belong to domain of f. If f is even, $-x$ is also in the domain of f, and $f(-x) = f(x)$. If f is odd, $-x$ is also in the domain of f, and $f(-x) = -f(x)$.

EXAMPLE 1. The functions defined by $x^2, x^4, |x|, \cos kx,$ and $\sqrt{4 - x^2}$ are even. The functions defined by $x, x^3, x^{1/3}, \sin kx,$ and $-\tan x$ are odd. The functions defined by $x + 5, \sqrt{x},$ $|x - 1|, x^3 - 3x^2 + 1,$ and $x + \cos x$ are neither even nor odd.

Figure 10.4 displays the graphs of two functions defined on $[-p, p]$. The function E in Figure 10.4(a) is even, the function O in Figure 10.4(b) is odd. From area considerations, it is easy to see that the integral from $-p$ to p of an even function, if it exists, is twice the integral of the same function from 0 to p. Also, the integral from $-p$ to p of an odd function, if it exists, is zero.

Let f be an even function of period $2p$ satisfying the Dirichlet conditions, let $\phi(x) = f(x) \cos(n\pi x/p)$, and let $\psi(x) = f(x) \sin(n\pi x/p)$.

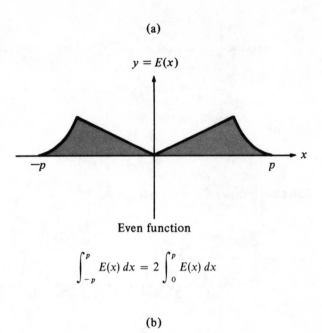

(a)

$y = E(x)$

$-p$ p x

Even function

$$\int_{-p}^{p} E(x)\, dx = 2 \int_{0}^{p} E(x)\, dx$$

(b)

$y = O(x)$

$-p$ p x

Odd function

$$\int_{-p}^{p} O(x)\, dx = 0$$

Fig. 10.4

Then,

$$\phi(-x) = f(-x) \cos\left(\frac{-n\pi x}{p}\right) = f(x) \cos\frac{n\pi x}{p} = \phi(x)$$

and hence ϕ is even. Also,

$$\psi(-x) = f(-x) \sin\left(\frac{-n\pi x}{p}\right) = f(x)\left(-\sin\frac{n\pi x}{p}\right) = -\psi(x)$$

and hence ψ is odd. Therefore,

$$a_n = \frac{1}{p}\int_{-p}^{p} f(x) \cos\frac{n\pi x}{p}\,dx = \frac{2}{p}\int_{0}^{p} f(x) \cos\frac{n\pi x}{p}\,dx, \quad n = 0, 1, 2, 3, \ldots$$

$$(10.27)$$

and

$$b_n = \frac{1}{p}\int_{-p}^{p} f(x) \sin\frac{n\pi x}{p}\,dx = 0, \quad n = 1, 2, 3, \ldots \qquad (10.28)$$

Thus, the Fourier series of an even function contains only cosine terms; the constant term $a_0/2$ is regarded as a cosine term with $n = 0$. A series of this type is called a *Fourier cosine series of f*.

Similarly, if f is odd

$$\phi(-x) = f(-x) \cos\left(\frac{-n\pi x}{p}\right) = -f(x) \cos\frac{n\pi x}{p} = -\phi(x)$$

and hence ϕ is odd.

Also,

$$\psi(-x) = f(-x) \sin\left(\frac{-n\pi x}{p}\right) = [-f(x)]\left[-\sin\frac{n\pi x}{p}\right]$$

$$= f(x) \sin\frac{n\pi x}{p} = \psi(x)$$

and hence ψ is even. Therefore, for f odd,

$$a_n = \frac{1}{p}\int_{-p}^{p} f(x) \cos\frac{n\pi x}{p}\,dx = 0, \quad n = 0, 1, 2, 3, \ldots \qquad (10.29)$$

and

$$b_n = \frac{1}{p}\int_{-p}^{p} f(x) \sin\frac{n\pi x}{p}\,dx = \frac{2}{p}\int_{0}^{p} f(x) \sin\frac{n\pi x}{p}\,dx \qquad (10.30)$$

Thus, the Fourier series of an odd function contains only sine terms. A series of this type is called a *Fourier sine series of f*. Fourier cosine and sine series are also termed *half-range Fourier series*.

Example 2. Find a Fourier cosine series of the function f defined by $f(x) = x$ on $0 < x < \pi$.

Solution: By (10.27) and (10.28), with $p = \pi$,

$$a_0 = \frac{2}{\pi} \int_0^\pi x \, dx = \frac{2}{\pi} \frac{x^2}{2} \Big]_0^\pi = \pi$$

$$a_n = \frac{2}{\pi} \int_0^\pi x \cos nx \, dx = \frac{2}{n^2 \pi} [\cos nx + nx \sin nx]_0^\pi$$

$$= \frac{2}{n^2 \pi} (\cos n\pi - 1) = \frac{2}{n^2 \pi} [(-1)^n - 1] = \begin{cases} \dfrac{-4}{n^2 \pi} & \text{for } n \text{ odd} \\ 0 & \text{for } n \text{ even, } n \geq 2 \end{cases}$$

and $b_n = 0$ for $n = 1, 2, 3, \ldots$. Therefore,

$$f(x) = \frac{\pi}{2} - \frac{4}{\pi} \left(\frac{\cos x}{1^2} + \frac{\cos 3x}{3^2} + \frac{\cos 5x}{5^2} + \cdots \right) \tag{10.31}$$

Example 3. Find a Fourier sine series of the function g defined by $g(x) = x$ on $0 < x < \pi$.

Solution: By (10.29) and (10.30), with $p = \pi$, $a_n = 0$ for $n = 0, 1, 2, 3, \ldots$ and

$$b_n = \frac{2}{\pi} \int_0^\pi x \sin nx \, dx = \frac{2}{n^2 \pi} [\sin nx - nx \cos nx]_0^\pi$$

$$= \frac{-2n\pi}{n^2 \pi} \cos n\pi = -\frac{2(-1)^n}{n} = \frac{2(-1)^{n+1}}{n}$$

for $n = 1, 2, 3, \ldots$. Therefore,

$$g(x) = 2 \left(\frac{\sin x}{1} - \frac{\sin 2x}{2} + \frac{\sin 3x}{3} + \frac{\sin 4x}{4} + \cdots \right) \tag{10.32}$$

Examples 2 and 3 illustrate the remarkable fact that different Fourier series can converge to the same function in a particular interval. The Fourier cosine series (10.31) and the Fourier sine series (10.32) both converge to the function F given by $F(x) = x$ in $0 < x < \pi$. When $x = 0$, both series converge to 0, but when $x = \pi$, (10.31) converges to π whereas (10.32) converges to 0. By changing the period from $2p = 2$ to $2p = 2\pi$ in Example 3 of Section 10.4, we would obtain still another Fourier series converging to x for $0 < x < \pi$.

In summary, if f is defined on $0 < x < p$, we can obtain an infinite number of Fourier series, all converging to f on $0 < x < p$, by defining f on $-p < x < 0$ in various ways. The Fourier sine and cosine series are the simplest Fourier series that can be so obtained. When we define f on $(-p, 0)$ by $f(-x) = -f(x)$, we are, we say, using the *odd periodic extension* of f, and when we define f on $(-p, 0)$ by $f(-x) = f(x)$, we are using the *even periodic extension* of f.

In many applications we are interested in a function f defined on an interval $[0, p]$, where $p > 0$. For example, the function might describe the temperature of a slender rod p units in length. To satisfy a given boundary condition, we may wish to expand f in an infinite series of sine terms. This can be done using the odd periodic extension of f to $(-p, 0)$, and then obtaining a Fourier sine series of f. The set of values to which the series converges in $(-p, 0)$ is of no concern, since this interval has no significance in the physical problem. Similarly, if we want to expand f in an infinite series of cosine terms, we use the even periodic extension of f to $(-p, 0)$ and obtain a Fourier cosine series of f. The application of half-range Fourier series will be illustrated in the next section on the vibrating string.

The principal uses of Fourier series in the field of DE arise in applying partial DE to boundary-value problems. Several of these applications will be presented in the remaining sections of this chapter. We conclude this section by considering briefly an application of Fourier series in the field of ordinary DE.

In Section 5.1, we considered physical systems governed by a DE of the form

$$\frac{w}{g}\frac{d^2x}{dt^2} + c\frac{dx}{dt} + kx = F(t) \tag{5.11}$$

In many mechanical and electrical systems, the forcing or input function F is periodic, as in an off-on type of control mechansim or current supply. To solve a DE of type (5.11) for an F that is periodic, it is often advantageous to obtain a Fourier series for F.

EXAMPLE 4. Find a particular solution of $x'' - 4x = F(t)$ if F is the periodic function defined in one period by $F(t) = |t|$ on $-\pi \le t < \pi$.

Solution: The Fourier series for F is given by (see Example 2)

$$F(t) = \frac{\pi}{2} - \frac{4}{\pi}\sum_{n=1}^{+\infty}\frac{\cos(2n-1)t}{(2n-1)^2}$$

We seek a particular solution X given by

$$X(t) = \frac{-\pi}{8} - \frac{4}{\pi}\sum_{n=1}^{+\infty} a_n \cos(2n-1)t$$

Then,

$$X'(t) = \frac{4}{\pi}\sum_{n=1}^{+\infty} a_n(2n-1)\sin(2n-1)t$$

and

$$X''(t) = \frac{4}{\pi}\sum_{n=1}^{+\infty} a_n(2n-1)^2\cos(2n-1)t$$

10.5 Fourier Sine and Cosine Series

403

Substituting $X(t)$ and $X''(t)$ into the DE yields

$$\frac{4}{\pi} \sum_{n=1}^{+\infty} a_n(2n-1)^2 \cos(2n-1)t + 4\left(\frac{\pi}{8}\right) + \frac{16}{\pi} \sum_{n=1}^{+\infty} a_n \cos(2n-1)t$$

$$= \frac{\pi}{2} - \frac{4}{\pi} \sum_{n=1}^{+\infty} \frac{\cos(2n-1)t}{(2n-1)^2}$$

Equating the coefficients of $\cos(2n-1)t$ on both sides of the equation, we obtain

$$a_n(2n-1)^2 + 4a_n = \frac{-1}{(2n-1)^2}$$

and hence

$$a_n = \frac{-1}{(2n-1)^4 + 4(2n-1)^2}$$

The particular solution X given by

$$X(t) = -\frac{\pi}{8} + \frac{4}{\pi} \sum_{n=1}^{+\infty} \frac{\cos(2n-1)t}{(2n-1)^4 + 4(2n-1)^2}$$

yields the steady-state output corresponding to the periodic input F.

It can be shown that F satisfies conditions under which term-by-term differentiation of the two Fourier series is valid. For a treatment of the differentiation and integration of Fourier series see Reference 10.2.

Problem List 10.3

1. Show that:

(a) $\int_{-p}^{p} \cos\frac{n\pi x}{p}\, dx = \int_{-p}^{p} \sin\frac{n\pi x}{p}\, dx = 0$ for $n = 1, 2, 3, \ldots$

(b) $\int_{-p}^{p} \cos^2\frac{n\pi x}{p}\, dx = \int_{-p}^{p} \sin^2\frac{n\pi x}{p}\, dx = p$ for $n = 1, 2, 3, \ldots$

(c) $\int_{-p}^{p} \cos\frac{m\pi x}{p} \cos\frac{n\pi x}{p}\, dx = 0$ for $m \geq 1, n \geq 1,$ and $m \neq n$

(d) $\int_{-p}^{p} \sin\frac{m\pi x}{p} \sin\frac{n\pi x}{p}\, dx = 0$ for $m \geq 1, n \geq 1,$ and $m \neq n$

(e) $\int_{-p}^{p} \cos\frac{m\pi x}{p} \sin\frac{n\pi x}{p}\, dx = 0$ for $m \geq 1, n \geq 1$

Hint: In (c), (d), and (e), use

$$\cos Ax \cos Bx = \frac{1}{2}[\cos(A+B)x + \cos(A-B)x]$$

$$\sin Ax \cos Bx = \frac{1}{2}[\sin(A+B)x + \sin(A-B)x]$$

and

$$\sin Ax \sin Bx = \frac{1}{2}[\cos(A - B)x - \cos(A + B)x]$$

2. Find the Fourier series of the periodic function defined in one period by:

(a) $f(x) = \begin{cases} 0 & \text{for } -1 < x < 0 \\ 1 & \text{for } 0 < x < 1 \end{cases}$

(b) $f(x) = \begin{cases} 1 & \text{for } -\pi < x < 0 \\ 0 & \text{for } 0 < x < \pi \end{cases}$

(c) $f(x) = \begin{cases} 0 & \text{for } -2 < x < 0 \\ 1 & \text{for } 0 < x < 2 \end{cases}$

(d) $f(x) = \begin{cases} 0 & \text{for } -\pi < x < 0 \\ 1 & \text{for } 0 < x < \dfrac{\pi}{2} \\ 0 & \text{for } \dfrac{\pi}{2} < x < \pi \end{cases}$

(e) $f(x) = \begin{cases} 1 & \text{for } 0 < x < \dfrac{\pi}{2} \\ 0 & \text{for } \dfrac{\pi}{2} < x < 2\pi \end{cases}$

(f) $f(x) = \begin{cases} -\pi & \text{for } -\pi < x < 0 \\ x & \text{for } 0 < x < \pi \end{cases}$

(g) $f(x) = 3 \sin 2x \qquad \text{for } \dfrac{-\pi}{2} < x < \dfrac{\pi}{2}$

(h) $f(x) = x^2 \qquad \text{for } 0 < x < 2\pi$

(i) $f(x) = \begin{cases} x & \text{for } -\pi < x < 0 \\ \pi - x & \text{for } 0 < x < \pi \end{cases}$

3. Find the Fourier series of the periodic function whose definition over one period is

$$f(x) = \begin{cases} -x & \text{for } -\pi < x < 0 \\ 0 & \text{for } 0 < x < \pi \end{cases}$$

State the value to which the series converges when
(a) $x = -\pi$; (b) $x = -\pi/2$; (c) $x = 0$; (d) $x = \pi/2$; (e) $x = \pi$

4. Use the result of Problem 2(h) to evaluate $\sum_{n=1}^{+\infty} (1/n^2)$.

5. A periodic function is defined in one period by

$$f(x) = \begin{cases} 1 & \text{for } 0 < x < 1 \\ 2 & \text{for } 1 < x < 2 \end{cases}$$

Sketch the graph of the function to which the Fourier series of f will converge for $-2 < x < 2$. To what value will the series converge for $x = 0$?

6. The periodic function defined in one period by

$$f(x) = \begin{cases} -\pi & \text{for } -\pi < x < 0 \\ x & \text{for } 0 < x < \pi \end{cases}$$

10.5 Fourier Sine and Cosine Series

has the Fourier series expansion

$$f(x) = \frac{-\pi}{4} + \sum_{n=1}^{+\infty} \left[\frac{\cos n\pi - 1}{n^2 \pi} \cos nx + \frac{1 - 2\cos n\pi}{n} \sin nx \right]$$

The sum $S_n(x)$ of $2n + 1$ terms of the Fourier series is given by

$$S_n(x) = \frac{-\pi}{4} + \sum_{k=1}^{n} \left[\frac{\cos k\pi - 1}{k^2 \pi} \cos kx + \frac{1 - 2\cos k\pi}{k} \sin kx \right]$$

(a) Write $S_5(x)$ without using a summation symbol.
(b) Evaluate $\lim_{n \to +\infty} S_n(\pi/4)$, the value to which the series converges when $x = \pi/4$.
(c) Evaluate $\lim_{n \to +\infty} S_n(0)$.
(d) Evaluate $\sum_{n=1}^{+\infty} [1/(2n - 1)^2]$.

7. Which of the following expressions define functions that are (i) even? (ii) odd? (iii) neither even nor odd?
   (a) $x + 1$ (b) $x^2 - 6$
   (c) $-x^5$ (d) x^{-1}
   (e) $\sin x \cos 3x$ (f) $x + |x|$
   (g) $\tan 2x$ (h) $\cos 2x$
   (i) $[x]$ (j) $\sin x$

8. Show that if for all x in the domain of f, $-x$ is also in the domain of f, then $\frac{1}{2}[f(x) + f(-x)]$ defines an even function and $\frac{1}{2}[f(x) - f(-x)]$ defines an odd function.

9. (a) Show that if for all x in the domain of f, $-x$ is also in the domain of f, then f is the sum of an even and an odd function. See Problem 8.
   (b) Write the function given by $f(x) = e^x$ as the sum of an even and an odd function.

10. A function f is periodic, with period 2, and $f(x) = 2x - x^2$ for $0 < x < 2$. Sketch the graph of f between $x = -2$ and $x = +2$. State whether f is odd, even, or neither.

11. A function f is periodic, with period 4, and $f(x) = 2x$ for $0 < x < 4$. Is it correct to determine the Fourier coefficients b_n for f by

$$b_n = \frac{1}{2} \int_{-2}^{2} 2x \sin \frac{n\pi x}{2} \, dx$$

If not, give the correct integral for b_n.

12. Find a Fourier sine series of the function f defined by $f(x) = -\pi$ on $0 < x < \pi$.

13. Find a Fourier sine series of the function f defined by $f(x) \equiv 1$ on $0 < x < \pi$.

14. Given a Fourier sine series of the function f defined on $0 < x < 2$ by

$$f(x) = \begin{cases} 1 & \text{for } 0 < x < 1 \\ 1.5 & \text{for } x = 1 \\ 2 & \text{for } 1 < x < 2 \end{cases}$$

sketch the graph of the function to which the series will converge for $-2 < x < 2$. To what value will the series converge when (a) $x = -1$? (b) $x = 0$?

15. Find a Fourier sine series of the function f defined by $f(x) = x$ on $0 < x < \pi/2$ and by $f(x) \equiv 0$ on $\pi/2 < x < \pi$. (Find only the terms involving $\sin x$, $\sin 2x$, and $\sin 3x$.) To what value does the series converge when (a) $x = \pi/4$? (b) $x = \pi/2$? (c) $x = 9\pi/4$?

16. Determine a half-range Fourier sine expansion converging to $f(x) = x$ on $0 < x < 2$. Sketch the odd extension of f on $-4 < x < 4$.

17. Find a Fourier sine series of the function f defined by $f(x) = 2 - x$ on $0 < x < 2$.

18. Write but do not evaluate an integral for the coefficients b_n of the Fourier sine series of the function f defined by $f(x) = e^{-x}$ on $0 < x < 2$.

19. Find a Fourier sine series of the function f defined by $f(x) = \cos x$ on $0 < x < \pi$. Write the terms of the series involving $\sin 2x$, $\sin 4x$, and $\sin 6x$.

20. Find a Fourier sine series of the function f defined by $f(x) = (\pi - x)/2$ on $0 < x < \pi$. Use the series to evaluate

$$\sum_{n=1}^{+\infty} \frac{\sin n}{n} = \frac{\sin 1}{1} + \frac{\sin 2}{2} + \frac{\sin 3}{3} + \cdots$$

21. Find a Fourier cosine series of the function f defined by $f(x) \equiv 0$ on $0 < x < 1$ and $f(x) \equiv 1$ on $1 < x < 2$. Sketch the graph of the function to which the series converges on $-2 < x < 2$.

22. Find a Fourier cosine series of the function f defined by $f(x) = x$ on $0 < x < 10$. To what value does the series converge when $x = 10$?

23. Find a Fourier cosine series of the function f defined by $f(x) = -x$ on $0 < x < 3$.

24. Find a Fourier cosine series of the function f defined by $f(x) = 2x$ on $0 < x < 1$. Evaluate $\sum_{n=1}^{+\infty} [1/(2n - 1)^2]$.

25. Let g be given by $g(x) = x^2$ on $0 < x < 3$ and let $f(x)$ denote the Fourier cosine series of g. State the value of (a) $f(2)$; (b) $f(-2)$; (c) $f(0)$.
    Let $h(x)$ denote the Fourier series of the periodic function of period 3 defined on $0 < x < 3$ by $h(x) = x^2$. State the value of (d) $h(2)$; (e) $h(-2)$; (f) $h(0)$.

26. (a) Find a Fourier cosine series of the function f defined by $f(x) = x(\pi - x)$ on $0 < t < \pi$.

    (b) Evaluate $\sum_{n=1}^{+\infty} \frac{1}{n^2}$.

    (c) Evaluate $\sum_{n=1}^{+\infty} \frac{(-1)^{n+1}}{n^2}$.

27. Find a Fourier cosine series of the function f defined by $f(x) \equiv 100$ on $0 < x < \pi$.

28. Find a Fourier cosine series of the function f defined by $f(x) = 2\cos^2 x$ on $0 < x < \pi$.

29. Find a Fourier cosine series of the function f defined by $f(x) = \sin x$ on $0 < x < 2\pi$.

30. Given the Fourier series

$$\sum_{n=1}^{+\infty} \frac{\sin nx}{n}$$

of the function f defined by $f(x) = (\pi - x)/2$ on $0 < x < \pi$ (see Problem 20), evaluate $\sum_{n=1}^{k} [(\sin nx)/n]$ for $k = 3, 5,$ and 10 for $x = 0, \pi/10, 2\pi/10, 3\pi/10, \ldots$

(step size $h = \pi/10$). Plot the three curves on one graph and compare with the graph of f. Repeat, using step size $h = \pi/100$ (computer problem).

31. Find a particular solution of the following ordinary DE:

$$x'' + 4x = \sum_{n=1}^{+\infty} \frac{\sin(2n-1)\pi t}{(2n-1)^3}$$

Assume that the right member is the Fourier series of a function F for which term-by-term differentiation of both F and F' is valid.

10.6 The Vibrating String

A perfectly elastic string is fastened at $(0, 0)$ and at $(L, 0)$ in Figure 10.5. The string is caused to vibrate by giving each of its points a small (possibly zero) initial vertical displacement in the xy plane and an initial velocity (possibly zero) perpendicular to the x axis.

A partial DE governing the vibration of the string is derived by making a few physical assumptions. We assume that the tension T in the string is sufficiently large that the effects of air resistance and gravity can be neglected. We assume also that the string is sufficiently flexible that the tension T acts tangent to the string at every point. The particles of the string are assumed to move vertically in the xy plane, and the maximum displacement of any point of the string from the x axis is considered small compared with L. The string is assumed to be homogeneous; therefore ρ, the mass of the string per unit length, is taken to be constant. This implies that the total mass of string between the lines $x = x$ and $x = x + \Delta x$ remains equal to $\rho(\Delta x)$ as the string vibrates.

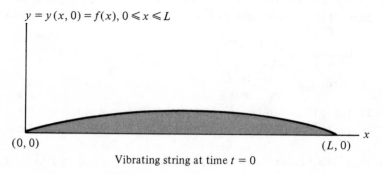

Vibrating string at time $t = 0$

Fig. 10.5

Figure 10.6 shows the configuration of the string at arbitrary time t. The height y of a point of the string x units from the y axis depends not only on x but also on the time t. We seek a partial DE satisfied by $y = y(x, t)$ where $0 < x < L$ and $t > 0$.

As shown in Figure 10.6, the portion of the string between $x = x$ and $x = x + \Delta x$ is acted on by the tension $T = T(x)$, acting at an angle $\theta = \theta(x)$

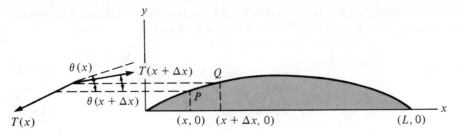

Vibrating string at time $t = t$

Fig. 10.6

at point P, and by the tension $T = T(x + \Delta x)$, acting at an angle $\theta = \theta(x + \Delta x)$ at point Q. The symbol θ denotes the inclination of the tangent to the string. Applying Newton's second law in the y and x directions, we obtain

$$T(x + \Delta x) \sin [\theta(x + \Delta x)] - T(x) \sin [\theta(x)] = \rho(\Delta x) \frac{\partial^2 \bar{y}}{\partial t^2} \quad \textbf{(10.33)}$$

and

$$T(x + \Delta x) \cos [\theta(x + \Delta x)] - T(x) \cos [\theta(x)] = 0 \quad \textbf{(10.34)}$$

where $\partial^2 \bar{y}/\partial t^2$ denotes the acceleration in the y direction of the center of mass of the small portion of the string under consideration. The right member of (10.34) is zero since, we are assuming, the string has no horizontal motion. Hence,

$$T(x + \Delta x) \cos [\theta(x + \Delta x)] = T(x) \cos [\theta(x)] = T \quad \textbf{(10.35)}$$

where T denotes the constant horizontal tension in the string. From (10.33) and (10.35),

$$\frac{T(x + \Delta x) \sin [\theta(x + \Delta x)]}{T(x + \Delta x) \cos [\theta(x + \Delta x)]} - \frac{T(x) \sin [\theta(x)]}{T(x) \cos [\theta(x)]} = \frac{\rho(\Delta x)}{T} \frac{\partial^2 \bar{y}}{\partial t^2}$$

or

$$\tan [\theta(x + \Delta x)] - \tan [\theta(x)] = \frac{\rho(\Delta x)}{T} \frac{\partial^2 \bar{y}}{\partial t^2} \quad \textbf{(10.36)}$$

Since $\tan \theta = \partial y/\partial x$, (10.36) can be written in the form

$$\frac{(\partial y/\partial x)|_{x = x + \Delta x} - (\partial y/\partial x)|_{x = x}}{\Delta x} = \frac{\rho}{T} \frac{\partial^2 \bar{y}}{\partial t^2}$$

Letting $\Delta x \to 0$, $\bar{y}$ approaches y at $x = x$, the left member approaches $\partial^2 y/\partial x^2$ at $x = x$, and we obtain the partial DE

$$\frac{\partial^2 y}{\partial x^2} = \frac{1}{c^2} \frac{\partial^2 y}{\partial t^2} \quad \textbf{(10.37)}$$

where $c = \sqrt{T/\rho}$. Equation (10.37) is called the *one-dimensional wave equation*.

10.6 The Vibrating String

We now proceed to find the height y of the vibrating string in terms of x and t. We must solve the boundary-value problem

$$y_{xx} = \frac{1}{c^2} y_{tt}; \quad 0 < x < L, t > 0 \tag{10.38}$$

$$y(0, t) = 0, \quad t \geq 0 \tag{10.39}$$

$$y(L, t) = 0, \quad t \geq 0 \tag{10.40}$$

$$y(x, 0) = f(x), \quad 0 \leq x \leq L \tag{10.41}$$

$$y_t(x, 0) = g(x), \quad 0 \leq x \leq L \tag{10.42}$$

The function f defines the initial configuration of the string and the function g prescribes the initial velocity of each point on the string. We assume that f and g are continuous on $[0, L]$, and we observe that $f(0) = f(L) = g(0) = g(L) = 0$ since the string is fixed at both ends. The graph of f is shown in Figure 10.5. Possible expressions for $f(x)$ are

$$a \sin \frac{\pi x}{L}, \quad a \sin \frac{2\pi x}{L}, \quad a \sin^2 \frac{\pi x}{L}, \quad ax(L - x)$$

where a is a nonzero constant. Another possibility is that in which $f(x) \equiv 0$, a situation that might happen when $g(x)$ is not identically zero. This occurs when a stationary piano string is struck by a felt hammer. When a string is given an initial displacement $f(x)$ and released from rest, $g(x) \equiv 0$.

Applying the method of separation of variables, we seek a solution of the boundary-value problem of the form

$$y(x, t) = X(x)T(t) \tag{10.43}$$

Substituting from (10.43) into (10.37), we obtain

$$X''T = \frac{1}{c^2} XT''$$

which for $XT \neq 0$ can be written in the form

$$\frac{X''}{X} = \frac{1}{c^2} \frac{T''}{T} \tag{10.44}$$

By the argument presented in Section 10.3, both sides of (10.44) have the same constant value k. We are thus led to the two ordinary DE

$$X'' - kX = 0 \tag{10.45}$$

and

$$T'' - kc^2 T = 0 \qquad \text{(10.46)}$$

It is left to Problem 1 to show that if $k = 0$

$$y(x, t) = (A + Bx)(C + Dt)$$

and that if $k = \lambda^2 > 0$,

$$y(x, t) = (A \sinh \lambda x + B \cosh \lambda x)(C \sinh \lambda ct + D \cosh \lambda ct)$$

and that in each of these cases, the boundary conditions are satisfied only if $y(x, t) \equiv 0$, that is, only if the trivial solution prevails.

We next consider the only remaining possibility $k = -\lambda^2 < 0$. From (10.39), (10.40), and (10.43) we conclude that if $y(x, t) \not\equiv 0$, then $X(0) = X(L) = 0$.

The two-point boundary-value problem

$$X'' + \lambda^2 X = 0; \quad X(0) = 0, \quad X(L) = 0 \qquad \text{(10.47)}$$

yields $X(x) = A \cos \lambda x + B \sin \lambda x$.

From $X(0) = 0$ we obtain $A = 0$, and from $X(L) = 0$ we obtain $B \sin \lambda L = 0$. We assume that $B \neq 0$, since otherwise $X(x) \equiv 0$ and $y(x, t) \equiv 0$. Thus, for boundary conditions (10.39) and (10.40) to be satisfied, $\sin \lambda L = 0$ must hold. This implies that

$$\lambda L = n\pi \text{ or } \lambda = n\pi/L \quad \text{for } n = \pm 1, \pm 2, \ldots$$

Since $\sin(-\alpha) \equiv -\sin \alpha$, the only nontrivial solutions of (10.47) have the form

$$X_n(x) = B_n \sin \lambda_n x = B_n \sin \frac{n\pi x}{L}$$

where n is a *positive* integer and each arbitrary constant $B_n \neq 0$.

Similarly, the ordinary DE $T'' + (\lambda c)^2 T = 0$ has solutions of the form

$$T_n(t) = C_n \sin c\lambda_n t + D_n \cos c\lambda_n t = C_n \sin \frac{n\pi c}{L} t + D_n \cos \frac{n\pi c}{L} t$$

$$\text{(10.48)}$$

where $n = 1, 2, 3, \ldots$.

In forming the product $X_n(x)T_n(t)$, we take each $B_n = 1$ without loss of generality, since C_n and D_n are arbitrary. The values of λ_n for which

$$y(x, t) = \sin \lambda_n x (C_n \sin c\lambda_n t + D_n \cos c\lambda_n t) \qquad \text{(10.49)}$$

satisfies (10.38), (10.39), and (10.40) are called the *eigenvalues* of the boundary-value problem, and the corresponding nonzero solutions of (10.47) given by $\sin(n\pi x/L)$ are called the *eigenfunctions* of the boundary-value problem. The objective at this stage is to select values of C_n, D_n, and permissible values of λ_n so that the additional boundary (initial) conditions (10.41) and (10.42) will be satisfied. To satisfy (10.41), Equation (10.49) must yield

$$y(x, 0) = D_n \sin \lambda_n x = D_n \sin \frac{n\pi x}{L} = f(x)$$

If $f(x)$ has the form $D_n \sin(n\pi x/L)$ or even $\sum_{k=1}^{n} D_k \sin(k\pi x/L)$, we can proceed as in Section 10.3 but generally this will not be the case. Instead, let us try a solution of the form

$$y(x, t) = \sum_{n=1}^{+\infty} \sin \lambda_n x (C_n \sin c\lambda_n t + D_n \cos c\lambda_n t) \tag{10.50}$$

where $\lambda_n = n\pi/L$; that is, let us superimpose an infinite number of solutions of the form (10.49). Then if (10.41) is to hold, we must have

$$y(x, 0) = \sum_{n=1}^{+\infty} D_n \sin \lambda_n x = f(x) \quad \text{on } 0 \le x \le L \tag{10.51}$$

We recognize (10.51) as the Fourier sine series of f, where the period $2p = 2L$. Hence, the Fourier coefficients are given by

$$D_n = \frac{2}{L} \int_0^L f(x) \sin \frac{n\pi x}{L} \, dx, \quad n = 1, 2, 3, \ldots \tag{10.52}$$

To satisfy the final boundary condition (10.42), we assume that term-by-term differentiation of (10.50) is valid. This assumption yields

$$y_t(x, t) = \sum_{n=1}^{+\infty} c\lambda_n \sin \lambda_n x (C_n \cos c\lambda_n t - D_n \sin c\lambda_n t) \tag{10.53}$$

where $\lambda_n = n\pi/L$.

Condition (10.42) reduces (10.53) to

$$y_t(x, 0) = \sum_{n=1}^{+\infty} c\lambda_n C_n \sin \lambda_n x = g(x) \quad \text{on } 0 \le x \le L \tag{10.54}$$

Letting $c\lambda_n C_n = F_n$, we recognize (10.54) as the Fourier sine series of g, where the period $2p = 2L$. Hence, the Fourier coefficients are given by

$$F_n = \frac{2}{L} \int_0^L g(x) \sin \frac{n\pi x}{L} \, dx, \quad n = 1, 2, 3, \ldots$$

and therefore,

$$C_n = \frac{F_n}{c\lambda_n} = \frac{LF_n}{n\pi c} = \frac{2}{n\pi c} \int_0^L g(x) \sin\frac{n\pi x}{L}\, dx, \quad n = 1, 2, 3, \ldots \quad \textbf{(10.55)}$$

Note: Some authors refer to the functions defined by (10.49) as the eigenfunctions.

Our formal computations have led to the possible solution (10.50) of the boundary-value problem, where C_n and D_n are given by (10.55) and (10.52). Two questions now present themselves. Is the function defined by (10.50) the unique solution? If so, does it provide a reasonable model of the physical motion of the string?

In References 10.3 and 10.5, it is shown that the function given by (10.50) defines the unique solution of the vibrating string problem whenever f, f', f'', g, and g' are continuous on $[0, L]$ with

$$f(0) = f(L) = 0, \quad g(0) = g(L) = 0, \quad f''(0) = f''(L) = 0$$

Even these general restrictions on f and g do not cover all cases of interest. Reference 10.6 uses advanced methods to show that (10.50) defines the unique solution of the "plucked string" boundary-value problem. In this problem, a point on the string at $x = x_0$ is "plucked," or raised, a distance h above the x axis and released from rest. The graph of f on $[0, L]$ consists of two straight line segments and f is not differentiable at $x = x_0$. See Problem 3(e).

The solution given by (10.50), although obtained by making numerous physical assumptions, does provide a useful model for the vibrating string. As in the case of simple harmonic motion, the motion of the string defined by (10.50) represents perpetual motion, since damping forces have been neglected. References 10.2 and 10.3 discuss weighted strings and strings with variable tension and density.

Let us consider the case in which the string is released from rest. Then $g(x) \equiv 0$, $C_n \equiv 0$, and (10.50) reduces to

$$y(x, t) = \sum_{n=1}^{+\infty} D_n \sin\frac{n\pi x}{L} \cos\frac{n\pi ct}{L}$$

If $D_2, D_3, \ldots$ are all zero,

$$y(x, t) = D_1 \sin\frac{\pi x}{L} \cos\frac{\pi ct}{L} \quad \text{and} \quad f(x) = D_1 \sin\frac{\pi x}{L}$$

See Figure 10.7.

In this mode of vibration, known as the first *normal mode*, the string vibrates as a single arch, and the frequency $F = \pi c L^{-1}/(2\pi) = c/(2L)$. The vibration of the string generates sound waves that are audible as a musical note if the tension T in the string is large enough ($c = \sqrt{T/\rho}$). The tone

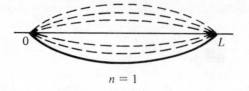

$n = 1$

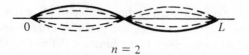

$n = 2$

$n = 3$

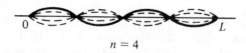

$n = 4$

Normal modes of a vibrating string

Fig. 10.7

corresponding to $n = 1$ is known as the *fundamental tone* of the vibration. It has the smallest frequency of the normal modes of vibration and is also known as the *first harmonic*.

If $D_n = 0$ except for D_2,

$$y(x, t) = D_2 \sin \frac{2\pi x}{L} \cos \frac{2\pi ct}{L} \quad \text{and} \quad f(x) = D_2 \sin \frac{2\pi x}{L}$$

In this second normal mode of vibration, the point $x = L/2$, called a *node*, remains fixed and the string vibrates as two half waves with frequency $2c/(2L) = 2[c/(2L)]$. This produces the second harmonic, called the *first overtone*.

Similarly, if

$$y(x, t) = D_n \sin \frac{n\pi x}{L} \cos \frac{n\pi ct}{L}$$

the vibration produces the nth harmonic, or $(n - 1)$st overtone. The frequencies of the higher harmonics are integral multiples of the fundamental frequency (frequency of the fundamental tone). The nth normal mode of vibration has $n - 1$ nodes and n half waves. The frequency of a vibration determines the important property of the resulting musical sound known as *pitch*.

In the general case, for somewhat arbitrary $f(x)$, the vibration consists of a superposition of the fundamental mode ($n = 1$) and various overtones. Musicians produce notes of pleasant quality by combining overtones with the fundamental note in various manners. The frequencies of the overtones of a vibrating membrane such as a drum are not integral multiples of the fundamental frequency.

The one-dimensional wave equation is a useful model for describing various types of vibrations and wave phenomena. For example, if a thin cylindrical shaft with axis along the x axis is fixed at one end and subjected to a twisting torque at the free end, each circular cross section will turn through an angle θ. If the torque is released, the shaft will vibrate and $\theta = \theta(x, t)$ will satisfy the one-dimensional wave equation. See Reference 10.12.

Problem List 10.4

1. (a) Show that if

$$\frac{X''}{X} = \frac{1}{c^2} \frac{T''}{T} = k = 0$$

the vibrating string problem has solutions of the form

$$y(x, t) = (A + Bx)(C + Dt)$$

which reduce to $y(x, t) \equiv 0$ if boundary conditions (10.39) and (10.40) are satisfied.

(b) Show that if

$$\frac{X''}{X} = \frac{1}{c^2} \frac{T''}{T} = k = \lambda^2 > 0$$

the vibrating string problem has solutions of the form

$$y(x, t) = (A \sinh \lambda x + B \cosh \lambda x)(C \sinh \lambda ct + D \cosh \lambda ct)$$

which reduce to $y(x, t) \equiv 0$ if boundary conditions (10.39) and (10.40) are satisfied.

2. Show that the constant c in the one-dimensional wave equation has the dimension of velocity.

3. The height y of a vibrating string L ft long is governed by the one-dimensional wave equation $y_{xx} = (1/c^2)y_{tt}$ and the boundary conditions

$$y(0, t) = 0, \quad y(L, t) = 0 \quad \text{for } t \geq 0$$

and

$$y(x, 0) = f(x), \quad y_t(x, 0) = g(x) \quad \text{for } 0 \leq x \leq L$$

Find y in terms of x and t if the initial position and velocity of the string are given by:

(a) $f(x) = 4 \sin \dfrac{3\pi}{L} x, g(x) \equiv 0$

(b) $f(x) = 2 \sin \dfrac{4\pi}{L} x - 6 \sin \dfrac{5\pi}{L} x, \; g(x) \equiv 0$

(c) $f(x) = 0.01 \sin \dfrac{2\pi x}{L}, \; g(x) = 0.01 \sin \dfrac{2\pi}{L} x$

(d) $f(x) = 0.01x(L - x), \; g(x) \equiv 0$

(e) $f(x) = \begin{cases} \dfrac{x}{40} & \text{on } 0 \le x < \dfrac{L}{2} \\[2mm] \dfrac{L - x}{40} & \text{on } \dfrac{L}{2} \le x \le L \end{cases}, \quad g(x) \equiv 0$

(f) $f(x) \equiv 0, \; g(x) = x(L - x)$

(g) $f(x) = \dfrac{1}{80} \sin^2 \dfrac{\pi x}{L}, \; g(x) \equiv 0$

4. Show by direct substitution that $y(x, t) = \sin(x + 3t) + \sin(x - 3t)$ defines a solution of $y_{xx} = \frac{1}{9} y_{tt}$.

5. Show that under the substitutions $u = x - ct$, $v = x + ct$, the one-dimensional wave equation $y_{xx} = (1/c^2)y_{tt}$ becomes $y_{vu} = 0$. Show that $y(x, t) = f(x - ct) + g(x + ct)$ defines a general solution.

6. Assume in the vibrating string problem that $f''(x)$ exists on $[0, L]$ and that $g(x) \equiv 0$ on $[0, L]$. Show that the solution

$$y(x, t) = \sum_{n=1}^{+\infty} D_n \sin \frac{n\pi x}{L} \cos \frac{n\pi ct}{L}$$

can be written in the form

$$y(x, t) = \frac{1}{2}[f(x + ct) + f(x - ct)]$$

Use this form to show that $y(x, t)$ is a solution of the boundary-value problem of the vibrating string under the given conditions. See Problem 5. *Hint:* Use the identity

$$\sin \alpha \cos \beta = \frac{1}{2}[\sin(\alpha + \beta) + \sin(\alpha - \beta)]$$

7. Show that the frequency F of the nth normal mode of vibration of a vibrating string is given by $F = [n/(2L)]\sqrt{T/\rho}$. What change in the horizontal tension T will double the frequency (pitch) of the fundamental tone?

10.7 Heat Flow

The axis of the thin uniform cylindrical rod in Figure 10.8 extends from $x = 0$ to $x = L$, where L denotes the length of the rod. The lateral surface of the rod is completely insulated, and consequently heat can enter and leave the rod only through the ends. The cross section of the rod has constant area A, and the temperature u of the rod at a fixed time $t = t_0$ is the same at each point

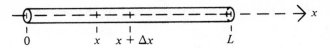

Fig. 10.8

of a cross section $x = x_0$. Thus, the temperature $u = u(x, t)$ is a function of x and t.

The partial DE governing the flow of heat in the rod is obtained by equating two expressions for the increase (decrease) ΔQ of the heat Q in the portion of the rod between $x = x$ and $x = x + \Delta x$ during the time interval Δt. One physical law states that at any instant, the amount of heat in the portion of length Δx equals the density ρ times the volume $A(\Delta x)$ times a constant s (the *specific heat* of the material) times the average temperature $\bar{u} = \bar{u}(x, t)$ of the portion at that instant. Therefore,

$$\Delta Q = s\rho A(\Delta x)\bar{u}(x, t + \Delta t) - s\rho A(\Delta x)\bar{u}(x, t) \tag{10.56}$$

A second physical law states that the amount of heat flowing across a section $x = x$ during a time interval Δt equals the cross-sectional area A times a constant κ (the *heat conductivity* of the rod) times the interval Δt times the average $\bar{u}_x = \bar{u}_x(x, t)$ of $u_x(x, t)$ during the time interval Δt. Hence,

$$\Delta Q = A\kappa(\Delta t)\bar{u}_x(x + \Delta x, t) - A\kappa(\Delta t)\bar{u}_x(x, t) \tag{10.57}$$

Written as a vector, the temperature gradient is given by $u_x(x, t)\mathbf{i}$. If $\bar{u}_x(x + \Delta x, t) > 0$ in (10.57), heat flows *into* the portion of length Δx at $x = x + \Delta x$ during Δt, since heat flows in the direction of decreasing temperature. Similarly, if $\bar{u}_x(x, t) > 0$, heat flows *out of* the portion of length Δx at $x = x$ during Δt. It follows that the expressions (10.56) and (10.57) have the same sign. Equating these two expressions for ΔQ and dividing by $s\rho A(\Delta x)(\Delta t)$, we obtain

$$\frac{\bar{u}(x, t + \Delta t) - \bar{u}(x, t)}{\Delta t} = \frac{\kappa}{s\rho}\left[\frac{\bar{u}_x(x + \Delta x, t) - \bar{u}_x(x, t)}{\Delta x}\right]$$

We assume that u, u_x, u_t, and u_{xx} are continuous. Letting $\Delta x \to 0$ and $\Delta t \to 0$, we obtain

$$u_t = a^2 u_{xx} \tag{10.58}$$

where $a^2 = \kappa/(s\rho)$ is called the *thermal diffusivity* of the material of the rod. Equation (10.58) is called the *one-dimensional heat equation*.

The flow of heat in a thin plane lamina is governed by the *two-dimensional heat equation*

$$u_t = a^2(u_{xx} + u_{yy}) \tag{10.59}$$

10.7 Heat Flow

and the flow of heat in a solid is governed by the *three-dimensional heat equation*

$$u_t = a^2(u_{xx} + u_{yy} + u_{zz}) \tag{10.60}$$

For derivations of (10.59) and (10.60), see Reference 10.12.

When the steady-state temperature prevails, $u_t = 0$; (10.58) reduces to $u_{xx} = 0$ (Laplace's equation in one dimension), (10.59) reduces to $u_{xx} + u_{yy} = 0$ (Laplace's equation in two dimensions), and (10.60) reduces to $u_{xx} + u_{yy} + u_{zz} = 0$ (Laplace's equation in three dimensions).

We now consider the following boundary-value problem for a uniform rod:

$$u_t = a^2 u_{xx}; \qquad 0 < x < L, \quad t > 0 \tag{10.61}$$

$$u(0, t) = 0, \quad t \geq 0 \tag{10.62}$$

$$u(L, t) = 0, \quad t \geq 0 \tag{10.63}$$

$$u(x, 0) = f(x), \quad 0 < x < L \tag{10.64}$$

Equations (10.62) and (10.63) express the conditions that both ends of the rod remain at zero degrees. Equation (10.64) expresses the condition that the initial temperature of the rod is given by a prescribed function f.

Letting $u(x, t) = X(x)T(t)$, substituting into (10.61), and separating the variables, we obtain

$$XT' = a^2 X''T$$

and

$$\frac{T'}{a^2 T} = \frac{X''}{X} = k \tag{10.65}$$

where k is constant.

If $k = 0$, then $T' = 0$, $X'' = 0$, and

$$u(x, t) = Ex + F \tag{10.66}$$

where E and F are arbitrary constants.

If $k = \lambda^2 > 0$, then $u(x, t) \to +\infty$ as $t \to +\infty$ (see Problem 1). We exclude this possibility that in a concrete physical application the temperature distribution is unbounded.

If $k = -\lambda^2 < 0$,

$$T' + a^2\lambda^2 T = 0, \qquad X'' + \lambda^2 X = 0$$

$$T = C_1[\exp(-a^2\lambda^2 t)], \qquad X = A_1 \cos \lambda x + B_1 \sin \lambda x$$

and

$$u(x, t) = [\exp(-a^2\lambda^2 t)](A \cos \lambda x + B \sin \lambda x) \tag{10.67}$$

where $A = C_1 A_1$ and $B = C_1 B_1$.

From (10.62), we obtain $A = 0$, and hence

$$u(x, t) = B[\exp(-a^2\lambda^2 t)] \sin \lambda x$$

From (10.63), we obtain

$$B[\exp(-a^2\lambda^2 t)] \sin \lambda L = 0$$

If $B = 0$, then $u(x, t) \equiv 0$ (the trivial solution).
If $B \neq 0$, then $\sin \lambda L = 0$, and

$$\lambda L = n\pi \quad \text{or} \quad \lambda = \lambda_n = \frac{n\pi}{L}, \quad n = \pm 1, \pm 2, \ldots$$

Thus, eigenfunctions of the heat flow problem are given by $\sin(n\pi x/L)$, and (10.61), (10.62), and (10.63) are satisfied by

$$u_n(x, t) = B_n \exp\left[-a^2(n\pi/L)^2 t\right] \sin \frac{n\pi x}{L} \tag{10.68}$$

for $n = 1, 2, 3, \ldots$.
To satisfy (10.64), we try

$$u(x, t) = \sum_{n=1}^{+\infty} B_n \exp\left[-a^2(n\pi/L)^2 t\right] \sin \frac{n\pi x}{L} \tag{10.69}$$

Then,

$$u(x, 0) = \sum_{n=1}^{+\infty} B_n \sin \frac{n\pi x}{L} = f(x)$$

and we choose

$$B_n = \frac{2}{L} \int_0^L f(x) \sin \frac{n\pi x}{L}\, dx \tag{10.70}$$

the coefficients of the Fourier sine series for f. We state without proof that (10.69) with B_n given by (10.70) defines the unique solution of the boundary-value problem whenever f is continuous on $[0, L]$. For a discussion of the existence and uniqueness of this solution when f and f' are piecewise continuous on $[0, L]$ and $f(x) = \frac{1}{2}[f(x+) + f(x-)]$ everywhere on $[0, L]$, see Reference 10.5.

EXAMPLE 1. Solve the following boundary-value problem for heat flow in a 10-ft rod:

$$u_t = 4u_{xx}; \quad 0 < x < 10, \quad t > 0 \tag{10.71}$$

$$u(0, t) = u(10, t) = 0, \quad t \geq 0 \tag{10.72}$$

$$u(x, 0) = 6 \sin 3\pi x, \quad 0 < x < 10 \tag{10.73}$$

Solution: From (10.68), Equations (10.71) and (10.72) are satisfied by

$$u_n(x, t) = B_n \exp\left[-4(n\pi/10)^2 t\right] \sin\frac{n\pi x}{10}$$

From (10.73),

$$B_n \sin\frac{n\pi x}{10} = 6 \sin 3\pi x$$

The solution is obtained by choosing $B_n = 6$ and $n = 30$:

$$u(x, t) = 6 \exp(-36\pi^2 t) \sin 3\pi x$$

The steady-state solution, which could have been anticipated from (10.72), is given by $U(x, t) = \lim_{t \to +\infty} u(x, t) \equiv 0$.

EXAMPLE 2. Solve Example 1 if (10.73) is replaced by

$$u(x, 0) = \begin{cases} 60 & \text{for } 0 < x < 5 \\ 0 & \text{for } 5 < x < 10 \end{cases} \tag{10.74}$$

Solution: Using (10.69), we try

$$u(x, t) = \sum_{n=1}^{+\infty} B_n \exp\left[-4(n\pi/10)^2 t\right] \sin\frac{n\pi x}{10}$$

From (10.74),

$$u(x, 0) = \sum_{n=1}^{+\infty} B_n \sin\frac{n\pi x}{10} = \begin{cases} 60 & \text{for } 0 < x < 5 \\ 0 & \text{for } 5 < x < 10 \end{cases}$$

and from (10.70),

$$B_n = \frac{2}{10}\int_0^5 60 \sin\frac{n\pi x}{10}\, dx = \frac{120}{n\pi}\left[1 - \cos\frac{n\pi}{2}\right]$$

Hence,

$$u(x, t) = \frac{120}{\pi}\sum_{n=1}^{+\infty}\frac{1}{n}\left[1 - \cos\frac{n\pi}{2}\right]\exp\left[-4(n\pi/10)^2 t\right]\sin\frac{n\pi x}{10} \tag{10.75}$$

NOTE: Although $u(5, 0)$ was not defined in (10.74), $u(5, 0) = \frac{1}{2}[0 + 60] = 30$ by (10.75).

EXAMPLE 3. Solve the heat flow problem in a rod of uniform cross section π ft long; the lateral surface and both ends of the rod are perfectly insulated and the initial temperature distribution in the rod is given by $u(x, 0) = x$.

Solution: We must solve the boundary-value problem:

$$u_t = a^2 u_{xx}; \quad 0 < x < \pi, \quad t > 0 \tag{10.76}$$

$$u_x(0, t) = 0, \quad t \geq 0 \tag{10.77}$$

$$u_x(\pi, t) = 0, \quad t \geq 0 \qquad\qquad \textbf{(10.78)}$$

$$u(x, 0) = x, \quad 0 < x < \pi \qquad\qquad \textbf{(10.79)}$$

From (10.67), $u(x, t) = \exp(-a^2\lambda^2 t)(A \cos \lambda x + B \sin \lambda x)$, we obtain

$$u_x(x, t) = \exp(-a^2\lambda^2 t)(-\lambda A \sin \lambda x + \lambda B \cos \lambda x)$$

By (10.77), $u_x(0, t) = \lambda B \exp(-a^2\lambda^2 t) = 0$, and hence $B = 0$.
By (10.78), $u_x(\pi, t) = \exp(-a^2\lambda^2 t)(-\lambda A \sin \lambda\pi) = 0$.
  Since $A = 0$ yields the trivial solution $u(x, t) \equiv 0$, we must have $\lambda\pi = n\pi$ or $\lambda = \lambda_n = n$.
  The eigenfunctions of the boundary-value problem are given by $\cos nx$ for $n = 0, 1, 2, 3, \ldots$.
  Trying $u(x, t) = \sum_{n=0}^{+\infty} A_n \exp(-a^2 n^2 t) \cos nx$, we must have, from (10.79),

$$u(x, 0) = \sum_{n=0}^{+\infty} A_n \cos nx = x$$

Using the Fourier cosine series for $f(x) = x$ on $(0, \pi)$, we obtain $A_n = (2/\pi)\int_0^\pi x \cos nx \, dx$. Thus,

$$A_0 = \frac{2}{\pi} \int_0^\pi x \, dx = \pi$$

and

$$A_n = \frac{2}{\pi}\left[\frac{1}{n^2}\cos nx + \frac{x}{n}\sin nx\right]_0^\pi = \frac{2}{n^2\pi}[(-1)^n - 1]$$

for $n = 1, 2, 3, \ldots$.
  Finally,

$$u(x, t) = \frac{\pi}{2} + \frac{2}{\pi}\sum_{n=1}^{+\infty}\frac{(-1)^n - 1}{n^2}\exp(-a^2 n^2 t)\cos nx$$

In Example 1 the solution for a fixed value $t = T$ of t is given by

$$u(x, T) = 6 \exp(-36\pi^2 T)\sin 3\pi x$$

The temperature profile at $t = T$ is easy to visualize; for every $T \geq 0$ the graph of $u(x, T)$ is an arch of a sine curve passing through the two ends of the rod. As $T \to +\infty$ the maximum height of the arch, achieved at $x = 5$ ft, decreases and approaches zero as the temperature approaches the steady-state temperature $U(x, t) \equiv 0$.
  In Examples 2 and 3, in which the solutions are Fourier series solutions, it is difficult to visualize the temperature profile for a fixed $t = T$, or to determine the manner in which the temperature profiles change as T assumes values on an increasing sequence, say, $T = 10, T = 20, T = 50, T = 100$, etc

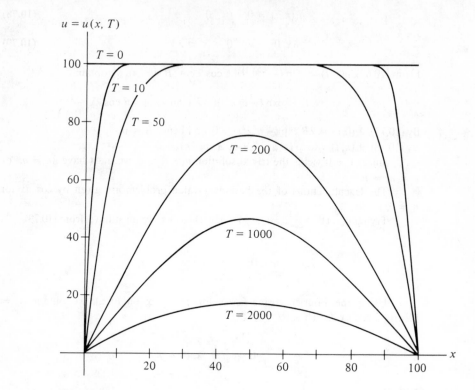

Fig. 10.9

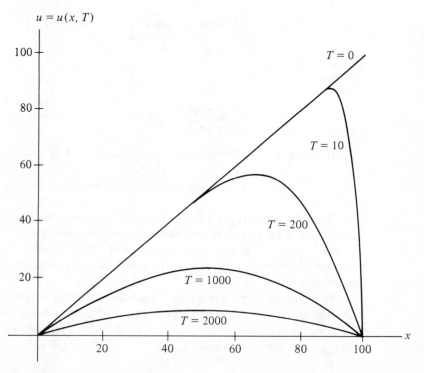

Fig. 10.10

In Examples 4 and 5 we present two problems in which the temperature profiles $u(x, T)$ are graphed by a computer for a few values of $t = T$. The author is indebted to Professor Howard Lewis Penn of the United States Naval Academy for these examples and for the accompanying computer graphics displayed in Figures 10.9 and 10.10. Dr. Penn employed these two boundary-value problems in a set of television tapes he developed to assist students in relating Fourier series solutions of boundary-value problems to the physical problems that led to the analytical solutions.

EXAMPLE 4. Solve the following boundary-value problem for heat flow in a 100-cm rod:

$$u_t = u_{xx}; \quad 0 < x < 100, \quad t > 0$$

$$u(0, t) = u(100, t) = 0, \quad t \geq 0$$

$$u(x, 0) = 100, \quad 0 < x < 100$$

Solution: The Fourier series solution (see Problem 13) is given by

$$u(x, t) = \sum_{n=1}^{+\infty} \frac{400}{(2n - 1)\pi} \exp\left[-(2n - 1)^2\pi^2 t/10{,}000\right] \sin\left[(2n - 1)\pi x/100\right]$$

In Figure 10.9 the temperature profiles are graphed by computer for $t = T = 10, 50, 200, 1000$, and 2000. For a fixed $t = T$, the computer sums a sufficient number of terms of the solution $u = u(x, t)$ so that the graph printed out by the computer and the graph of the actual solution $u = u(x, t)$ are visually indistinguishable.

It is seen that at $T = 10$, the temperature, except near the ends of the rod, is 100 throughout the rod. The temperature at the center of the rod remains 100 until some time between $T = 50$ and $T = 200$, after which the center temperature approaches zero as $T \to +\infty$.

EXAMPLE 5. Solve the following boundary-value problem for heat flow in a 100-cm rod:

$$u_t = u_{xx}; \quad 0 < x < 100, \quad t > 0$$

$$u(0, t) = u(100, t) = 0, \quad t \geq 0$$

$$u(x, 0) = x, \quad 0 < x < 100$$

Solution: The Fourier series solution (see Problem 13) is given by

$$u(x, t) = \sum_{n=1}^{+\infty} \frac{200}{n\pi} (-1)^{n+1} \exp\left(-n^2\pi^2 t/10{,}000\right) \sin\left(n\pi x/100\right)$$

In Figure 10.10 the temperature profiles are graphed by computer for $t = T = 10, 200, 1000$, and 2000. It is seen that, for a considerable period of time, the temperature decreases in the right half of the rod but remains unchanged in the left half of the rod. The heat flowing out the left end is replaced by heat flowing to the left from the warmer right end. The hottest point in the rod keeps moving to the left until the temperature profile resembles the profile in Example 4.

10.7 Heat Flow

We conclude this section with a brief discussion of the heat flow problem in which (10.62) and (10.63) are replaced by

$$u(0, t) = u_0, \quad u(L, t) = u_L, \quad t \geq 0 \tag{10.80}$$

where u_0 and u_L are different from zero. These boundary conditions are called *nonhomogeneous*, since the sum of two or more eigenfunctions no longer yields solutions satisfying (10.80). For example, $u_1(0, t) + u_2(0, t) = 2u_0$ instead of u_0. To handle this situation, we assume the steady-state solution

$$u(x, t) = Ex + F \tag{10.66}$$

corresponding to the eigenvalue $k = 0$. To determine E and F, we use (10.80) (see Problem 2) to obtain

$$u(x, t) = U(x) = u_0 + (u_L - u_0)\frac{x}{L} \tag{10.81}$$

We then try as a solution the sum of the steady-state solution (10.81) and a transient solution of the form (10.69); that is, we try

$$u(x, t) = U(x) + \sum_{n=1}^{+\infty} B_n \exp \frac{-a^2 n^2 \pi^2 t}{L^2} \sin \frac{n\pi x}{L}$$

Then $u(0, t) = U(0) = u_0$ and $u(L, t) = U(L) = u_L$ as required and

$$u(x, 0) = U(x) + \sum_{n=1}^{+\infty} B_n \sin \frac{n\pi x}{L} = f(x)$$

or

$$\sum_{n=1}^{+\infty} B_n \sin \frac{n\pi x}{L} = f(x) - U(x)$$

We then use the Fourier sine series of $f - U$ to obtain

$$B_n = \frac{2}{L} \int_0^L [f(x) - U(x)] \sin \frac{n\pi x}{L} \, dx$$

The procedure is analogous to the procedure used in Chapter 4 in solving nonhomogeneous linear ordinary DE.

Problem List 10.5

1. Show that if $k = \lambda^2 > 0$ in (10.66), then $u(x, t) \to +\infty$ as $t \to +\infty$.
2. Show that if $u(0, t) = U(0) = u_0$ and $u(L, t) = U(L) = u_L$, then

$$u(x, t) = U(x) = Ex + F \tag{10.66}$$

is given by

$$u(x, t) = U(x) = u_0 + (u_L - u_0)\frac{x}{L} \qquad (10.81)$$

3. The flow of heat in a rod is governed by the partial DE $u_t = 9u_{xx}$. Find u in terms of x and t if

$$u(0, t) = u(12, t) = 0 \quad \text{and} \quad u(x, 0) = 5 \sin \frac{\pi x}{6}$$

4. Solve the following boundary-value problem for heat flow in a 1-ft rod:

$$u_t = u_{xx}; \quad 0 < x < 1, \quad t > 0$$

$$u(0, t) = u(1, t) = 0, \quad t \geq 0$$

$$u(x, 0) = 50, \quad 0 < x < 1$$

5. (a) Solve the following boundary-value problem for heat flow in a 4-ft rod:

$$u_t = 4u_{xx}; \quad 0 < x < 4, \quad t > 0$$

$$u(0, t) = u(4, t) = 0, \quad t \geq 0$$

$$u(x, 0) = 5 \sin \pi x, \quad 0 < x < 4$$

(b) Solve part (a) with $u(x, 0) = 5 \sin \pi x$, $0 < x < 4$, changed to $u(x, 0) = 60$, $0 < x < 4$.

6. The curved surface of a thin rod of length L is perfectly insulated against the flow of heat so that the temperature $u(x, t)$ satisfies the one-dimensional heat equation $u_t = a^2 u_{xx}$. The rod is initially at a uniform temperature of $50°$ $(0 < x < L)$. Find $u(x, t)$ if both ends are kept at the temperature $0°$.

7. The partial DE governing heat flow in a 10-ft bar is $u_t = u_{xx}$. The initial temperature of the bar is given by

$$u(x, 0) = \begin{cases} 0 & \text{for } 0 < x < 5 \\ 60 & \text{for } 5 < x < 10 \end{cases}$$

The bar is insulated, except that both ends are kept at temperature $0°$. Find u in terms of x and t. Write the first four nonzero terms of the series for $u(x, t)$.

8. Solve the heat flow boundary-value problem:

$$u_t = \frac{1}{4}u_{xx}; \quad 0 < x < 6, \quad t > 0$$

$$u(0, t) = u(6, t) = 0, \quad t \geq 0$$

$$u(x, 0) = f(x) = \sum_{n=1}^{+\infty} \frac{1}{n^2 + 1} \sin \frac{n\pi x}{6}, \quad 0 < x < 6$$

9. Solve the one-dimensional heat equation $u_t = u_{xx}$ with the boundary conditions $u(0, t) = u(L, t) = 0, t \geq 0; u(x, 0) = x, 0 < x < L$.

10. Solve the one-dimensional heat equation $u_t = u_{xx}$ with boundary conditions $u_x(0, t) = u_x(L, t) = 0, t \geq 0; u(x, 0) = x/2, 0 < x < L$.

11. Show in illustrative Example 3 that the steady-state temperature of the rod is the average of the initial temperature distribution given by $f(x) = u(x, 0) = x$ on $0 \le x \le \pi$.

12. Solve the one-dimensional heat equation $u_t = u_{xx}$ with the boundary conditions

$$u(0, t) = u(1, t) = 0, \quad t \ge 0$$

$$u(x, 0) = \begin{cases} x & \text{for } 0 < x < \dfrac{1}{2} \\ 1 - x & \text{for } \dfrac{1}{2} \le x < 1 \end{cases}$$

13. Solve the boundary-value problem in (a) illustrative Example 4 and (b) illustrative Example 5.

10.8 Two-Dimensional Heat Flow

Let $u = u(x, y, t)$ denote the temperature distribution in a thin plate, or lamina, with insulated faces. Under certain physical assumptions that are closely approximated in numerous important applications, the temperature distribution is governed by the two-dimensional heat equation

$$u_t = a^2(u_{xx} + u_{yy}) \tag{10.59}$$

When the lamina covers a circle, a sector of a circle, or an annular ring, it is usually advantageous to use the polar form

$$u_t = a^2(u_{rr} + r^{-1}u_r + r^{-2}u_{\theta\theta}) \tag{10.82}$$

of (10.59). See Problem 1.

EXAMPLE 1. A circular lamina with insulated faces covers the circle $r = 1$. The temperature on the circumference of the lamina is given by $u = f(\theta)$ and the interior temperature is governed by (10.82). Find the steady-state temperature given by $u = u(r, \theta)$.

Solution: Setting $u_t = 0$ in (10.82), we solve the following boundary-value problem, known as a *Dirichlet problem*:

$$u_{rr} + r^{-1}u_r + r^{-2}u_{\theta\theta} = 0; \quad 0 \le r < 1, \quad 0 \le \theta < 2\pi \tag{10.83}$$

$$u(1, \theta) = f(\theta), \quad 0 \le \theta < 2\pi \tag{10.84}$$

where f is continuous and $f(\theta + 2\pi) = f(\theta)$ for all θ.

Letting $u = u(r, \theta) = R(r)\Theta(\theta)$ in (10.83), we obtain $R''\Theta + r^{-1}R'\Theta + r^{-2}R\Theta'' = 0$. Separation of variables yields

$$\frac{1}{R}(r^2R'' + rR') = -\frac{\Theta''}{\Theta} = k \tag{10.85}$$

If $k = 0$,

$$u(r, \theta) = (A + B \ln r)(C\theta + D) \tag{10.86}$$

See Problem 2.

We consider it physically impossible for $\lim_{r \to 0^+} u(r, \theta) = +\infty$ to hold. Therefore, (10.86) can represent a solution only if $B = 0$. With $B = 0$, (10.86) defines a linear function of θ and can represent a solution only if f is a linear function of θ.

In Problem 3 the student is asked to show that no solution of the boundary-value problem can be obtained when $k < 0$.

If $k = \lambda^2 > 0$, we must solve the ordinary DE

$$r^2 R'' + r R' - \lambda^2 R = 0 \tag{10.87}$$

and

$$\Theta'' + \lambda^2 \Theta = 0 \tag{10.88}$$

Equation (10.87), Euler's DE, is solved by the substitution (see Problem 4) $r = e^z$, to obtain $R(r) = Ar^\lambda + Br^{-\lambda}$.

Equation (10.88) yields $\Theta = C \cos \lambda\theta + D \sin \lambda\theta$, and hence

$$u(r, \theta) = (Ar^\lambda + Br^{-\lambda})(C \cos \lambda\theta + D \sin \lambda\theta)$$

Without loss of generality we assume $\lambda > 0$. We also assume that it is physically impossible for $\lim_{r \to 0^+} u(r, \theta) = +\infty$ to hold, and hence, since $\lim_{r \to 0^+} r^{-\lambda} = +\infty$, we conclude that $B = 0$.

If we set $A = 1$ and let $\lambda = n$, where $n = 0, 1, 2, 3, \ldots$, we see that

$$u_n(r, \theta) = r^n(C_n \cos n\theta + D_n \sin n\theta)$$

defines eigenfunctions of the boundary-value problem for $n = 0, 1, 2, 3, \ldots$. When $n = 0$, an eigenfunction corresponding to the eigenvalue $k = 0$ is used. See (10.68).

To satisfy boundary condition (10.84), we try

$$u(r, \theta) = C_0 + \sum_{n=1}^{+\infty} r^n(C_n \cos n\theta + D_n \sin n\theta) \tag{10.89}$$

Then

$$u(1, \theta) = C_0 + \sum_{n=1}^{+\infty} (C_n \cos n\theta + D_n \sin n\theta) = f(\theta)$$

Since f has a Fourier series, C_0, C_n, and D_n are given by

$$C_0 = \frac{1}{2\pi} \int_0^{2\pi} f(\theta)\, d\theta \tag{10.90}$$

$$C_n = \frac{1}{\pi} \int_0^{2\pi} f(\theta) \cos n\theta\, d\theta, \quad n = 1, 2, 3, \ldots \tag{10.91}$$

and

$$D_n = \frac{1}{\pi} \int_0^{2\pi} f(\theta) \sin n\theta\, d\theta, \quad n = 1, 2, 3, \ldots \tag{10.92}$$

It can be shown (see Reference 10.3) that the solution given by (10.89), with C_0, C_n, and D_n given by (10.90), (10.91), and (10.92), is the unique solution of the boundary-value problem. To generalize the solution to the case $r = a > 0$, see Problem 5.

10.8 Two-Dimensional Heat Flow

EXAMPLE 2. A circular lamina with insulated faces covers the circle $r = 1$. The temperature u depends on r and the time t, and is given initially inside the circle by a prescribed function f of r. Find u in terms of r and t if the circumference of the lamina is kept at zero degrees.

Solution: Since u is independent of θ, $u_{\theta\theta} = 0$ in (10.82). Thus, we consider the boundary-value problem:

$$u_t = a^2(u_{rr} + r^{-1}u_r); \quad 0 \leq r < 1, \quad t > 0 \tag{10.93}$$

$$u(1, t) = 0, \quad t \geq 0 \tag{10.94}$$

$$u(r, 0) = f(r), \quad 0 \leq r < 1 \tag{10.95}$$

Letting $u = u(r, t) = R(r)T(t)$ in (10.93), we obtain

$$RT' = a^2(R''T + r^{-1}R'T)$$

Separation of variables yields

$$\frac{T'}{a^2T} = \frac{R''}{R} + \frac{R'}{rR} = k \tag{10.96}$$

The constant k must be negative so that a nontrivial solution of the boundary-value problem can be obtained. See Problem 7.

Letting $k = -\lambda^2$ in (10.96), we obtain the ordinary DE

$$T' + a^2\lambda^2 T = 0 \tag{10.97}$$

and

$$r^2R'' + rR' + \lambda 2r^2R = 0 \tag{10.98}$$

The solution of (10.97) is given by

$$T = T(t) = Ce^{-a^2\lambda^2 t} \tag{10.99}$$

Noting that Equation (10.98) resembles Equation (8.26) with the parameter $p = 0$ (Bessel's DE), we let $z = \lambda r$ in (10.98). Then

$$R' = \frac{dR}{dr} = \frac{dR}{dz} \cdot \frac{dz}{dr} = \lambda \frac{dR}{dz}$$

$$R'' = \frac{d^2R}{dr^2} = \frac{d}{dz}\left(\lambda \frac{dR}{dz}\right)\frac{dz}{dr} = \lambda^2 \frac{d^2R}{dz^2}$$

and (10.98) becomes

$$\frac{z^2}{\lambda^2}\left(\lambda^2 \frac{d^2R}{dz^2}\right) + \frac{z}{\lambda}\left(\lambda \frac{dR}{dz}\right) + z^2R = 0$$

or

$$z^2 \frac{d^2R}{dz^2} + z\frac{dR}{dz} + z^2R = 0 \tag{10.100}$$

This is Bessel's DE of order zero, which has by (8.36) a complete solution given by $R(z) = c_1 J_0(z) + c_2 Y_0(z)$. Thus

$$R = R(r) = c_1 J_0(\lambda r) + c_2 Y_0(\lambda r) \tag{10.101}$$

As $r \to 0^+$, $Y_0(\lambda r) \to +\infty$, and hence if $c_2 \neq 0$, then $u(r, t) \to +\infty$ (or $-\infty$). We assume that this result is physically impossible; that is, we assume that the temperature of the lamina is bounded. We assume therefore that $c_2 = 0$ in (10.101).

Letting $C\, c_1 = A$, we conclude from (10.99) and (10.101) that if the boundary-value problem has a solution of the form $u(r, t) = R(r)T(t)$, then

$$u(r, t) = A[\exp(-a^2 \lambda^2 t)] J_0(\lambda r) \tag{10.102}$$

To satisfy boundary condition (10.94), we must have

$$u(1, t) = A[\exp(-a^2 \lambda^2 t)] J_0(\lambda) = 0$$

Since $e^{-a^2 \lambda^2 t} > 0$, and $A = 0$ yields the trivial solution $u(r, t) \equiv 0$, we set $J_0(\lambda) = 0$.

The zeros of J_0, shown in Figure 8.1, yield the eigenvalues of the problem. The positive zeros $\lambda_1 < \lambda_2 < \lambda_3 < \cdots < \lambda_n < \cdots$ yield the linearly independent eigenfunctions $J_0(\lambda_n r)$ and hence (10.93) is satisfied by

$$u_n(r, t) = A_n[\exp(-a^2 \lambda_n^2 t)] J_0(\lambda_n r) \tag{10.103}$$

for $n = 1, 2, 3, \ldots$.

To satisfy boundary condition (10.95), it is tempting to try superposition of an infinite number of the functions given by (10.103). This is suggested by the success of this approach using Fourier series. That is, we might try

$$u(r, t) = \sum_{n=1}^{+\infty} A_n[\exp(-a^2 \lambda_n^2 t)] J_0(\lambda_n r) \tag{10.104}$$

and hope to determine A_n in

$$u(r, 0) = \sum_{n=1}^{+\infty} A_n J_0(\lambda_n r) = f(r) \tag{10.105}$$

In the theory of Fourier series, the set of functions

$$\left\{ \cos \frac{n\pi x}{L}, n = 0, 1, 2, \ldots; \quad \sin \frac{n\pi x}{L}, n = 1, 2, \ldots \right\}$$

has the property that the integral of the product of any two distinct members of the set, taken over the interval of interest in the problem, is zero; whereas the integral of the square of any member of the set, taken over the same interval, is positive. It is this important property that makes possible the determination of the Fourier coefficients after multiplication of both sides of

$$f(x) = \frac{A_0}{2} + A_1 \cos \frac{\pi x}{L} + A_2 \cos \frac{2\pi x}{L} + \cdots + B_1 \sin \frac{\pi x}{L} + B_2 \sin \frac{2\pi x}{L} + \cdots$$

by $\cos(n\pi x/L)$ or $\sin(n\pi x/L)$.

Definition A sequence $\{\phi_n(x)\}$ of real functions is said to be *orthogonal* on an interval $[a, b]$ if and only if

$$\int_a^b \phi_m(x)\phi_n(x)\,dx = \begin{cases} 0 & \text{for } m \neq n \\ P_n > 0 & \text{for } m = n \end{cases}$$

If the sequence $\{J_0(\lambda_n r)\}$ was orthogonal on $[0, 1]$, the Fourier method could be applied to (10.105) to compute the coefficients A_n. Although this is too much to expect, a remarkable extension of the definition of orthogonality enables us to extend the Fourier method.

Definition A sequence $\{\phi_n(x)\}$ of real functions is said to be *orthogonal with respect to the weight function* $p(x)$ on an interval $[a, b]$ if and only if

$$\int_a^b p(x)\phi_m(x)\phi_n(x)\,dx = \begin{cases} 0 & \text{for } m \neq n \\ P_n > 0 & \text{for } m = n \end{cases}$$

Note: It is assumed that p is continuous on $[a, b]$ and that $p(x) > 0$ on (a, b).

It is shown in Reference 10.12 that the sequence $\{J_0(\lambda_n r)\}$ is orthogonal with respect to the weight function given by $p(r) = r$ on the interval $[0, 1]$. Specifically, it is shown that

$$\int_0^1 \int_0^1 rf(r)J_0(\lambda_n r)\,dr = \begin{cases} 0 & \text{for } m \neq n \\ \dfrac{1}{2}J_1^2(\lambda_n) & \text{for } m = n \end{cases} \tag{10.106}$$

where $J_1(\lambda_n) = -J_0'(\lambda_n)$.

We now multiply both sides of (10.105) by $rJ_0(\lambda_n r)$ and integrate over $[0, 1]$ to obtain

$$\int_0^1 rf(r)J_0(\lambda_n r)\,dr = \sum_{n=1}^{+\infty} A_n \int_0^1 rJ_0(\lambda_m r)J_0(\lambda_n r)\,dr$$

$$= A_n \int_0^1 rJ_0^2(\lambda_n r)\,dr$$

from which, by (10.106),

$$A_n = \frac{\displaystyle\int_0^1 rf(r)J_0(\lambda_n r)\,dr}{\displaystyle\int_0^1 rJ_0^2(\lambda_n r)\,dr} = \frac{2\displaystyle\int_0^1 rf(r)J_0(\lambda_n r)\,dr}{J_1^2(\lambda_n)}$$

The series (10.105), with A_n given by

$$A_n = \frac{2\displaystyle\int_0^1 rf(r)J_0(\lambda_n r)\,dr}{[J_1(\lambda_n)]^2} \tag{10.107}$$

is known as a *Fourier-Bessel series* of f. If f satisfies the Dirichlet conditions, the series converges to $f(r_0)$ if f is continuous at r_0 and to $\frac{1}{2}[f(r_0+) + f(r_0-)]$ if f is discontinuous at r_0. See References 10.1 and 10.2.

The solution of the boundary-value problem is given by (10.104) with A_n given by (10.107).

Sequences of orthogonal functions often arise as solutions of ordinary DE. Each of the sequences $\{J_0(\lambda_n x)\}$, $\{J_1(\lambda_n x)\}$, $\{J_2(\lambda_n x)\}$, ... of Bessel functions of positive integral order with $J_m(\lambda_n) = 0$ is a set of orthogonal functions over an appropriate interval. The sequence $\{P_n(x)\}$ of Legendre polynomials encountered in Chapter 8 forms a set of orthogonal functions over $[-1, 1]$. See Reference 10.12. Given a sequence $\{\phi_n(x)\}$ orthogonal on $[a, b]$, we often can find a series of the form

$$f(x) = \sum_{n=1}^{+\infty} A_n \phi_n(x) \tag{10.108}$$

with A_n given by

$$A_n = \frac{\displaystyle\int_a^b f(x)\phi_n(x) \, dx}{\displaystyle\int_a^b [\phi_n(x)]^2 \, dx} \tag{10.109}$$

that represents a fairly arbitrary function f on $[a, b]$. Series (10.108) is called an *orthogonal*, or *generalized Fourier*, *series* of f, and the A_n in (10.109) are called the *generalized Fourier coefficients* of f. In the theory of orthogonal functions, a theorem stating sufficient conditions on a function f under which f can be represented by an orthogonal series is called an *expansion theorem*. Orthogonal series are important in solving boundary-value problems. That Fourier series are special cases of orthogonal series indicates the importance of Fourier's discovery.

In order to prove that a series of orthogonal functions converges to a function f on a given interval, the sequence of orthogonal functions must be *complete* with respect to a set of functions containing f. This important property of completeness is discussed in References 10.2 and 10.12. In a Fourier sine series for f, the term involving $\sin 3\pi x/L$ could not be omitted; in the Fourier-Bessel series of Example 2, the term involving $J_0(\lambda_2 r)$ could not be omitted; and so on. If these omissions were made, the solutions would proceed formally with no difficulty, but the resulting series would not converge to the required function f.

EXAMPLE 3. Let $\lambda_1 \approx 2.405$, $\lambda_2 \approx 5.520$, $\lambda_3 \approx 8.654$, $\lambda_4 \approx 11.791$, ... denote the positive zeros of $J_0(x)$. Using the formulas

$$\int_0^1 x J_0^2(\lambda_n x) \, dx = \frac{1}{2} J_1^2(\lambda_n)$$

$$\int u J_0(u) \, du = u J_1(u) + C$$

find the Fourier-Bessel series of $f(x) \equiv 1$ over the interval $(0, 1)$ in terms of the functions $J_0(\lambda_n x)$.

Solution: Multiplying both sides of $1 = \sum_{n=1}^{+\infty} A_n J_0(\lambda_n x)$ by $x J_0(\lambda_n x)$ and integrating from 0 to 1, we obtain

$$\int_0^1 x J_0(\lambda_n x) \, dx = A_n \int_0^1 x J_0^2(\lambda_n x) \, dx$$

Hence,

$$\frac{1}{\lambda_n^2} \int_0^1 (\lambda_n x) J_0(\lambda_n x) \lambda_n \, dx = \frac{1}{\lambda_n^2} \left(\lambda_n x\right) J_1(\lambda_n x)]_0^1$$

$$= \frac{J_1(\lambda_n)}{\lambda_n} = A_n \left[\frac{J_1^2(\lambda_n)}{2} \right]$$

from which

$$A_n = \frac{2}{\lambda_n J_1(\lambda_n)} \quad \text{and} \quad 1 = 2 \sum_{n=1}^{+\infty} \frac{J_0(\lambda_n x)}{\lambda_n J_1(\lambda_n)}$$

Problem List 10.6

1. Derive (10.82) from (10.59).
2. Show that $u = u(r, \theta)$ is given by (10.86) when $k = 0$ in (10.85).
3. Show that no solution of the boundary-value problem given by (10.83) and (10.84) can be obtained when $k < 0$.
4. Use the substitution $r = e^z$ in $r^2 R'' + r R' - \lambda^2 R = 0$ to obtain the solution given by $R(r) = Ar^\lambda + Br^{-\lambda}$.
5. Show that if the condition $u(1, \theta) = f(\theta)$ in Example 1 is replaced by $u(a, \theta) = f(\theta)$, where $a > 0$, the solution is then given by

$$u(r, \theta) = C_0 + \sum_{n=1}^{+\infty} \left(\frac{r}{a}\right)^n (C_n \cos n\theta + D_n \sin n\theta)$$

with C_0, C_n, and D_n given by (10.90), (10.91), and (10.92).

6. A plane lamina has the shape of an annular ring with inner radius 1 and outer radius 2. Find the steady-state temperature if the temperature of the inner boundary remains at $30°$ and the temperature of the outer boundary at $70°$. Find $u(1.5)$. Find r if $u(r) = 50°$.

7. Show that k must be negative in (10.96) to obtain a nontrivial solution of the boundary-value problem in illustrative Example 2. Assume that $u(r, t)$ is bounded.

8. Let the sequence of functions $\{\phi_n(x)\}$ be orthogonal on $[a, b]$ with respect to the weight function given by $p = p(x)$ where $p(x) \geq 0$ on $[a, b]$. Show that the sequence of functions $\{\sqrt{p(x)}\phi_n(x)\}$ is orthogonal on $[a, b]$.

9. Assume that the function given by $u = u(r, t)$ satisfies the partial DE $u_{tt} = u_{rr} + r^{-1} u_r$.

Show that if $u = R(r)T(t)$, and $T(t) = A \cos \lambda t + B \sin \lambda t$, then the function R satisfies the DE $r R'' + R' + \lambda^2 r R = 0$, known as Bessel's DE of order zero with parameter λ.

10. If $\{\phi_n(x)\}$ is a sequence of orthogonal functions on $[a, b]$ and if $\int_a^b \phi_n^2(x)\, dx = 1$, $n = 1, 2, 3, \ldots$, then the functions are said to be *orthonormal* on $[a, b]$.

Given that the sequence $\{\phi_n(x)\} = \{\cos n\pi x\}$ is orthogonal on $[0, 2]$, prove that $\{\cos n\pi x\}$ is orthonormal on $[0, 2]$.

11. Use the formulas

$$2u^{-1}J_1(u) = J_0(u) + J_2(u)$$

and

$$\int_0^c u^3 J_0(u)\, du = c^3 J_1(c) - 2c^2 J_2(c)$$

to show that the Fourier-Bessel expansion of the function f given by $f(x) = x^2$ on $(0, 1)$ with $J_0(\lambda_n) = 0$, $n = 1, 2, 3, \ldots$, is given by

$$x^2 = 2 \sum_{n=1}^{+\infty} \frac{\lambda_n^2 - 4}{\lambda_n^3 J_1(\lambda_n)} J_0(\lambda_n x)$$

References

10.1 Carslaw, H. S. 1952. *Introduction to the Theory of Fourier's Series and Integrals*, 3rd ed. New York: Dover.

10.2 Churchill, R. V. 1963. *Fourier Series and Boundary Value Problems*, 2nd ed. New York: McGraw-Hill.

10.3 Davis, Harry F. 1963. *Fourier Series and Orthogonal Functions*. Boston: Allyn and Bacon.

10.4 Greenspan, Donald. 1961. *Introduction to Partial Differential Equations*. New York: McGraw-Hill.

10.5 Kreider, D. L., R. G. Kuller, D. R. Ostberg, and F. W. Perkins. 1966. *An Introduction to Linear Analysis*. Reading, Mass.: Addison-Wesley.

10.6 Sagan, H. 1961. *Boundary and Eigenvalue Problems in Mathematical Physics*. New York: John Wiley & Sons.

10.7 Scarborough, J. B. 1965. *Differential Equations and Applications*. Baltimore: Waverly Press.

10.8 Scarborough, J. B. 1966. *Numerical Mathematical Analysis*, 6th ed. Baltimore: Johns Hopkins University Press.

10.9 Sokolnikoff, I. S., and R. M. Redheffer. 1966. *Mathematics of Physics and Modern Engineering*. New York: McGraw-Hill.

10.10 Stakgold, Ivar. 1967. *Boundary-Value Problems of Mathematical Physics*, vol. 1. New York: Macmillan.

10.11 Tierney, John A. 1979. *Calculus and Analytic Geometry*, 4th ed. Boston: Allyn and Bacon.

10.12 Wylie, C. R., Jr. 1966. *Advanced Engineering Mathematics*, 3rd ed. New York: McGraw-Hill.

Answers to Selected Odd-Numbered Problems

Chapter 1

List 1.1

3. (a) $y = e^x + c$

(b) $y = e^x + cx + k$

(c) $y = -\cos x + cx + k$

(d) $y = \dfrac{(x-1)^4}{24} + \dfrac{c_1 x^2}{2} + c_2 x + c_3$

(e) $y = \dfrac{x^2}{2} \ln x - \dfrac{x^2}{4} + c$

7. (a) second-order, linear

(b) first-order, linear

(c) third-order, linear

(d) third-order, nonlinear

(e) second-order, nonlinear

(f) second-order, linear

(g) first-order, nonlinear

(h) first-order, linear

9. $y \equiv -2$

11. $y' = e^t - e^{-t}$, not unique

13. $y = D \sin^2 x + E \cos^2 x$

15. $y \equiv 0$

19. $y \equiv 0$ **25.** (b) $y = 3x - 9$, $y = -3x - 9$

27. (a) $y = 1 + \tan x$

(b) $y = -e^x + 1 + e$

(c) $y = \dfrac{x^2}{2} + 2x + 1$

(d) $y = -x + 3$

List 1.2

1. (a) $y = 2x^3 - x^2 + C$ (b) $y = 2 \tan^2 x + C$ (c) $y = x \ln x - x + C$

3. The function defined by $f(x) = e^{1/x}$ is not continuous on $(-1, +1)$.

7. (a) $y = \begin{cases} \dfrac{x^2}{2} + 4 & \text{for } x \geq 0 \\[2mm] -\dfrac{x^2}{2} + 4 & \text{for } x < 0 \end{cases}$ (b) $y = \begin{cases} \dfrac{x^2}{2} + 8 & \text{for } x \geq 0 \\[2mm] -\dfrac{x^2}{2} + 8 & \text{for } x < 0 \end{cases}$

List 1.3

1. (a) $y = 3x^2 + 4$ (b) $y = -x^2 + 7$

(c) $y = 4x - 3$ (d) $y = \dfrac{x^2}{2} - 3x + 8$

(e) $y = \frac{1}{3}x^3 - x^2 - 3x + 10$ (f) $y = x^3 + 4x + 19$

3. $y = 2 \ln 3x$ for $x > 0$ (a) $\ln 36$ (b) $\frac{1}{3}$

5. (a) $y = x^2 + 5x - 1$ (b) $y = 2x^3 + x - 3$

(c) $y = -x^3 + x^2 - 3$ (d) $y = x^4 + 3x - 1$

7. $x(2) = 6$ ft, $x'(2) = 7$ ft/sec

11. $\dfrac{125}{6}$ ft

13. $s(16) = 960$ ft

15. A overtakes B at $(27, 0)$.

17. 576 ft

19. $20\sqrt{6}$ ft/sec

21. $x(1) = 6 + 3e^{-1}$ ft, $v(1) = x'(1) = 6 - 2e^{-1}$ ft/sec. As $t \to +\infty, x(t) \to +\infty$ and $x'(t) \to 6$ ft/sec.

23. (a) 4.5 (b) 84

Chapter 2

List 2.1

3. (a) $xy^2 = c$ (b) $xe^y = c$

(c) $e^x \ln y = c$ (d) $x^2y + y^3 = c$

(e) $x \sin y = c$

5. (a) $xy^2 + x^2y^3 = 12$ (b) $x^2y + x \cos y = 2$

(c) $x^3 - 3xy + y^3 = 0$

7. $A = 4; y(2) = \frac{1}{2}(\sqrt{5} - 1)$

List 2.2

1. (a) $y = \pm\sqrt{c + \ln|x|}$ (b) $y = \dfrac{x^2}{6} + k$

(c) $y = \pm\sqrt{\operatorname{Tan}^{-1}x - c}$ (d) $-e^{-x} = -e^{-y} + c$

(e) $x^2 + \operatorname{Tan}^{-1}y = c$ (f) $\sec^{-1}x - \sin^{-1}y = c$

(g) $\ln|y| = -x^{-1} + c$ (h) $2 + y^2 = c(3 + x^2)$

3. $xy = 4$

List 2.3

1. (a) $xy = c$ (b) $y = kx$ (c) $x^2 - \operatorname{Tan}^{-1}\dfrac{y}{x} = c$

(d) $\sqrt{x^2 + y^2} + e^y = c$ (e) $\sqrt{x^2 - y^2} - 2x^2 = c$

(f) $\ln(x^2 + y^2) + y = c$

7. $y = (x + 5)\cos x$

List 2.4

1. (a) $y = \dfrac{x^3}{5} + \dfrac{x}{3} + \dfrac{c}{x^2}$ (b) $y = 1 + c\exp(-x^2)$

   (c) $y = \dfrac{x^3}{4} - \dfrac{k}{x}$ (d) $y = \dfrac{-(\cos x + k)}{x}$

   (e) $r = \dfrac{\theta^2}{6} + \dfrac{c}{\theta^4}$ (f) $y = 1 + c\exp(-e^x)$

   (g) $y = (3x + c)(x + 1)^2$ (h) $y = (\csc x)\ln|\csc x - \cot x| + c\csc x$

3. (a) $y = e^{-x}(x + 5)$ (b) $y = 2 + 4x^{-2}$

   (c) $y = x^{-1}e^x(x - 1) + (ex)^{-1}$ (d) $y = \sin x + 2\csc x$

   (e) $y = x^4 + x^2\ln x + 3$

5. $y = \dfrac{b}{a} + \left(k - \dfrac{b}{a}\right)e^{-ax}$

9. $y = b\exp\left(-\displaystyle\int_a^x \dfrac{B(t)\,dt}{A(t)}\right)$

11. $y = \begin{cases} 1 - x + 5e^{-x} & \text{on } (-1, 0] \\ x - 1 + 7e^{-x} & \text{on } (0, 1) \end{cases}$

13. $y = \begin{cases} (C + e)e^{-x} & \text{on } (-\infty, 1] \\ 1 + Ce^{-x} & \text{on } (1, +\infty) \end{cases}$

   $y = \begin{cases} 3e^{1-x} & \text{on } (-\infty, 1] \\ 1 + 2e^{1-x} & \text{on } (1, +\infty) \end{cases}$

List 2.5

1. (a) $y = c\ln|x| + k$ (b) $y = 4x - ce^{-x} + k$

   (c) $y = -ke^{-x} + A$ (d) $y^2 = Ax + B$

3. (a) $y = [x^2\ln|cx^{-1}|]^{-1}$ (b) $y^4x^4 = 2x^2 + c$

   (c) $y^2 = \dfrac{-b}{a} + ce^{2ax}$

5. (b) $y = 1 - x + \tan(x + k)$

7. (b) $y = 1 + (x + c)^{-1}$

List 2.6

1. (a) 3   (b) 1   (c) 3   (d) 0   (e) 0

3. (a) $y = x\ln\dfrac{x}{2}$ on $(0, +\infty)$ (b) $y = x^2/(3 - x)$ (c) $y = -\sqrt{x^2 + 8x}$

List 2.7

1. (a) $y = 2x^2 + c$ (b) $y = 4x + c$

   (c) $y = c$ (d) $y = -3x^2 + c$

(e) $y = -x^2 + 3x + c$ (f) $y = \dfrac{x^3}{3} + c$

(g) $y = -x^3 + c$ (h) $y = 2\sqrt{x} + c$

3. $y = 2x^2 - 3$

7. $y = \dfrac{x^2 + 7}{4}$

9. (a) $\mathrm{Tan}^{-1} \dfrac{y}{x} + \dfrac{1}{2} \ln (x^2 + y^2) = \ln c, r = ce^{-\theta}, r = e^{-\theta}$

11. Let
$$y = \begin{cases} (x - 2)^3 & \text{on } [2, +\infty) \\ 0 & \text{on } (k, 2)] \\ (x - k)^3 & \text{on } (-\infty, k) \end{cases}$$
and let $k = -1, -2, -3, \ldots, -n, \ldots$.

13. In Example 10 we solved the DE by dividing by y^2; hence we must check the possible solution $y \equiv 0$. This is a solution and is unique since Picard's theorem applies just as it did in Example 10.

15. (a) $y = x^2/4, x \geq 0$

A second solution is given by $y \equiv 0,\ -\infty < x < +\infty$. Picard's theorem does not apply since $F_y(x, y) = \frac{1}{2} y^{-1/2}$ is not defined at $(0, 0)$.

(b) $y = (x + 2)^2/4, x \geq -2$

Both $F(x, y) = \sqrt{y}$ and $F_y(x, y) = \frac{1}{2} y^{-1/2}$ are continuous in the rectangle having vertices $(-1, \frac{1}{2}), (1, \frac{1}{2}), (1, 2),$ and $(-1, 2)$.

Chapter 3

List 3.1

1. 2 hr

3. 2000

5. $8\sqrt{2}A_0 \approx 11.3A_0$, between 11 and 12 times the original amount

7. $\dfrac{\ln 2}{\ln 10} \approx 0.30$ yr

9. 1.25 g

11. 15,512 yr

13. $14.7(0.5)^{1/18} \approx 14.1$ psi

15. $q = q_0 e^{kt}, q \to 0$ as $t \to +\infty$ since $k < 0$

List 3.2

1. $\dfrac{\ln 2}{0.07} \approx 9.9$ yr   3. $r \approx 10.986\%$ 5. 171.828%

7. (a) \$20,161.65   (b) \$21,386.70

List 3.3

1. 0.206   3. $T_m = (T_0 T_1)^{1/2}$ 5. $141.8°$

List 3.4

1. 8.84 min   3. 3.1 min   5. $y = y_0 e^{kt}$
7. (a) 30 min   (b) 34.8^+ min
9. $t \approx 8.1$ hr, 0.0051 mg/cm^3

List 3.5

1. 36.8 lb
3. (a) 107.6 lb   (b) 116.5 lb
5. $Q(4) \approx 7.39$ g, $t \approx 2.63$ hr

List 3.6

1. 64.3 g, 80 min
3. $x = ax_0[(a - x_0)e^{-akt} + x_0]^{-1}$; as $t \to +\infty$, $x \to a$;
$$x = \frac{a}{2}, \max \frac{dx}{dt} = \frac{ka^2}{4}, t = \frac{1}{ak} \ln \frac{(a - x_0)}{x_0}$$

List 3.7

1. 31.1 yr
3. $t \approx 124.6$ yr, just after the middle of 1914
    (Using the Verhulst-Pearl formula, Lotka gives the date 1 April 1914.)
5. $N(60) \approx 23.0$ in 1850, $N(110) \approx 76.5$ in 1900, $N(160) \approx 148.4$ in 1950
    (The census figures for these years are 23.2, 76.2, and 151.3.)
    $$\frac{A}{B} \approx 197 \text{ million}$$
7. N decreases steadily toward AB^{-1}.

List 3.8

1. 917 companies

List 3.9

1. $i = 5(1 - e^{-10t})$
3. $i = 5 - 5e^{-5t}$, $i(0.2) \approx 3.16$ A
5. $i = 2e^{-2t} \sin 25t + 20e^{-2t}$; $i(0.5) \approx 7.309$ A

List 3.10

1. (a) 38.2 min   (b) 11.2 min   (c) 2.04 ft
3. 3.25 min

List 3.11

1. -2.94 cal/sec
3. $T = 450r^{-1} - 25$; $T(7) = 39\frac{2}{7}$, $T(8) = 31\frac{1}{4}$, $T(9) = 25$ (T in degrees Celsius)

List 3.12

1. (a) $I = I_0 \exp\left(\dfrac{2a}{3} t^{3/2}\right)$ (b) $I = -k^{-1} \ln\left(-akt + e^{-kI_0}\right)$

3. $\bar{Q} = \dfrac{ad - bc}{b + d}$

List 3.13

1. (a) $y' = 3yx^{-1}$ (b) $y' = -xy^{-1}$ (c) $y' = y$
  (d) $y' = 1 + y - x$ (e) $y' = xy^{-1}$ (f) $y' = y(2x)^{-1}$
  (g) $y' = -yx^{-1}$ (h) $y' = xy' + f(y')$ [Clairaut's equation]
3. (a) $y'' - y = 0$ (b) $y'' = 0$ (c) $y'' = 2$ (d) $y'' + y = 0$
  (e) $y'' + 4y = 0$
5. $[(y')^2 + 1](y - x)^2 = (x + yy')^2$
7. $y' = \dfrac{-y(2x^3 - y^3)}{x(2y^3 - x^3)}$

List 3.14

3. $x^2 + y^2 = kx$
5. $y^2 + 2x^2 = k,\ 2x^2 + y^2 = 6,\ y^2 = 4x$
9. $r = c(1 - \cos\theta)$

Chapter 4

List 4.1

5. $y(x) = x^3 + 2x + 5$
9. $W(y_1, y_2) = -4x^{-1}$
11. $y = c_1 \sin 2x + c_2 \cos 2x$
13. $W(y_1, y_2) = -2x^{-1},\ y = c_1 x + c_2 x^{-1}$
15. $y \equiv 2$
17. $y'' - 9y = 0,\ y'' - 3y' = 0,\ y'' - y' - 6y = 0, \ldots$ (not unique)
19. $y(t) = 2\sin 2t + \cos 2t + \cos 3t$
21. Result does not contradict Theorem 4-II, since the DE is nonlinear.
23. Result does not contradict Theorem 4-V, since y_1 and y_2 are not solutions of a second-order linear DE of the form (4.8).
27. (a) $y = c_1 x + c_2 x^{-1}$ (b) $y = c_1 \sin x + c_2 \cos x$
  (c) $y = c_1 e^{4x} + c_2 x c^{4x}$ (d) $y = c_1 x^{-1/2} \sin x + c_2 x^{-1/2} \cos x$

  (e) $y = c_1 x + c_2[\exp(x^2) + x \int \exp(x^2)\,dx]$

  (f) $y = c_1 x + c_2\left(-1 + \dfrac{x}{2} \ln\dfrac{1 + x}{1 - x}\right)$

List 4.2

1. (a) $y = c_1 e^{4x} + c_2 e^{3x}$ (b) $y = c_1 e^{-3x} + c_2 e^{2x}$
  (c) $y = c_1 e^{-x} + c_2 e^{-6x}$ (d) $y = c_1 e^{2x} + c_2 e^{-2x}$
  (e) $y = c_1 + c_2 e^{9x}$ (f) $y = c_1 + c_2 e^{-5x}$
  (g) $y = c_1 e^{(2+\sqrt{3})x} + c_2 e^{(2-\sqrt{3})x}$ (h) $y = c_1 e^{3x/4} + c_2 e^{-x/4}$
  (i) $y = c_1 e^{3x} + c_2 x e^{3x}$ (j) $y = c_1 e^{-2x} + c_2 x e^{-2x}$

(k) $y = c_1 \sin 5x + c_2 \cos 5x$ (l) $y = e^{3x}(c_1 \sin 4x + c_2 \cos 4x)$

(m) $y = e^{-5x}(c_1 \sin 2x + c_2 \cos 2x)$ (n) $y = e^{-x}(c_1 \sin x + c_2 \cos x)$

3. $y = ce^{-3x}$

5. $W(y_1, y_2) \equiv 1 \neq 0$ on an interval I

7. (a) $c_1 \sin \beta x + c_2 \cos \beta x$ (b) $c_2 e^{\alpha x}$

9. (a) $y = c_1 e^{2x} + c_2 e^{-5x}$ (b) $y = c_1 e^{3x} c_2 x e^{3x}$

(c) $y = c_1 + c_2 e^{-4x}$ (d) $y = e^{3x}(c_1 \sin 5x + c_2 \cos 5x)$

(e) $y = c_1 \sin 2x + c_2 \cos 2x$

(f) $y = c_1 \exp(3 + \sqrt{5})x + c_2 \exp(3 - \sqrt{5})x$

11. $y(x) = A(\sin x + \cos x)$

List 4.3

1. (a) $Y = 3e^x$ (b) $Y = -5e^{-x}$ (c) $Y = \frac{8}{5}e^{2x}$

(d) $Y = 3x^2 - 6x + 1$ (e) $Y \equiv 4$ (f) $Y = 4 \cos x$

(g) $Y = \sin 2x + \cos 2x$ (h) $Y = 2e^x + x^2$ (i) $Y = \frac{5}{2}x e^{2x}$

(j) $Y = -\frac{3}{2}x \cos x$

3. (a) $y = -1 + 2e^{2x} + \frac{1}{3}e^{3x}$ (b) $y = 4 \sin x + 2e^{-3x}$

(c) $y = e^{-x} + \dfrac{x^3}{3} + x$ (d) $y = \frac{1}{2}e^x + \frac{1}{2}e^{3x} - e^{2x}$

(e) $y = \cos x + 2x$ (f) $y = \sin 4x - 4x \cos 4x$

7. $y'' + y = 2e^x$

9. $y = \dfrac{x^7}{14} + \dfrac{c_1 x^5}{5} + c_2$

11. $y = e^x(\sin x - 2 \cos x)/5$

List 4.4

1. (a) $Y = (\sin x) \ln(\csc x - \cot x)$ (b) $Y = x \sin x + (\cos x) \ln(\cos x)$

(c) $Y = \frac{1}{2} \sin x \tan x$ (d) $Y = 3e^x$

(e) $Y = (\sin x) \ln(\sin x) - x \cos x$ (f) $Y = -2x \cos x$

3. $y = \frac{1}{9}(x^4 - x - 3x \ln x)$

5. $y = \exp(-\int P\,dx) \int [Q \exp(\int P\,dx)]\,dx + C \exp(-\int P\,dx)$

7. $Y = x^{-1/2}$

11. (b) $y_p = -\cos x \ln(\sec x + \tan x)$

(c) $y_p = x \sin x + \cos x \ln \cos x$

(d) $y = e^{-x}(1 - x + x \ln x)$

List 4.6

1. (a) $6x$ (b) $-30e^{2x}$

(c) $(r_2 - r_1)(r_3 - r_1)(r_3 - r_2) \exp(r_1 + r_2 + r_3)x$

(d) $-2e^{-x}$ (e) 12 (f) 0

7. (a) $y(x) = c_1 + c_2 e^x + c_3 e^{-x}$ (b) $y(x) = c_1 e^x + c_2 e^{3x} + c_3 e^{-2x}$

9. (a) $y = A + Bx$ (b) $y = Ax e^{2x}$

(c) $y = Ae^{-3x}$ (d) $y = x(A \sin 3x + B \cos 3x)$

(e) $y = A \sin 4x + B \cos 4x$ (f) $y = A \sin x + B \cos x$

(g) $y = A \sin 4x + B \cos 4x$ (h) $y = A \sin 2x + B \cos 2x$

11. (a) $y = 1 - x + \frac{1}{2}x^2 - e^{-x}$, valid on $(-\infty, +\infty)$

 (b) $y = 6 - 8 \cos x + 2 \cos 2x$, valid on $(-\infty, +\infty)$

 (c) $y = \frac{1}{2}(-1 + x^2 - 2x \ln x)$, valid on $(0, +\infty)$

 (d) $y = 18 + 12x - 18e^x + 6xe^x + 3x^2$, valid on $(-\infty, +\infty)$

13. (a) $y = -\frac{1}{2}e^{-x}$ 　　　　(b) $y = 5x^4 e^{-x}$

 (c) $y = \frac{1}{2}(\cos x - \sin x)$ 　(d) $y = e^{-x}(2x^2 + 4x + 3)$

15. $y = e^{ax}(k_n x^{n-1} + k_{n-1} x^{n-2} + \cdots + k_2 x + k_1)$

List 4.7

1. $y = c_1 + c_2 \ln |x|$

3. (a) $y(x) = c_1 x^2 + c_2 x^{-1}$ 　　　　(b) $y(x) = cx^{-4}$

 (c) $y(x) = c_1 x^{1/2} + c_2 x^{1/2} \ln x$ 　(d) $y(x) = c_1 \sin (\ln x) + c_2 \cos (\ln x)$

 (e) $y(x) = c_1 + c_2 x + c_3 x^{-2}$ 　　　(f) $y(x) = c_1 x^{-3} + c_2 \sin (\ln x)$

　　　　　　　　　　　　　　　　　　　　$+ c_3 \cos (\ln x)$

Chapter 5

List 5.1

1. $a = -\frac{1600}{3} \cos 40t, v = -\frac{40}{3} \sin 40t,$

 $x = \frac{1}{3} \cos 40t; T = \dfrac{\pi}{20}$ sec, am. $= \frac{1}{3}$ ft

3. $x = \frac{1}{2} \cos 48t - \frac{1}{4} \sin 48t, v = -24 \sin 48t - 12 \cos 48t,$

 max $|v| = 12\sqrt{5}$ ft/sec, am. $= \dfrac{\sqrt{5}}{4}$ ft, $T = \dfrac{\pi}{24}$ sec

7. $v = 160 - 160e^{-t/5}$

9. $v = 8 - 8e^{-4t}, \lim_{t \to +\infty} v = 8$ ft/sec

11. $x = -\frac{1}{3}e^{-12t} + \frac{2}{3}e^{-6t}, v = 4e^{-12t} - 4e^{-6t}$

13. $x = \frac{1}{3}e^{-6t}(\sin 6t + \cos 6t), v = -4e^{-6t} \sin 6t$

15. $x = e^{-t}[\frac{1}{2}\cos \sqrt{15}t - \frac{1}{6}(19\sqrt{15}) \sin \sqrt{15}t] + 12 \sin 4t,$

 $v = e^{-t}[-48 \cos \sqrt{15}t + \frac{1}{3}(8\sqrt{15}) \sin \sqrt{15}t] + 48 \cos 4t$

17. $x = \frac{1}{4} \cos 5t, v = -\frac{5}{4} \sin 5t, T = \dfrac{2\pi}{5}$ sec

19. $x = \frac{1}{4} \cos 8\sqrt{3}t, v = -2\sqrt{3} \sin 8\sqrt{3}t$, am. $= \frac{1}{4}$ ft, $T = \dfrac{\pi\sqrt{3}}{12}$ sec

21. $\beta = 16$

23. $x = 3 \sin 4t$

25. (a) $T = 1.418$ hr 　　(b) $|v| = 17{,}725$ mph

27. $t = \dfrac{25}{16} \displaystyle\int_{30}^{40} \dfrac{dv}{v^{5/4} - 50} \approx 0.48$ sec

List 5.2

1. $\omega = 9$

3. $X = \dfrac{F_0}{2m\omega_0} t \sin \omega_0 t$

7. $X = A \cos \omega t + B \sin \omega t$;

$$A = \frac{(k - \omega^2 m)F_0}{(k - \omega^2 m) + \omega^2 c^2}, \quad B = \frac{\omega c F_0}{(k - \omega^2 m)^2 + \omega^2 c^2}$$

9. (a) 12.04   (b) 120.00

List 5.3

1. 781 ft

3. 4000i

5. (a) $v(10) \approx 5978$ ft/sec, $x(10) \approx 26{,}931$ ft
   (b) $v(20) \approx 23{,}026$ ft/sec, $x(20) \approx 148{,}831$ ft

7. (a) $v = \left(\dfrac{f^2 c^2}{f^2 + c^2 m_0^2 t^{-2}}\right)^{1/2}$ (b) c

   (c) $x = \dfrac{c}{f}(f^2 t^2 + c^2 m_0^2)^{1/2} - \dfrac{c^2 m_0}{f}$

9. (a) 0.815 ft   (b) 3.260 ft

11. $b = 3w \cos \theta - 2w \cos \theta_0$

13. 8.02 ft/sec

List 5.4

1. (a) $2\dfrac{d^2 q}{dt^2} + 50\dfrac{dq}{dt} + 20q = 40$

   (b) $2\dfrac{d^2 q}{dt^2} + 40\dfrac{dq}{dt} + 1000q = 220 \cos 10t$

   (c) $2\dfrac{d^2 q}{dt^2} + 12\dfrac{dq}{dt} + 100q = 24 \sin 10t$

   (d) $10\dfrac{dq}{dt} + 100q = 200\sqrt{t}e^{-4t}$

3. $q = 1 + \cos 10t$, $i = -10 \sin 10t$

List 5.5

1. (a) 0   (b) $\dfrac{x_0}{a}$ (c) $+\infty$ (d) $+\infty$

3. $y = \dfrac{x^3}{6x_0^2} + \dfrac{x_0^2}{2x} - \dfrac{2x_0}{3}; y\Big]_{x=0.5x_0} = \dfrac{17x_0}{48}$

5. $y = \frac{1}{4}[x^2 - 2x - 2\ln(1 - x)]$

List 5.6

1. $x = -100$ ft, $y \approx 225.53$ ft; $x = 200$ ft, $y \approx 308.62$ ft

5. $y = 60 \cosh\dfrac{x}{60}$, span $= 120 \ln 2 \approx 83.18$ ft, sag $= 15$ ft

List 5.7

3. 2.4 in.

5. (a) 0.1152 in.   (b) 0.1617 in.

7. $y = \dfrac{-w}{48EI}(2x^4 - 5lx^3 + 3l^2x^2)$

9. $y = \dfrac{-wx^2}{24EI}(x - l)^2, d = \dfrac{wl^4}{384EI}$

List 5.8

3. $c_i - c_e = \dfrac{Q\rho_0}{3h}$

List 5.9

3. (a) $\bar{P} = 8, P = -13e^{-t} + 7e^{-2t} + 8$, stable

(b) $\bar{P} = 1, P = e^t(-\tfrac{3}{2}\sin 2t + 2\cos 2t) + 1$, unstable

Chapter 6

List 6.1

5. (a) $\dfrac{dx_1}{dt} = x_2, \dfrac{dx_2}{dt} = -4x_2 - -3tx_1 + e^{-t}$ where $x_1 = y$

(b) $\dfrac{dx_1}{dt} = x_2, \dfrac{dx_2}{dt} = x_3, \dfrac{dx_3}{dt} = t^2x_3 + x_2 - x_1^2$ where $x_1 = x$

7. $y = e^{-4t}, x = 3e^{-4t}$

9. $x = 2 - 2(1 + t^2)^{-1/2}, y = 2t - 2t(1 + t^2)^{-1/2} + 5$

11. $x = -\tfrac{9}{10}\sin 2t - \tfrac{1}{5}\cos 2t + \tfrac{9}{5}e^{-t}, y = \tfrac{1}{2}\sin 2t - e^{-t}$

List 6.2

1. (a) $(0, 0)$ (b) $(\tfrac{1}{2}, -\tfrac{1}{2})$

3. $\dfrac{dx}{dt} = v, \dfrac{dv}{dt} = -\dfrac{cg}{w}v - \dfrac{kg}{w}x$

The critical points of the system are the points at which the velocity and the acceleration are both zero.

5. $\dfrac{dx}{dt} = y, \dfrac{dy}{dt} = -\dfrac{g}{1}x$

$x = A\sin(\sqrt{g/1}\,t + \phi), \quad y = \dfrac{dx}{dt} = A\sqrt{g/1}\cos(\sqrt{g/1}\,t + \phi)$

Orbits are ellipses that are traversed clockwise. Ellipse through $P(x, y)$ can be made arbitrarily close to $(0, 0)$ by choosing P sufficiently close to $(0, 0)$. Origin is stable but is not asymptotically stable since $P(x, y)$ does not approach $(0, 0)$ as $t \to +\infty$.

7. $x = \tfrac{1}{2}[(1 - t)^{-1} - (1 - t)], y = \tfrac{1}{2}[(1 - t)^{-1} + (1 - t)]; y^2 - x^2 = 1$

9. (a) $r^2 - 3r + 2 = (r - 2)(r - 1) = 0; r = 2, 1$

(b) $r^2 + 3r + 2 = (r + 2)(r + 1) = 0; r = -2, -1$

(c) $r^2 + 1 = 0; r = 0 \pm i$

(d) $r^2 - 1 = 0; r = +1, -1$

(e) $r^2 - 2r + 1 = (r - 1)^2 = 0; r = 1, 1$

List 6.3

3. (a) $R_2 = 480i + 576j$, $V_2 = 240i + 256j$, $A_2 = -32j$
   (b) $R_{10} = 2400i + 1600j$, $V_{10} = 240i$, $A_{10} = -32j$

5. $y_{max} = 2048$ ft, $x = 4096$ ft, $R = 4096i + 2048j$

7. $\dfrac{v_0^2 \sin 2\alpha}{g}$ is max when $\alpha = 45°$.

9. $v = v_0 \cos \alpha$ is attained at vertex of parabola since $t = (v_0 \sin \alpha)/g$ there.

List 6.4

1. $M = (-\sin \theta)i + (\cos \theta)j$

   $\dfrac{dM}{d\theta} = (-\cos \theta)i + (-\sin \theta)j = -L$

3. $v_0 = \dfrac{h}{r_0}$; since r_0 is min, v_0 is max when $\theta = 0$.

5. $f(r) = 2mh^2 r^{-3}$

7. $e \approx 0.278 < 1$, $\beta(\pi) \approx 7010$ miles, $T \approx 138$ min

9. $r_0 \approx 4289$ miles, altitude ≈ 329 miles

11. orbital speed $\approx 16,860$ mph, escape speed $\approx 23,844$ mph

13. $v_0 \approx 6.950$ miles/sec

List 6.5

3. $x = \phi(t) = \cos \omega_1 t + \sin \omega_1 t = \sqrt{2} \sin \left(\omega_1 t + \dfrac{\pi}{4} \right)$

   $y = \psi(t) = \sqrt{3} \cos \omega_1 t + \sqrt{3} \sin \omega_1 t = \sqrt{6} \sin \left(\omega_1 t + \dfrac{\pi}{4} \right)$

7. $\lambda_1 = \sqrt{3}\sqrt{1 + w_2^2/w_1^2} = \sqrt{3}\sqrt{1 + 2 + \sqrt{3}}$
   $= \sqrt{3}\sqrt{3 + \sqrt{3}} = \sqrt{9 + 3\sqrt{3}} \approx 3.768$
   $\lambda_2 = \sqrt{3}\sqrt{1 + 1} = \sqrt{6} \approx 2.449$

9. $-k_1 x + k_2(y - x) = m_1 \dfrac{d^2 x}{dt^2}$

   $-k_2(y - x) + k_3(z - y) = m_2 \dfrac{d^2 y}{dt^2}$

   $-k_3(z - y) = m_3 \dfrac{d^2 z}{dt^2}$

List 6.6

3. $i_1 = i_2 + i_3$ and any two of

   $4i_1 + 20 \dfrac{di_3}{dt} - \sin 2t = 0$

   $4i_2 + 4 \dfrac{d^2 i_3}{dt^2} + 20 \dfrac{di_3}{dt} - \sin 2t = 0$

   $20 \dfrac{di_3}{dt} - 2 \dfrac{di_2}{dt} = 0$

5. $i_3 = i_1 + i_2$ and any two of

$$30i_1 + 20q - 10i_2 - 2\frac{di_2}{dt} = 0$$

$$2\frac{di_2}{dt} + 10i_2 - 120 = 0$$

$$30i_1 + 20q - 120 = 0$$

where $i_1 = \dfrac{dq}{dt}$

List 6.7

1. $t = \frac{25}{3}\ln 2 \approx 5.78$ min, $y_{max} = 4.5$ lb

3. $y^a x^c e^{-by-dx} - k = 0$

5. $\dfrac{dx}{dt} = (A - B)x + Cy$

$$\frac{dy}{dt} = Bx - (D + C)y + Ez$$

$$\frac{dz}{dt} = Dy - (E + F)z$$

List 6.8

1. (a) $Y = Y_0, D = D_0 + kY_0t,$

$B = \dfrac{iD_0 + ikY_0t}{Y_0}, \quad \lim\limits_{t \to +\infty} B(t) = +\infty$

(b) $Y = at + Y_0, D = \dfrac{kat^2}{2} + kY_0t + D_0,$

$B = \dfrac{(ika/2)t^2 + ikY_0t + iD_0}{at + Y_0}, \quad \lim\limits_{t \to +\infty} B(t) = +\infty$

(c) $Y = \dfrac{at^2}{2} + Y_0, D = \dfrac{kat^3}{6} + kY_0t + D_0,$

$B = \dfrac{(ika/6)t^3 + ikY_0t + iD_0}{(a/2)t^2 + Y_0}, \quad \lim\limits_{t \to +\infty} B(t) = +\infty$

(d) $Y = \dfrac{at^2}{2} + bt + Y_0, D = D_0 + \dfrac{kat^3}{6} + \dfrac{kbt^2}{2} + kY_0t,$

$B = \dfrac{iD_0 + ikat^3/6 + ikbt^2/2 + ikY_0t}{at^2/2 + bt + Y_0},$

$\lim\limits_{t \to +\infty} B(t) = +\infty$

(e) $Y = \dfrac{at^3}{3} + \dfrac{bt^2}{2} + ct + Y_0,$

$D = \dfrac{kat^4}{12} + \dfrac{kbt^3}{6} + \dfrac{kct^2}{2} + kY_0t + D_0,$

$B = \dfrac{ikat^4/12 + ikbt^3/6 + ikct^2/2 + ikY_0t + iD_0}{at^3/3 + bt^2/2 + ct + Y_0},$

$\lim\limits_{t \to +\infty} B(t) = +\infty$

3. $D(t) = D_0 - kY_0c^{-1} + kc^{-1}Y_0e^{ct}$

Chapter 7

List 7.1

1. (a) 0, all s (b) $\dfrac{2}{s}$, $s > 0$ (c) $(s - 1)^{-1}$, $s > 1$

(d) $(s + 1)^{-1}$, $s > -1$ (e) $(s - 3)^{-1}$, $s > 3$

(f) $\dfrac{s}{s^2 + a^2}$, $s > 0$ (g) $\dfrac{a}{s}$, $s > 0$

3. $\mathscr{L}[t] = \dfrac{1!}{s^{1+1}} = s^{-2}$

7. $\displaystyle\lim_{s \to 0} \dfrac{1 - e^{-s}}{s} = \lim_{s \to 0} \dfrac{e^{-s}}{1} = 1 = g(0)$

9. $\dfrac{1}{s} - \dfrac{e^{-s}}{s^2}$

List 7.2

5. $\displaystyle\int_0^1 \dfrac{e^{-st}\, dt}{t} \geq e^{-s} \int_0^1 \dfrac{dt}{t}$ and $\displaystyle\int_0^1 \dfrac{dt}{t}$ diverges

List 7.3

1. (a) $\dfrac{6}{s^3} + \dfrac{2s}{s^2 + 4}$ (b) $\dfrac{1}{s^2} - \dfrac{12}{s^2 + 9}$ (c) $\dfrac{6}{s^4} - \dfrac{1}{s^2 + 1} + \dfrac{4}{s}$

(d) $\dfrac{s}{s^2 - a^2}$, $s > |a|$ (e) $\dfrac{s^2 + 2a^2}{s(s^2 + 4a^2)}$ (f) $\dfrac{72}{s^5} - \dfrac{\Gamma(\frac{3}{2})}{s^{3/2}}$

(g) $\dfrac{\sqrt{2}}{2} \dfrac{s + 1}{s^2 + 1}$ (h) $\dfrac{2}{s^3} + \dfrac{2}{s^2} + \dfrac{1}{s}$ (i) $\dfrac{6}{s^4} + \dfrac{2}{s^2} - \dfrac{4}{s}$

(j) $\dfrac{3}{s} + \dfrac{5}{s + 1} + \dfrac{2}{s^2 + 1} - \dfrac{6s}{s^2 + 4}$

3. $\mathscr{L}[1] = \dfrac{1}{s}$, $\mathscr{L}[e^{at}(1)] = \dfrac{1}{s - a}$

5. (a) $\dfrac{2a(3s^2 - a^2)}{(s^2 + a^2)^3}$, $s > 0$ (b) $\dfrac{2s(s^2 - 3a^2)}{(s^2 + a^2)^3}$, $s > 0$

(c) $\mathscr{L}[t \cos t] = \dfrac{s^2 - 1}{(s^2 + 1)^2}$

$\mathscr{L}[e^{-t} \cos t] = \dfrac{s^2 + 2s}{(s^2 + 2s + 2)^2}$

(d) $\mathscr{L}[\sin^2 at] = \dfrac{1}{2}\left(\dfrac{1}{s} - \dfrac{s}{s^2 + 4a^2}\right)$

$\mathscr{L}[t \sin^2 at] = \dfrac{1}{2}\left[\dfrac{1}{s^2} - \dfrac{s^2 - 4a^2}{(s^2 + 4a^2)^2}\right]$, $s > 0$

7. $\ln \dfrac{s - 3}{s - 2}$, $s > 3$

9. $\mathscr{L}[\sin t] = s\mathscr{L}[-\cos t] - (-\cos 0) = s\left(\dfrac{-s}{s^2 + 1}\right) + 1 = \dfrac{1}{s^2 + 1}$

11. $u = at, \dfrac{1}{a}\displaystyle\int_0^{+\infty} e^{-su/a} F(u)\, du = \left(\dfrac{1}{a}\right) f\left(\dfrac{s}{a}\right)$ for $\dfrac{s}{a} > s_0$

List 7.4

3. (a) $1 - e^{-t}$ (b) $t + e^{-t} - 1$ (c) $e^t + e^{-t} - 2$
     (d) $2e^t - 2e^{-t} - 4t$ (e) $1 - \cos t$ (f) $t - \sin t$
     (g) $6e^t - 6e^{-2t}$ (h) $-6e^{-t} + 4e^{-2t} + 2e^t$
     (i) $-3e^{-t} + 2te^{-t} + 5e^{2t}$ (j) $-2 + 2\cos 2t + 2\sin 2t$
     (k) $-4 - t + 4e^t$ (l) $2e^{3t} + 3\cos t$
     (m) $2 - 2\cos 3t$ (n) $\frac{5}{3}e^t - \frac{2}{3}e^{-2t}$

5. $\dfrac{1}{a}\displaystyle\int_0^t \sin au\, du = \dfrac{1 - \cos at}{a^2}$

7. $\dfrac{d}{dt}\left(\dfrac{1}{a}\sin at\right) = \cos at$

9. $\cos^3 t$

List 7.5

1. (a) $Y(t) = 2e^{2t} + e^{3t}$ (b) $Y(t) = 4te^t$
     (c) $Y(t) = e^t$ (d) $Y(t) = e^{3t} + 2e^{-t}$
     (e) $Y(t) = e^{3t} + 2e^t$ (f) $Y(t) = 10te^{-5t}$
     (g) $Y(t) = 4 - 2e^{-4t}$ (h) $Y(t) \equiv 1$
         $(\cos 3t - 2\sin 3t)$
     (i) $Y(t) = 2 - 3e^t + 3e^{2t}$ (j) $Y(t) = 3\cos 2t - 5\sin 2t + 4t$
     (k) $Y(t) = 2\sin 2t - 2t\cos 2t$ (l) $Y(t) = 2e^t - 2\cos t - 2\sin t$
     (m) $Y(t) = \frac{3}{4}e^t - \frac{1}{2}te^t + \frac{3}{2}t^2 e^t - \frac{3}{4}e^{-t}$

3. $Y(t) = Y_0 \cos \omega t + \dfrac{Y_0'}{\omega}\sin \omega t$

5. $Y(t) = e^{2-2t} + 2t - 1$

7. $Y(t) = 2t^2$

List 7.6

1. Set $a = 0$. Then $\mathscr{L}[1] = \dfrac{1}{s}$.

3. (a) $H(t - 3)$ (b) $(3t - 6)H(t - 2)$
     (c) $3e^{2(t-4)}H(t - 4)$ (d) $H(t - 1) + (t - 3)H(t - 3)$

5. $\lim_{t \to 1^-} Y''(t) = \cos 3$, $\lim_{t \to 1^+} Y''(t) = (\cos 3) - 1$

List 7.7

1. (a) $\dfrac{s}{s^2 + 1}$ (b) $\dfrac{1}{s[1 + \exp(-ps/2)]}$

     (c) $\dfrac{1}{(s^2 + 1)(1 - e^{-\pi s})}$ (d) $\dfrac{\coth(\pi s/2)}{s^2 + 1}$

     (e) $\dfrac{k}{ps^2} - \dfrac{ke^{-ps}}{s(1 - e^{-ps})}$ (f) $\dfrac{1}{s^2}\tanh\dfrac{s}{2}$

3. $\dfrac{\coth(\pi s/2)}{s^2 + 1}$

5. $Y(t) = 1 - \cos t + \displaystyle\sum_{k=1}^{+\infty} (-1)^k [1 - \cos (t - k)] H(t - k)$

9. (a) $Y(t) = \sin t - \cos t + e^{-t}$

(b) $Y(t) = \frac{1}{2}(\sin 2t - 2 \cos 2t + 4)$

(c) $Y(t) = 2(-e^{-t} \sin t - e^{-t} \cos t + 1)$

11. (a) $1 - e^{-t}$　　　　　　(b) $t + e^{-t} - 1$

(c) $\frac{1}{2}(e^t - e^{-t} - 2t)$　　　(d) $e^{-t}(-t - 1) + 1$

(e) $\frac{1}{2}(1 - \cos 2t)$　　　　(f) $e^{3t} - e^{2t}$

(g) $\dfrac{-1 + e^{3t}}{3}$　　　　　(h) $2 (\sin 2t + \cos 2t - 1)$

(i) $\dfrac{1 - e^{-5t}}{5}$　　　　　(j) $\dfrac{e^{-4t}}{16} - \dfrac{1}{16} + \dfrac{t}{4}$

(k) $\dfrac{t \sin t}{2}$　　　　　　(l) $\dfrac{t \sin at}{2a}$

(m) $\frac{1}{2}(e^t - \sin t - \cos t)$　(n) $\dfrac{t^2}{2} - 1 + \cos t$

(o) $\cosh at$　　　　　　(p) $\dfrac{\sin at + at \cos at}{2a}$

(q) $\dfrac{e^{-at}}{2b^3} (\sin bt - bt \cos bt)$　(r) $e^{at} \displaystyle\int_0^t e^{-au} F(u) \, du$

List 7.8

1. $X = 2e^{-t}, Y = e^{-t}$

3. $X = e^{2t} + \frac{1}{2}, Y = e^{2t} - \frac{1}{2}$

5. $X = e^t(t + 2), Y = e^t(t + 1)$

7. $V = 16 + 2e^{-2t}$

9. $X = \frac{1}{3} \cos 8t + \frac{1}{4} \sin 8t$, am. $= \frac{5}{12}$, per. $= \dfrac{\pi}{4}, f = \dfrac{4}{\pi}$

11. $X = \frac{1}{2} \cos 2t - 4 \sin 2t + 5 \sin t$

13. $A = A_0 e^{0.01rt}$

15. $Y = \dfrac{-w}{48EI} (2x^4 - 5lx^3 + 3l^2x^2)$

17. $I = 20e^{-2t} + 2e^{-2t} \sin 25t$

19. $I_1 = 5 - \dfrac{5e^{-2t}}{\sqrt{6}} \sinh \dfrac{\sqrt{6}t}{2} - 5e^{-2t} \cosh \dfrac{\sqrt{6}t}{2}$

$I_2 = 5 - \dfrac{10}{\sqrt{6}} e^{-2t} \sinh \dfrac{\sqrt{6}t}{2} - 5e^{-2t} \cosh \dfrac{\sqrt{6}t}{2}$

$I_3 = \dfrac{5}{\sqrt{6}} e^{-2t} \sinh \dfrac{\sqrt{6}t}{2}$

Chapter 8

List 8.1

1. (a) $y = c_0 \left(1 + x + \dfrac{x^2}{2!} + \cdots\right) = c_0 e^x$

(b) $y = c_0 \left(1 - x + \dfrac{x^2}{2!} - \cdots \right) = c_0 e^{-x}$

(c) $y = c_0 \left[1 + 2x + \dfrac{(2x)^2}{2!} + \cdots \right] = c_0 e^{2x}$

(d) $y = c_0 e^x + c_1 e^{-x}$

(e) $y = c_0 \left(1 + \dfrac{x^2}{2} + \dfrac{x^4}{8} + \cdots + \dfrac{x^{2n}}{2^n n!} + \cdots \right) = c_0 e^{x^2/2}$

(f) $y = c_0 + c_1 \displaystyle\sum_{n=0}^{+\infty} \dfrac{(-1)^n}{2n+1} x^{2n+1} = c_0 + c_1 \operatorname{Tan}^{-1} x$

(g) $y = c_0 \left[1 + \displaystyle\sum_{n=1}^{+\infty} \dfrac{(-1)^n x^{2n}}{2 \cdot 4 \cdot 6 \cdots (2n)}\right] + c_1 \left[x + \displaystyle\sum_{n=1}^{+\infty} \dfrac{(-1)^n x^{2n+1}}{3 \cdot 5 \cdot 7 \cdots (2n+1)}\right]$

(h) $y = c_0 \left[1 - \dfrac{x^2}{2!} - \displaystyle\sum_{n=2}^{+\infty} \dfrac{1 \cdot 3 \cdot 5 \cdots (2n-3)}{(2n)!} x^{2n}\right] + c_1 x$

(i) $y = c_0 e^{-x^3}$

3. $y = 2 + 2x + 0 + \frac{2}{3}x^3 + \frac{1}{2}x^4 + 0 + \frac{1}{9}x^6 + \cdots$

9. $P_0(x) \equiv 1,\ P_1(x) = x,\ P_2(x) = \frac{1}{2}(3x^2 - 1),\ P_3(x) = \frac{1}{2}(5x^3 - 3x),$
$P_4(x) = \frac{1}{8}(35x^4 - 30x^2 + 3),\ P_5(x) = \frac{1}{8}(63x^5 - 70x^3 + 15x)$

11. See Problem 9 for P_1, P_2, and P_3.

13. $y = 1 + \displaystyle\sum_{n=1}^{+\infty} \dfrac{(-1)^n x^{3n}}{2 \cdot 5 \cdot 8 \cdots (3n-1)3^n n!}$

15. $y = 1 + x - \dfrac{x^2}{2} - \dfrac{x^4}{12} + \dfrac{x^5}{24} - \dfrac{x^6}{144} + \cdots$

17. $y(x) = 1 - x^3/6 + x^5/120 + x^6/180$

List 8.2

1. (a) (i) 0, 4   (ii) 4
   (b) (i) 0, -1 (ii) 0
   (c) (i) none   (ii) none
   (d) (i) 0   (ii) none
   (e) (i) 0, 1, -1 (ii) 0, 1
   (f) (i) 0   (ii) 0

3. $0, -1$

5. $y = a_0 \left(1 - \dfrac{x^2}{4} + \dfrac{x^4}{64} - \cdots \right)$

7. $0, \frac{3}{2}$

9. $r = 0, 1;\ y = c_0 \left(x + x^2 + \dfrac{x^3}{3} + \dfrac{x^4}{18} + \cdots \right)$

17. Let $u = \lambda x$.

19. $y(x) = A + Bx^{-1}$

21. $y = c_0 x^{-1},\ x \neq 0$

23. $J_0(1) \approx 0.7652,\ J_1(1) \approx 0.4401$

List 8.3

1. (a) $y = ce^{-x}$ (b) $y = c_1 \sin x + c_2 \cos x$
   (c) $y = c_0 e^x + c_1 e^{-x}$

3. (a) $y = P_0(x) \equiv 1$ for $y_0 = 1$

(b) $y = P_3(x) = \frac{1}{2}(5x^3 - 3x)$ for $y_0' = -\frac{3}{2}$

5. $y = 1 - 2x^2$

Chapter 9

List 9.1

1. (a) $y_1(x) = 1 - x$, $y_2(x) = 1 - x + \frac{x^2}{2}$, $y_3(x) = 1 - x + \frac{x^2}{2} - \frac{x^3}{6}$,

$y(x) = e^{-x}$

$y_3(0.2) = 0.82133$, $y_3(0.5) = 0.60417$, $y_3(1) = \frac{1}{3} \approx 0.33333$

$y(0.2) = 0.81873$, $y(0.5) = 0.60653$, $y(1) = e^{-1} \approx 0.36788$

(b) $y_1(x) = 1 + \frac{x^2}{2}$, $y_2(x) = 1 + \frac{x^2}{2} + \frac{x^4}{8}$,

$y_3(x) = 1 + \frac{x^2}{2} + \frac{x^4}{8} + \frac{x^6}{48}$, $y(x) = \exp\frac{x^2}{2}$

$y_3(0.2) = 1.02020$, $y_3(0.5) = 1.13314$, $y_3(1) = 1.64583$

$y(0.2) = 1.02020$, $y(0.5) = 1.13315$, $y(1) = 1.64872$

(c) $y_1(x) = 1 - x$, $y_2(x) = 1 - x + x^2 - \frac{x^3}{3}$,

$y_3(x) = 1 - x + x^2 - x^3 + \frac{2}{3}x^4 - \frac{x^5}{3} + \frac{x^6}{9} - \frac{x^7}{63}$, $y(x) = (1 + x)^{-1}$

$y_3(0.2) = 0.83297$, $y_3(0.5) = 0.65786$, $y_3(1) = 0.42857$

$y(0.2) = 0.83333$, $y(0.5) = 0.66667$, $y(1) = 0.5$

(d) $y_1(x) = \frac{x^2}{2}$, $y_2(x) = \frac{x^2}{2} - \frac{x^3}{6}$,

$y_3(x) = \frac{x^2}{2} - \frac{x^3}{6} + \frac{x^4}{24}$, $y(x) = e^{-x} + x - 1$

$y_3(0.2) = 0.01873$, $y_3(0.5) = 0.10677$, $y_3(1) = 0.37500$

$y(0.2) = 0.01873$, $y(0.5) = 0.10653$, $y(1) = 0.36788$

3. $y_1(x) = 1 + \frac{x^3}{3} + x$, $y_2(x) = 1 + x + x^2 + \frac{2x^3}{3} + \frac{x^4}{6} + \frac{2x^5}{15} + \frac{x^7}{63}$,

$y(x) = 1 + x + x^2 + \frac{4}{3}x^3 + \cdots$

$y_2(0.1) = 1.11068$, $y_2(0.2) = 1.24564$

$y(0.1) = 1.11133$, $y(0.2) = 1.25067$

5. $y_1(x) = \frac{x^3}{3}$, $y_2(x) = \frac{x^3}{3} - \frac{x^7}{63}$,

$y_3(x) = \frac{x^3}{3} - \frac{x^7}{63} + \frac{2x^{11}}{2079} - \frac{x^{15}}{59,535}$

7. $y_n(x) = 1 + x + 2\left(\frac{x^2}{2!} + \frac{x^3}{3!} + \cdots + \frac{x^n}{n!}\right) + \frac{x^{n+1}}{(n + 1)!}$

$\lim_{n \to +\infty} y_n(x) = 1 + x + 2(e^x - x - 1) + 0 = 2e^x - x - 1 = y(x)$

List 9.2

1. $y(1) \approx 2.051$, $y(x) = e^{x^2}$, $y(1) = e \approx 2.718$

rel. error ≈ 0.245

3. $y(0.2) \approx 1.24$ with $h = 0.025$

List 9.3

1. $y_6 \approx 1.8204$, $y(0.6) = e^{0.6} \approx 1.8221$
   rel. error $\approx 0.0009 = 0.09\%$

3. $y_5 \approx 4.7027$, $y(1) = e + 2 \approx 4.7183$
   rel. error $\approx 0.0033 \approx 0.3\%$

5. $y_5 \approx 0.7838$, $y(1) = \text{Tan}^{-1}(1) = \dfrac{\pi}{4} \approx 0.7853$

   rel. error $\approx 0.002 = 0.2\%$

7. $y_2 \approx 1.2515$, $u(x) = (1 - x)^{-1}$, $v(x) = \tan\left(x + \dfrac{\pi}{4}\right)$

   $u(0.2) = 1.25 < 1.2515 < v(0.2) = \tan\left(0.2 + \dfrac{\pi}{4}\right) \approx 1.5085$

List 9.4

1. $y_1 \approx 1.6484$, $y(0.5) = e^{0.5} \approx 1.6487$

3. $y_1 \approx 0.77881$, $y(0.25) = e^{-0.25} \approx 0.77880$

5. computer answer 1.94616
   $y(x) = 3e^x - 2x - 2$
   $y(0.5) = 3\sqrt{e} - 3 \approx 1.94616$

7. computer answers $h = 0.5, 4.77125$
   $\qquad\qquad\qquad\qquad h = 0.1, 4.77259$

   $y(x) = x(1 + 2 \ln x)$
   $y(2) = 2(1 + \ln 4) \approx 4.77259$

9. computer answers $h = 0.2, 1.702$
   $\qquad\qquad\qquad\qquad h = 0.1, 1.702$

11. computer answers $h = 0.1, 0.5054$
   $\qquad\qquad\qquad\qquad h = 0.05, 0.5054$

13. $y = \text{Tan}^{-1} x$, $y(1) = \dfrac{\pi}{4}$, $\pi = 4y(1)$

   computer answers
   $h = 0.5, y(1) \approx 0.785392, 4y(1) \approx 3.142$
   $h = 0.25, y(1) \approx 0.785398, 4y(1) \approx 3.142$

15. $y_2 \approx 0.9800$
   $y(x) = \cos x$, $y(0.2) = \cos(0.2) \approx 0.9801$

Chapter 10

List 10.1

5. $z = \phi(x + y)$
9. The left member is always positive.
11. $Pz_x + Qz_y - R = 0$ (Lagrange's equation)

List 10.2

1. (a) $u = 4e^{-3x - 2y}$
   (b) $u = 2e^{4(x - t)}$
   (c) $u = 5[\exp(-4\pi^2 t)] \sin \pi x$

(d) $u = 2\left[\exp\left(\dfrac{-\pi^2 t}{25}\right)\right]\sin \pi x + 3\left[\exp\left(\dfrac{-4\pi^2 t}{25}\right)\right]\sin 2\pi x$

(e) $u = \dfrac{4 \sin \pi x \sinh \pi y}{\sinh \pi}$

(f) $u = 2 \sin \pi x \cos \pi t$

3. (a) $u = x^2 y^2 + \dfrac{y^2}{2} + 2$

(b) $u = -y \sin x + xy + 3y^2$

(c) $u = 2xy + y^2 + x^3$

(d) $u = x^3 y - 3xy + 2y$

(e) $u = \dfrac{x^2 y^2}{4} + x + e^y$

(f) $u = xy^3 + 2y^2 + \ln x$

List 10.3

3. $f(x) = \dfrac{\pi}{4} - \dfrac{2}{\pi}\left[\dfrac{\cos x}{1^2} + \dfrac{\cos 3x}{3^2} + \dfrac{\cos 5x}{5^2} + \cdots\right]$
$\qquad + \left[\dfrac{-\sin x}{1} + \dfrac{\sin 2x}{2} - \dfrac{\sin 3x}{3} + \cdots\right]$

(a) $\dfrac{\pi}{2}$ (b) $\dfrac{\pi}{2}$ (c) 0 (d) 0 (e) $\dfrac{\pi}{2}$

5. 1.5

7. (i) b, h (ii) c, d, e, g, j (iii) a, f, i

9. (b) $f(x) = \cosh x + \sinh x$

11. incorrect, $b_n = \dfrac{1}{2}\displaystyle\int_0^4 2x \sin \dfrac{n\pi x}{2}\, dx$

13. $f(x) = \dfrac{4}{\pi}\left(\dfrac{\sin x}{1} + \dfrac{\sin 3x}{3} + \dfrac{\sin 5x}{5} + \cdots\right)$

15. $f(x) = \dfrac{2}{\pi}\left(\sin x + \dfrac{\pi}{4}\sin 2x - \dfrac{1}{9}\sin 3x + \cdots\right)$

(a) $\dfrac{\pi}{4}$ (b) $\dfrac{0 + \pi/2}{2} = \dfrac{\pi}{4}$ (c) $\dfrac{9\pi}{4} - 2\pi = \dfrac{\pi}{4}$

17. $f(x) = \dfrac{4}{\pi}\left(\sin \dfrac{\pi x}{2} + \dfrac{1}{2}\sin \dfrac{2\pi x}{2} + \dfrac{1}{3}\sin \dfrac{3\pi x}{2} + \cdots\right)$

19. $f(x) = \dfrac{1}{\pi}\left(\dfrac{8}{3}\sin 2x + \dfrac{16}{15}\sin 4x + \dfrac{24}{25}\sin 6x + \cdots\right)$

21. $f(x) = \dfrac{1}{2} + \dfrac{2}{\pi}\left(-\cos \dfrac{\pi x}{2} + \dfrac{1}{3}\cos \dfrac{3\pi x}{2} - \dfrac{1}{5}\cos \dfrac{5\pi x}{2} + \dfrac{1}{7}\cos \dfrac{7\pi x}{2} - \cdots\right)$

23. $f(x) = \dfrac{-3}{2} + \dfrac{12}{\pi^2}\displaystyle\sum_{n=1}^{+\infty}\dfrac{1}{n^2}\cos \dfrac{n\pi x}{3}$, n odd

25. (a) 4 (b) 4 (c) 0 (d) 4 (e) 1 (f) 4.5

27. $f(x) \equiv 100$ on $(-\infty, +\infty)$

29. $\sin x = \dfrac{8}{\pi}\left(\dfrac{1}{3}\cos \dfrac{x}{2} - \dfrac{1}{5}\cos \dfrac{3x}{2} - \dfrac{1}{21}\cos \dfrac{5x}{2} - \cdots\right)$

31. $X(t) = \sum\limits_{n=1}^{+\infty} \dfrac{\sin(2n-1)\pi t}{4(2n-1)^3 - \pi^2(2n-1)^5}$

List 10.4

3. (a) $y(x,t) = 4\sin\dfrac{3\pi x}{L}\cos\dfrac{3\pi ct}{L}$

(b) $y(x,t) = 2\sin\dfrac{4\pi x}{L}\cos\dfrac{4\pi ct}{L} - 6\sin\dfrac{5\pi x}{L}\cos\dfrac{5\pi ct}{L}$

(c) $y(x,t) = 0.01\sin\dfrac{2\pi x}{L}\left(\cos\dfrac{2\pi ct}{L} + \dfrac{L}{2\pi c}\sin\dfrac{2\pi ct}{L}\right)$

(d) $y(x,t) = \dfrac{0.08L^2}{\pi^3}\sum\limits_{n=1}^{+\infty}\dfrac{1}{(2n-1)^3}\sin\dfrac{(2n-1)\pi x}{L}\cos\dfrac{(2n-1)\pi ct}{L}$

(e) $y(x,t) = \dfrac{L}{10\pi^2}\sum\limits_{n=1}^{+\infty}\dfrac{1}{n^2}\sin\dfrac{n\pi}{2}\sin\dfrac{n\pi x}{L}\cos\dfrac{n\pi ct}{L}$

(f) $y(x,t) = \dfrac{8L^3}{\pi^{4c}}\sum\limits_{n=1}^{+\infty}\left(\dfrac{1}{2n-1}\right)^4\sin\dfrac{(2n-1)\pi x}{L}\sin\dfrac{(2n-1)\pi ct}{L}$

(g) $y(x,t) = \dfrac{1}{10\pi}\sum\limits_{n=1}^{+\infty}\dfrac{1}{(2n-1)[4-(2n-1)^2]}\sin\dfrac{(2n-1)\pi x}{L}\cos\dfrac{(2n-1)\pi ct}{L}$

7. T must be multiplied by 4 to double F_1.

List 10.5

3. $u(x,t) = 5\left[\exp\left(\dfrac{-\pi^2 t}{4}\right)\right]\sin\dfrac{\pi x}{6}$

5. (a) $u(x,t) = 5[\exp(-4\pi^2 t)]\sin\pi x$

(b) $u(x,t) = \sum\limits_{n=1}^{+\infty}\dfrac{120}{n\pi}(1-\cos n\pi)\left[\exp\left(\dfrac{-\pi^2 n^2 t}{4}\right)\right]\sin\dfrac{n\pi x}{4}$

7. $u(x,t) = \dfrac{120}{\pi}\left\{\left[\exp\left(\dfrac{-\pi^2 t}{100}\right)\right]\sin\dfrac{\pi x}{10} - \left[\exp\left(\dfrac{-\pi^2 t}{25}\right)\right]\sin\dfrac{\pi x}{5}\right.$
$\left. + \dfrac{1}{3}\left[\exp\left(\dfrac{-9\pi^2 t}{100}\right)\right]\sin\dfrac{3\pi x}{100} + \dfrac{1}{5}\left[\exp\left(\dfrac{-\pi^2 t}{4}\right)\right]\sin\dfrac{\pi x}{2} + \cdots\right\}$

9. $u(x,t) = \dfrac{2L}{\pi}\sum\limits_{n=1}^{+\infty}\dfrac{(-1)^{n+1}}{n}\sin\dfrac{n\pi x}{L}\left[\exp\left(\dfrac{-n^2\pi^2 t}{L^2}\right)\right]$

Index

Boundary-value problem, 375, 385
two-point, 15
Branch points, 248
Burnout, velocity, 190
height, 190
Bush, Vannevar, 345

Cables, 197–200
parabolic, 200
Cantilever beam, 203, 303
Capacitance, 192
Capacitor, 191
Catenary, 199
Cauchy, Augustin-Louis, 9, 160, 351
Cauchy polygon, 352–353
Central force system, 231, 232
Change of variable, 41–43
Characteristic equation, 128, 154, 227
Characteristic polynomial, 146
Chemical reaction, 82
Clairaut's equation, 8
Closed form solution, 312
Closed system, 252
Coefficient of capacitance, 192
Coefficient of contraction, 96
Coefficient of utilization, 103
Compartmental systems, 252–254
Complementary function, 123, 138
Complete solution, 5, 123, 381
Completeness, 431
Complex roots of auxiliary
equation, 131
Compound interest law, 70
Computer solutions, 10, 61
Concentration, 80, 205, 206
Conservation law of mechanical
energy, 184
Conservation of linear momentum, 185
Conservative force field, 183
Constant slope method, 351
Constants, arbitrary, 4, 216
essential, 4, 216
of integration, 4
Continuing method, 350
Continuity, equation of, 79, 97, 208, 252
Convolution, 289–295

formula, 290, 291
theorem, 290, 291
Cooling, Newton's law of, 74, 301
Corrector formula, 362
Cosine series, Fourier, 399, 401, 421
Coulomb, 192
Coupled systems, 242, 306
Critically damped motion, 167
Critical points, 222, 223
asymptotically stable, 224
stable, 224
unstable, 224
Current, 92
Curves of pursuit, 194–196

D'Alembert, Jean, 124
Damping, 167
factor, 169
Debt burden, 256
Decay, 65
constant, 65
of radium, 65
Deflection of beams, 202–205
Degree of a DE, 6
Differential, 21
exact (total), 21
form, 20
operator, 117, 215
Differential analyzer, 345
Differential equation(s), 1
equivalent, 29
exact, 20
of a family of functions, 103
first-order, 20
linear, 4, 116
with one variable missing, 41
of organic decay, 65
of organic growth, 63, 256
partial, 2, 379
second-order, 118, 162
solution of, 2, 4, 5
Diffusion, 76
coefficient of, 207
equation, 79, 383
Diffusivity, thermal, 417
Direction field, 48
Dirichlet, Peter Gustave, 395
Dirichlet, conditions, 396
problem, 426

Discharge coefficient, 96
Discontinuity, 38, 263, 395
  jump, 38, 263, 395
  ordinary, 263, 395
  removable, 263, 395
Discretization error, 353, 356
Displacement function, 15
Dissolution, rate of, 80
Domar, E., 100
Domar model, 100, 255
Driving force, 170
Dynamically stable equilibrium,
    102, 210

Economic applications, 100, 209
Economic equilibrium, 100, 209
  dynamically stable, 102, 210
Eigenfunctions, 412, 413, 419, 427
Eigenvalues, 412, 427
Elastic curve, 201, 303
Elbow path, 23
Electric circuits, 92, 191–194, 248,
    304–306
Elliptic integral of the first kind,
    189
Equation of continuity, 79, 97,
    208, 252
Equilibrium, populations, 255
  point, 223
  position, 163
  price, 101, 210
Equivalent, DE, 29
  systems of DE, 219
Error, analysis, 357
  discretization, 353, 356
  function, 11
  relative, 355
  roundoff, 356
  truncation, 353, 357
Escape speed, 240
Essential arbitrary constants, 4,
    216
Euler, Leonard, 32, 128, 133, 140,
    158, 160, 333, 351, 358
Euler differential equation, 158–
    160, 328
Euler identity, 133
Euler's method, 351–356
Even function, 399
Even periodic extension of a
    function, 402

Exact differential, 21
Exact equation, 20–21
Excess demand, 101
Existence, of solution, 9
Existence theorem, 10, 53, 218
Expansion theorem, 431
Explicit method, 362
Exponential order, 263
Exponents, 326

Family, DE of, 103
Farad, 192
Fick's law of diffusion, 76, 208
First harmonic, 414
First integral, 42, 166, 189
First-order equations, 20, 60
  exact, 20
  homogeneous, 44
  linear, 35
First-order reaction, 82
Flexural rigidity constant, 204
Flow of heat, 74, 97, 426
Flow of liquid from an orifice, 94
Force field, 183, 184
Forcing function, 170
Fourier, J. B. J., 390–391
Fourier, coefficients, 394, 395
  cosine series, 399, 401, 421
  half-range series, 401
  series, 392, 394
  sine series, 399, 401
Fourier–Bessel series, 431
Frequency, 164
Friction, 60–61, 71
  belt, 71
  coefficient of, 72
Frobenius, Ferdinand Georg, 324
  method of, 323–332
  series, 324
Full-wave rectification of the sine
    function, 296
Fundamental period, 287
Fundamental tone, 414

Gamma function, 261, 264, 334
General solution, 5, 381
Geometric interpretation of DE,
    48–58
Gravitation, Newton's universal
    law of, 231, 236